Lecture Notes in Physics

The Lecture Notes in Physics

The series Lecture Notes in Physics (LNP), founded in 1969, reports new developments in physics research and teaching – quickly and informally, but with a high quality and the explicit aim to summarize and communicate current knowledge in an accessible way. Books published in this series are conceived as bridging material between advanced graduate textbooks and the forefront of research and to serve three purposes:

- to be a compact and modern up-to-date source of reference on a well-defined topic

- to serve as an accessible introduction to the field to postgraduate students and nonspecialist researchers from related areas

- to be a source of advanced teaching material for specialized seminars, courses and schools

Both monographs and multi-author volumes will be considered for publication. Edited volumes should, however, consist of a very limited number of contributions only. Proceedings will not be considered for LNP.

Volumes published in LNP are disseminated both in print and in electronic formats, the electronic archive being available at springerlink.com. The series content is indexed, abstracted and referenced by many abstracting and information services, bibliographic networks, subscription agencies, library networks, and consortia.

Proposals should be sent to a member of the Editorial Board, or directly to the managing editor at Springer:

Christian Caron
Springer Heidelberg
Physics Editorial Department I
Tiergartenstrasse 17
69121 Heidelberg / Germany
christian.caron@springer.com

E. Papantonopoulos (Ed.)

Physics of Black Holes

A Guided Tour

 Springer

Eleftherios Papantonopoulos
National Technical University
Dept. of Physics
Zografou Campus
157 80 Athens
Greece
lpapa@central.ntua.gr

Papantonopoulos, E., (Ed.), *Physics of Black Holes: A Guided Tour*, Lect. Notes Phys.
769 (Springer, Berlin Heidelberg 2009), DOI 10.1007/ 978-3-540-88460-6

ISBN: 978-3-540-88459-0   e-ISBN: 978-3-540-88460-6

DOI 10.1007/978-3-540-88460-6

Lecture Notes in Physics ISSN: 0075-8450   e-ISSN: 1616-6361

Library of Congress Control Number: 2008936825

Cover design: Integra Software Services Pvt. Ltd.

Printed on acid-free paper

9 8 7 6 5 4 3 2 1

springer.com

Preface

This book is an edited version of the review talks given in the Fourth Aegean School on Black Holes held in Mytilene on Lesvos Island, Greece, from 17 to 22 September 2007. The aim of this book is not to present another proceedings volume, but rather an advanced multiauthored textbook which meets the needs of both the postgraduate students and the young researchers in the fields of gravity, relativity, modern cosmology and astrophysics.

Black holes are the most mysterious and fascinating objects of our universe. They were predicted by the Einstein's theory of general relativity and their existence is the triumph of this theory. A possible detection of gravitational waves by the gravity experiments will give vital information on the nature and properties of black holes. On the other hand, the recent advances in string theory offered a new understanding of the classical and quantum properties of black holes. This book is a guided tour, by world experts, on the new developments in the physics of the black holes.

In the first part of the book, Samir Mathur discusses the information paradox of black holes. The information paradox connects quantum mechanics with gravity. As he says in his article "if quantum gravity effects are confined to within a given length scale and the vacuum is assumed to be unique, then there will be information loss". However, he goes one step further explaining how quantum effects in string theory resolve this problem.

Another important problem is discussed by Elizabeth Winstanley: do hairy black holes exist? She reviews the properties of hairy black holes in $SU(2)$ Einstein–Yang–Mills (EYM) theory in asymptotically anti-de Sitter space and she discusses recent work in which it is shown that stable hair also exists in $SU(N)$ EYM for arbitrary N. Next, the thermodynamics of black holes is discussed in Steven Carlip's article. He reviews what we currently know about black hole thermodynamics and statistical mechanics, which suggests a rather speculative "universal" characterization of the underlying states, and he describes some key open questions.

The first part of the book ends with the discussion of astrophysical black holes. In Ulrich Sperhake's article the basic techniques of numerical relativity and black holes simulations are presented, while Nikolaos Stergioulas describes the way black holes are formed through gravitational collapse of rotating stars. It is demonstrated, in the case of rotating neutron stars which are unstable to quasi-radial oscillations, how a complete transition from one stationary solution of Einstein's equations to another occurs, including the formation of horizons and gravitational wave emission.

The second part of the book presents the new ideas in black hole physics coming from string theory and braneworlds. One of the basic ingredient of these theories is the existence of black holes in higher than four dimensions. Niels Obers reviews some of the recent progress in uncovering the phase structure of black hole solutions in higher-dimensional vacuum Einstein gravity. Ruth Gregory gives an overview of braneworlds and she discusses black holes on the brane, the obstructions to finding exact solutions and ways of tackling these difficulties. She describes also some known solutions and concludes with some open questions and controversies.

The next article by Christos Charmousis describes a higher-order gravity theory, the Lovelock theory, that generalizes in higher dimensions than four, general relativity. He discusses a generic staticity theorem, quite similar to Birkhoff's theorem in general relativity, which gives charged static black hole solutions. He also presents Lovelock exact black hole solutions in the context of braneworlds.

This part of the book also includes Sanjeev Seahra's article on a black string model of a braneworld black hole. He develops the perturbation formalism for Randall–Sundrum model and discusses the weak field limit of the model. He solves numerically the equations of motion for the gravitational waves in the black string background and discusses their behaviour. Finally, Panagiota Kanti in her article addresses the topic of the creation of small black holes during particle collisions in a ground-based accelerator, such as Large Hadron Collider at CERN, in the context of a higher-dimensional theory. She points out that the most important observable effect associated with their creation is likely to be the emission of Hawking radiation during their evaporation process.

The last part of the book deals with the very important issue of perturbations and stability of black holes in various dimensions. Hideo Kodama in his article explains the gauge-invariant formulation for perturbations of background spacetimes. He derives his famous master equations for a variety of important spacetimes such as static black holes, static black branes and rotating black holes in various dimensions. As applications, he discusses the stability of static black holes in higher dimensions and flat black branes. The article by George Siopsis discusses the analytic calculation of quasi-normal modes of various types of perturbations of black holes in both asymptotically flat and anti-de Sitter spaces. He pays special attention to low-frequency modes in anti-de Sitter space because, as it is known, they may have experimental consequences for the quark-gluon plasma formed in heavy ion collisions.

The Fourth Aegean School and consequently this book became possible with the kind support of many people and organizations. The school was organized by the physics department of the National Technical University of Athens and supported by the physics department of the University of Tennessee. We also received financial support from the following sources and organizations and this is gratefully acknowledged: Ministry of National Education and Religious Affairs, the Ministry of the Aegean, Alexander Onassis Foundation, the University of the Aegean, Prefecture of Lesvos, Municipality of Mytilene. The administrative support of the Fourth Aegean

School was taken up with great care by Mrs. Fani Siatra. We acknowledge the help of Vasilis Zamarias who designed and maintained the website of the school.

Last, but not the least, we are grateful to the staff of Springer-Verlag, responsible for the Lecture Notes in Physics, whose abilities and help contributed greatly to the appearance of this book.

Athens, June 2008   *Lefteris Papantonopoulos*

Contents

Part III Perturbations of Black Holes

Part I
Black Holes and their Properties

Chapter 1
What Exactly is the Information Paradox?

S.D. Mathur

Abstract The black hole information paradox tells us something important about the way quantum mechanics and gravity fit together. In these lectures I try to give a pedagogical review of the essential physics leading to the paradox, using mostly pictures. Hawking's argument is recast as a 'theorem': if quantum gravity effects are confined to within a given length scale and the vacuum is assumed to be unique, then there will be information loss. We conclude with a brief summary of how quantum effects in string theory violate the first condition and make the interior of the hole a 'fuzzball'.

1.1 Introduction

The black hole information paradox is probably the most important issue for fundamental physics today. If we cannot understand its resolution, then we cannot understand how quantum theory and gravity work together. Yet very few people seem to understand how robust the original Hawking arguments are and what exactly it would take to resolve the problem.

In this review I try to explain the power of this paradox using mostly pictures. In Sect. 1.7, I formulate the paradox as a 'theorem': if quantum gravity effects are confined to within the planck length and the vacuum is unique, then there *will* be information loss. I conclude with a brief outline of how the paradox is resolved in string theory: quantum gravity effects are not confined to a bounded length (due to an effect termed 'fractionation'), and the information of the hole is spread throughout its interior, creating a 'fuzzball'.

S.D. Mathur (✉)
Department of Physics, The Ohio State University, Columbus, OH 43210, USA
mathur@mps.ohio-state.edu

Mathur, S.D.: *What Exactly is the Information Paradox?*. Lect. Notes Phys. **769**, 3–48 (2009)
DOI 10.1007/978-3-540-88460-6_1

1.2 Puzzles with Black Holes

There are two closely connected problems that arise when we consider quantum theory in the context of black holes: the 'entropy puzzle' and the 'information paradox'.

1.2.1 The Entropy Puzzle

Take a box containing some gas and throw it into a black hole. The gas had some entropy, so after the gas has vanished into the singularity, we have decreased the entropy of the Universe, and violated the second law!

Of course this sounds silly: if we threw the box into a trash can, then its entropy would be inside the trash can, whether we wanted to look in there or not. The black hole case is a little different however, since it is not clear how we would look into the hole to see the entropy of the gas. Nevertheless, physical intuition tells us that the entropy of the hole should have gone up when it swallowed the box of gas.

When the box falls into the hole, it increases the mass of the hole, and therefore the size of its horizon. Careful work with thermodynamics shows that we should attribute a 'Bekenstein entropy' [1]

$$S_{bek} = \frac{A}{4G} \tag{1.1}$$

to the black hole, where A is the area of the horizon and we have set $c = \hbar = 1$. Then we have

$$\frac{dS_{total}}{dt} = \frac{dS_{matter}}{dt} + \frac{dS_{bek}}{dt} \geq 0 \tag{1.2}$$

and the second law of thermodynamics is saved.

This looks nice, but thinking a bit more, we find a deeper puzzle. From statistical physics we know that the entropy of any system is given by $S = \ln \mathcal{N}$, where $\mathcal{N}$ is the number of states of the system for the given macroscopic parameters. Applying this to the black hole, we should find

$$\mathcal{N} = e^{S_{bek}} \tag{1.3}$$

states for a black hole of given mass. Note that (1.1) is the area of the horizon measured in planck units. Thus for a solar mass black hole with a horizon radius ~ 3 km, we would have

$$\mathcal{N} \sim 10^{10^{77}} \tag{1.4}$$

states, an enormous number! Where should we look for these microstates? Since S_{bek} is proportional to the horizon area, people tried to look for small 'deformation modes' of the horizon. But it turns out that there is no such deformation in general; any excitation near the horizon either falls to the singularity or flows off to infinity, leaving a spherically symmetric horizon again. This observation came to be called 'black holes have no hair', signifying that the horizon cannot hold any information in its vicinity. But if we find a unique geometry for the hole then the entropy would be $S = \ln 1 = 0$, in sharp contrast to (1.4).

One may therefore think that the entropy is somehow at the singularity; after all the matter that made the hole in the first place disappeared into this singularity. In that case we would not see the differences between microstates in the classical geometry of the hole, and quantum effects at the singularity would differentiate the different states. But as we will see now, this possibility leads to an even more serious problem: the information paradox.

1.2.2 The Information Paradox

We have seen above that a black hole has entropy S_{bek}. It has an energy $E = M$, where M is the mass of the hole. One may therefore ask if we could have the usual thermodynamic relation

$$T dS_{bek} = dE \ . \tag{1.5}$$

This would imply that the black hole has a temperature

$$T = \left(\frac{dS}{dE} \right)^{-1} = \left(\frac{d}{dM} \left(\frac{4\pi(2GM)^2}{4G} \right) \right)^{-1} = \frac{1}{8\pi GM} \ . \tag{1.6}$$

If the black hole has a temperature, should it radiate? Temperature by itself does not imply radiation, but by the law of detailed balance in thermodynamics what we *can* say is that if the black hole can absorb quanta of a certain wavenumber with cross section $\sigma(k)$, then it should radiate the same quanta at a rate

$$\Gamma = \int \frac{d^3 k}{(2\pi)^3} \sigma(k) \frac{1}{e^{\frac{\omega(k)}{T}} - 1} \ . \tag{1.7}$$

But we know that $\sigma(k)$ is nonzero, since quanta can fall into the hole. Thus we must get radiation from the hole.

But the classical geometry of the hole does not allow any worldlines to emerge from the horizon! How then will we get this radiation? The answer, discovered by Hawking [2, 3], is that we must consider quantum processes, more precisely quantum fluctuations of the vacuum. In the vacuum pairs of particles and antiparticles are continuously being created and annihilated. Consider such fluctuations for electron–positron pairs. Suppose we apply a strong electric field in a region which is pure vacuum. When an electron–positron pair is created, the electron gets pulled one way by the field and the positron gets pulled the other way. Thus instead of annihilation of the pair, we can get creation of real (instead of virtual) electrons and positrons which can be collected on opposite ends of the vacuum region. Thus we get a current flowing through the space even though there is no material medium filling the region where the electric field is applied. This is called the 'Schwinger effect'.

A similar effect happens with the black hole, with the effect of the electric field now replaced by the gravitational field. We do not have particles that are charged in opposite ways under gravity. But the attraction of the black hole falls off with radius, so if one member of a particle–antiparticle pair is just outside the horizon it can flow off to infinity, while if the other member of the pair is just inside the horizon then it can get sucked into the hole. The particles flowing off to infinity represent the 'Hawking radiation' coming out of the black hole. Doing a detailed computation, one finds that the rate of this radiation is given by (1.7). Thus we seem to have a very nice thermodynamical physics of the black hole. The hole has entropy, energy, and temperature and radiates as a thermal body should.

But there is a deep problem arising out of the *way* in which this radiation is created by the black hole. As we will discuss in detail in the coming sections, the radiation which emerges from the hole is not in a 'pure quantum state'. Instead, the emitted quanta are in a 'mixed state' with excitations which stay inside the hole. There is nothing wrong in this by itself, but the problem comes at the next step. The hole loses mass because of the radiation and eventually disappears. Then the quanta in the radiation outside the hole are left in a state that is 'mixed', but we cannot see anything that they are mixed with! Thus the state of the system has become a 'mixed' state in a fundamental way. This does not happen in usual quantum mechanics, where we start with a pure state $|\Psi\rangle$ and evolve it by some Hamiltonian H as $|\Psi'\rangle = e^{-iHt}|\Psi\rangle$ to get another pure state at the end. We will describe mixed states in detail later, but for now we note that mixed states arise in usual physics when we coarse-grain over some variables and thereby discard some information about a system. This coarse-graining is done for convenience, so that we can extract the gross behavior of a system without keeping all its fine details, and is a standard procedure in statistical mechanics. But there is always a 'fine-grained' description available with all information about the state, so that underlying the full system there is always a pure state. With black holes we seem to be getting a loss of information in a fundamental way. We are not throwing away information for convenience; rather we cannot get a pure state even if we wanted. This implies a fundamental change in quantum theory, and Hawking advocated that in the presence of gravity (which will make black holes) we should not formulate quantum mechanics with pure states and unitary evolution operators. Rather, we should think of mixed states as being basic and describe these in terms of their 'density matrices'. The evolution of these density matrices will be given not by the S matrix but by a dollar matrix $ [2, 3].

This was a radical proposal, and most physicists were not happy to abandon ordinary quantum mechanics when it works so well in all other contexts. But if we are to bypass this 'information paradox' then we have to see how exactly this radiation is emitted and what changes to the physics could make this radiation emerge in a pure state. Enormous effort has been spent on this problem. With string theory, we will see that we can now obtain a resolution of the paradox. Perhaps it should not be surprising that this resolution itself comes with a radical change in our understanding of how quantum gravity works. Earlier attempts at resolving the paradox had assumed that quantum gravity effects operate over distances of order the planck length or less. This seems natural since the only fundamental length scale that we can make out of the fundamental constants $c, \hbar$, and G is

$$l_p = \left(\frac{\hbar G}{c^3} \right)^{\frac{1}{2}} \sim 10^{-33} \text{ cm}. \qquad (1.8)$$

As we will see below, if it were indeed true that all quantum gravity effects were confined to within a length scale like l_p (or any other fixed length scale) then we would get information loss and quantum mechanics would need to be changed. But how can we get any other natural length scale for quantum gravity effects? If we collide two gravitons then it is true that quantum gravity effects should start when the wavelengths of the gravitons become order l_p. But a black hole is made up of a large number of quanta N; the larger the black hole the larger this number N. Then we have to ask if quantum gravity effects extend over distances $\sim l_p$ or over distances $N^\alpha l_p$ where α is some appropriate constant. In string theory we find that the latter is true and that N, α are such that the length scale of quantum gravity effects becomes of the order of the radius of the horizon. This changes the process by which the radiation is emitted, and the radiation can emerge in a pure state.

1.2.3 The Plan of the Review

There exist many reviews on the subject of black holes, and there are also reviews of the 'fuzzball' structure emerging from string theory [4–6]. What I will do here is a bit different: I will try to give a detailed pictorial description of the information problem. We will study the black hole geometry in detail and see how wavemodes evolve to create Hawking radiation. Then we will discuss the 'mixed' nature of the quantum state that is created in this radiation process. Most importantly, we will discuss why the argument of Hawking showing information loss is *robust* and can only be bypassed by a radical change in one of the fundamental assumptions that we usually make about quantum gravity. We will close with a brief summary of black holes in string theory and the fuzzball nature of the black hole interior.

1.3 Particle Creation in Curved Space

The story of Hawking radiation really begins with the understanding of particle creation in curved spacetime (for reviews see [7, 8]). Particles are described in terms of an underlying quantum field, say a scalar field ϕ. We can write a covariant action for this field and do a path integral. But how do we define particles? In flat space we expand the field operator as

$$\hat{\phi} = \sum_{k} \frac{1}{\sqrt{V}} \frac{1}{\sqrt{2\omega}} \left(\hat{a}_k e^{ik \cdot x - i\omega t} + \hat{a}_k^{\dagger} e^{-ik \cdot x + i\omega t} \right) , \qquad (1.9)$$

where V is the volume of the spatial box where we have taken the field to live and $\omega = \sqrt{|k|^2 + m^2}$ for a field with mass m. The vacuum is the state annihilated by all the $\hat{a}$:

$$\hat{a}_k |0\rangle = 0, \qquad (1.10)$$

and the $\hat{a}_k^{\dagger}$ create particles.

In *curved* spacetime, on the other hand, there is no canonical definition of particles. We can choose any coordinate t for time and decompose the field into positive and negative frequency modes with respect to this time t. Let the positive frequency modes be called $f(x)$; then their complex conjugates give negative frequency modes $f^*(x)$. The field operator can be expanded as

$$\hat{\phi}(x) = \sum_{n} \left(\hat{a}_n f_n(x) + \hat{a}_n^{\dagger} f_n^*(x) \right) . \qquad (1.11)$$

Then we can define a vacuum state as one that is annihilated by all the annihilation operators

$$\hat{a}_n |0\rangle_a = 0 . \qquad (1.12)$$

The creation operators generate particles; for example a 1-particle state would be

$$|\psi\rangle = \hat{a}_n^{\dagger} |0\rangle_a . \qquad (1.13)$$

We have added the subscript a to the vacuum state to indicate that the vacuum is defined with respect to the operators $\hat{a}_n$. But since there is no unique choice of the time coordinate t, we can choose a different one $\tilde{t}$. We will then have a different set of positive and negative frequency modes and an expansion

$$\hat{\phi}(x) = \sum_{n} \left(\hat{b}_n h_n(x) + \hat{b}_n^{\dagger} h_n^*(x) \right) . \qquad (1.14)$$

Now the vacuum would be defined as

$$\hat{b}_n |0\rangle_b = 0 \qquad (1.15)$$

and the $\hat{b}_n^{\dagger}$ would create particles.

The main point now is that a person using the operators $\hat{a}, \hat{a}^{\dagger}$ would think that $|0\rangle_a$ was a vacuum, but he would not think that the state $|0\rangle_b$ was a vacuum – he would find it to contain particles of the type created by the $\hat{a}_n^{\dagger}$. Let us see how one finds exactly how many $\hat{a}^{\dagger}$ particles there are in the state $|0\rangle_b$. The mode functions f_n are normalized using an inner product defined as follows. Take any spacelike hypersurface, with volume element $d\Sigma^{\mu}$ (thus the vector $d\Sigma^{\mu}$ points normal to the hypersurface and has a value equal to the volume of the surface element). Then

$$(f, g) \equiv -i \int d\Sigma^{\mu} \left(f \partial_{\mu} g^* - g^* \partial_{\mu} f \right) . \qquad (1.16)$$

Under this inner product we will have

$$(f_m, f_n) = \delta_{mn}, \quad (f_m, f_n^*) = 0, \quad (f_m^*, f_n^*) = -\delta_{mn} . \tag{1.17}$$

Now from the two different expansions of $\hat{\phi}$ we have

$$\sum_n \left(\hat{a}_n f_n(x) + \hat{a}_n^\dagger f_n^*(x) \right) = \sum_n \left(\hat{b}_n h_n(x) + \hat{b}_n^\dagger h_n^*(x) \right) . \tag{1.18}$$

Taking the inner product with f_m on each side, we get

$$\hat{a}_m = \sum_n (h_n, f_m)\hat{b}_n + \sum_n (h_n^*, f_m)\hat{b}_n^\dagger \equiv \sum_n \alpha_{mn}\hat{b}_n + \sum_n \beta_{mn}\hat{b}_n^\dagger . \tag{1.19}$$

Thus the vacuum $|0\rangle_a$ satisfies

$$0 = \hat{a}_m |0\rangle_a = \left(\sum_n \alpha_{mn}\hat{b}_n + \sum_n \beta_{mn}\hat{b}_n^\dagger \right) |0\rangle_a . \tag{1.20}$$

Let us see how to solve this equation. Suppose we had just one mode, with a relation

$$(b + \gamma b^\dagger)|0\rangle_a = 0 . \tag{1.21}$$

The solution to this equation is of the form

$$|0\rangle_a = C e^{\mu \hat{b}^\dagger \hat{b}^\dagger} |0\rangle_b , \tag{1.22}$$

where C is a normalization constant and μ is a number that we have to determine. Expand the exponential in a power series

$$e^{\mu \hat{b}^\dagger \hat{b}^\dagger} = \sum_n \frac{\mu^n}{n!} (\hat{b}^\dagger \hat{b}^\dagger)^n . \tag{1.23}$$

With a little effort using the commutator $[\hat{b}, \hat{b}^\dagger] = 1$, we find that

$$\hat{b}(\hat{b}^\dagger \hat{b}^\dagger)^n = (\hat{b}^\dagger \hat{b}^\dagger)^n \hat{b} + 2n\hat{b}^\dagger (\hat{b}^\dagger \hat{b}^\dagger)^{n-1} . \tag{1.24}$$

Putting this in the series for the exponential, we find that

$$\hat{b} e^{\mu \hat{b}^\dagger \hat{b}^\dagger} |0\rangle_b = 2\mu \hat{b}^\dagger e^{\mu \hat{b}^\dagger \hat{b}^\dagger} |0\rangle_b . \tag{1.25}$$

Looking at (1.21) we see that we should choose $\mu = -\frac{\gamma}{2}$, and we get

$$|0\rangle_a = C e^{-\frac{\gamma}{2} \hat{b}^\dagger \hat{b}^\dagger} |0\rangle_b . \tag{1.26}$$

This state has the form

$$|0\rangle_a = C|0\rangle_b + C_2 \hat{b}^\dagger \hat{b}^\dagger |0\rangle_b + C_4 \hat{b}^\dagger \hat{b}^\dagger \hat{b}^\dagger \hat{b}^\dagger |0\rangle_b + \cdots , \tag{1.27}$$

so it looks like a part that is the b vacuum, a part that has two particles of type b, a part with four such particles, and so on.

Returning to our full equation (1.20) we have the solution

$$|0\rangle_a = C e^{-\frac{1}{2}\sum_{m,n}\hat{b}_m^\dagger \gamma_{mn} \hat{b}_n^\dagger}|0\rangle_b \, , \tag{1.28}$$

where the matrix γ is symmetric and is given by

$$\gamma = \frac{1}{2}\left(\alpha^{-1}\beta + (\alpha^{-1}\beta)^T\right) \, . \tag{1.29}$$

To summarize, there are many ways to define time and therefore many ways to define the vacuum and particles in curved space. The vacuum in one definition looks, in general, full of particle pairs in other definitions. How then are we going to do any physics with these particles?

What helps is that we will usually detect particles in some region which is far away from the region where spacetime is curved, for example at asymptotic infinity in a black hole geometry. There is a natural choice of coordinates at infinity, in which the metric looks like $\eta_{\mu\nu}$. We can still make boosts that keep the metric in this form, but the change of time coordinate under these boosts does not change the vacuum. What happens is that positive frequency modes change to other positive frequency modes, giving the expected change of the energy of a quantum when it is viewed from a moving frame.

But even though this may be a natural choice of coordinates, giving a natural definition of particles, we may still ask why we cannot use some other curvilinear coordinate system and its corresponding particles. The point is that we have to know the following physics at some point: what is the energy carried by these particles? This information is not given by the definition of the particle modes; rather, we need to know the energy–momentum tensor for these particle states. For the physical fields that we consider, we assume that the particles defined in the flat coordinate system with metric $\eta_{\mu\nu}$ are the ones which give the expected physical energy of the state, an energy which shows up, for example, in the gravitational attraction between these particles.

So there is no ambiguity in how particles are defined at infinity, but if there is some region of spacetime which is curved, then wavemodes that travel through that region and back out to spatial infinity can have a nontrivial number of particles at the end, even though they may have started with no particles excited in them at the start. What we need now is to get some physical feeling for the length and time scales involved in this process of particle creation.

1.3.1 Particle Creation: Physical Picture

Let us first get a simpler picture of why particles can get created when spacetime is curved. We know that each Fourier mode of a quantum field behaves like a harmonic oscillator, and if we are in the excited state $|n\rangle$ for this oscillator then we have n particles in this Fourier mode. Thus the amplitude of this Fourier mode, which we call a, has a Lagrangian of the form

$$L = \frac{1}{2}\dot{a}^2 - \frac{1}{2}\omega^2 a^2 \ . \tag{1.30}$$

But as we move to later times, the spacetime can distort, and the frequency of the mode can change, so that we get

$$L' = \frac{1}{2}\dot{a}^2 - \frac{1}{2}\omega'^2 a^2 \ . \tag{1.31}$$

We picture this situation in Fig. 1.1. Figure 1.1(a) shows the potential where the frequency is ω. Let us require that here no particles are present in this Fourier mode. Then we will have the vacuum wavefunction $|0\rangle$ for this harmonic oscillator. Now suppose we change the potential to the one for frequency ω'; this potential is shown in Fig. 1.1(b). For this new potential, the vacuum state is a different wavefunction from the one for frequency ω, and we sketch it in Fig. 1.1(b).

First suppose that the change of frequency from ω to ω' was very slow. Then we will find that the vacuum wavefunction will keep changing as the potential changes in such a way that it remains the vacuum state for whatever potential we have at any given time. In particular when we reach the final potential with frequency ω', the vacuum wavefunction of Fig. 1.1(a) will have become the vacuum wavefunction of Fig. 1.1(b). This fact follows from the 'adiabatic theorem', which describes the evolution of states when the potential changes slowly.

Now consider the opposite limit, where the potential changes from the one in Fig. 1.1(a) to the one in Fig. 1.1(b) very *quickly*. Then the wavefunction has had hardly any time to evolve, and we get the situation in Fig. 1.1(c). The potential is that for frequency ω', but the wavefunction is still the vacuum wavefunction for frequency ω. This is not the vacuum wavefunction for frequency ω', but we can expand it in terms of the wavefunctions $|n\rangle_{\omega'}$ which describe the level n excitation of the harmonic oscillator for frequency ω':

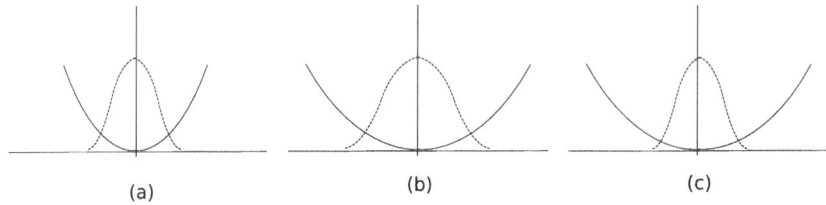

(a)　　　　　　　　(b)　　　　　　　　(c)

Fig. 1.1 (**a**) The potential characterizing a given Fourier mode, and the vacuum wavefunction for this potential. (**b**) If the spacetime distorts, the potential changes to a new one, with its own vacuum wavefunction. (**c**) If the potential changes suddenly, we have the new potential but the old wavefunction, which will not be the vacuum wavefunction for this changed potential; thus we will see particles

$$|0\rangle_\omega = c_0|0\rangle_{\omega'} + c_1|1\rangle_{\omega'} . + c_2|2\rangle_{\omega'} + \cdots . \tag{1.32}$$

Actually since the wavefunction that we have is symmetric under reflections $a \to -a$, we will get only the even levels $|n\rangle$ in our expansion

$$|0\rangle_\omega = c_0|0\rangle_{\omega'} + c_2|2\rangle_{\omega'} + c_4|4\rangle_{\omega'} + \cdots . \tag{1.33}$$

This is like the expansion (1.27), and a little more effort shows that the coefficients c_n will be of the form that will give the exponential form (1.26).

Thus under slow changes of the potential the Fourier mode remains in a vacuum state, while if the changes are fast then the Fourier mode gets populated by particle pairs. But what is the timescale that distinguishes slow changes from fast ones? The only natural timescale in the problem is the one given by frequency of the oscillator:

$$\Delta T \sim \omega^{-1} \sim \omega'^{-1} , \tag{1.34}$$

where we have assumed that the two frequencies involved are of the same order. If the potential changes over times that are *small* compared to ΔT, then in general particle pairs will be produced.

We can now put this discussion in the context of curved spacetime. Let the variations of the metric be characterized by the length scale L; i.e., the length scale for variations of g_{ab} is $\sim L$ in the space and time directions, and the region under consideration also has length $\sim L$ in the space and time directions. We assume that the metric varies significantly (i.e., $\delta g \sim g$) in this region. Then the particles produced in this region will have a wavelength $\sim L$ and the number of produced particles will be of order unity. Thus there is no other 'large dimensionless number' appearing in the physics, and the length scale L governs the qualitative features of particle production.

An example of such a metric variation would be if we take a star with radius $6GM$ (so it is not close to being a black hole), and then this star shrinks to a size $4GM$ (still not close to a black hole) over a time of order $\sim GM$. Then in this process we would produce order unity number of quanta for the scalar field, and these quanta will have wavelengths $\sim GM$. After the star settles down to its new size, the metric becomes time independent again, and there is no further particle production.

As it stands, this particle production is a very small effect, from the point of view of energetics. In the above example, the length GM is of order kilometers or more, so the few quanta we produce will have wavelengths of the order of kilometers. The energy of these quanta will be very small, much smaller than the energy M present in the star which created the changing metric. So particle production can be ignored in most cases where the metric is changing on astrophysical length scales.

We will see that a quite different situation emerges for the black hole, where particle production keeps going on until all the mass of the black hole is exhausted.

1.3.2 Particle Production in Black Holes

The metric of a Schwarzschild hole is often written as

$$ds^2 = -\left(1 - \frac{2GM}{r}\right) dt^2 + \frac{dr^2}{1 - \frac{2GM}{r}} + r^2(d\theta^2 + \sin^2\theta d\phi^2) . \tag{1.35}$$

This metric looks time independent, so we might think at first that there should be *no* particle production. If we had a time-independent geometry for a star, there would indeed be no particle production. What is different in the black hole case? The point is that the coordinate system in the above metric covers only a part of the spacetime – the part outside the horizon $r = 2GM$. Once we look at the full metric we will not see a time-independent geometry. The full geometry is traditionally described by a Penrose diagram, which we sketch in Fig. 1.2. The region of this diagram where the particle production will take place is indicated by the box with dotted outline around the horizon.

From the Penrose diagram we can easily see which point is in the causal future of which point, but since lengths have been 'conformally scaled' we cannot get a good idea of relative lengths at different locations on the diagram. Thus in Fig. 1.3 we make a schematic sketch of the shaded region in Fig. 1.2. The horizontal axis is r, which is a very geometric variable in the problem – the value of r at any radius is given by writing the area of the 2-sphere at that point as $4\pi r^2$. The line at $r = 0$ is the 'center' of the black hole; thus this is a region of high curvature (the singularity) after the black hole forms. The line $r = 2GM$ is the horizon. Spatial infinity is on the right, at $r \to \infty$.

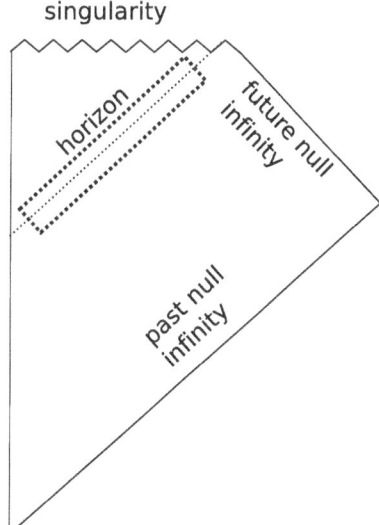

Fig. 1.2 The Penrose diagram for a black hole (without the backreaction effects of Hawking evaporation). Null rays are straight lines at 45°. Thus we see that the horizon is a null surface. Hawking radiation collects at future null infinity

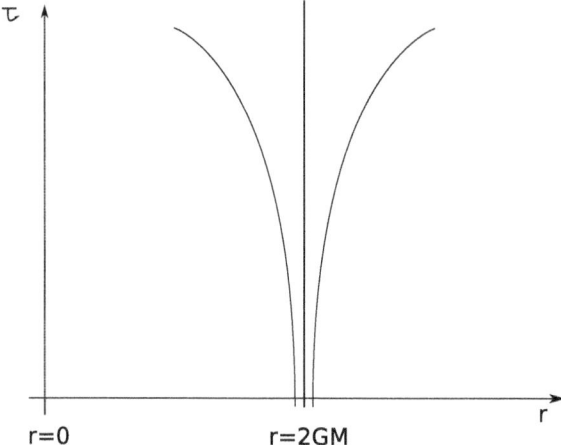

Fig. 1.3 A schematic picture of the dotted box in Fig. 1.2. The horizon has been rotated to be vertical. One coordinate is r. The other axis has been called τ, but there is no canonical choice of τ (the metric will degenerate at the horizon anyway if we try to make it independent of τ). We see that the null geodesics on the two sides of the horizon move away from $r = 2GM$ as they evolve

The vertical axis in Fig. 1.3 is called τ; it is some time coordinate that we have introduced to complement r. At large r we let $\tau \to t$, where t is the Schwarzschild time. The metric will *not* be good everywhere in the coordinates (r, τ); it will degenerate at the horizon. This will not matter since all we want to do with the help of this figure is show how geodesics near the horizon evolve to smaller or larger r values.

A massless particle that is at the horizon and trying its best to fly out never manages to escape, but stays on the horizon. This can be seen as follows. The massless particle follows a null geodesic. Let us allow no angular part to its momentum to ensure that all the momentum is directed radially outward in the attempt to escape. Thus from the metric (1.35) we will have

$$0 = ds^2 = -\left(1 - \frac{2GM}{r}\right) dt^2 + \frac{dr^2}{1 - \frac{2GM}{r}}, \tag{1.36}$$

which gives

$$dr = \left(1 - \frac{2GM}{r}\right) dt . \tag{1.37}$$

So if we are on the horizon $r = 2GM$, we get $dr = 0$, i.e., the particle stays on the horizon.

What if the particle started slightly outside the horizon and tried to fly radially outward? Now it can escape, so after some time the particle will reach out to a larger radius, say $r \sim 3GM$. This null geodesic starts out near the horizon, but 'peels off' toward infinity.

Similarly, consider a massless particle that starts a little *inside* the horizon and tries to fly radially outward. This time it cannot escape the hole or even remain where it started; this null geodesic 'peels off' and falls in toward smaller r. The figure shows the geodesic reaching the radius $r \sim GM$ which is inside the hole, though still comfortably away from the singularity.

Now we see that in this vicinity of the horizon, there is a 'stretching' of spacetime going on. A small region near the horizon gets 'pulled apart', with the part inside the horizon moving deeper in and the part outside the horizon moving out. We will make this more precise later, but we can now see that the metric indeed has a time dependence which can cause particle creation. Moreover, this stretching goes on as long as the black hole lasts, since whenever we have the horizon we will see such a 'peeling off' of null geodesics from the two sides of the horizon. Thus if there will be particle production from this stretching of spacetime, it will keep going on till the black hole disappears and there is no more horizon.

1.4 Slicing the Black Hole Geometry

We have seen that the Schwarzschild coordinates cover only the exterior of the horizon, and so do not give a useful description of the spacetime for the purposes of understanding Hawking radiation. What we need is a set of spacelike slices that 'foliate' the spacetime geometry, covering both the outside and the inside of the hole. Let us see how to make such spacelike slices.

Consider the slices sketched in Fig. 1.4. Far outside the horizon, we would like to have the spacelike slice look quite like a spacelike slice in ordinary flat spacetime. Thus we let it be the surface $t = constant$, all the way from infinity to say $r = 3GM$, a point that is comfortably far away from the vicinity of the horizon. We call this part of the spacelike surface S_{out}.

What should we do inside the horizon? From the metric (1.35) we see that inside the horizon $r = 2GM$ space and time interchange roles; i.e., the t direction is spacelike while the r direction is timelike. Thus for the part of the slice inside the horizon we use a $r = constant$ slice. Let us take this slice at $r = GM$, comfortably far away from the horizon at $r = 2GM$ and also from the singularity at $r = 0$. Let us call this part of the spacelike surface S_{in}.

We must now connect these two parts of our spacelike surface. It is not hard to convince oneself that this can be done with a smooth 'connector' segment, which is everywhere spacelike. Let us call this segment of the spacelike surface S_{con}.

One might be worried that this spacelike slice is not covering the region near $r = 0$. Let us assume that the black hole formed at some time $t \sim t_{initial}$. Then for $t \ll t_{initial}$, there was no singularity at $r = 0$. Thus imagine extending the part S_{in} of the slice down to a time before this singularity, whereupon we bend it smoothly to reach $r = 0$. (This part of the slice is not depicted in the figure, since it will not be of immediate use to us in the discussions that follow.)

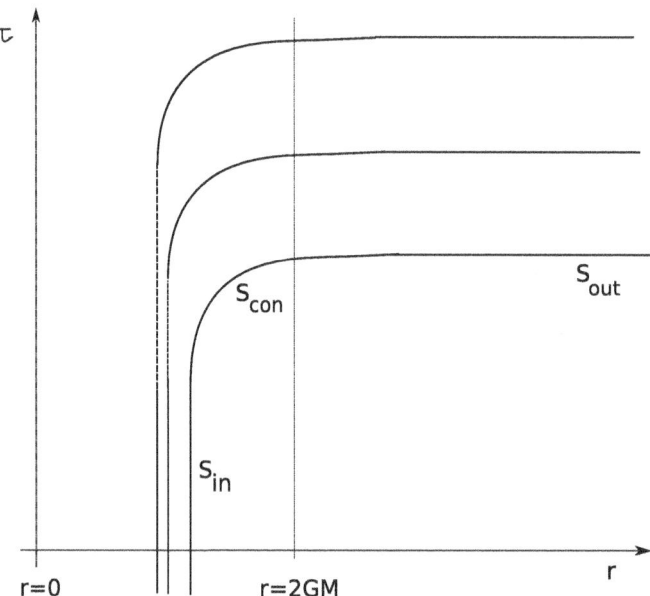

Fig. 1.4 Constructing a slicing of the black hole geometry. For $r > 3GM$ we have the part S_{out} as a $t = constant$ slice. The 'connector' part S_{con} is almost the same on all slices and has a smooth intrinsic metric as the surface crosses the horizon. The inner part of the slice S_{in} is a $r = constant$ surface, with the value of r kept away from the singularity at $r = 0$. The coordinate τ is only schematic; it will degenerate at the horizon

All this makes one spacelike slice, but what we need is a family of such slices to foliate the region of spacetime that is of interest. Let us try to make a 'later' slice in our foliation. For the part S_{out} we know what this means: we should take $t = constant$ with a larger value of t. What do we do for the inside part S_{in}? If we wish to advance this part forward in the direction that is locally timelike, then we have to move it *inward* toward smaller r (recall that r is the timelike direction inside the horizon). If we keep moving our successive slices toward smaller r, we will soon reach the vicinity of the singularity at $r = 0$, which we did not want to do. So we will move each successive slice to a smaller r value but only by a very *small* amount; we will make this amount smaller and smaller so that the slices asymptote to say the surface $r = \frac{GM}{2}$, still comfortably away from the singularity at $r = 0$.

So what is the essential difference between one slice and a later slice? The outer part S_{out} has just moved up in time τ, but not changed its intrinsic geometry. The 'connector' part S_{con} has not changed its intrinsic geometry much either. The nontrivial change has been in the inner part S_{in}, which has not changed in its r location very much, but it has become *longer*; there is an extra part indicated by the dotted segment that has emerged to allow S_{in} to connect to the rest of the slice.

In the $r - \tau$ plane of Fig. 1.4 this may look a strange set of slices, so we redraw them a bit differently in Fig. 1.5. The lowest slice corresponds to the time before

Fig. 1.5 The slices of Fig. 1.4
redrawn in a different way to
show the changes from one
slice to the next

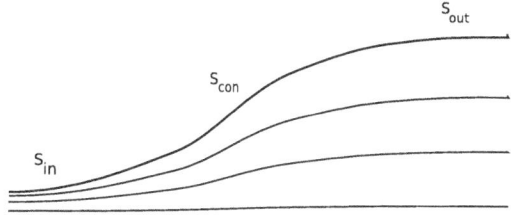

the black hole is formed. Thus it is essentially a flat slice $t = constant$ all through. On later slices, the part on the right, which is in the 'outer' region, keeps advancing forward in time. The part on the 'inside' advances very little. As a consequence there is a lot of stretching in the part that connects the part on the left to the part on the right. Later and later slices have to stretch more and more in this region.

In any spacetime we always have the freedom of pushing forward our spacelike slice at different rates at different locations (Wheeler terms this 'many-fingered' time in general relativity). But note that in flat spacetime, for example, we could not have done what we see in Fig. 1.5. Thus consider flat spacetime, and let the first slice be $t = constant$. Now for later slices we try to keep the left side of the slice fixed (or advancing very slightly) and we make the right side move up to later times. Then after a while we will find that the part of the slice joining these two parts is no longer spacelike; it will become null somewhere and then become timelike. Thus the kind of slices that we see in Fig. 1.5 is particular to the black hole geometry, and the infinite stretching that we see in these slices can be traced to the presence of a horizon.

Finally, in Fig. 1.6 we depict the slices on the Penrose diagram. The later the time slice, the more it moves up near future null infinity before coming into the horizon. Thus the later and later time slices will be able to capture more and more of the Hawking radiation emitted from the hole.

The important fact about slices in the black hole geometry is the following. In Schwarzschild coordinates both g_{tt} and g_{rr} become singular at the horizon: one vanishes and one diverges. With the slices we have chosen the spatial metric along the slices remains regular as we cross the horizon. If we allow our slices to reach the singularity at $r = 0$, then we can foliate the geometry by slices which are spacelike, and which are all similar to each other as far as their intrinsic geometry is concerned. Then why will there be particle production? The point is that there is no *timelike killing vector in the geometry*. Suppose we draw a vector connecting a point at $r = r_0$ on one slice with the point at $r = r_0$ on the next slice. If this vector was timelike everywhere we could use it to define time evolution, and everything would look time independent: the slices do not change, and the metric with this choice of time direction will look time independent. But this vector will *not* be timelike everywhere; it will become null on the horizon and be spacelike inside the horizon.

We have taken extra care to make our slices not approach the singularity – we let them follow a $r = constant$ path to an early enough stage where the singularity had not formed, and then take them in to $r = 0$. This feature of the slices is not

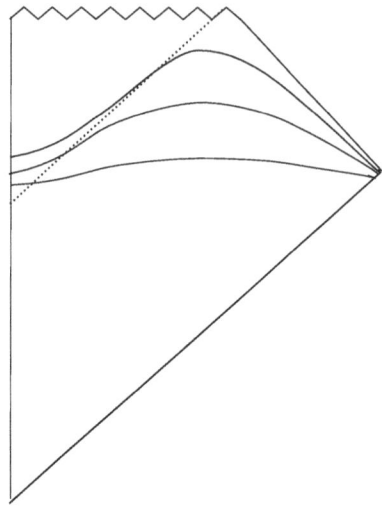

Fig. 1.6 The slices drawn on the Penrose diagram. Later slices go up higher near future null infinity and will thus capture more of the Hawking radiation

directly related to the production of particles in Hawking radiation, but we have done the slicing in this way so that the evolution stays in a domain where curvature is everywhere low and so classical gravity would appear to be trustworthy.

To summarize, the central point that we see with these different ways of exhibiting the slices is that the geometry of the black hole is not really a time-independent one, and particle production can therefore be expected to happen.

1.4.1 The Wavemodes

Let us now look at the wavemodes of the scalar field in the black hole geometry.

We will look at non-rotating holes, so the metric is spherically symmetric, and we can decompose the modes of the scalar field ϕ into spherical harmonics. Most of the Hawking radiation turns out to be in the lowest harmonic, the s-wave, so we will just focus our attention on this $l = 0$ mode; the physics extends in an identical way to the other harmonics. We will suppress the θ, ϕ coordinates, drawing all waves only in the r, t plane.

To study the emission from the hole, in principle we should solve the wave equation in the metric of the black hole. This is complicated, though many approximations have been developed to carry out this computation and a lot of numerical work has been done as well. But the basic ideas involved in the computation of Hawking radiation can be understood by using a very simple description of the wavemodes: solving them by the 'eikonal approximation', which we now describe.

Since we have taken the harmonic $l = 0$, all we have to do is describe the wave in the r, t plane. In flat space, we have two kinds of modes: ingoing modes and outgoing modes. For higher l, there is a 'centrifugal barrier' from angular momentum, and there will not be such a clean separation between ingoing and outgoing waves at small r. But if we are looking at a high-frequency mode then this angular momentum term is ignorable, and the physics again splits into an ingoing and an outgoing mode. From the wave equation $\Box\phi = 0$ we find that these ingoing and outgoing modes travel at the speed of light.

Consider an outgoing wavemode, and look at it on a spacelike hypersurface. Then we would see a sinusoidal oscillation of its phase with some wavelength which we call λ. We assume that

$$\lambda \ll GM \ . \tag{1.38}$$

Here GM is the scale over which the metric of the black hole varies, so we are asking that the wavelength be much smaller than the scale over which the metric is curved. Thus the wave oscillations will locally look like oscillations on a piece of flat spacetime. It will turn out that as our wavemodes evolve their wavelength will increase, and will finally become order $\sim GM$, but by the time that happens they would be waves traveling near infinity where we understand their physics very well. Thus while we may find the overall radiation rate to be incorrect by a factor of order unity because $\lambda \sim GM$ near the end of the evolution, the basic problem created by the 'entanglement of Hawking pairs' will be very robust and will not be affected by the errors caused by our approximation.

In Fig. 1.7 we sketch the wavemode as seen on a spacelike surface. At each point on the surface, the wavemode is a complex number given by an amplitude and a phase. Let this be an outgoing mode, of the type $e^{ik(r-t)}$ at infinity. Take a point A on this spacelike surface, and suppose the phase of the wavemode is $e^{i\phi_0}$ at this point. Draw a radial null geodesic through A, going out to infinity. Assign the phase $e^{i\phi_0}$ to all points on this null geodesic. Do the same for all points on the initial surface. The amplitude of the wavemode at point A also determines the amplitude at all points along the null geodesic through A, but we should note that in $d+1$ spacetime

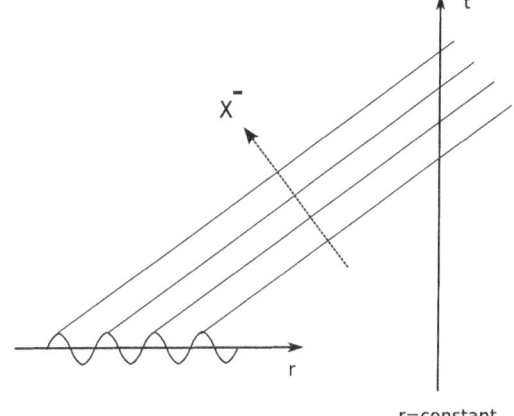

Fig. 1.7 The wavemode on an initial spacelike surface is evolved by letting the phase be constant on outgoing null rays. At infinity we can describe the mode by its intersection with a space-like surface, which gives a function $\sim e^{ikr}$. Alternatively we can give its intersection with a timelike surface which gives a function $\sim e^{-ikt}$. Lastly, we can describe the mode by giving the phase on different outgoing null rays, which gives a function e^{-ikX^-}

dimensions the amplitude of a spherical wave falls off as $\frac{1}{r^{\frac{d-1}{2}}}$, so we put in this decrease with r when finding the amplitude at all points along the null geodesic.

This process gives us a wavemode evolved to all points to the future of our space-like hypersurface. If the wavelength was everywhere small compared to the curvature of the manifold, this would be a very good approximation to the actual solution of the wave equation; as it is, it will be an approximate solution that will serve our purpose in what follows. To summarize, we have evolved the outgoing wavemode by assuming that the phase of the mode stays constant along the outgoing null rays.

We can describe the wavemode in a few different ways. First, we can 'catch' it on a spacelike surface as we did in Fig. 1.7; in this case we see a waveform e^{ikr} on the spacelike surface. We can also 'catch' the wavemode by looking at its intersection with a timelike surface $r = constant$. Then on this timelike surface we will see a phase like $e^{-i\omega t}$; this can be seen from the way the null lines intersect the timelike surface $r = constant$. Third we can describe the wavemode by giving the phase on each null ray. For outgoing waves the null rays are of the form

$$t - r \equiv X^- = constant \; . \tag{1.39}$$

Thus we can write outgoing modes as e^{-ikX^-}, or mode generally as $f(X^-)$. We will generally chose the function f so as to make a localized wavepacket.

1.5 The Evolution of Modes in the Black Hole

In this section we will put together a lot of the tools we developed in the above discussion. Our goal is to look at wavemodes in the black hole background and to see how they evolve. At the end of this evolution the initial vacuum modes will be populated with particles. What we wish to understand is the nature of this state with particles, in particular, how the various particles are correlated or 'entangled' with each other. The entire essence of the information paradox lies in understanding this entanglement.

Let us begin our discussion with a look at the Penrose diagram again, sketched in Fig. 1.8(a). We have drawn a circle around the region that is of immediate interest to us. In Fig. 1.8(b) we have drawn an expanded view of this region. The important thing about this region of spacetime is that in the traditional black hole picture this is a region of 'empty space'. Thus there is no large curvature here or any other matter that our scalar field ϕ could interact with. To understand better the state of the scalar field here, consider the evolution of field modes depicted in Fig. 1.8(a). Vacuum modes start off at past null infinity as ingoing modes. They reach $r = 0$ and scatter back as *outgoing* modes. At this stage there is no singularity at $r = 0$, so we just get outgoing vacuum modes after this scattering. These outgoing modes then show up in the circled region of Fig. 1.8(a). We are interested in the further evolution of these modes.

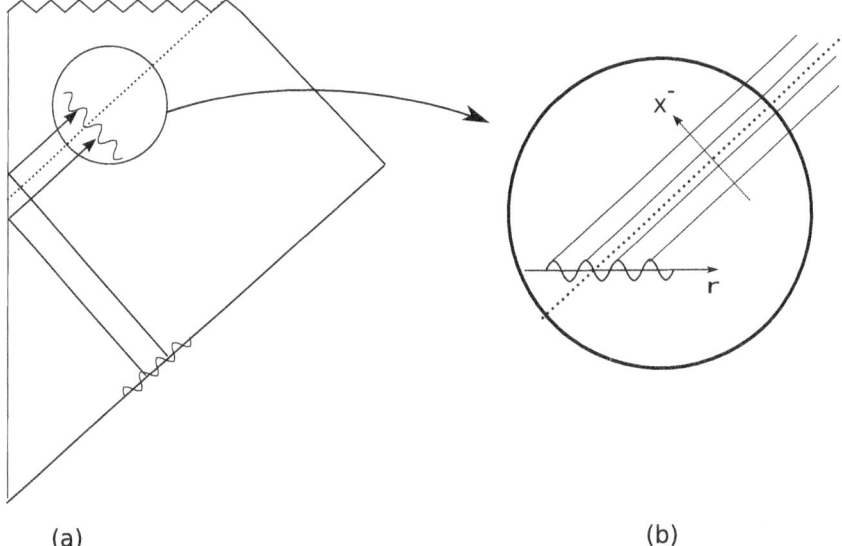

Fig. 1.8 (**a**) The region around the horizon is a vacuum. (**b**) An outgoing wavemode on an initial spacelike surface is evolved by letting the phase be constant on outgoing null geodescis

The outgoing field mode is drawn in more detail in Fig. 1.8(b), where we have caught the mode on a spacelike surface which we will call our 'initial slice'. We follow our above described method of evolving the wavemodes by letting the phase be constant along outgoing radial null geodesics. These null geodesics look like straight lines on the Penrose diagram, so at first it might seem that the wavelength of the mode is not changing as we follow the mode out toward infinity. This is not true, since in the Penrose diagram the actual distances between points are large when the points are near infinity. (In drawing the Penrose diagram we squeeze the spacetime in a 'conformal' way so that all of spacetime fits in a finite box; this automatically squeezes points near infinity by a large amount.)

What we really want to see is how the wavelength of the mode changes as the mode is evolved. So in Fig. 1.9 we sketch the evolution in the $r - \tau$ diagram that we discussed above. The initial slice is drawn again, with the outgoing wavemode on it. The lines of constant phase are drawn too, but now they do not look like straight lines. We had seen that the horizon itself is an outgoing null geodesic that stays at all times at $r = 2GM$. The rays starting slightly outside the horizon eventually 'peel off' and go to spatial infinity, while those starting slightly inside 'peel off' and fall in toward small r. Thus the wavemode will get distorted as it evolves.

What we want to do now is to 'catch' the wavemode on a later spacelike slice. By following the null rays, we can obtain the phase of the wavemode all along this later slice. We can see that there is quite a distortion between the wavemode as seen on the initial spacelike slice and the wavemode as it is 'caught' on this later slice, and the changes come because the null geodesics just inside and outside the

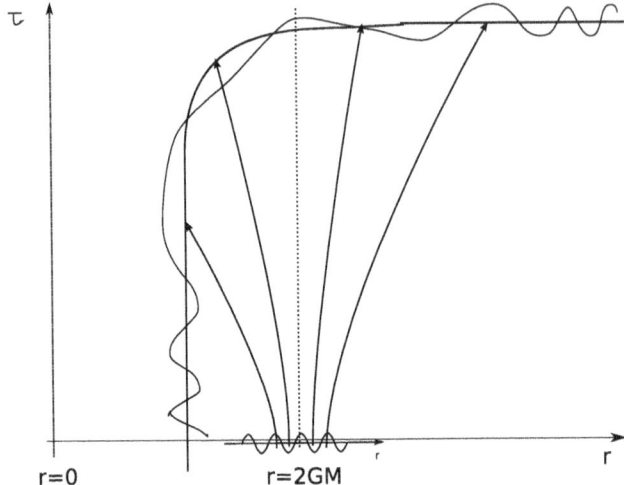

Fig. 1.9 A wavemode which is a positive frequency mode on the initial spacelike surface gets distorted when it evolves to a later spacelike surface; the mode will not be made of purely positive frequencies after the distortion

horizon evolve in quite different ways. But if the wavemode is distorted, there can be particle creation. We will now look at the distortion in much more detail and discuss the nature of this particle creation.

1.5.1 The Coordinate Map Giving the Expansion

Consider the vicinity of the horizon sketched in Fig. 1.8(a). The local geometry is approximately flat space, and the field modes are in the vacuum state. Let us use null coordinates y^+, y^- to describe the spacetime here (recall that the angular S^2 is suppressed throughout). The outgoing modes, which are of interest to us, are then of the form

$$\psi_{initial} \sim e^{iky^-} . \tag{1.40}$$

We will assume that $k > 0$. In the expansion of the field $\hat{\phi}$ the positive frequency modes multiply the annihilation operators $\hat{a}_k$. We will write the negative frequency modes as e^{-iky^-}; these will multiply the creation operators.

Now let us see what coordinates would be good on the late time spacelike slice, sketched in Fig. 1.9. Consider the outer part of the slice S_{out}. This part is in a region which is close to flat Minkowski spacetime. We had discussed above that particles were well defined in such a region of spacetime, and this definition required us to use positive frequency modes based on the usual coordinates on Minkowski space. So we just use the standard definition of null coordinates here:

$$X^+ = t + r, \quad X^- = t - r, \tag{1.41}$$

and the positive frequency modes are of the form

$$\psi_{out} \sim e^{iKX^-}, \tag{1.42}$$

with $K > 0$. Since we evolve our field modes by keeping the phase of the mode constant on the outgoing null rays, all we need to know now is the relation between the outgoing null coordinates on S_{out} and the outgoing null coordinates on the initial slice:

$$X^- = X^-(y^-). \tag{1.43}$$

Note that X^+ is not involved in this relation, so modes that start off as functions of only y^- become modes involving only X^-.

What we need to know now is the nature of the function in (1.43). This requires us to study the black hole metric and its geodesics. We will not carry out those computations here, but instead just quote the results and focus on the qualitative physics which emerges. Detailed derivations of the results we will use can be found in [2, 9, 10]. A good review in the 2D context can be found in [11].

First consider points very close to the horizon. Let $y^- = 0$ be the horizon itself. Then points $y^- < 0$ will be outside the horizon, and points $y^- > 0$ will be inside the horizon.

First consider null rays that are close to and just outside the horizon. It turns out that the null coordinate X^- describing the rays at infinity is related to the label used near the horizon by a relation which is logarithmic:

$$X^- \sim -\ln(-y^-). \tag{1.44}$$

Note that $y^- < 0$ for these rays, so we are taking the log of a positive number, as we should. Since $|y^-|$ is very small, the log is negative, so X^- is actually positive. But as it stands this relation does not have the right units. The coordinate y^- has units of length, so we must first make a dimensionless variable and then take the log. The only natural length scale in the black hole geometry is GM, and the relation actually looks like

$$X^- = -(GM) \ln\left(-\frac{y^-}{GM}\right). \tag{1.45}$$

This is a very interesting relation. A simple Fourier mode e^{iky^-} will get distorted by the logarithmic map. But to completely understand this logarithmic map we also need to understand what happens when the rays are *not* very close to the horizon. Thus look at $y^- \lesssim -GM$. For such values of y^- we are no longer close to the horizon. The rays are thus almost like rays in flat spacetime, and so there is no serious deformation of the wavemodes. Thus the relation (1.45) will change over to a relation like

$$X^- = -y^-, \quad \text{for } y^- \lesssim -GM, \tag{1.46}$$

and there will be *no* distortion of modes in this region which is away from the horizon.

Now let us look at the range $y^- > 0$, i.e., the part of the mode inside the horizon. There are no natural coordinates to describe the inside of the black hole. But since we have chosen a way of drawing spacelike slices across the entire geometry, we can use wavemodes here that are natural to this slicing; the actual choice of wavemodes inside the horizon will not matter at the end. Thus consider the part of the slice S_{in} inside the horizon. We can introduce a coordinate Y on this part of the slice which is linear in the distance measured along the slice. The null rays for $y^- > 0$ will intersect this slice at various points. We assign to each ray a null coordinate Y^- which is equal to Y at the point where the null ray intersects S_{in}. Thus this coordinate assignment is similar to the coordinate X^- defined on S_{out}, with the difference that in the case of S_{out} there was a natural physical choice and particles defined using X^- had the correct energy–momentum tensor to be the real particles at infinity.

With such a coordinate choice Y^-, using the black hole geometry we find a relation like

$$Y^- \sim -\ln y^- \qquad (1.47)$$

or with the correct dimensionful parameters inserted

$$Y^- = -GM \ln\left(\frac{y^-}{GM}\right) . \qquad (1.48)$$

Thus there will be a distortion of the wavemodes on this side of the horizon as well.

The last point to note is that if $|y^-|$ is *very small* (i.e., the null ray is very close to the horizon) then the ray intersects the late time slice on the 'connector region' S_{con}, rather than on S_{out} or S_{in}.

We now come to a crucial point. The wavemode on the initial slice straddles both sides of the horizon. Indeed, the horizon is not a 'special place' in the geometry from a local point of view; this can be seen from the circle drawn in Fig. 1.8(a), which circles a region of spacetime much like any other. Thus wavemodes near the horizon naturally continue from one side of the horizon to the other. But the subsequent expansion of the geometry, encoded in the behavior of the null geodesics, treats the parts of the wavemode outside and inside the horizon quite differently. The part of the initial wavemode for $y^- \ll -GM$ does not 'stretch', the part for y^- negative and small in magnitude (but not too small) reaches S_{out} with a logarithmic stretching, the part for *very* small $|y^-|$ ends up on the connector region S_{con}, and the part for $y^- > 0$ (but not too small) ends up on S_{in}. The consequent distortion of the wavemode is sketched in Fig. 1.9.

From this figure we can observe some basic facts about the distortion of the wavemode. The distortion is large around the point where the rays move from being inside the horizon to being outside the horizon. As we will see in more detail below, there is very little distortion away from this region. Because the wavemode gets distorted, a given Fourier mode on the initial slice becomes a combination of modes on the later slice. Note that all modes involved are outgoing modes; we have functions of y^-, X^-, Y^-. This fact is a consequence of our 'ray approximation' where we evolve the mode by letting the phase of ϕ be constant on outgoing rays.

The most important thing here is that the part of the mode straddling the horizon splits into a part on S_{out} and a part on S_{in}. This will make the state of the created particles a 'mixed state' of the outside and inside quanta, as we shall discuss in more detail below.

1.5.2 Detailed Nature of the Wavemode

Consider the part of the wavemode that escapes to $r \to \infty$. In Fig. 1.10 we have drawn, on the $r - \tau$ plane, the lines of constant phase for this part of the wavemode. We have drawn a timelike surface ($r = constant$) on which we 'catch' the mode outside the hole; it will be easier to understand the mode on this surface first and then read off its behavior on any spacelike surface near infinity.

Our goal is to see where the distortion is large enough to create particles. On the initial slice near the horizon, we take a Fourier mode e^{iky^-}. Consider this mode for the range $-GM < y^- < 0$. Recall that y^- is negative outside the horizon and zero on the horizon. Also, for $y^- \lesssim -GM$, there is no significant distortion of null rays,

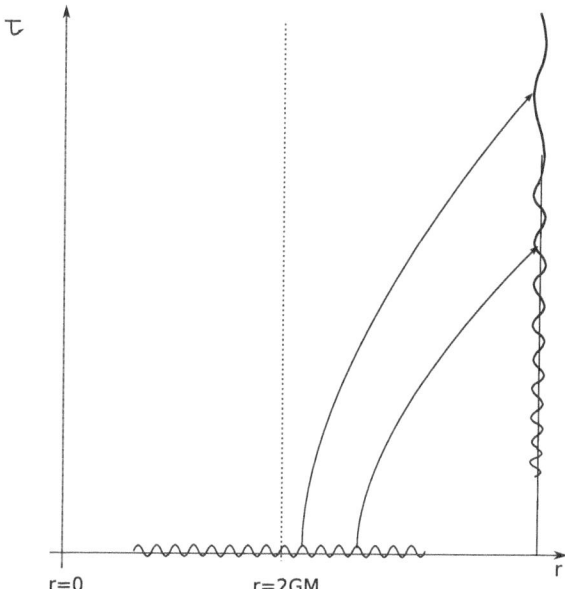

Fig. 1.10 The Fourier mode on the initial spacelike slice is evolved in the eikonal approximation and 'caught' on the timelike surface $r = constant$ near infinity. (From the behavior of the mode on this surface we can immediately obtain what it looks like on any spacelike surface near infinity.) The wavelength of oscillations becomes longer and longer as we go up the surface, with the last oscillation to emerge from the horizon extending all the way to $t = \infty$

so that we get $y^- \approx X^-$. What we now wish to show is that even though there is a logarithmic distortion of coordinates for smaller $-GM \lesssim y^- < 0$, there is no *particle production* for most of this range of y^-; in fact particle production will be relevant only for the few oscillations of the wavemode near $y^- = 0$.

Thus we now look at the range $-GM < y^- < 0$, where we assume that the logarithmic maps (1.45) are a good approximation. Consider the Fourier mode e^{iky^-} on the initial slice. Let the wavelength of this mode be much smaller than GM:

$$\lambda = \frac{2\pi}{k} = \epsilon GM, \qquad \epsilon \ll 1 . \tag{1.49}$$

Thus the number of oscillations of the wavemode in our range $-GM < y^- < 0$ is large:

$$\text{\# oscillations} = \frac{1}{\epsilon} \gg 1 . \tag{1.50}$$

After the mode evolves to the late time slice we have to look at the wavelength in the X^- coordinate system. Consider one oscillation of the wavemode, which in the y^- coordinate system extends over the range

$$-\alpha < y^- < -\alpha + \epsilon \tag{1.51}$$

(here $\alpha > 0$). The wavelength in the X^- frame will be

$$\lambda_1 = |\delta X^-| = \left| \left(\frac{dX^-}{dy^-} \right) \delta y^- \right| = \frac{GM}{\alpha} \epsilon = \frac{GM}{\alpha} \epsilon . \tag{1.52}$$

Now comes an important question: what about the *next* oscillation of the wavemode? On the initial slice this spans the range

$$-\alpha - \epsilon < y^- < -\alpha , \tag{1.53}$$

where we have chosen to look at the oscillation that is the neighboring one on the side closer to the horizon. This evolves to have a wavelength

$$\lambda_2 = \frac{GM}{|y^-|} \epsilon = \frac{GM}{\alpha - \epsilon} \epsilon . \tag{1.54}$$

How different is (1.54) from (1.52)? Let us first suppose that we are *not* looking at the first few oscillations of the wavemode near the horizon. Then we have $|y^-| >> \epsilon$, and

$$\frac{\lambda_2}{\lambda_1} \approx \frac{\alpha}{\alpha - \epsilon} \approx 1 . \tag{1.55}$$

This is the important fact: we can take several adjacent oscillations of the wavemode on the initial slice and find they evolve to almost the same final wavelength. Thus the stretching they suffer can be called an almost *uniform* rescaling of coordinates. But under a uniform rescaling we do *not* create particles, a fact that we can see as follows. Suppose an initial mode e^{iky^-} evolves to $e^{ik(\mu y^-)}$, with μ a (positive) constant. To check for particle creation we would compute (f,g) where $f = e^{iky^-}$ is the positive frequency mode defining the initial vacuum and $g = (e^{ik\mu y^-})^*$ is the *negative* frequency mode for the final vacuum. But

$$(f,g) = -i \int dy^- \, e^{iky^-} \, e^{ik\mu y^-} \to 0 \tag{1.56}$$

since both Fourier modes involved in the integral have the same sign of the exponent. (We get a nonzero integral only if we have e^{iky} and e^{-iky} in the integrand.)

So what we seem to be finding is that the part of the wavemode that is not too close to the horizon undergoes deformations due to the logarithmic stretching, but this does not create particles because under this stretching there is no significant mixing of positive and negative frequency modes. The underlying reason for why we failed to create particles is the same as the analysis of scales that we did in the toy model with harmonic oscillators. In the latter case there was no particle creation if the change of the potential was too slow compared to the time period of the oscillator. In the present case, *there is very little change in the stretching factor over the period of oscillation of the wave*, and so we again get no significant particle creation.

To make the above conclusion more precise we recall the notion of *wavepackets*.

1.5.3 Wavepackets

In Fig. 1.11(a) we depict a wavemode with a definite wavenumber k_0. This wavemode has an infinite spatial extent. For physical arguments it is more convenient to have a wavemode that is localized in some region of space. Such a wavemode can be obtained by appropriately superposing wavemodes of different k. But we also wish to retain some properties of the mode arising from the fact that the wavenumber was k_0. Thus we use only a small band of k around the value k_0:

$$k_0 - \Delta k < k < k_0 + \Delta k, \qquad \frac{\Delta k}{|k_0|} << 1 . \tag{1.57}$$

This makes a wavetrain that 'sort of' has the wavenumber k_0 but which decays after a certain number of oscillations and is thus localized. Our discussions are mostly qualitative, so we will allow ourselves to use wavetrains that are only a

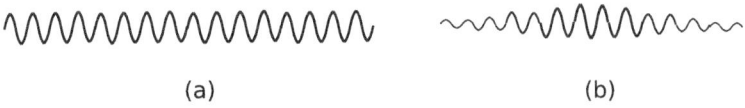

(a)   (b)

Fig. 1.11 (a) A Fourier mode with given wavelength $\lambda = \frac{2\pi}{k_0}$. (b) Appropriately superposing fourier modes with wavenumbers near k_0 we can make a wavepacket

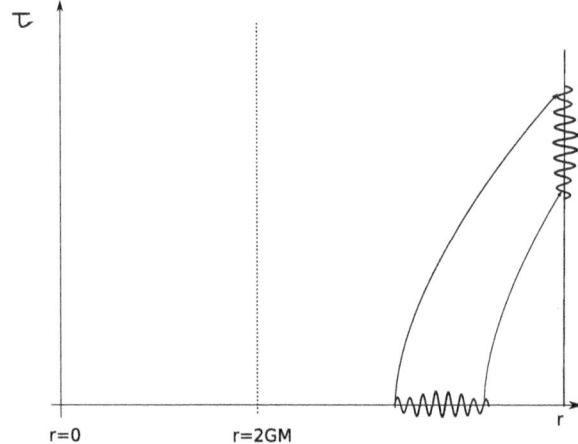

Fig. 1.12 If we look at the oscillations that are not too close to the horizon, then we can make a wavepacket out of them that evolves to a wavepacket at infinity. Suppose we can make a localized wavepacket such that in the region occupied by the wavepacket the 'stretching' of space is approximately uniform. Then there will be no mixing of positive and negative frequencies and therefore no particle production

few oscillations long; this means that we will not take $\frac{\Delta k}{k_0}$ to be very small, but for our pictorial understanding it will be enough to have k in the rough neighborhood of k_0.

Let us now use the above discussion together to make the point that we are after. In Fig. 1.12 we make a wavepacket out of a few oscillations that are not too close to the horizon. This wavepacket evolves to a wavepacket near spatial infinity without significant distortion, since the oscillations making the wavepacket suffer an almost uniform stretching under the evolution. Thus there is no significant particle production from the part of the wavemode where $|y^-| \gg \epsilon$.

1.5.4 Modes Straddling the Horizon

So far we have seen what part of the wavemode does *not* create particles. The part at $y^- \lesssim -GM$ does not get deformed. The part $-GM \lesssim y^- \ll -\epsilon$ deforms logarithmically but can be broken up into wavepackets, each of which suffers 'nearly uniform stretching', so again we do not get particle creation. A similar analysis can be performed for the domain $y^- > 0$ which is inside the horizon. We can now turn to the part of the wavemode that *does* create particle pairs.

Consider the wavemode on the initial surface and look at the domain of y^- which covers a few oscillations on either side of the horizon $y^- = 0$. Thus we have

$$|y^-| \sim \epsilon . \tag{1.58}$$

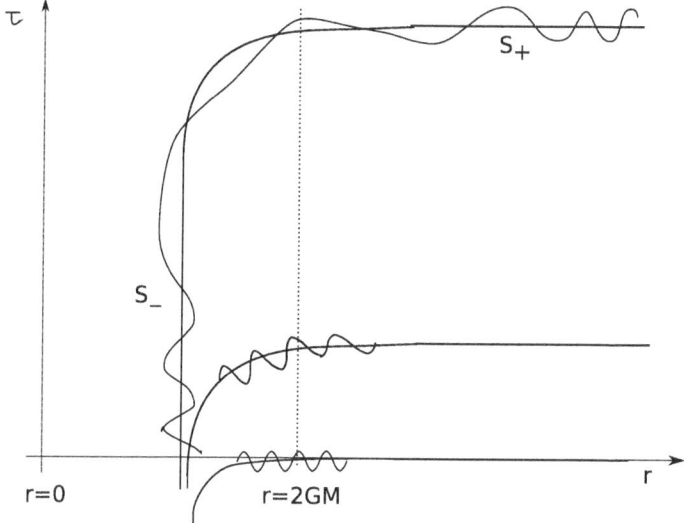

Fig. 1.13 A Fourier mode on the initial spacelike surface is evolved to later spacelike surfaces. In the initial part of the evolution the wavelength increases but there is no significant distortion of the general shape of the mode. At this stage the initial vacuum state is still a vacuum state. Further evolution leads to a distorted waveform, which results in particle creation

With just a few oscillations in this range, we cannot break this part of the wavemode further into wavepackets. Thus we must evolve it as a whole to the late time surface and see what it becomes. The evolution is described in Fig. 1.13. On the initial slice we have regularly spaced oscillations. If we look at surface just a little later, they are still pretty much like regularly spaced oscillations, since there has not been much deformation; thus so far there is no significant particle production. On slices that are much later, we see that the mode has deformed significantly: there are a few oscillations on the part S_- of the surface that is inside the horizon, then a large gap until we reach a region on S_+, where we find oscillations again.

Note that on this late time slice the deformation of these oscillations of the wavemode is very *nonuniform*. We have a positive frequency mode on the initial surface e^{iky^-}, but on the late time surface we will get an admixture of positive frequency modes e^{iKX^-} and negative frequency modes e^{-iKX^-}. The same happens for the part of the mode on S_-. Thus there will be particle creation.

The most important part of our entire discussion comes now. We know from (1.28) that when we create particles by deforming spacetime the vacuum state changes to a state of the form $e^{-\frac{1}{2}\Sigma_{ij}\gamma_{ij}\hat{b}_i^\dagger\hat{b}_j^\dagger}|0\rangle$. But in the present case we can break the creation operators $\hat{b}_k^\dagger$ into two sets: those on S_+ which we call $\hat{b}_k^\dagger$ and those on S_- which we call $\hat{c}_k^\dagger$. When we compute the state on the late time surface it turns out to have the form

$$e^{\Sigma_k \gamma \hat{b}_k^\dagger \hat{c}_k^\dagger}|0\rangle \ . \tag{1.59}$$

We do not derive this result here; the derivation can be found, for example, in [2, 3, 9–11]. But this is the crucial result for the physics of information, so we will now spend some time in understanding it.

1.5.5 The Nature of the Created Pairs

Consider again Fig. 1.13. On the initial surface the wavemode had a very short wavelength. On later time surfaces the wavelength has been stretched to a longer one, though there is no particle production because the stretching is almost uniform over the oscillations under consideration. The wavelength keeps getting longer as we go to later time slices, till the deformation becomes nonuniform and particles are created. But there is only one length scale in the geometry – the scale GM – and one can see easily that when particles are produced the wavelength of the mode has become $\sim GM$. At this point the wavemode has also moved to distances $\gtrsim GM$ from the horizon, and further deformation stops. Thus the wavelength of the produced quanta is $\sim GM$. These are the Hawking radiation quanta, so we see that this radiation has a temperature $\sim \lambda^{-1} \sim \frac{1}{GM}$. The exact temperature is [2, 3]

$$T = \frac{1}{8\pi GM} .$$
(1.60)

So the wavemode ends its evolution with a wavelength $\sim GM$, but what was its wavelength on the initial slice that we had drawn? On this initial slice there are modes of all possible wavelengths. Consider a wavemode with wavelength *shorter* than the one shown in Fig. 1.13. Then this mode will evolve for a *longer* time before it suffers a nonlinear deformation.

This situation in depicted in Fig. 1.14. On the initial slice we have drawn two wavemodes of different wavelengths. The one with the longer wavelength becomes distorted first and creates the quanta labeled b_1 and c_1 on the late time slice. The wavemode with shorter wavelength evolves for a longer time before becoming distorted and creates the quanta labeled b_2, c_2.

The state of the first pair b_1, c_1 is of the form

$$|\psi\rangle_1 = C e^{\gamma \hat{b}_1^\dagger \hat{c}_1^\dagger} |0\rangle .$$
(1.61)

Here $\hat{b}_1^\dagger$ is an operator that creates a quantum in the localized wavepacket depicted as b_1 in Fig. 1.14, and similarly $\hat{c}_1^\dagger$ creates the quantum of the wavepacket labeled c_1. Because we have broken up wavemodes into localized wavepackets, we can define a sort of local vacuum $|0\rangle_{b_1}$ in the region occupied by this mode b_1. If we are in this vacuum state then there are no quanta in this region, if we act with $\hat{b}_1^\dagger$ once then we have one quantum with this wavepacket, if we act with $\hat{b}_1^\dagger \hat{b}_1^\dagger$ then we have two quanta of this type, and so on. Doing the same for the modes on S_- we can write the state (1.61) as

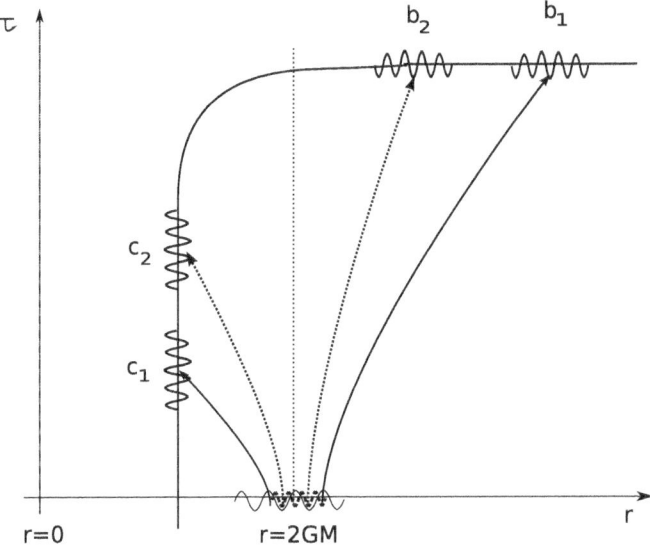

Fig. 1.14 On the initial spacelike slice we have depicted two Fourier modes: the longer wavelength mode is drawn with a *solid line* and the shorter wavelength mode is drawn with a *dotted line*. The mode with longer wavelength distorts to a nonuniform shape first and creates an entangled pair b_1, c_1. The mode with shorter wavelength evolves for some more time before suffering the same distortion, and then it creates an entangled pair b_2, c_2

$$|\psi\rangle_1 = Ce^{\gamma \hat{b}_1^\dagger \hat{c}_1^\dagger}|0\rangle_{b_1}|0\rangle_{c_1} . \tag{1.62}$$

A similar state is produced by the wavemode which started off with a shorter wavelength on the initial slice. We get particle pairs described by

$$|\psi\rangle_2 = Ce^{\gamma \hat{b}_2^\dagger \hat{c}_2^\dagger}|0\rangle_{b_2}|0\rangle_{c_2} . \tag{1.63}$$

The pairs b_k, c_k for different k lie in regions that do not overlap, so the overall state on the late time slice is the direct product of the states $|\psi\rangle_k$:

$$|\psi\rangle = |\psi\rangle_1 \otimes |\psi\rangle_2 \otimes |\psi\rangle_3 \otimes \cdots . \tag{1.64}$$

We have presented a simplified discussion of the created pairs; more technical details can be found in [2, 3, 9–11]. For a more accurate description we should use a large number of oscillations in making each wavepacket (we have used just a few), and then we will have to consider many wavenumbers in each of the intervals on $S_\pm$ over which the wavepackets extend. But the above approximate description has all the essence of what we need to understand the entanglement of quanta.

1.5.6 The Entangled Nature of $|\psi\rangle$

Consider the state $|\psi\rangle_1$

$$|\psi\rangle_1 = C\left(|0\rangle_{b_1} \otimes |0\rangle_{c_1} + \gamma\hat{b}_1^\dagger|0\rangle_{b_1} \otimes \hat{c}_1^\dagger|0\rangle_{c_1} + \frac{\gamma^2}{2}\hat{b}_1^\dagger\hat{b}_1^\dagger|0\rangle_{b_1} \otimes \hat{c}_1^\dagger\hat{c}_1^\dagger|0\rangle_{c_1} + \cdots\right)$$

$$= C\left(|0\rangle_{b_1} \otimes |0\rangle_{c_1} + \gamma|1\rangle_{b_1} \otimes |1\rangle_{c_1} + \gamma^2|2\rangle_{b_1} \otimes |2\rangle_{c_1} + \cdots\right), \qquad (1.65)$$

where $|n\rangle_{b_1}$ means that we have n quanta of type b_1 in the state, etc.

The important feature of this state is that the b_1 and c_1 excitations are 'entangled'. To understand this in more detail, let us take a simple example of an entangled state.

1.5.7 Entanglement and the Idea of 'Mixed States'

Consider two electrons, kept at two different locations, and let each of them have a 'spin-up' state and a 'spin-down' state. Then this system can have 'factored states' of the form

$$|\psi\rangle = |\psi\rangle_1 \otimes |\psi\rangle_2 . \qquad (1.66)$$

Examples are

$$|\psi\rangle = |\uparrow\rangle_1 \otimes |\downarrow\rangle_2$$

$$|\psi\rangle = \frac{1}{\sqrt{2}}(|\uparrow\rangle_1 + |\downarrow\rangle_1) \otimes \frac{1}{\sqrt{2}}(|\uparrow\rangle_2 + |\downarrow\rangle_2), \qquad (1.67)$$

etc. But we can also have "entangled" states which cannot be written as a product of the type (1.66), for example

$$|\psi\rangle = \frac{1}{\sqrt{2}}(|\uparrow\rangle_1 \otimes |\downarrow\rangle_2 + |\downarrow\rangle_1 \otimes |\uparrow\rangle_2) . \qquad (1.68)$$

Suppose we ask, what is the state of electron 1? For states of type (1.66) we can answer this question: we ignore the state of electron 2 and just give the answer $|\psi\rangle_1$. But for states of type (1.68) we cannot do this, and only the state of the entire system makes sense. Suppose we nevertheless want to ignore electron 2 in some way. Then we can make a 'density matrix'

$$\rho = |\psi\rangle\langle\psi| . \qquad (1.69)$$

For the two-electron system we get

$$\rho = \frac{1}{2} \ |{\uparrow}\rangle_1 \otimes |{\downarrow}\rangle_2 \quad {}_1\langle{\uparrow}| \otimes {}_2\langle{\downarrow}|$$

$$+ \frac{1}{2} \ |{\uparrow}\rangle_1 \otimes |{\downarrow}\rangle_2 \quad {}_1\langle{\downarrow}| \otimes {}_2\langle{\uparrow}|$$

$$+ \frac{1}{2} \ |{\downarrow}\rangle_1 \otimes |{\uparrow}\rangle_2 \quad {}_1\langle{\uparrow}| \otimes {}_2\langle{\downarrow}|$$

$$+ \frac{1}{2} \ |{\downarrow}\rangle_1 \otimes |{\uparrow}\rangle_2 \quad {}_1\langle{\downarrow}| \otimes {}_2\langle{\uparrow}| \ . \tag{1.70}$$

We can now 'trace over' the states of system 2, which for the above case means that the bra and ket states of system 2 must be the same in the terms that we keep. Then we get a 'reduced density matrix' describing system 1:

$$\rho_1 = \frac{1}{2} \ |{\uparrow}\rangle_1 \ {}_1\langle{\uparrow}| \ + \ \frac{1}{2} \ |{\downarrow}\rangle_1 \ {}_1\langle{\downarrow}| \ . \tag{1.71}$$

In general we get a density matrix of the form $\rho_1 = \sum_{m,n} C_{mn} |m\rangle_1 \ {}_1\langle n|$. The probability to find system 1 in state k is given by the coefficient C_{kk}. These probabilities must add up to unity, so we have $\mathrm{tr}\rho = 1$. The *entropy* that results from ignoring system 2 is given by

$$S = -\mathrm{tr}\,\rho \ln\rho \ . \tag{1.72}$$

For the density matrix (1.71) we can compute S easily since it is a diagonal density matrix

$$S = - \left[\frac{1}{2} \ln \frac{1}{2} + \frac{1}{2} \ln \frac{1}{2} \right] = \ln 2 \ . \tag{1.73}$$

If the state $|\psi\rangle$ in (1.69) is 'factorized' as in (1.66) then when we make ρ_1 and compute S we get $S = 0$. Roughly speaking, S gives the log of the number of terms in a sum like (1.68). The entropy is thus a measure of how 'entangled' the systems 1 and 2 are.

1.5.8 Entropy of the Hawking Radiation

Let us now return to the black hole. The state (1.65) is not factorized between the b_1 and c_1 excitations. The number γ is order unity, so the first few terms in the sum will be of relevance. To explain the significance of the entangled nature of the state we will for convenience replace the state (1.65) by the simpler state

$$|\psi\rangle_1 = \frac{1}{\sqrt{2}} \left(|0\rangle_{b_1} \otimes |0\rangle_{c_1} + |1\rangle_{b_1} \otimes |1\rangle_{c_1} \right) \ . \tag{1.74}$$

The quanta of type b_1 lie on the part S_+ of the spacelike surface which is outside the horizon, while the quanta of type c_1 lie on the part S_- which is inside the horizon. Due to the entanglement between b_1 and c_1 quanta, we cannot restrict ourselves to the Hawking radiation quanta b_1 and still describe them by a 'pure' quantum state. If we wish to ignore the quanta c_1 then we have to find the density matrix for the quanta b_1. For the state (1.74) we will get an entanglement entropy $S = \ln 2$. (The state (1.65) would have given an S of the same order.)

Now we can look at the other pairs of quanta $(b_2, c_2), (b_3, c_3)$, etc. We had seen that each of these sets (b_k, c_k) lives at a location different from the other pairs, so the overall state (1.64) was a direct product of states for each of these pairs. A little thought shows that the total entanglement entropy S will then be the sum of the entropies from each pair (b_k, c_k). Let us see how many such pairs there will be. The temperature of the Hawking radiation is (1.60), so the energy of the typical emitted quantum is $\sim (GM)^{-1}$. The mass of the hole is M, so the number of quanta that will be emitted when the hole has evaporated is

$$\# \, quanta \sim M(GM) \sim \frac{(GM)^2}{G} \,. \tag{1.75}$$

With an entropy of order unity from each set (b_k, c_k) we see that the entropy of the radiation is

$$S_{rad} \sim \frac{(GM)^2}{G} \sim S_{bek} \,, \tag{1.76}$$

so we see that the radiation has an 'entanglement entropy' of the order of the entropy of the black hole.

1.5.9 The Problem with the Entangled State

Consider the two-electron state (1.68), and suppose that we want to concentrate on the first electron. We have seen that we cannot write a quantum state for this electron alone. We can make a density matrix ρ_1, but this is not a 'pure' quantum state. Rather it is a statistical construct that allows us to get probabilities for different states of electron 1, and one cannot see the usual quantum principles of linear superposition or phase interference by looking at ρ_1.

Of course there is no fundamental problem with such an entangled state; all we have to do is realize that it is only the complete two-electron system that can be described by a quantum state. The situation is a bit different for the black hole case. As long as we are willing to look at both sets of quanta, b_k and c_k, we have an entangled quantum wavefunction. But if the black hole eventually disappears, then we will be left with the quanta of type b_k floating at infinity. We know that they cannot be described by a pure quantum state, and now we cannot write a mixed state either, for there is nothing for them to mix with! Thus the only way we can describe the b_k quanta is by the reduced density matrix ρ_b describing the b_k, and this description is inherently statistical, rather than a usual quantum mechanical one. This is what led Hawking to postulate that quantum mechanics in the presence of gravity is not a consistent theory by itself; he suggested that general configurations can only be described by density matrices, and we must make a quantum theory based on such a description.

Attempts to modify quantum theory in this way have not made much progress. Others have argued that the black hole does not completely evaporate away, but instead stabilizes after reaching planck size because of quantum gravity effects. In this case the quanta c_k are never removed from the system, and we have a pure state overall. But one is then forced to accept that there can be an infinite number of possible states of such a planck-sized remnant (since the remnant can result from an arbitrarily large black hole). Allowing the theory to have infinitely many states within a bounded spatial region and within a bounded energy range is unnatural, and creates many problems for the theory. It would therefore seem best if somehow we could get the black hole to disappear and yet have the quanta b_k left in a pure state. Let us now discuss what would be needed for this to be possible.

1.6 Common Misconceptions About Information Loss

We will find it helpful to start by considering several common misconceptions about how information can come out of the black hole.

1.6.1 Is the Emitted Radiation Exactly Thermal?

A common argument about Hawking radiation is the following. The above discussed computations give 'thermal radiation', but there could be corrections (from the gravitational backreaction of the created pairs, for example) which generate small deviations from 'thermality', and these deviations can encode the information that should escape from the hole.

The problem here is the word 'thermal'. What is thermal radiation? One might think that 'thermal' means the spectrum of radiation should be planckian; this spectrum is depicted by the solid curve in Fig. 1.15. Small deviations from this spectrum are shown by the dotted curve in Fig. 1.15. Can such a change in spectrum bring out information from the black hole? We will now see that the shape of the spectrum itself does not have much to do with whether the information comes out.

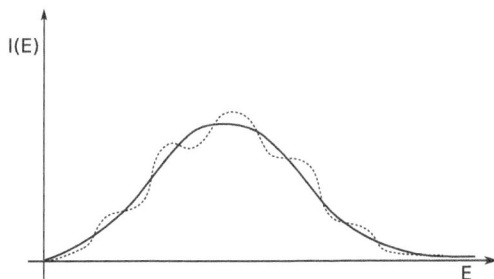

Fig. 1.15 The planck distribution; small deviations from this distributions are indicated by the *dotted curve*

For one thing, the spectrum of the semiclassical radiation from the black hole is not of the planck shape; the spectrum is modified by graybody factors. This is a general feature of radiation from any warm body – there is a modification to the spectrum if the emitted wavelength is comparable to the size of the body. For black holes, this wavelength is $\sim GM$, which is the same order as the black hole size $r \sim 2GM$. Thus the spectrum is not planckian anyway.

A more correct definition of 'thermal' radiation is that if the body has an absorption cross section $\sigma(k)$ for quanta of a certain wavenumber, then the emission rate for the same wavenumber is

$$\Gamma = \sigma(k) \frac{d^3k}{(2\pi)^3} \frac{1}{e^{\frac{\omega}{T}} - 1} . \tag{1.77}$$

The semiclassical radiation from the hole is 'thermal' in this sense. But the essential problem that we have is *not* created by this 'thermality', but by the entangled nature of the state. Whether we have the entangled state (1.65) (which can be shown to be 'thermal' in the above sense) or the entangled state (1.74), which is very different from 'thermal', we face the *same* problem. There is order unity entropy of entanglement from the state created by each pair of operators $(\hat{b}_k^\dagger, \hat{c}_k^\dagger)$, and so there is an entanglement entropy (1.76) for the radiation which is order S_{bek}. It is this entanglement that will eventually lead to information loss. By contrast, if a piece of coal burns away completely to radiation, then this radiation is in a pure state, even though it looks much more 'thermal' than a state which has the form (1.74) for each of the $(\hat{b}_k^\dagger, \hat{c}_k^\dagger)$.

Thus 'thermality' is not really the issue; the issue is the entangled nature of the state created in the process of black hole evaporation.

1.6.2 Can Small Quantum Gravity Effects Encode Information in the Radiation?

Consider the derivation of Hawking radiation discussed in the above sections. We have used a classical metric and a quantum field ϕ on this 'curved space', but gravity itself has not been treated as quantized; this is called the *semiclassical approximation*. Thus the semiclassical computation of radiation does not use the physics of quantum gravity anywhere. Since spacetime curvature was low in the regions where the wavemodes deformed and created particles, this would seem to be a good approximation. But one can still wonder if the *small* corrections that would arise from quantum gravity effects could change the state of the radiation to a pure state. There are two aspects to this question:

(a) The first point to note is that a *small* change in the state of the quantum field will *not* succeed in making the state of the b quanta a pure state. Focusing again on a given set (b_1, c_1) we see that their state is a mixed one like (1.65). To get no entanglement of the b_1 quanta with the c_1 quanta we would need a state like

$$|\psi\rangle_1 = \left(C_0|0\rangle_{b_1} + C_1|1\rangle_{b_1} + \cdots\right) \otimes \left(D_0|0\rangle_{c_1} + D_1|1\rangle_{c_1} + \cdots\right). \qquad (1.78)$$

But the state (1.78) is *not* a small perturbation on a state like (1.65). The two states are completely different, so we need an order *unity* change in the state of each set (b_k, c_k) before the state can become pure. Thus if quantum gravity is to help us, *then it must completely change the evolution of the wavemodes that we have been drawing in the above sections.*

(b) The second point is that even if we had a state like (1.78), and thus the radiation quanta b_k formed a pure state by themselves, it would not solve the information problem. Consider the Penrose diagram in Fig. 1.16(a). There are not two but *three* kinds of matter involved in the problem. There is the matter that fell in to make the hole, marked Q. Then there are the Hawking radiation quanta b_k (we have labeled them B) and their entangled partners, the c_k (labeled C in the figure).

The problem is that not only do the quanta B have to form a pure state, but they have to carry the information of the matter Q. This is because in quantum mechanics the evolution of states is one to one and onto, and so different states of the initial matter Q have to give different states of the final radiation B. In Fig. 1.16(b) we have drawn the slices as shown in Fig. 1.5, with Q, B, C indicated. We see that the quanta Q reach small r first, and exist on each slice. The way we have drawn our slices keeps Q always in a region of low curvature; to achieve this we have evolved the small r region very little as we move from slice to slice. As the evolution proceeds the b_k and c_k quanta start appearing out of the vacuum modes. But these vacuum modes were localized in the region between the b and c quanta, far away from where Q sits on the slice. So *how can the matter Q transfer its information to the b_k?* This is the essence of the information problem.

Note that all the evolution depicted in Fig. 1.16(b) has been in a low-curvature region, with slices that are smooth and carrying matter that is always of low density. Thus it would appear that the situation is like the low-curvature physics encountered in the solar system, and no unexpected quantum effects can occur. The only unusual thing is that through the course of the evolution the slices *stretch* by a large amount, as discussed in Sect. 1.4. In conventional relativity the total stretching from initial to final slice does *not* matter; quantum gravity effects will not come in as long as the *rate* of change is small. This fact may not be true in string theory; for a discussion see [12, 13].

1.6.3 What Is the Difference Between Hawking Radiation and Radiation from a Burning Piece of Coal?

Suppose a piece of coal burns away completely, leaving behind only the radiation it emitted. This time we know that subtle correlations in the emitted quanta encode the entire information about the state of the coal. But because these correlations are subtle, we cannot see them easily. How does this radiation differ from the Hawking radiation emitted by the black hole?

Consider the first photon emitted by the coal. This photon *can* be in a mixed state with the matter left behind in the coal. Let us assume that an atom emits this

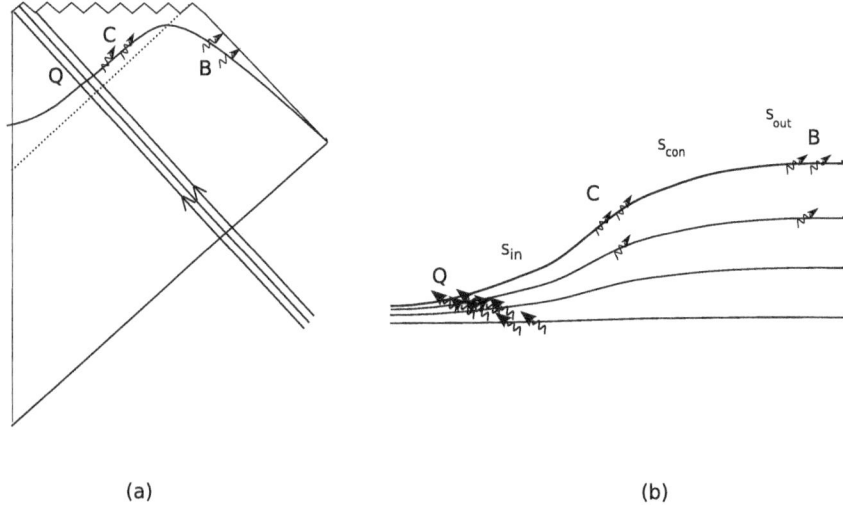

(a)	(b)

Fig. 1.16 (a) The infalling matter Q and the entangled pairs C,B shown on the spacelike slices in the Penrose diagram. (b) Q,C,B sit at different locations on the spacelike slices. To catch all three of these on the slices while staying in a low-curvature region we have evolved the small r side less and the large r side more, something that we are certainly allowed to do in classical gravity

photon and that after the emission the spin of the atom and the spin of the photon are correlated in an entangled wavefunction as follows

$$|\psi\rangle_1 = \frac{1}{\sqrt{2}} (|\uparrow\rangle_a \otimes |\downarrow\rangle_p + |\downarrow\rangle_a \otimes |\uparrow\rangle_p) , \qquad (1.79)$$

where $|\uparrow\rangle_a$ stands for the spin-up state of the atom, $|\downarrow\rangle_p$ stands for the spin-down state of the emitted photon, etc. Thus far, the situation looks just like the case of entangled b,c in the black hole. But the crucial difference is that when later photons are emitted from the coal, they can bounce off the atom left behind in the coal, and thus the spins of these later photons can carry the information left behind in this atom. If this atom drifts out itself (as a piece of ash) then it can also carry the information of its spin. Thus at the end the quanta collecting at infinity are entangled only with themselves and form a pure state carrying all the information in the initial piece of coal.

Contrast this with the state of the radiated quanta b_k in the black hole case, shown in Fig. 1.14. The quanta of type b_1 are correlated with the quanta of type c_1, which are located at a certain region on the part S_- of the spacelike slice. *But this place where c_1 is located is not involved any further in the process of radiation from the black hole.* For example, consider a later pair, say (b_{10}, c_{10}), and look at the region where this mode is suffering its nonuniform deformation. This region is *not* causally connected to the location where the earlier quanta c_1 is located, so c_1 cannot have any influence on the later-emitted quantum b_{10}. In the case of the coal the atom left behind after the first emission *was* in causal contact with later quanta leaving the coal. The black hole is different because each pair (b_k, c_k) is created at a point on a spacelike surface, and then this surface *stretches* so that the b_k, c_k quanta are moved away in different directions. New quanta are again created in the middle (i.e., at the horizon); these are again moved away by stretching, and so on. Thus all the created quanta b_k, c_k are located along different points of a very long spatial slice, with no overlap in their locations.

Since the quanta are prevented from influencing each other by being spread out along this very long spatial slice, we should ask the basic question: how did we get this very long spatial slice when the black hole only had a given size $\sim GM$? Recall from Fig. 1.4 that the spacelike slice inside the horizon was of the form $r = constant$, and it could be made arbitrarily long while remaining in the region $r < 2GM$. This possibility is unique to the black hole geometry, since it needs the light cones to 'turn over' and make the $r = constant$ direction spacelike. This does not happen for the coal, and so later quanta can (and do) carry the information left in entangled pairs from earlier quanta.

1.7 The Hawking 'Theorem'

There is one more common misconception about Hawking's computation of radiation which is very important to address. Look at the evolving mode drawn in Fig. 1.13. On the late time surface this mode was deformed, but if we follow the mode to the far *past* then it is just a simple Fourier mode with no particles in that mode; i.e., $\hat{a}_k|0\rangle = 0$. The further back we look, the smaller the wavelength. In fact if we follow the mode to times before the black hole formed then we find that its wavelength was much shorter than planck length; such modes are called 'transplanckian'. But perhaps we do not really know how to do quantum field theory when transplanckian wavelengths are concerned. In normal physics we take a field, break it into Fourier modes, make operators $\hat{a}_k, \hat{a}_k^\dagger$, and define a vacuum annihilated by the $\hat{a}_k$. Maybe all this is incorrect when describing transplanckian modes, and quantum gravity must be brought in some essential way?

If this argument were correct, then we have no information paradox, since Hawking's semiclassical computation would be invalid. In this section we argue that we do *not* need to know the physics of transplanckian modes to make Hawking's claim; we can formulate his argument using only physics at scales that we understand. More precisely, we will formulate his argument in the form of the following 'theorem':

Suppose we are given that

(a)   The effects of quantum gravity are confined to within a fixed length like planck l_p or string length l_s.

(b)   The vacuum is unique. Then when a black hole forms and evaporates, we *will* have information loss.

The meaning of these conditions will become clearer as we go through the argument.

1.7.1 The Local Vacuum

Let us first chose a length scale where we believe that we *do* understand quantum field theory and its vacuum structure. This could be $\lambda \sim 1$ fermi, since experiments on nuclear scales agree well with computations of Feynman graphs in field theory. Or we could take $\lambda \sim 1$ Å, since we understand atomic physics well, including the effects of vacuum fluctuations in effects like the Lamb shift. It does not matter what scale we choose; we will just keep it fixed henceforth as a scale $\lambda \sim \lambda_{known}$. The black hole itself will be taken as big, so we have

$$l_p \ll \lambda_{known} \ll GM . \tag{1.80}$$

A given Fourier mode starts off with very small wavelength $\lambda \ll l_p$, evolves to longer wavelengths $\lambda \sim \lambda_{known}$, and then continues to evolve to $\lambda \sim GM$, where its distortion becomes nonuniform and particle pairs are created. The important point is that since $\lambda_{known} \ll GM$, no particle pairs have been created when $\lambda \sim \lambda_{known}$. Thus we will look at the physics at this intermediate scale λ_{known} which is much larger than planck length and where the wavemode is still in the vacuum state.

1.7.2 The Consequences of Conditions (a) and (b)

Look at the region circled in Fig. 1.8(a). If we assume condition (a) of our 'theorem', then since the circled region is far from the singularity we have 'normal' physics' in this region, with no quantum gravity effects. That is, the metric is that of empty, almost flat, spacetime. Now focus on a mode which in this region has $\lambda \sim \lambda_{known}$. By condition (b) of the 'theorem' the vacuum is unique, which means that there is only 'one kind of empty space' possible in the theory; this empty space must therefore be described by the usual quantum vacuum that we use in field theory. Since there is nothing strange about the state of the spacetime region under consideration, the Fourier mode that we are studying (with $\lambda \sim \lambda_{known}$) will have to behave the way we expect a mode to behave in usual field theory.

Since we have 'normal physics' for this mode, the possible states of this mode are the vacuum $|0\rangle$, 1-particle $|1\rangle$, 2-particle $|2\rangle$, etc. There are now two possibilities:

(i)   First assume that the state of the field mode is the vacuum $|0\rangle$. Then the state will evolve in the way shown in Fig. 1.13. So the mode will become distorted and create entangled particle pairs described by a state like (1.65), and we would have the information problem created by such an entangled state.

(ii)   What happens if we assume that the state of the mode λ was *not* the vacuum
        state in this circled region? It is in principle possible that we get such an
        excited state for the mode because as we have argued above, we do not really
        know the evolution of the mode at the time when it was transplanckian. So
        suppose the mode is in a 1-particle state $|1\rangle$ when it reaches $\lambda \sim \lambda_{known}$. Then
        because for this mode we have 'normal physics', we will have the energy
        density expected from quanta of $\lambda \sim \lambda_{known} \sim 1$ fermi in the circled region
        of Fig. 1.8. So there would be matter of *nuclear* density filling this region.
        This would *not* agree with this region being low-curvature 'empty space', as
        required by postulate (a). More generally, the state of the wavemode $\lambda \sim$
        λ_{known} can be

$$|\psi\rangle = C_1|0\rangle + C_1|1\rangle + C_2|\rangle + \cdots . \tag{1.81}$$

If $C_i, i > 0$ are not small, then we get the nuclear density matter distribution
around the horizon. (It does not help to ask that the C_i be small but nonzero,
since then the evolved state will be close to (1.65), and we have already seen
that we need an order *unity* change in this state to remove the entanglement.)

1.7.3 The Consequence of a Non-unique Vacuum

It may appear that there is one way that we can have the classical geometry of the
hole depicted in the circled region of Fig. 1.8(a) and yet avoid information loss. This
way would be to drop condition (b) from our set of natural physics assumptions. Let
us see what dropping this condition would imply.

Consider again the state of the quantum field in the circled region of Fig. 1.8(a).
Suppose that the state here is *not* the usual vacuum, and yet it has *no* energy density.
This sounds strange, and indeed there are no such states in usual field theory. But
it could be that the transplanckian modes, which we do not understand, have some
complicated states which are not the usual vacuum and yet have no extra energy over
that of the vacuum. Then the evolution of modes with $\lambda \sim \lambda_{known}$ *can* be different
from the normally expected evolution because of interaction with these 'hidden'
transplanckian excitations. The allowed states for modes $\lambda \sim \lambda_{known}$ may not be of
the form (1.81), and the evolution of these modes may not be the usual free wave
evolution depicted pictorially in Fig. 1.13.

But if such a situation were permitted in our full quantum gravity theory, then we
would have to say that the vacuum of the theory is non-unique. There would be an
arbitrarily large number of states possible in a given region, with energy arbitrarily
close to the vacuum. For each such state we would find a totally different evolution
for modes with $\lambda \sim \lambda_{known}$. In this situation the theory loses all predictive power. In
the lab we would not know which of these 'vacuum' states we have, so we would not
know how modes with $\lambda \sim \lambda_{known}$ would behave. We could never do the physics at
any length scale, because modes with shorter length scales could be 'corrupting the
vacuum' and modifying evolution, without being detectable since they contribute

no net energy. Thus we normally assume condition (b) of our theorem that the vacuum *is* unique. (For an example of a theory with a non-unique vacuum created by nonlocal identifications, see [14].)

1.7.4 Summary of the Information Paradox

Thus we see that if we assume the two very reasonable sounding assumptions (a) and (b) of the Hawking 'theorem', then we are forced into a situation where the outgoing radiation will not be a pure state carrying the information of the black hole. To evade the information paradox we will therefore need some radical change in our basic understanding of quantum mechanics and gravity. Let us first summarize the main ideas that have led to the information paradox.

The central point is that vacuum modes evolve over smooth spacetime in the manner sketched in Fig. 1.13, and thus create entangled particle pairs. Entangled states are not a problem by themselves. The problem arises because gravity is an attractive force with a negative potential energy, and this makes the quanta c_k inside the horizon have a net negative energy. Thus the matter Q and the quanta C in Fig. 1.16 can have a net mass zero. Then all the energy will go to the b_k quanta and there is no net mass left in the hole. If we assume that there cannot be an infinite number of light 'remnants' in our theory then we are forced to assume that the black hole disappears. Now the radiation quanta b_k are 'entangled with nothing', and we cannot describe them by any wavefunction.

To save this situation we need some way to *change significantly* the evolution depicted in Fig. 1.13. In fact what we need is not only that the matter labeled B in Fig. 1.16 be in a pure state (so that it should not be entangled with C), but that it should reflect all the information in the matter Q. In the derivation of the Hawking 'theorem' we saw that we could restrict attention to wavemodes with $\lambda \gtrsim \lambda_{known}$, where the physics of evolution is well understood. The evolution of these modes, depicted in Fig. 1.13, would seem to be governed by physics that we know very well – the physics of quantum fields on gently curved space. *Yet, to save quantum theory we need that this evolution be changed by order unity, leading to a completely different state than the entangled pair state that we got!* A small change in the evolution, leading to a small change in the final state, will not help.

We will see in the next section that in string theory it is condition (a) that fails; quantum gravity effects can change the entire interior of the hole and resolve the information paradox.

1.8 Black Holes in String Theory: Fuzzballs

String theory provides a consistent theory of perturbative quantum gravity, so we can hope that the theory might also be able to avoid contradictions when it comes to nonperturbative things like black holes. The theory has no free parameters, and no fields can be added or removed from the theory. To make the black hole we must use the objects present in the theory. Let us compactify the 10D spacetime of string theory as follows:

$$M_{9,1} \rightarrow M_{4,1} \times T^4 \times S^1 . \tag{1.82}$$

We can wrap a string around the S^1; this will look like a point mass from the viewpoint of the noncompact directions. We can take a large number n_1 of these strings and ask what metric they produce. The important thing is that we take a *bound* state of the strings, otherwise we will make 'many small black holes' rather than the one massive hole that we are seeking. The bound state of these strings is easy to picture: the string just wraps n_1 times around S^1 before closing. There is just one such state of the string, since the string is an 'elastic band' and settles down to its shortest length for the given winding. Thus the microscopic count of states would suggest an entropy $S_{micro} = \ln 1 = 0$. What about the 'black hole' that it creates? The string carries 'winding charge' and radiates a corresponding 2-form gauge field $B_{\mu\nu}$. When we make the metric with the mass and charge of the string we find that the horizon coincides with the singularity, and so the horizon area is zero. Thus the Bekenstein entropy $S_{bek} = A/4 = 0$, and so we get $S_{bek} = S_{micro}$.

Alternatively we can take the massless gravitons of the theory and allow them to circle around the S^1; this would also look like a mass point from the viewpoint of the noncompact directions, but now the mass point will carry 'momentum charge' due the momentum carried by the gravitons. To get a 'bound state' of these gravitons we would have to put all the momentum into one energetic graviton, so the microscopic entropy would be again $S_{micro} = \ln 1 = 0$. The metric produced by this graviton carrying energy and 'momentum charge' again ends up with no horizon area, and we get $S_{bek} = 0 = S_{micro}$.

To get something more interesting let us *combine* the winding and momentum charges. To make a bound state of winding and momentum we simply let the momentum be carried as traveling waves on the string. But now we see that there are many states for a given winding n_1 and a given momentum n_p: we can put all the energy in the lowest harmonic, or some in the first and some in the second harmonic, or take any other distribution of the energy into harmonics. The number of such states turns out to give an entropy [15–19]

$$
\begin{aligned}
T^4: \quad & S = 2\sqrt{2}\pi\sqrt{n_1 n_p} , \\
K3: \quad & S = 4\pi\sqrt{n_1 n_p} ,
\end{aligned}
\tag{1.83}
$$

where we have also included the answer for a case where the T^4 in (1.82) has been replaced with another 4D manifold called $K3$.

We can compute the geometry produced by a point source carrying the energy and gauge fields produced by the string winding and momentum. In this computation we should note that the string action contains R^2 corrections to the leading Einstein action R. This modifies the expression for the Bekenstein entropy (to the 'Bekenstein–Wald entropy' [20]). With these needed corrections this entropy has been computed for the case of K3 compactification, and one finds that [21]

$$S_{bek} = 4\pi\sqrt{n_1 n_p} = S_{micro} \,, \tag{1.84}$$

so the microscopic count exactly reproduces the entropy from the geometry of the horizon.

We can make more complicated holes, by adding n_5 5-branes wrapped on $T^4 \times S^1$ (or on $K3 \times S^1$). This time the horizon area is large enough that we do not need the R^2 corrections to the action, and one finds an exact agreement again with the microscopic count of states [22, 23]

$$S_{bek} = \frac{A}{4} = 2\pi\sqrt{n_1 n_p n_5} = S_{micro} \,. \tag{1.85}$$

So we seem to understand something about black hole entropy, but what about the information problem? To understand what can change in Hawking's derivation of information loss, we need to understand what is going on inside the black hole. Let us return to the 2-charge hole made with string winding and momentum. The crucial point is that the elementary string of string theory has no *longitudinal waves*; it admits only transverse oscillations. Thus when carrying the momentum as traveling waves it spreads over some transverse region, instead of just sitting at a point in the noncompact space. Instead of the spherically symmetric hole with a central singularity at $r = 0$ we get a 'fuzzball', with different states of the string creating different fuzzballs. Interestingly, the boundary of the typical fuzzball has an area that satisfies

$$\frac{A}{G} \sim \sqrt{n_1 n_p} \sim S_{micro} \,. \tag{1.86}$$

So we see that the region occupied by the vibrating string is of order the entire horizon interior; in fact a horizon never forms [24, 25]. We depict this situation in Fig. 1.17. Now there is no information problem: any matter falling onto the fuzzball gets absorbed by the fuzz and is eventually re-radiated with all its information, which is just how any other body would behave. The crucial point is that we do not have a horizon whose vicinity is 'empty space'. The matter making the hole, instead of sitting at $r = 0$, spreads all the way to the horizon. So it can send its information out with the radiation, just like a piece of coal would do.

Similar constructions have been done with many states of the 3-charge hole carrying winding, momentum, and 5-brane charges, and more complicated holes with four kinds of charges [26–31]. Some states of non-extremal holes have been made as well [32]. Radiation from these non-extremal gravity states has been computed [33] and found to agree exactly with the radiation expected from the corresponding state of strings and branes [34].

We can still ask, why does all this work? What feature of string theory led to this large change in the picture of the hole and allowed the interior of the horizon to depart from the naive classical expectation? The answer would seem to be 'fractionation', a phenomenon peculiar to string theory which is a theory of *extended* objects. Consider spacetime with a compact circle of length L. Suppose we want to make an excitation of this system, while adding no net charge. What is the lowest energy ΔE that we will need? We can take one graviton in the lowest allowed harmonic

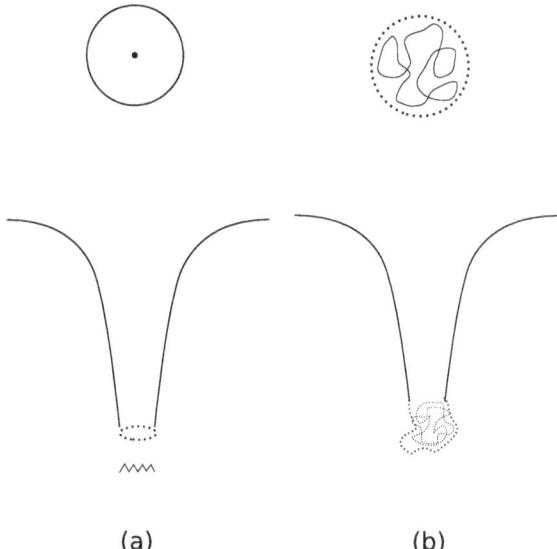

Fig. 1.17 (**a**) If the string winding and momentum excitations could sit at a point, then we would get the usual black hole; in the lower diagram the geometry is shown with flat space at infinity, then a 'throat', ending in a horizon with a singularity inside. (**b**) The string cannot carry the momentum without transverse vibrations, and thus spreads over a horizon-sized transverse area. The geometry depicted in the lower diagram has no horizon; instead the throat ends in a 'fuzzball'

running clockwise on the S^1, and one running; this would give an energy $\Delta E = \frac{4\pi}{L}$. Now suppose on this circle we already had a wrapped string with winding n_1. Now we can excite a clockwise momentum mode of energy $\frac{2\pi}{n_1 L}$ on the string, and with a similar contribution from the anticlockwise mode we get $\Delta E = \frac{4\pi}{n_1 L}$. If $n_1 \gg 1$ then this ΔE is much smaller than the energy gap in the absence of the strings. We say that in the presence of the strings the momentum comes in *fractional* units, which are $\frac{1}{n_1}$th of a full unit [35].

This looks like a simple physical effect, so what can it have to do with black holes? In string theory we have duality, which allows us to map different objects in the theory to each other. Thus we can map the n_1 times wound string to a bound state of n_1 5-branes. At the same time the momentum mode would map to a string winding along the S^1. Now the 'fractional momentum mode' becomes a 'fractional string'. But what is a fractional string? The original string had a tension of string scale, which is order planck scale. But the fractional string has a tension which is $\frac{1}{n_1}$th of this value, and so for n_1 large it will be a very low tension object [36].

One can extend such constructions further to bound states of many kinds of branes. Let us take the black hole described in (1.85). One finds that there exist very low tension 'floppy fractional objects' that stretch over distances of order [37]

$$D \sim \left[\frac{(n_1 n_p n_5)^{\frac{1}{2}} g^2 \alpha'^4}{VL} \right]^{\frac{1}{3}}, \qquad (1.87)$$

where V is the volume of T^4, L is the length of S^1, and g and α' are the string coupling and tension. But this turns out to be just the order of the horizon radius of the black hole with these charges! This argument tells us that fractionation can generate quantum effects over horizon scales. We can then return to simpler holes like the 2-charge hole (1.83), where we can construct the internal state of the hole, and see that we indeed get a 'fuzzball' instead of the traditional hole.

This solves the information paradox but raises many natural questions about the behavior of black holes. While the dynamics of fuzzballs is in its infancy, we can make some simple observations and conjectures relevant to such questions.

If a shell of dust is collapsing, will it suddenly change its dynamics when it reaches horizon size?

No, the fuzzball proposal does not require that. The essential point is that there are *two* timescales in the black hole problem. One is the 'crossing timescale' of order $\sim GM$, over which the collapse occurs. The other is the much longer Hawking evaporation timescale, $t_{evap} \sim GM(\frac{M}{m_{plank}})^2$. The collapsing matter was in a low-entropy state, and will take some time to come to statistical equilibrium and reach a generic state (which we expect to be a fuzzball-type state). It is known that the entropy of radiation from the hole S_{rad} is somewhat larger than S_{bek}, since the radiation free-streams out of the hole rather than leave in a 'quasi-static' way [38, 39]. Thus the matter can collapse as classically expected on the crossing timescale and even use some fraction of t_{evap} to stabilize to the fuzzball configuration; we can still carry the information out in the remaining radiation.

After the black hole has stabilized to the fuzzball configuration, will an infalling body feel a very different environment from that of the usual black hole geometry?

Not necessarily, since the 'fuzz' is a very low density 'web', at least in the simple 2-charge examples that we can explicitly study [40]. If a body is heavy (compared to the energy of a Hawking radiation quantum) *and* we follow it only over the short 'crossing' timescale $\sim GM$, then we may not see a dynamics that departs significantly from the classical one. But over the long Hawking evaporation timescale the information in the heavy body should get incorporated in the fuzz and eventually get radiated away.

After the black hole has stabilized to the fuzzball configuration, will the evolution of Hawking radiation quanta be very different from that expected in the classical geometry?

Yes, and that *should* happen. If we do not modify the evolution of $\lambda \sim GM$ Fourier modes in the vicinity of the horizon, we will have information loss, as argued in the above sections. The fuzzball structure of the hole ensures that the information of the hole reaches out to the boundary of the hole and so the mode evolution of Fig. 1.13 is altered, not slightly, but rather by order unity effects. This is what is needed to prevent information loss.

1.9 Conclusion

So what is the information paradox? We would like ordinary quantum theory to be valid, even when black holes form and evaporate. But with the traditional picture of the black hole, the explicit computation of Hawking radiation generates entangled pairs, and the state of the outgoing quanta is not a pure quantum state when the black hole disappears. Furthermore, the state of these outgoing quanta b_k has no relation to the matter that made the hole; they just made a specific entangled state with their partners c_k. *To resolve the paradox we have to find some way to change the evolution of vacuum modes depicted in Fig. 1.13, so that the b_k form a pure state containing the information of the initial matter.* Small changes in the evolution will not help; it has to be an order unity change since we want a completely different outcome. But if we make some very reasonable sounding assumptions – that quantum effects are confined to within planck distances and that the vacuum is unique – then we can establish that there *cannot* be any such change to the evolution of Fig. 1.13.

String theory resolves the problem by telling us that the first assumption is false: quantum gravity effects are *not* confined to a given distance, but instead range over distances that increase with the number of quanta making up the bound state corresponding to the hole. We find an effect called 'fractionation' which shows that in a bound state of strings and branes the quantum effects stretch to distances of order horizon scale (1.87). This is a crude estimate, but we can then return to simple black hole states and construct them explicitly, finding in each case that there is no horizon; instead the interior of the hole is a 'fuzzball'.

The information paradox was important because its resolution would have to challenge some basic assumptions that we have held about quantum gravity. We do indeed find a change in our basic idea of how quantum gravity acts when we have large dense systems of strings and branes. The goal is now to formalize this understanding and apply it to other basic problems like the early Universe where quantum gravity can be important.

Acknowledgments I would like to thank the organizers for a wonderful school at Mytilene. I am grateful to Steve Avery, Borun Chowdhury, and Jeremy Michelson for many helpful comments on this manuscript. This work was supported in part by DOE grant DE-FG02-91ER-40690.

References

1. J. D. Bekenstein, Phys. Rev. D **7**, 2333 (1973).
2. S. W. Hawking, Commun. Math. Phys. **43**, 199 (1975) [Erratum-ibid. **46**, 206 (1976)].
3. S. W. Hawking, Phys. Rev. D **14**, 2460 (1976).
4. S. D. Mathur, Fortsch. Phys. **53**, 793 (2005) [arXiv:hep-th/0502050].
5. S. D. Mathur, Class. Quant. Grav. **23**, R115 (2006) [arXiv:hep-th/0510180].
6. I. Bena and N. P. Warner, arXiv:hep-th/0701216.
7. N. D. Birrell and P. C. W. Davies, *Quantum Fields In Curved Space*, (Cambridge University Press, Cambridge 1982), 340p.

8. S. A. Fulling, Aspects of quantum field theory in curved space-time, London Math. Soc. Student Texts **17**, 1 (1989).
9. R. M. Wald, Commun. Math. Phys. **45**, 9 (1975).
10. L. Parker, Phys. Rev. D **12**, 1519 (1975).
11. S. B. Giddings and W. M. Nelson, Phys. Rev. D **46**, 2486 (1992) [arXiv:hep-th/9204072].
12. S. D. Mathur, Int. J. Mod. Phys. A **15**, 4877 (2000) [arXiv:gr-qc/0007011].
13. S. D. Mathur, Int. J. Mod. Phys. D **11**, 1537 (2002) [arXiv:hep-th/0205192].
14. A. Chamblin and J. Michelson, Class. Quant. Grav. **24**, 1569 (2007) [arXiv:hep-th/0610133].
15. L. Susskind, arXiv:hep-th/9309145.
16. J. G. Russo and L. Susskind, Nucl. Phys. B **437**, 611 (1995) [arXiv:hep-th/9405117].
17. A. Sen, Nucl. Phys. B **440**, 421 (1995) [arXiv:hep-th/9411187].
18. A. Sen, Mod. Phys. Lett. A **10**, 2081 (1995) [arXiv:hep-th/9504147].
19. C. Vafa, Nucl. Phys. B **463**, 435 (1996) [arXiv:hep-th/9512078].
20. R. M. Wald, Phys. Rev. D **48**, 3427 (1993) [arXiv:gr-qc/9307038].
21. A. Dabholkar, Phys. Rev. Lett. **94**, 241301 (2005) [arXiv:hep-th/0409148].
22. A. Strominger and C. Vafa, Phys. Lett. B **379**, 99 (1996) [arXiv:hep-th/9601029].
23. C. G. Callan and J. M. Maldacena, Nucl. Phys. B **472**, 591 (1996) [arXiv:hep-th/9602043].
24. O. Lunin and S. D. Mathur, Nucl. Phys. B **623**, 342 (2002) [arXiv:hep-th/0109154].
25. O. Lunin and S. D. Mathur, Phys. Rev. Lett. **88**, 211303 (2002) [arXiv:hep-th/0202072].
26. S. Giusto, S. D. Mathur and A. Saxena, Nucl. Phys. B **710**, 425 (2005) [arXiv:hep-th/0406103].
27. O. Lunin, JHEP **0404**, 054 (2004) [arXiv:hep-th/0404006].
28. I. Bena and N. P. Warner, arXiv:hep-th/0701216.
29. I. Bena and N. P. Warner, Adv. Theor. Math. Phys. **9**, 667 (2005) [arXiv:hep-th/0408106].
30. V. Balasubramanian, E. G. Gimon and T. S. Levi, arXiv:hep-th/0606118.
31. I. Kanitscheider, K. Skenderis and M. Taylor, arXiv:0704.0690 [hep-th].
32. V. Jejjala, O. Madden, S. F. Ross and G. Titchener, Phys. Rev. D **71**, 124030 (2005) [arXiv:hep-th/0504181].
33. V. Cardoso, O. J. C. Dias, J. L. Hovdebo and R. C. Myers, Phys. Rev. D **73**, 064031 (2006) [arXiv:hep-th/0512277].
34. B. D. Chowdhury and S. D. Mathur, arXiv:0711.4817 [hep-th].
35. S. R. Das and S. D. Mathur, Phys. Lett. B **375**, 103 (1996) [arXiv:hep-th/9601152].
36. J. M. Maldacena and L. Susskind, Nucl. Phys. B **475**, 679 (1996) [arXiv:hep-th/9604042].
37. S. D. Mathur, Nucl. Phys. B **529**, 295 (1998) [arXiv:hep-th/9706151].
38. T. M. Fiola, J. Preskill, A. Strominger and S. P. Trivedi, Phys. Rev. D **50**, 3987 (1994) [arXiv:hep-th/9403137].
39. E. Keski-Vakkuri and S. D. Mathur, Phys. Rev. D **50**, 917 (1994) [arXiv:hep-th/9312194].
40. S. D. Mathur, arXiv:0706.3884 [hep-th].

Chapter 2
Classical Yang–Mills Black Hole Hair in Anti-de Sitter Space

E. Winstanley

Abstract The properties of hairy black holes in Einstein–Yang–Mills (EYM) theory are reviewed, focusing on spherically symmetric solutions. In particular, in asymptotically anti-de Sitter space (adS) stable black hole hair is known to exist for $\mathfrak{su}(2)$ EYM. We review recent work in which it is shown that stable hair also exists in $\mathfrak{su}(N)$ EYM for arbitrary N, so that there is no upper limit on how much stable hair a black hole in adS can possess.

2.1 Introduction

We begin by very briefly reviewing the "no-hair" conjecture and motivating the study of hairy black holes.

2.1.1 The "no-hair" Conjecture

The black hole "no-hair" conjecture [142] states that (see, for example, [51, 52, 77–79, 118] for detailed reviews and comprehensive lists of references): *All stationary, asymptotically flat, four-dimensional black hole equilibrium solutions of the Einstein equations in vacuum or with an electromagnetic field are characterized by their mass, angular momentum, and (electric or magnetic) charge.*

According to the no-hair conjecture, black holes are therefore extraordinarily simple objects, whose geometry (exterior to the event horizon) is a member of the Kerr–Newman family and completely determined by just three quantities (mass, angular momentum and charge). Furthermore, these quantities are *global charges* which can (at least in principle) be measured at infinity, far from the black hole event horizon. If a black hole is formed by the gravitational collapse of a dying star, the

E. Winstanley (✉)

School of Mathematics and Statistics, The University of Sheffield, Hicks Building, Hounsfield Road, Sheffield S3 7RH, UK

E.Winstanley@sheffield.ac.uk

Winstanley, E.: *Classical Yang–Mills Black Hole Hair in Anti-de Sitter Space*. Lect. Notes Phys. **769**, 49–87 (2009)

DOI 10.1007/978-3-540-88460-6_2   © Springer-Verlag Berlin Heidelberg 2009

initial star will be a highly complex object described by many different parameters. The final, equilibrium, black hole is, by contrast, rather simple and described by a very small number of quantities. During the process of the formation of a black hole, an enormous amount of (classical) information about the star which collapsed has therefore been lost. Similarly, if a complicated object is thrown down a black hole event horizon, once the system settles down, the only changes in the final state will be changes in the total mass, total angular momentum and total charge. Advances in astrometry [174] and future gravitational wave detectors [5] may even be able to probe the validity of the "no-hair" conjecture for astrophysical black holes by verifying that the mass, angular momentum and quadrupole moment Q_2 of the black hole satisfy the relation $Q_2 = J^2/M$ which holds for Kerr black holes.

The "no-hair" conjecture, stated above, has been proved by means of much complicated and beautiful mathematics (as reviewed in, for example, [51, 52, 77–79, 118]), subject to the assumptions of stationarity, asymptotic flatness, four-dimensional spacetime and the electrovac Einstein equations. It is perhaps unsurprising that if one or more of these assumptions is relaxed, then the conjecture does not necessarily hold. For example, if a negative cosmological constant is included, so that the spacetime is no longer asymptotically flat but instead approaches anti-de Sitter (adS) space at infinity, then the event horizon of the black hole is not necessarily spherical, giving rise to "topological" black holes (see, for example, [18, 64, 97, 98, 103, 112, 165]). More recently, the discovery of "black ring" solutions in five spacetime dimensions ([60], see [61] for a recent review) and the even more complicated "black Saturn" [59] solutions indicates that Einstein–Maxwell theory has a rich space of black solutions in higher dimensions, which are not given in terms of the Myers-Perry [121] metric (which is the generalization of the Kerr–Newman geometry to higher dimensions).

2.1.2 Hairy Black Holes

In this article we consider what happens when the other condition in the "no-hair" conjecture, namely that the Einstein equations involve electrovac matter only, is relaxed. The "generalized" version of the no-hair conjecture [79] states that all stationary black hole solutions of the Einstein equations with any type of self-gravitating matter field are determined uniquely by their mass, angular momentum and a set of global charges. Even in asymptotically flat space, this conjecture does not hold, even for the simplest type of self-gravitating matter, a scalar field. The first such counterexample is the famous BBMB black hole [12, 13, 27] which has the same metric as the extremal Reissner–Nordström black hole but possesses a conformally coupled scalar field. However, this solution is controversial due to the divergence of the scalar field on the event horizon [158] and is also highly unstable [48]. Therefore, in some ways the first "hairy" black hole is considered to be the Gibbons solution [71], which describes a Reissner–Nordström black hole with a non-trivial dilaton field. While there are many results which rule out scalar field hair in quite general models, particularly in asymptotically flat spacetimes (see, for example, [14] for a review), in recent years many other examples of black holes with non-trivial scalar field hair have been found. For example, minimally coupled scalar field hair has been found when the cosmological constant is positive [161] or negative [162] and non-minimally coupled scalar field hair has also been considered (see, for example, [176, 177] and references therein).

In this short review, we will focus on another particular matter model, Einstein–Yang–Mills theory (EYM), where the matter is described by a non-Abelian (Yang-Mills) gauge field. It is now well-known that this theory possesses "hairy" black hole solutions, whose metric is not a member of the Kerr–Newman family (see [171] for a detailed review). Furthermore, unlike the Kerr–Newman black holes, the geometry exterior to the event horizon is not determined uniquely by global charges measureable at infinity, although only a small number of parameters are required in order to describe the metric and matter field (see Sect. 2.3 for further details). All the asymptotically flat black hole solutions of pure EYM theory discovered to date are unstable [47] (however, there are examples of asymptotically flat, stable hairy black holes in variants of the EYM action, such as Einstein–Skyrme [22, 58, 80, 81], Einstein-non-Abelian-Proca [73, 110, 159, 160, 163] and Einstein–Yang–Mills–Higgs [1] theories). This means that, while the "letter" of the no-hair theorem is violated in this case (as there exist solutions which are not described by the Kerr–Newman metric), its "spirit" is intact, as stable equilibrium black holes remain simple objects, described by a few parameters if not exactly of the Kerr–Newman form (see [21] for a related discussion along these lines).

The situation is radically different if one considers EYM solutions in asymptotically adS space, rather than asymptotically flat space. For $\mathfrak{su}(2)$ EYM, at least some black hole solutions with hair are stable [25, 26, 175]. These stable black holes require one new parameter (see Sect. 2.4) to completely describe the geometry exterior to their event horizons. Therefore, one might still argue that the true "spirit" of the "no-hair" conjecture remains intact and that stable equilibrium black holes are comparatively simple objects, described by just a few parameters.

One is therefore led to a natural question: are there hairy black hole solutions in adS which require an infinite number of parameters to fully describe the geometry and matter exterior to the event horizon? In other words, is there a limit to how much hair a black hole in adS can be given? This is the question we will be seeking to address in this article.

2.1.3 Scope of this Article

The subject of hairy black holes in EYM theory and its variants is very active, with many new solutions appearing each year. The review [171], written in 1998, is very detailed and thorough and contains a comprehensive list of references to solutions known at that time. We have therefore not sought to be complete in our references prior to that date, and have, instead, chosen to highlight a few solutions (the selection being undoubtedly personal). Even considering just work after 1998, we have been unable to do justice to the huge body of work in this area (for example, the seminal paper [7] has 172 arXiv citations between 1999 and the time of writing) and have instead chosen some examples of solutions. As well as [171], reviews of various aspects of solitons and black holes in EYM can be found in [21, 66, 72, 152, 153, 166].

The outline of this article is as follows. In Sect. 2.2 we will outline $\mathfrak{su}(N)$ EYM theory, including our ansatz for the gauge field and the form of the field equations. We will then, in Sect. 2.3, briefly review some of the properties of the well-known asymptotically flat solutions of this theory. Our main focus in this article are asymptotically adS black holes, and we begin our discussion of these in Sect. 2.4 by reviewing the key features of the $\mathfrak{su}(2)$ EYM black holes in adS, before moving on to describe very recent work on $\mathfrak{su}(N)$, asymptotically adS, EYM black holes in Sect. 2.5. Our conclusions are presented in Sect. 2.6. Throughout this article the metric has signature $(-,+,+,+)$ and we use units in which $4\pi G = c = 1$.

2.2 $\mathfrak{su}(N)$ Einstein–Yang–Mills Theory

In this section we gather together all the formalism and field equations we shall require for our later study of black hole solutions.

2.2.1 Ansatz, Field Equations and Boundary Conditions

In this article we shall be interested in four-dimensional $\mathfrak{su}(N)$ EYM theory with a cosmological constant, described by the following action, given in suitable units:

$$S_{\text{EYM}} = \frac{1}{2} \int d^4x \sqrt{-g} \left[R - 2\Lambda - \text{Tr} F_{\mu\nu} F^{\mu\nu} \right], \tag{2.1}$$

where R is the Ricci scalar of the geometry and Λ the cosmological constant. Here we have chosen the simplest type of EYM-like theory, many variants have been studied in the literature (see, for example, [171] for a selection of examples).

Varying the action (2.1) gives the field equations

$$T_{\mu\nu} = R_{\mu\nu} - \frac{1}{2}R g_{\mu\nu} + \Lambda g_{\mu\nu};$$
$$0 = D_\mu F_\nu{}^\mu = \nabla_\mu F_\nu{}^\mu + \left[A_\mu, F_\nu{}^\mu \right]; \tag{2.2}$$

where the YM stress–energy tensor is

$$T_{\mu\nu} = \text{Tr} F_{\mu\lambda} F_\nu{}^\lambda - \frac{1}{4} g_{\mu\nu} \text{Tr} F_{\lambda\sigma} F^{\lambda\sigma}. \tag{2.3}$$

In this article we consider only static, spherically symmetric black hole geometries, with metric given, in standard Schwarzschild-like co-ordinates, as

$$ds^2 = -\mu S^2 \, dt^2 + \mu^{-1} \, dr^2 + r^2 \, d\theta^2 + r^2 \sin^2 \theta \, d\phi^2, \qquad (2.4)$$

where the metric functions μ and S depend on the radial co-ordinate r only. In the presence of a negative cosmological constant $\Lambda < 0$, we write the metric function μ as

$$\mu(r) = 1 - \frac{2m(r)}{r} - \frac{\Lambda r^2}{3}. \qquad (2.5)$$

The most general, spherically symmetric, ansatz for the $\mathfrak{su}(N)$ gauge potential is [99]:

$$A = \mathscr{A} \, dt + \mathscr{B} \, dr + \frac{1}{2} \left(C - C^H \right) d\theta - \frac{i}{2} \left[\left(C + C^H \right) \sin \theta + D \cos \theta \right] d\phi, \qquad (2.6)$$

where $\mathscr{A}$, $\mathscr{B}$, C and D are all $(N \times N)$ matrices and C^H is the Hermitian conjugate of C. The matrices $\mathscr{A}$ and $\mathscr{B}$ are purely imaginary, diagonal, traceless and depend only on the radial co-ordinate r. The matrix C is upper triangular, with non-zero entries only immediately above the diagonal:

$$C_{j,j+1} = \omega_j(r) e^{i \gamma_j(r)}, \qquad (2.7)$$

for $j = 1, \cdots, N - 1$. In addition, D is a constant matrix:

$$D = \mathrm{Diag} \left(N - 1, N - 3, \cdots, -N + 3, -N + 1 \right). \qquad (2.8)$$

Here we are primarily interested only in purely magnetic solutions, so we set $\mathscr{A} \equiv 0$. We may also take $\mathscr{B} \equiv 0$ by a choice of gauge [99]. From now on we will assume that all the $\omega_j(r)$ are non-zero (see, for example, [69, 94–96] for the possibilities in asymptotically flat space if this assumption does not hold). In this case one of the Yang–Mills equations becomes [99]

$$\gamma_j = 0 \qquad \forall j = 1, \cdots, N - 1. \qquad (2.9)$$

Our ansatz for the Yang–Mills potential therefore reduces to

$$A = \frac{1}{2} \left(C - C^H \right) d\theta - \frac{i}{2} \left[\left(C + C^H \right) \sin \theta + D \cos \theta \right] d\phi, \qquad (2.10)$$

where the only non-zero entries of the matrix C are

$$C_{j,j+1} = \omega_j(r). \qquad (2.11)$$

The gauge field is therefore described by the $N - 1$ functions $\omega_j(r)$. We comment that our ansatz (2.10) is by no means the only possible choice in $\mathfrak{su}(N)$ EYM. Techniques for finding *all* spherically symmetric $\mathfrak{su}(N)$ gauge potentials can be found in [6], where all irreducible models are explicitly listed for $N \le 6$.

With the ansatz (2.10), there are $N - 1$ non-trivial Yang–Mills equations for the $N - 1$ functions ω_j:

$$r^2 \mu \omega_j'' + \left(2m - 2r^3 p_\theta - \frac{2\Lambda r^3}{3}\right) \omega_j' + W_j \omega_j = 0 \tag{2.12}$$

for $j = 1, \ldots, N - 1$, where a prime $'$ denotes d/dr,

$$p_\theta = \frac{1}{4r^4} \sum_{j=1}^{N} \left[\left(\omega_j^2 - \omega_{j-1}^2 - N - 1 + 2j\right)^2 \right], \tag{2.13}$$

$$W_j = 1 - \omega_j^2 + \frac{1}{2} \left(\omega_{j-1}^2 + \omega_{j+1}^2\right), \tag{2.14}$$

and $\omega_0 = \omega_N = 0$. The Einstein equations take the form

$$m' = \mu G + r^2 p_\theta, \qquad \frac{S'}{S} = \frac{2G}{r}, \tag{2.15}$$

where

$$G = \sum_{j=1}^{N-1} \omega_j'^2. \tag{2.16}$$

Altogether, then, we have $N + 1$ ordinary differential equations for the $N + 1$ unknown functions $m(r)$, $S(r)$ and $\omega_j(r)$. The field equations (2.12) and (2.15) are invariant under the transformation

$$\omega_j(r) \to -\omega_j(r) \tag{2.17}$$

for each j independently, and also under the substitution:

$$j \to N - j. \tag{2.18}$$

We are interested in black hole solutions of the field equations (2.12) and (2.15). We assume there is a regular, non-extremal, black hole event horizon at $r = r_h$, where $\mu(r)$ has a single zero. This fixes the value of $m(r_h)$ to be:

$$2m(r_h) = r_h - \frac{\Lambda r_h^3}{3}. \tag{2.19}$$

However, the field equations (2.12) and (2.15) are singular at the black hole event horizon $r = r_h$ and at infinity $r \to \infty$. We therefore need to impose boundary conditions on the field variables $m(r)$, $S(r)$ and $\omega_j(r)$ at these singular points. When the cosmological constant Λ is zero, local existence of solutions of the field equations in neighbourhoods of these singular points has been rigorously proved [100, 125]. This proof can be extended to the case when the cosmological constant is negative [8, 11].

We assume that the field variables $\omega_j(r)$, $m(r)$ and $S(r)$ have regular Taylor series expansions about $r = r_h$:

$$m(r) = m(r_h) + m'(r_h)(r - r_h) + O(r - r_h)^2;$$
$$\omega_j(r) = \omega_j(r_h) + \omega'_j(r_h)(r - r_h) + O(r - r_h)^2;$$
$$S(r) = S(r_h) + S'(r_h)(r - r_h) + O(r - r_h). \qquad (2.20)$$

Setting $\mu(r_h) = 0$ in the Yang–Mills equations (2.12) fixes the derivatives of the gauge field functions at the horizon:

$$\omega'_j(r_h) = -\frac{W_j(r_h)\omega_j(r_h)}{2m(r_h) - 2r_h^3 p_\theta(r_h) - \frac{2\Lambda r_h^3}{3}}. \qquad (2.21)$$

Therefore the expansions (2.20) are determined by the $N + 1$ quantities $\omega_j(r_h)$, r_h, $S(r_h)$ for fixed cosmological constant Λ. For the event horizon to be non-extremal, it must be the case that

$$2m'(r_h) = 2r_h^2 p_\theta(r_h) < 1 - \Lambda r_h^2, \qquad (2.22)$$

which weakly constrains the possible values of the gauge field functions $\omega_j(r_h)$ at the event horizon. Since the field equations (2.12) and (2.15) are invariant under the transformation (2.17), we may consider $\omega_j(r_h) > 0$ without loss of generality.

At infinity, we require that the field variables $\omega_j(r)$, $m(r)$ and $S(r)$ converge to constant values as $r \to \infty$ and have regular Taylor series expansions in r^{-1} near infinity:

$$m(r) = M + O\left(r^{-1}\right); \qquad S(r) = 1 + O\left(r^{-1}\right); \qquad \omega_j(r) = \omega_{j,\infty} + O\left(r^{-1}\right). \qquad (2.23)$$

If the spacetime is asymptotically flat, with $\Lambda = 0$, then the values of $\omega_{j,\infty}$ are constrained to be

$$\omega_{j,\infty} = \pm\sqrt{j(N - j)}. \qquad (2.24)$$

This condition means that the asymptotically flat black holes have no magnetic charge at infinity, or, in other words, these solutions have no global magnetic charge. Therefore, at infinity, they are indistinguishable from Schwarzschild black holes. However, if the cosmological constant is non-zero, so that the geometry approaches (a)dS at infinity, then there are no *a priori* constraints on the values of $\omega_{j,\infty}$. In general, therefore, the (a)dS black holes will be magnetically charged. It should be noted that the boundary conditions in the case when the cosmological constant Λ is positive are more complex, as there is a cosmological horizon between the event horizon and infinity.

2.2.2 Some "trivial" Solutions

Although the field equations (2.12) and (2.15) are highly non-linear and rather complicated, they do have some trivial solutions which can easily be written down:

Schwarzschild(-(a)dS)   Setting

$$\omega_j(r) \equiv \pm\sqrt{j(N-j)} \qquad (2.25)$$

for all j gives the Schwarzschild(-(a)dS) black hole with

$$m(r) = M = \text{constant} \qquad (2.26)$$

We note that, by setting $M = 0$, pure Minkowski ($\Lambda = 0$) or (a)dS ($\Lambda \neq 0$) space is also a solution.

Reissner–Nordström(-(a)dS)   Setting

$$\omega_j(r) \equiv 0 \qquad (2.27)$$

for all j gives the Reissner–Nordström(-(a)dS) black hole with metric function

$$\mu(r) = 1 - \frac{2M}{r} + \frac{Q^2}{r^2} - \frac{\Lambda r^2}{3}, \qquad (2.28)$$

where the magnetic charge Q is fixed by

$$Q^2 = \frac{1}{6}N(N+1)(N-1). \qquad (2.29)$$

Embedded $\mathfrak{su}(2)$ *solutions*   For our later numerical and analytic work, an additional special class of solutions turns out to be extremely useful. We begin by setting

$$\omega_j(r) = \pm\sqrt{j(N-j)}\,\omega(r) \qquad \forall j = 1,\ldots,N-1, \qquad (2.30)$$

then follow [100] and define

$$\lambda_N = \sqrt{\frac{1}{6}N(N-1)(N+1)}, \qquad (2.31)$$

and then rescale the field variables as follows:

$$R = \lambda_N^{-1}r; \qquad \tilde{\Lambda} = \lambda_N^2\Lambda; \qquad \tilde{m}(R) = \lambda_N^{-1}m(r);$$
$$\tilde{S}(R) = S(r); \qquad \tilde{\omega}(R) = \omega(r). \qquad (2.32)$$

Note that we rescale the cosmological constant Λ (this is not necessary in [100] as there $\Lambda = 0$). The field equations satisfied by $\tilde{m}(R)$, $\tilde{S}(R)$ and $\tilde{\omega}(R)$ are then

$$\frac{d\tilde{m}}{dR} = \mu\tilde{G} + R^2\tilde{p}_\theta;$$
$$\frac{1}{\tilde{S}}\frac{d\tilde{S}}{dR} = -\frac{2\tilde{G}}{R};$$
$$0 = R^2\mu\frac{d^2\tilde{\omega}}{dR^2} + \left[2\tilde{m} - 2R^3\tilde{p}_\theta - \frac{2\tilde{\Lambda}R^3}{3}\right]\frac{d\tilde{\omega}}{dR} + \left[1 - \tilde{\omega}^2\right]\tilde{\omega}; \quad (2.33)$$

where we now have

$$\mu = 1 - \frac{2\tilde{m}}{R} - \frac{\tilde{\Lambda}R^2}{3}, \tag{2.34}$$

and

$$\tilde{G} = \left(\frac{d\tilde{\omega}}{dR}\right)^2, \qquad \tilde{p}_\theta = \frac{1}{2R^4}\left(1 - \tilde{\omega}^2\right)^2. \tag{2.35}$$

The (2.33) are precisely the $\mathfrak{su}(2)$ EYM field equations. Furthermore, the boundary conditions (2.20) and (2.23) also reduce to those for the $\mathfrak{su}(2)$ case.

2.2.3 Dyonic Field Equations

As will be discussed in Sect. 2.4.3, if either $N > 2$ or we have a negative cosmological constant Λ, then we do not need to restrict ourselves to considering only purely magnetic equilibrium gauge potentials. If the electric part of the gauge potential (2.6), $\mathscr{A}$, is non-zero, there is still sufficient gauge freedom to set $\mathscr{B} = 0$ in (2.6) [99]. Then, provided none of the ω_j vanish identically, one of the Yang–Mills equations again tells us that all the γ_j are identically zero. Following [99] it is convenient to define new real variables $\alpha_j(r)$ by

$$\mathscr{A}_{jj} = i\left[-\frac{1}{N}\sum_{k=1}^{j-1} k\alpha_k + \sum_{k=j}^{N-1}\left(1 - \frac{k}{N}\right)\alpha_k\right] \tag{2.36}$$

so that the matrix $\mathscr{A}$ is automatically purely imaginary, diagonal and traceless. In this case the Yang–Mills equations (2.12) now take the form [99]

$$r^2\mu\omega_j'' + \left(2m - 2r^3 p_\theta - \frac{2\Lambda r^3}{3}\right)\omega_j' + W_j\omega_j + \frac{\mu}{r^2}\alpha_j^2\omega_j = 0, \tag{2.37}$$

and there are additional Yang–Mills equations for the α_j, namely [99]

$$\left[r^2 S^{-1}\left(\mu S\alpha_j\right)'\right]' = 2\alpha_j\omega_j^2 - \alpha_{j-1}\omega_{j-1}^2 - \alpha_{j+1}\omega_{j+1}^2. \tag{2.38}$$

The Einstein equations retain the form (2.15) but the quantities p_θ (2.13) and G (2.16) now read [99]

$$p_\theta = \frac{1}{4r^4}\sum_{j=1}^{N}\left[\left(\omega_j^2 - \omega_{j-1}^2 - N - 1 + 2j\right)^2 + \left(\frac{r^2}{S}\left(\mu S\mathscr{A}_{jj}\right)'\right)^2\right]$$

$$G = \sum_{j=1}^{N-1}\left[\omega_j'^2 + \alpha_j^2\omega_j^2\right]. \tag{2.39}$$

2.2.4 Perturbation Equations

We are also interested in the stability of the static, equilibrium solutions. For simplicity, we consider only linear, spherically symmetric perturbations of the purely magnetic solutions. We return to the general gauge potential of the form (2.6), and the metric (2.4), where now all functions depend on time t as well as r. There is still sufficient gauge freedom to enable us to set $\mathscr{A} \equiv 0$. This choice of gauge is particularly useful as then we shall shortly see that the perturbation equations decouple into two sectors, the "gravitational" and "sphaleronic" sectors [102]. We consider perturbations about the equilibrium solutions of the form

$$\omega_j(t,r) = \omega_j(r) + \delta\omega_j(t,r), \tag{2.40}$$

where $\omega_j(r)$ are the equilibrium functions and $\delta\omega_j(t,r)$ are the linear perturbations. There are similar perturbations for the other equilibrium quantities m and S, and in addition we have the perturbations $\delta\gamma_j(t,r)$ and $\delta\beta_j(t,r)$, the latter being the entries along the diagonal of the matrix $\mathscr{B}$ (2.6):

$$\mathscr{B} = \mathrm{Diag}\,(i\delta\beta_1, \cdots, i\delta\beta_N). \tag{2.41}$$

Note that the $\delta\beta_j$ are not independent because the matrix $\mathscr{B}$ is traceless, so

$$\delta\beta_1 + \cdots + \delta\beta_N = 0, \tag{2.42}$$

but it simplifies the derivation of the perturbation equations to retain all the $\delta\beta_j$ for the moment. We ignore all terms involving squares or higher powers of the perturbations. The full derivation of the perturbation equations is highly involved and the details will be presented elsewhere [11]. Instead here we summarize the key features of the perturbation equations. As usual, we will employ the "tortoise" co-ordinate r_*, defined by

$$\frac{dr_*}{dr} = \frac{1}{\mu S}, \tag{2.43}$$

where μ and S are the equilibrium metric functions.

2.2.4.1 Sphaleronic Sector

The sphaleronic sector consists of the $2N - 1$ perturbations $\delta\beta_j$, $j = 1, \ldots, N$ and $\delta\gamma_j$, $j = 1, \ldots, N-1$. We define new variables $\delta\Phi_j$ by

$$\delta\Phi_j = \omega_j\delta\gamma_j. \tag{2.44}$$

The perturbation equations for the sphaleronic sector arise solely from the Yang–Mills equations, and comprise

$$\delta\ddot{\beta}_j = \frac{S}{r^2}\left[\omega_{j-1}\partial_{r_*}\left(\delta\Phi_{j-1}\right) - \omega_j\partial_{r_*}\left(\delta\Phi_j\right)\right]$$

$$+\frac{S}{r^2}\left[\left(\partial_{r_*}\omega_j\right)\delta\Phi_j - \left(\partial_{r_*}\omega_{j-1}\right)\delta\Phi_{j-1}\right]$$

$$+\frac{\mu S^2}{r^2}\left[\omega_j^2\left(\delta\beta_{j+1} - \delta\beta_j\right) - \omega_{j-1}^2\left(\delta\beta_j - \delta\beta_{j-1}\right)\right]; \tag{2.45}$$

$$\delta\ddot{\Phi}_j = \partial_{r_*}^2\left(\delta\Phi_j\right) - \frac{1}{\omega_j}\left(\partial_{r_*}^2\omega_j\right)\delta\Phi_j + \mu S\omega_j\partial_{r_*}\left(\delta\beta_j - \delta\beta_{j+1}\right)$$

$$+\left[\mu\left(\partial_{r_*}S\right)\omega_j + \left(\partial_{r_*}\mu\right)S\omega_j + 2\mu S\left(\partial_{r_*}\omega_j\right)\right]\left(\delta\beta_j - \delta\beta_{j+1}\right); \tag{2.46}$$

together with the *Gauss constraint*

$$0 = \partial_{r_*}\left(\delta\dot{\beta}_j\right) + \left[\frac{2\mu S}{r} - \frac{\partial_{r_*}S}{S}\right]\delta\dot{\beta}_j + \frac{S}{r^2}\left[\omega_j\delta\dot{\Phi}_j + \omega_{j-1}\delta\dot{\Phi}_{j-1}\right], \tag{2.47}$$

where a dot denotes $\partial/\partial t$. It is important to note that the cosmological constant Λ only appears in these equations through the metric function μ (2.5), and therefore the perturbation equations (2.45) and (2.46) and the Gauss constraint (2.47) have exactly the same form as derived in [47] for arbitrary gauge groups in asymptotically flat space.

2.2.4.2 Gravitational Sector

The gravitational sector consists of the perturbations of the metric functions $\delta\mu$ and δS as well as the perturbations of the remaining gauge field functions $\delta\omega_j$. Both the Einstein equations and the remaining Yang–Mills equations are involved in this sector. For an arbitrary gauge group and asymptotically flat space, the perturbation equations in this sector have been considered in [47]. In asymptotically adS, we also find that the metric perturbations can be eliminated to give a set of equations governing the perturbations $\delta\omega_j$, which can be written in matrix form

$$\underline{\delta\ddot{\omega}} = \partial_{r_*}^2\left(\underline{\delta\omega}\right) + \mathcal{M}_G\underline{\delta\omega}, \tag{2.48}$$

where $\underline{\delta\omega} = \left(\delta\omega_1,\ldots,\delta\omega_{N-1}\right)^T$ and the $(N-1)\times(N-1)$ matrix $\mathcal{M}_G$ has entries

$$\mathcal{M}_{G,j,j} = \frac{\mu S^2}{r^2}\left[W_j - 2\omega_j^2\right] + \frac{4}{\mu Sr}\Upsilon\left(\partial_{r_*}\omega_j\right)^2 + \frac{8S}{r^3}W_j\omega_j\left(\partial_{r_*}\omega_j\right);$$

$$\mathcal{M}_{G,j,j+1} = \frac{\mu S^2}{r^2}\omega_j\omega_{j+1} + \frac{4}{\mu Sr}\Upsilon\left(\partial_{r_*}\omega_j\right)\left(\partial_{r_*}\omega_{j+1}\right)$$

$$+\frac{8S}{r^3}\left[W_j\omega_j\left(\partial_{r_*}\omega_{j+1}\right) + W_{j+1}\omega_{j+1}\left(\partial_{r_*}\omega_j\right)\right];$$

$$\mathcal{M}_{G,j,k} = \frac{4}{\mu Sr}\Upsilon\left(\partial_{r_*}\omega_j\right)\left(\partial_{r_*}\omega_k\right) + \frac{8S}{r^3}\left[W_j\omega_j\left(\partial_{r_*}\omega_k\right) + W_k\omega_k\left(\partial_{r_*}\omega_j\right)\right]; \tag{2.49}$$

where $k \neq j, j+1$, and Υ is given in terms of the equilibrium metric functions μ and S as follows:

$$\Upsilon = \frac{1}{\mu}\partial_{r_*}\mu + \frac{1}{S}\partial_{r_*}S + \frac{\mu S}{r}. \tag{2.50}$$

2.3 Asymptotically Flat/de Sitter Solutions for $\mathfrak{su}(N)$ EYM

We now turn to black hole solutions of the EYM field equations, beginning by briefly reviewing some of the key features of solutions in asymptotically flat or asymptotically de Sitter space.

2.3.1 Asymptotically Flat, Spherically Symmetric $\mathfrak{su}(2)$ Solutions

Apart from the trivial solutions given above (2.25) and (2.27), the first black hole solutions of the EYM field equations were found by Yasskin [182], and correspond to embedding the Reissner–Nordström electromagnetic gauge field into a higher–dimensional gauge group. The metric of these solutions is still Reissner–Nordström. Yasskin conjectured that his solutions were the only ones possible. This conjecture was only shown to be false 25 years later [19, 101, 168, 169]. That the discovery of hairy black holes in $\mathfrak{su}(2)$ EYM took so long may be attributed to the conjecture that there were no soliton solutions in this model. This conjecture is based on the fact that there are no solitons in pure gravity (see, for example, [78, 104]); no solitons in Einstein–Maxwell theory [77], no pure YM solitons in flat spacetime [53, 56] and no EYM solitons in three spacetime dimensions [57]. However, once Bartnik and McKinnon [7] had discovered non-trivial EYM solitons in four-dimensional spacetime, Yasskin's no-hair conjecture for EYM theory was quickly shown to be false [19].

For $\mathfrak{su}(2)$ EYM, it has been shown [23, 62, 67] that non-trivial solutions (i.e., solutions in which the gauge field is not essentially Abelian) must have a purely magnetic gauge potential, which is described by a single gauge field function $\omega(r)$ (2.10). Note that the ansatz (2.10) for $\mathfrak{su}(2)$ is not the same as the Witten ansatz [179] which was used in the original papers [7, 19], but it gives equivalent field equations. In this case the $\mathfrak{su}(2)$ EYM equations have the form

$$\frac{dm}{dr} = \left(1 - \frac{2m}{r}\right)\left(\frac{d\omega}{dr}\right)^2 + \frac{1}{2r^2}\left(1 - \omega^2\right)^2;$$

$$\frac{1}{S}\frac{dS}{dr} = -\frac{2}{r}\left(\frac{d\omega}{dr}\right)^2;$$

$$0 = r^2\left(1 - \frac{2m}{r}\right)\frac{d^2\omega}{dr^2} + \left[2m - \frac{\left(1 - \omega^2\right)^2}{r}\right]\frac{d\omega}{dr} + \left[1 - \omega^2\right]\omega. \tag{2.51}$$

It is the highly non-linear nature of these equations which allows for non-trivial soliton and hairy black hole solutions, which may be thought of heuristically as arising from a balancing of the gravitational and gauge field interactions (see [82] for a recent discussion). The non-linear nature of the equations also means, however, that (apart from the solutions for the Yang–Mills field on a fixed Schwarzschild metric [28, 34]) solutions can only be found numerically.

The numerical work in [7, 19, 101, 168, 169] found discrete families of solutions [156], indexed by the event horizon radius r_h (with $r_h = 0$ for solitons) and n, the number of zeros of the single gauge field function ω, each pair (r_h, n) identifying a solution of the field equations. A key feature of the solutions is that $n > 0$, so that the gauge field function must have at least one zero (or "node"). Later analytic work [29, 149–151] rigorously proved these numerical features. The black holes are "hairy" in the sense that they have no magnetic charge [23, 62, 67] and are therefore indistinguishable at infinity from a standard Schwarzschild black hole. However, the "hair", that is, the non-trivial structure in the matter fields, extends some way out from the event horizon, leading to the "no-short-hair" conjecture [122].

Although initially controversial [20, 24, 152, 173], rapidly it was accepted that both the soliton [154] and the black hole solutions [155] are unstable. This instability is not unexpected if we consider the solutions as arising from a balancing of the gauge field and gravitational interactions. Studies of the non-linear stability of the solutions [183, 184] reveal that the gauge field "hair" either radiates away to infinity or falls down the black hole event horizon, leaving, as the end-point, a bald Schwarzschild black hole. Due to this instability, the black holes, while they violate the "letter" of the no-hair conjecture, may be thought of as not contradicting its "spirit", and one might be led to conjecture that all *stable* black holes are fixed by their mass, angular momentum and conserved charges.

Originally these hairy black holes were shown to be unstable using numerical techniques [155] but the instability can also be shown analytically [68, 170]. In the $\mathfrak{su}(2)$ case, the perturbation equations (2.45), (2.46) and (2.48) simplify considerably. The sphaleronic sector reduces to a single equation (see Sect. 2.4.2 below for further details)

$$-\ddot{\zeta} = -\partial_{r_*}^2 \zeta + \left[\frac{\mu S^2}{r^2} \left(1 + \omega^2\right) + \frac{2}{\omega^2} \left(\frac{d\omega}{dr_*}\right)^2 \right] \zeta, \qquad (2.52)$$

while, on eliminating the metric perturbations, the gravitational sector also has just one equation:

$$-\delta\ddot{\omega} = -\partial_{r_*}^2 (\delta\omega) \qquad (2.53)$$
$$+ \frac{\mu S^2}{r^2} \left[3\omega^2 - 1 - 4r\omega'^2 \left(\frac{1}{r} - \frac{(1-\omega^2)^2}{r^3}\right) + \frac{8}{r} \omega\omega' \left(\omega^2 - 1\right) \right] \delta\omega.$$

The instability has been compared to that of the flat-space Yang–Mills sphaleron [170], which has a single unstable mode. The situation is slightly more complicated here, due to the two sectors of perturbations. The sphaleronic sector certainly, as its name suggests, mimics the perturbations of the flat-space sphaleron. It can be shown [167] that the number of instabilities in the sphaleronic sector equals n, the number of zeros of the gauge field function ω. The same is true in the gravitational sector, as conjectured in [102] and can be shown using catastrophe theory, by considering the more general EYM–Higgs solutions [115]. The above concerns only spherically symmetric perturbations. It is known that the flat-space sphaleron has instabilities only in the spherically symmetric sector [4]. Extending this to the $\mathfrak{su}(2)$ EYM black holes requires complicated analysis [143], using a curvature-based formalism developed in [43, 144, 145].

Using the isolated horizons formalism, these "hairy" black holes can be interpreted as bound states of ordinary black holes with the Bartnik–MacKinnon solitons [3, 54, 55]. In particular, the soliton masses are given in terms of the masses of the corresponding black holes [55], and the instability of the colored black holes arises naturally from the instability of the corresponding solitons [3, 54].

Since these initial discoveries a plethora of new, asymptotically flat, hairy black hole solutions have been found in Einstein–Yang–Mills theory and its variants (see [171] for a review of those solutions discovered prior to 1999). Most of these are, indeed, unstable. However, there are notable exceptions, including (a) the Skyrme black hole [22, 58, 80, 81] where the existence of an integer-valued topological winding number renders the solutions stable, (b) Einstein–Yang–Mills–Higgs black holes in the limit of infinitely strong coupling of the Higgs field [1] and (c) a particular branch of Einstein-non-Abelian-Proca black holes [73, 110, 159, 160, 163]. We will not consider additional matter fields further in this article.

2.3.2 Non-spherically Symmetric, Asymptotically Flat $\mathfrak{su}(2)$ Solutions

One of the surprising aspects of the failure of black hole uniqueness in EYM is that almost every step in the uniqueness theorem in Einstein–Maxwell theory has a counterexample in EYM (see [79] for detailed discussions on this topic, and [45, 128, 153, 156, 157] for examples of some results from Einstein–Maxwell theory which do generalize). An important example of this is Israel's theorem [86, 87], which states that the geometry outside the event horizon of a static black hole must be spherically symmetric. This is not true in EYM: there are static black hole solutions which are not spherically symmetric but only axisymmetric [90] (in more general matter models, static black holes do not necessarily possess any symmetries at all [138, 139]). These solutions are found numerically by writing the metric in isotropic co-ordinates

$$ds^2 = -f(r,\theta)\,dt^2 + \frac{m(r,\theta)}{f(r,\theta)}dr^2 + \frac{m(r,\theta)r^2}{f(r,\theta)}d\theta^2 + \frac{L(r,\theta)r^2\sin^2\theta}{f(r,\theta)}d\phi^2, \quad (2.54)$$

and using the following ansatz for the $\mathfrak{su}(2)$ gauge field [137]

$$A = \frac{1}{2r} \left\{ \tau_\phi^p \left[H_1(r,\theta)\,dr + (1 - H_2(r,\theta))\,r\,d\theta \right] \right.$$
$$\left. - p \left[\tau_r^p H_3(r,\theta) + \tau_\theta^p \left(1 - H_4(r,\theta)\right) \right] r\sin\theta\,d\phi \right\}, \qquad (2.55)$$

where

$$\tau_r^p = \underline{\tau}.\left(\sin\theta\cos p\phi, \sin\theta\sin p\phi, \cos\theta \right),$$
$$\tau_\theta^p = \underline{\tau}.\left(\cos\theta\cos p\phi, \cos\theta\sin p\phi, -\sin\theta \right),$$
$$\tau_\phi^p = \underline{\tau}.\left(-\sin p\phi, \cos p\phi, 0 \right), \qquad (2.56)$$

with

$$\underline{\tau} = \left(\tau_x, \tau_y, \tau_z \right), \qquad (2.57)$$

where τ_x, τ_y, τ_z are the usual generators of $\mathfrak{su}(2)$. Here, p is a winding number, with $p = 1$ corresponding to spherically symmetric solutions (with the gauge potential written in a different form to that we have used in (2.10)). Substituting the ansatz into the field equations gives a complicated set of partial differential equations, solutions of which are exhibited in [90]. Static, axisymmetric soliton solutions also exist [65, 85, 91].

It is less surprising that rotating black holes also exist in this model [92, 93], generalizing the Kerr–Newman metric (as predicted in [156]). These solutions are indexed by the winding number p (2.56) and a node number n. They carry no magnetic charge, but all have non-zero electric charge [156, 157]. The question of whether there are rotating solitons in pure $\mathfrak{su}(2)$ EYM has yet to be conclusively settled, however. Rotating soliton solutions have been found in EYM–Higgs theory [127], but not in pure EYM theory. Although rotating solitons are predicted perturbatively [44], the consensus in the literature is now that it seems unlikely that rotating soliton solutions do exist [17].

2.3.3 Asymptotically Flat $\mathfrak{su}(N)$ Solutions

We shall next consider generalizations of the $\mathfrak{su}(2)$ YM gauge group. The simplest such generalization is to consider $\mathfrak{su}(N)$ EYM. The results of [62, 67] do not extend to this larger gauge group, and it is possible to have solutions with electric charge [69], which correspond to a superposition of electrically charged Reissner–Nordström and the $\mathfrak{su}(2)$ EYM black holes. Numerical solutions of the field equations have been found in the following papers: [69, 94–96]. As N increases, the possible structures of the gauge field potential (2.6) become ever more complicated. A method for computing all spherically symmetric $\mathfrak{su}(N)$ gauge field potentials is given in [6], where all the irreducible possibilities are enumerated for $N \leq 6$. As in the $\mathfrak{su}(2)$ case, black hole solutions are found at discrete points in the parameter space $\{\omega_j(r_h), j = 1 \ldots N - 1\}$.

There is comparatively little analytic work for more general gauge groups. Local existence of solutions of the field equations (2.12) and (2.15) near the black hole event horizon and at infinity has been proven for gauge group $\mathfrak{su}(N)$ [100], and subsequently extended to arbitrary compact gauge group [124, 125]. The existence of non-trivial black hole solutions to the field equations has been proven rigorously only in the $\mathfrak{su}(3)$ case [140, 141], although there are arguments that hairy black hole solutions exist for all N [116]. In the $\mathfrak{su}(3)$ case, Ruan [140, 141] has proved that there are infinitely many hairy black hole solutions, indexed by the numbers of zeros (n_1, n_2), respectively, of the two gauge field functions (ω_1, ω_2). Furthermore, provided that the radius of the event horizon is sufficiently large, there is a black hole solution for any combination of (n_1, n_2). The global properties of the solutions for arbitrary compact gauge group are studied in [126]. However, it will come as no surprise to learn that all these solutions, in asymptotically flat space, and for any compact gauge group, are unstable [46, 47]. To show instability it is sufficient to find a single unstable mode, and therefore the work in [46, 47] studies the simpler, sphaleronic sector of perturbations (see Sect. 2.2.4).

2.3.4 Asymptotically de Sitter $\mathfrak{su}(2)$ EYM Solutions

Another natural generalization of asymptotically flat $\mathfrak{su}(2)$ EYM is the inclusion of a non-zero cosmological constant Λ. When the cosmological constant is positive, soliton [172] and black hole [164] $\mathfrak{su}(2)$ EYM solutions have been found (other numerical solutions are presented in [41, 119]). These solutions possess a cosmological horizon and approach de Sitter space at infinity (for a complete classification of the possible spacetime structures, see [30]). The phase space of solutions is again discrete, and the single gauge field function ω must have at least one zero. Unsurprisingly, these solutions again turn out to be unstable [42, 63, 164]. Given this instability, the asymptotically de Sitter solutions have received rather less attention in the literature, but some analytic work can be found in [105–107].

2.4 Asymptotically anti-de Sitter Solutions for $\mathfrak{su}(2)$ EYM

We now turn to the main focus of this article: asymptotically anti-de Sitter solutions. We begin by reviewing some of the properties of black holes in $\mathfrak{su}(2)$ EYM.

2.4.1 Spherically Symmetric, Asymptotically adS, $\mathfrak{su}(2)$ EYM Solutions

Black hole solutions of $\mathfrak{su}(2)$ EYM with a negative cosmological constant were first studied in [175], and subsequently in [25, 26]. The field equations now take the form

$$\frac{dm}{dr} = \left(1 - \frac{2m}{r} - \frac{\Lambda r^2}{3}\right)\left(\frac{d\omega}{dr}\right)^2 + \frac{1}{2r^2}\left(1 - \omega^2\right)^2;$$

$$\frac{1}{S}\frac{dS}{dr} = -\frac{2}{r}\left(\frac{d\omega}{dr}\right)^2;$$

$$0 = r^2\left(1 - \frac{2m}{r} - \frac{\Lambda r^2}{3}\right)\frac{d^2\omega}{dr^2} + \left[2m - \frac{2\Lambda r^3}{3} - \frac{\left(1 - \omega^2\right)^2}{r}\right]\frac{d\omega}{dr}$$

$$+ \left[1 - \omega^2\right]\omega. \tag{2.58}$$

The inclusion of a negative cosmological constant means that boundary conditions at infinity (2.23) are considerably less stringent than in the asymptotically flat case; it is therefore unsurprising that it is easier to find solutions in asymptotically adS.

The space of solutions in adS is very different to that in asymptotically flat space. Instead of finding solutions at discrete values of $\omega(r_h)$, solutions exist in continuous, open intervals. Furthermore, for sufficiently large $|\Lambda|$, we now find solutions in which the single gauge field function $\omega(r)$ has no zeros. A typical example of such a solution is shown in Fig. 2.1, further examples can be found in [175]. These properties of the space of solutions of the (2.58) are proved in [175].

We now examine the structure of the space of solutions, more details of which can be found in [8, 9, 175]. There are three parameters describing the solutions, r_h, Λ and $\omega(r_h)$. In order to plot two-dimensional figures, we fix either r_h or Λ and vary the other two quantities. For $\mathfrak{su}(2)$ black holes, the constraint (2.22) on the value of the gauge field function at the event horizon reads

$$\left(\omega(r_h)^2 - 1\right)^2 < r_h^2\left(1 - \Lambda r_h^2\right). \tag{2.59}$$

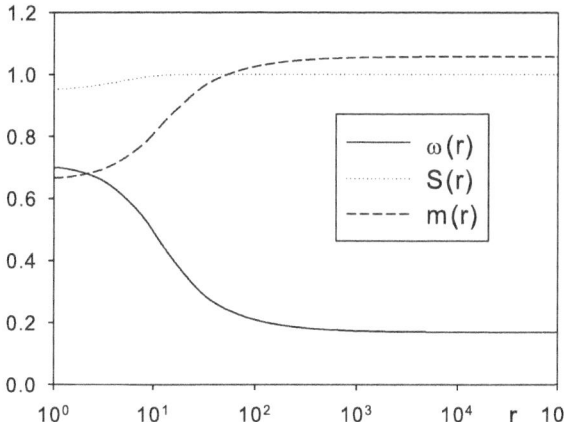

Fig. 2.1 An example of an $\mathfrak{su}(2)$ EYM black hole in adS in which the gauge field function $\omega(r)$ has no zeros. Here, $\Lambda = -1$, $r_h = 1$ and $\omega(r_h) = 0.7$

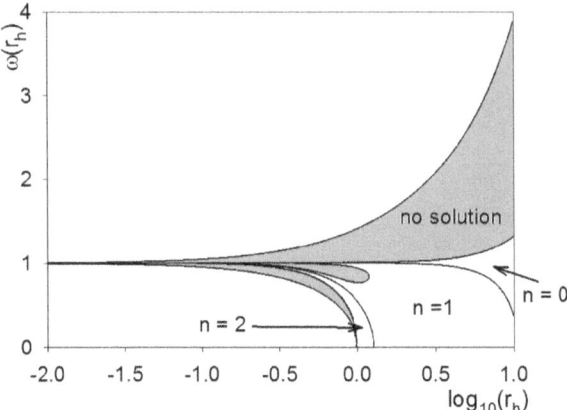

Fig. 2.2 The space of $\mathfrak{su}(2)$ black hole solutions when $\Lambda = -0.01$, for varying r_h. The *shaded region* indicates values of the gauge field function $\omega(r_h)$ at the event horizon for which the constraint (2.59) is satisfied, but for which we find no well-behaved black hole solution. The number of zeros n of the gauge field function ω are indicated in those regions of the phase space where we find black hole solutions. Elsewhere on the diagram, the constraint (2.59) is not satisfied. Between the region where $n = 2$ and the *shaded region* we find black hole solutions with $n = 3, 4$ and 5, but these regions are too small to indicate on the graph. Taken from [9]

Whether we are varying r_h or Λ, we perform a scan over all values of ω_h which satisfy (2.59). First, we show in Fig. 2.2 the space of black hole solutions for fixed $\Lambda = -0.01$ and varying event horizon radius r_h. The outermost curves in Fig. 2.2 are where the inequality (2.59) is saturated. Immediately inside these curves we have a shaded region, which represents values of $(r_h, \omega(r_h))$ for which the constraint (2.59) is satisfied, but for which we are unable to find black hole solutions which remain regular all the way out to infinity. Where we do find solutions, we indicate in Fig. 2.2 the number of zeros of the gauge field function $\omega(r)$. The solution for which $\omega(r_h) = 1$ is simply the Schwarzschild-adS black hole, while that for $\omega(r_h) = 0$ is the magnetically charged Reissner–Nordström-adS black hole (see Sect. 2.2.2). As $r_h \to 0$, the constraint (2.59) implies that $\omega(r_h) \to 1$, as can be seen in Fig. 2.2. The black hole solutions become solitons in this limit. However, for this value of Λ, there are different soliton solutions, with ω having different numbers of zeros [31], a feature which is not readily apparent from Fig. 2.2. We find similar behavior on varying r_h for different values of Λ.

If we now fix the event horizon radius to be $r_h = 1$ and vary Λ, the solution space is shown in Fig. 2.3, with a close-up for smaller values of $|\Lambda|$ in Fig. 2.4.

Again, in Figs. 2.3 and 2.4 we have shaded those regions where the constraint (2.59) is satisfied, but no regular black hole solutions could be found. Where we do find solutions, the number of zeros of the gauge field function $\omega(r)$ is indicated in the figures. As $\Lambda \to 0$, the phase space breaks up into discrete points, which correspond to the asymptotically flat "colored" $\mathfrak{su}(2)$ black holes described in Sect. 2.3.1 [19]. For sufficiently large $|\Lambda|$, we find solutions in which the gauge field function has no zeros.

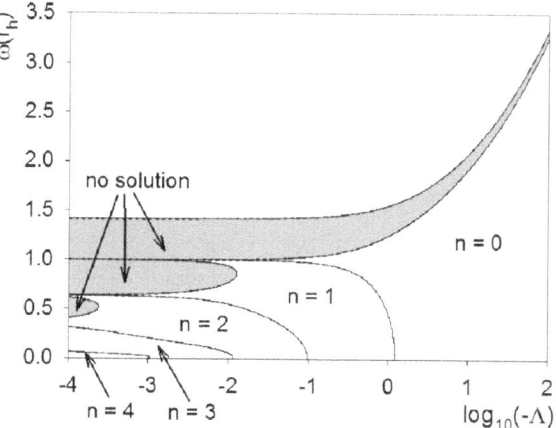

Fig. 2.3 Phase space of $\mathfrak{su}(2)$ black holes with $r_h = 1$ and varying Λ. The *shaded region* indicates values of the gauge field function $\omega(r_h)$ at the event horizon for which the constraint (2.59) is satisfied, but for which we find no well-behaved black hole solution. The number of zeros n of the gauge field function ω are indicated in those regions of the phase space where we find black hole solutions. Elsewhere on the diagram, the constraint (2.59) is not satisfied. As well as the regions where $n = 0, \ldots, 4$ as marked on the diagram, we find a *small region* in the *bottom left* of the plot where $n = 5$. This region is too small to indicate on the current figure, but can be seen in Fig. 2.4. Taken from [9]

The spectrum of black hole solutions (that is, the relationship between the mass M and magnetic charge Q of the black holes) was first studied in [26]. We plot in Fig. 2.5 the black hole mass versus magnetic charge for black holes with $r_h = 1$ and varying values of Λ (cf. Fig. 8 in [26]). For large values of $|\Lambda|$, there are only nodeless solutions and the spectrum is simple, with the black holes being uniquely

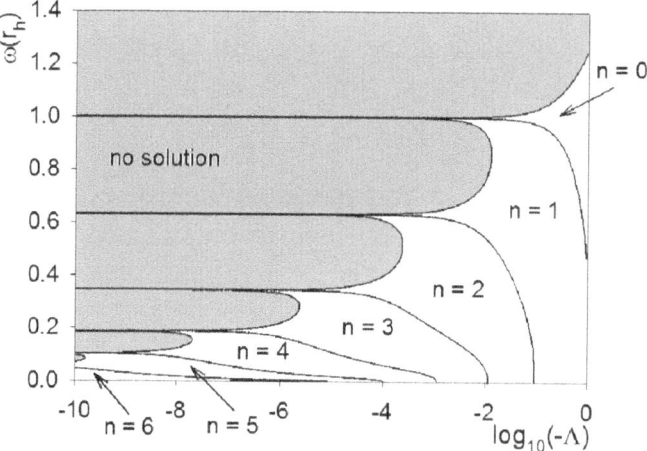

Fig. 2.4 Close-up of the phase space of $\mathfrak{su}(2)$ black holes with $r_h = 1$ and smaller values of Λ. In the *bottom left* of the plot there is a *small region* of solutions for which $n = 7$, but the region is too small to be visible. Taken from [9]

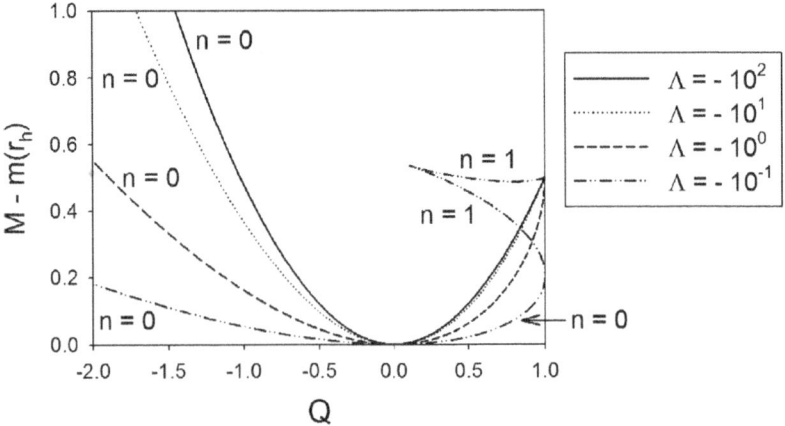

Fig. 2.5 Black hole mass M and magnetic charge Q for $\mathfrak{su}(2)$ EYM black holes with $r_h = 1$ and varying Λ (cf. Fig. 8 in [26])

specified by Λ, r_h and Q_M. As $|\Lambda|$ decreases, the spectrum becomes more complicated. For example, looking at the $\Lambda = -0.1$ curve in Fig. 2.5 we see that a branch structure emerges. The lower M curve for $\Lambda = -0.1$ consists of $n = 0$ (nodeless) solutions, and extends from negative Q up to $Q = 1$. When $Q = 1$, a branch of $n = 1$ solutions appears, which have larger mass. As Q decreases along this branch of solutions, the mass M increases, until a bifurcation point is reached and a second branch of $n = 1$ solutions appears, with even larger mass, and with the charge increasing as M increases. For smaller values of $|\Lambda|$, we find ever more complicated spectra, which appear to become "fractal" as $|\Lambda| \to 0$ [26, 114]. In view of the catastrophe theory analysis of other hairy black hole solutions [159, 160, 163], one might anticipate that the stability of the solutions changes at the points in the spectrum where two branches of solutions meet, but this has yet to be fully investigated in the literature (see [31] for an in-depth stability analysis of the soliton solutions). We therefore next consider the stability of these black holes.

2.4.2 Stability of the Spherically Symmetric Solutions

As discussed in Sect. 2.3.1, for the asymptotically flat $\mathfrak{su}(2)$ EYM black holes, it has been shown that the number of instabilities is twice the number of zeros of the gauge field function $\omega(r)$. Therefore, one might anticipate that at least some solutions when $\omega(r)$ has no zeros could be stable. For the $\mathfrak{su}(2)$ EYM case, the perturbation equations (2.45), (2.46) and (2.48) simplify considerably. In the sphaleronic sector, there is a single $\delta\Phi$ (2.44) and two further perturbations $\delta\beta_1$, $\delta\beta_2$, although these are not independent (2.42), so we may consider just $\delta\nu = \delta\beta_2 - \delta\beta_1$. The sphaleronic sector perturbations equations (2.45) and (2.46) then reduce to

$$\delta \ddot{v} = \frac{2S}{r^2} \left[\omega \partial_{r_*} (\delta \Phi) - (\partial_{r_*} \omega) \delta \Phi \right] - \frac{2\mu S^2}{r^2} \omega^2 \delta v; \tag{2.60}$$

$$\delta \ddot{\Phi} = \partial_{r_*}^2 (\delta \Phi) - \frac{1}{\omega} \left(\partial_{r_*}^2 \omega \right) \delta \Phi - \mu S \omega \partial_{r_*} (\delta v)$$
$$+ \left[\mu (\partial_{r_*} S) \omega + (\partial_{r_*} \mu) S \omega - 2\mu S (\partial_{r_*} \omega) \right] \delta v; \tag{2.61}$$

and the Gauss constraint (2.47) is now

$$0 = \partial_{r_*} (\delta \dot{v}) + \left[\frac{2\mu S}{r} - \frac{\partial_{r_*} S}{S} \right] \delta \dot{v} + \frac{S}{r^2} \omega \delta \Phi. \tag{2.62}$$

By introducing a new variable ζ (note our notation above is different from that used in [175])

$$\zeta = \frac{r^2}{S} \delta v, \tag{2.63}$$

the sphaleronic sector then reduces to a single equation [175]

$$-\ddot{\zeta} = -\partial_{r_*}^2 \zeta + \left[\frac{\mu S^2}{r^2} \left(1 + \omega^2 \right) + \frac{2}{\omega^2} \left(\frac{d\omega}{dr_*} \right)^2 \right] \zeta, \tag{2.64}$$

while the gravitational sector (2.48) also has just one equation:

$$-\delta \ddot{\omega} = -\partial_{r_*}^2 (\delta \omega) \tag{2.65}$$
$$+ \frac{\mu S^2}{r^2} \left[3\omega^2 - 1 - 4r\omega'^2 \left(\frac{1}{r} - \Lambda r - \frac{(1 - \omega^2)^2}{r^3} \right) + \frac{8}{r} \omega \omega' (\omega^2 - 1) \right] \delta \omega.$$

The sphaleronic sector equation (2.64) is exactly the same as that in the asymptotically flat $\mathfrak{su}(2)$ EYM case (2.52), but the gravitational sector equation (2.53) unsurprisingly is modified by the presence of non-zero Λ. Both (2.64) and (2.65) have the standard Schrödinger form

$$-\ddot{\Psi} = -\partial_{r_*}^2 \Psi + \mathcal{U} \Psi, \tag{2.66}$$

with potential $\mathcal{U}$. For the sphaleronic sector, when the gauge field function $\omega(r)$ has no zeros, it is immediately clear that the potential $\mathcal{U}$ is positive, so there are no instabilities in this sector (this result does not hold in the asymptotically flat case because the zeros of $\omega(r)$ in that case mean that $\mathcal{U}$ is not regular). The gravitational sector potential is more complex to analyze, but, for sufficiently large $|\Lambda|$ and $\omega(r_h) > 1/\sqrt{3}$, it can be shown that the potential is positive and there are no instabilities in this sector either. Therefore there are at least some hairy black holes which are stable under linear, spherically symmetric, perturbations. It can further be proved that at least some of these solutions remain stable when non-spherically symmetric perturbations are considered [146, 178] but the analysis is highly involved and so we do not attempt to summarize it here.

It should be remarked that it is unlikely that *all* nodeless black hole solutions are stable, although this has not been investigated in the literature. An in-depth study of the corresponding solitonic solutions [31] has revealed that some soliton solutions for which $\omega(r)$ has no zeros, although they do not have any instabilities in the sphaleronic sector, do possess unstable modes in the gravitational sector. A scaling behavior analysis of the solitonic solutions [83] has shown that the stable soliton solutions can be approximated well by the stable solitons which exist on pure adS space. On the other hand, the unstable solitons are interpreted as the unstable Bartnik–MacKinnon solitons [7] dressed with solitons on pure adS.

2.4.3 Other Asymptotically Anti-de Sitter $\mathfrak{su}(2)$ EYM Solutions

2.4.3.1 Dyonic Solutions

In asymptotically adS, it is no longer the case that the only genuinely non-Abelian solutions must have vanishing electric part in the gauge potential (2.6), so the results of [62, 67] do not extend to non-asymptotically flat solutions. As well as the magnetically charged solutions described above, dyonic black holes were discussed in [25, 26], which we shall not consider further here. The stability of the dyonic solutions remains an open question as the perturbation equations do not decouple into two sectors in this case, making analysis difficult.

2.4.3.2 Topological Black Holes

As in Einstein–Maxwell theory, topological black hole solutions exist for $\mathfrak{su}(2)$ EYM in adS [16]. The metric in this case reads

$$ds^2 = -\mu S^2 \, dt^2 + \mu^{-1} \, dr^2 + r^2 \, d\theta^2 + r^2 f^2(\theta) \, d\phi^2, \tag{2.67}$$

where

$$f(\theta) = \begin{cases} \sin\theta & \text{for} \quad k = 1, \\ \theta & \text{for} \quad k = 0, \\ \sinh\theta & \text{for} \quad k = -1, \end{cases} \tag{2.68}$$

and

$$\mu = k - \frac{2m(r)}{r} - \frac{\Lambda r^2}{3}. \tag{2.69}$$

The ansatz for the purely magnetic gauge field potential is now [16]

$$A = \tau_x \omega(r) \, d\theta + \left[\tau_y \omega(r) + \tau_z \frac{d\ln f}{d\theta} \right] f(\theta) \, d\phi. \tag{2.70}$$

When $\Lambda = 0$, only spherically symmetric solutions with $k = 1$ are possible, but for $\Lambda < 0$, solutions with both $k = 0$ and $k = -1$ have been found [16]. All the solutions are nodeless, which can be easily proved from the field equations [16]. It is found in [16] that all the $k = 0$ solutions are stable under spherically symmetric perturbations in both the sphaleronic and the gravitational sectors. The same is true for the $k = -1$ solutions for which $\omega > 1$ as $r \to \infty$ [16].

2.4.3.3 Non-spherically Symmetric Solutions

As in the asymptotically flat case, there are both soliton [129] and black hole [136] solutions which are static but not spherically symmetric, so that the metric and gauge potential take the form (2.54) and (2.55). Rotating black holes have also been found [113], and there are also rotating dyonic soliton solutions [131].

2.5 Asymptotically Anti-de Sitter Solutions for $\mathfrak{su}(N)$ EYM

In the previous section we found that stable hairy black holes exist in $\mathfrak{su}(2)$ EYM with a sufficiently large and negative cosmological constant. A natural question is therefore whether there are stable hairy black hole solutions of $\mathfrak{su}(N)$ EYM in adS, and we examine this question in this section.

2.5.1 Spherically Symmetric Numerical Solutions

For any fixed N, the field equations (2.12) and (2.15) can be solved numerically using standard techniques. We will outline briefly some of the key features of the black hole solutions for $\mathfrak{su}(3)$ EYM. Details of the corresponding soliton solutions and the solution space for $\mathfrak{su}(4)$ EYM can be found in [9].

For $\mathfrak{su}(3)$ EYM, there are two gauge field functions $\omega_1(r)$ and $\omega_2(r)$, and therefore four parameters describing black hole solutions: r_h, Λ, $\omega_1(r_h)$ and $\omega_2(r_h)$. Using the symmetry of the field equations (2.17), we set $\omega_1(r_h), \omega_2(r_h) > 0$ without loss of generality. The constraint (2.22) on the values of the gauge field functions at the horizon becomes, in this case

$$\left[\omega_1(r_h)^2 - 2\right]^2 + \left[\omega_1(r_h)^2 - \omega_2(r_h)^2\right]^2 + \left[2 - \omega_2(r_h)^2\right]^2 < 2r_h^2\left(1 - \Lambda r_h^2\right). \quad (2.71)$$

Two typical black hole solutions are shown in Figs. 2.6 and 2.7. The metric functions behave in a very similar way to the $\mathfrak{su}(2)$ solutions, smoothly interpolating between their values at the horizon and at infinity. We note that $S(r)$ in particular converges very rapidly to 1 as $r \to \infty$. In Fig. 2.6, we show an example of a black hole solution in which both gauge field functions have no zeros. We note that both

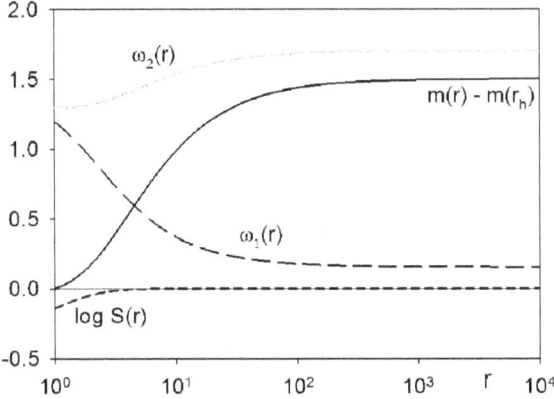

Fig. 2.6 Typical $\mathfrak{su}(3)$ black hole solution, with $r_h = 1$, $\Lambda = -1$, $\omega_1(r_h) = 1.2$ and $\omega_2(r_h) = 1.3$. In this example, both gauge field functions have no zeros. Taken from [9]

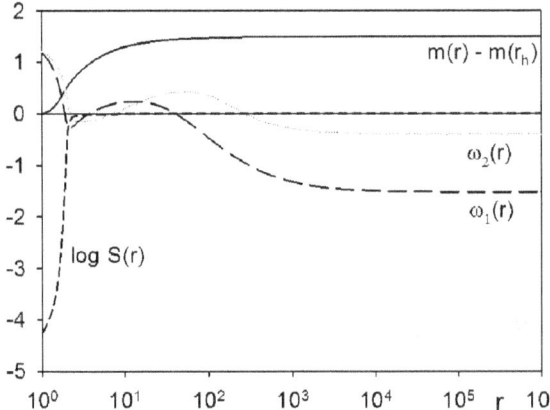

Fig. 2.7 Example of an $\mathfrak{su}(3)$ black hole solution, with $r_h = 1$, $\Lambda = -0.0001$, $\omega_1(r_h) = 1.184$ and $\omega_2(r_h) = 1.216$. In this case, both gauge field functions have three zeros. Taken from [9]

gauge field functions are monotonic, however, one is monotonically increasing and the other monotonically decreasing. In our second example (Fig. 2.7) both gauge field functions have three zeros. Although, in both our examples the two gauge field functions have the same number of zeros, we also find solutions where the two gauge field functions have different numbers of zeros (see Figs. 2.8 and 2.9).

We now examine the space of black hole solutions. Since we have four parameters, in order to produce two-dimensional figures, we need to fix two parameters in each case. We find that varying the event horizon radius produces similar behavior to the $\mathfrak{su}(2)$ case, so for the remainder of this section we fix $r_h = 1$ and consider the phase space for different, fixed values of Λ, scanning all values of $\omega_1(r_h)$, $\omega_2(r_h)$ such that the constraint (2.71) is satisfied. From the discussion in Sect. 2.2, we have embedded $\mathfrak{su}(2)$ black hole solutions when, from (2.30)

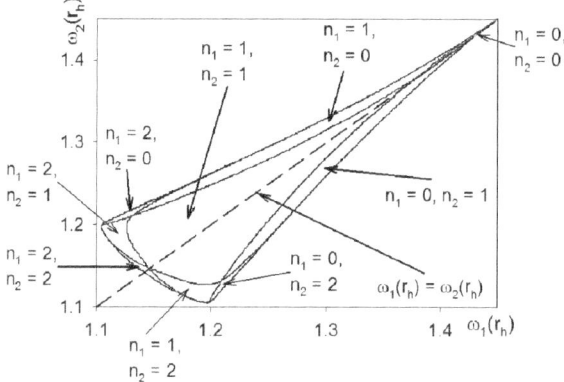

Fig. 2.8 Solution space for $\mathfrak{su}(3)$ black holes with $r_h = 1$ and $\Lambda = -0.1$. The numbers of zeros of the gauge field functions for the various regions of the solution space are shown. For other values of $\omega_1(r_h)$, $\omega_2(r_h)$ we find no solutions. There is a very small region containing solutions in which both gauge field functions have no zeros, in the *top-right-hand corner* of the plot. Taken from [9]

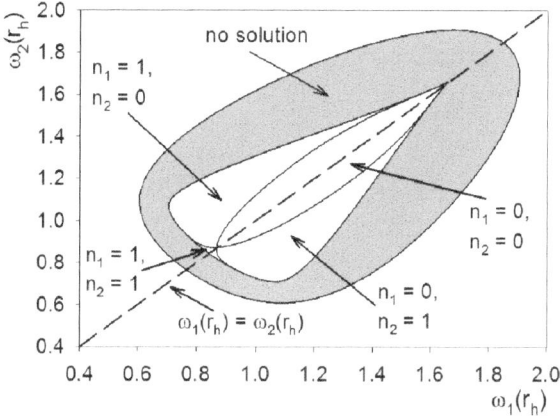

Fig. 2.9 Solution space for $\mathfrak{su}(3)$ black holes with $r_h = 1$ and $\Lambda = -1$. The *shaded region* indicates where the constraint (2.71) is satisfied but we do not find black hole solutions. Outside the *shaded region* the constraint (2.71) does not hold. Where there are solutions, we have indicated the number of zeros of the gauge field functions within the different regions. For this value of Λ there is a large region in which both gauge field functions have no zeros. Taken from [9]

$$\omega_1(r) = \sqrt{2}\omega(r) = \omega_2(r) \tag{2.72}$$

which occurs when $\omega_1(r_h) = \omega_2(r_h)$.

In Figs. 2.8, 2.9 and 2.10 we plot the phase space of solutions for fixed event horizon radius $r_h = 1$ and varying cosmological constant $\Lambda = -0.1$, -1 and -5, respectively. In each of Figs. 2.8, 2.9 and 2.10 we plot the dashed line $\omega_1(r_h) = \omega_2(r_h)$, along which lie the embedded $\mathfrak{su}(2)$ black holes. It is seen in all these figures that the solution space is symmetric about this line, as would be expected from the

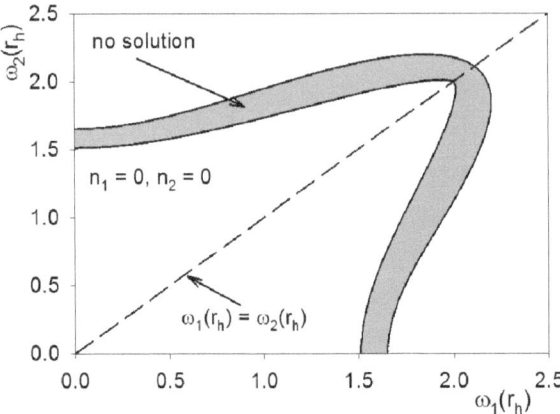

Fig. 2.10 Solution space for $\mathfrak{su}(3)$ black holes with $r_h = 1$ and $\Lambda = -5$. It can be seen that for the vast majority of the phase space for which the constraint (2.71) is satisfied, we have black hole solutions in which both gauge field functions have no zeros. Taken from [9]

symmetry (2.18) of the field equations. The solution space is found to be symmetric about the line $\omega_1(r_h) = \omega_2(r_h)$ not only in terms of where we find solutions but also in terms of the numbers of zeros of the gauge field functions. To state this precisely, suppose that at the point $\omega_1(r_h) = a_1$, $\omega_2(r_h) = a_2$ we find a black hole solution in which $\omega_1(r)$ has n_1 zeros and $\omega_2(r)$ has n_2 zeros. Then, at the point $\omega_1(r) = a_2$, $\omega_2(r) = a_1$, we find a black hole solution in which $\omega_1(r)$ has n_2 zeros and $\omega_1(r)$ has n_1 zeros. This is clearly seen in Figs. 2.8 and 2.9 and follows from the symmetry (2.18) of the field equations. As we increase $|\Lambda|$, we find (see Figs. 2.8, 2.9, and 2.10) that the solution space expands as a proportion of the space of values of $\omega_1(r_h)$, $\omega_2(r_h)$ satisfying the constraint (2.71). It can also be seen from Figs. 2.8, 2.9 and 2.10 that the number of nodes of the gauge field functions decreases as $|\Lambda|$ increases and that the space of solutions becomes simpler. For $\Lambda = -0.1$, there is a very small region of the solution space where both gauge field functions have no zeros. This region expands as we increase $|\Lambda|$, until for $\Lambda = -5$, both gauge field functions have no zeros for all the solutions we find.

The solution space becomes progressively more complicated as N increases, due to the increased number of parameters required to describe the solutions. However, the key feature described above is found; namely that for sufficiently large $|\Lambda|$, all the solutions we find are such that all the gauge field functions ω_j have no zeros. These solutions are of particular interest since one might hope that at least some of them might be stable.

As with the $\mathfrak{su}(2)$ black holes we may consider the spectra of black hole solutions by plotting the relationship between the mass M and the magnetic charge Q of the solutions (see Fig. 2.5 for the $\mathfrak{su}(2)$ case). As may be expected, for higher N the spectra are even more complicated than for $\mathfrak{su}(2)$. In Fig. 2.11 we plot some of the possible values of M and Q for $\mathfrak{su}(3)$ EYM black holes with $\Lambda = -0.1$ and $r_h = 1$. In Fig. 2.11 we have color coded the various possible numbers of zeros of

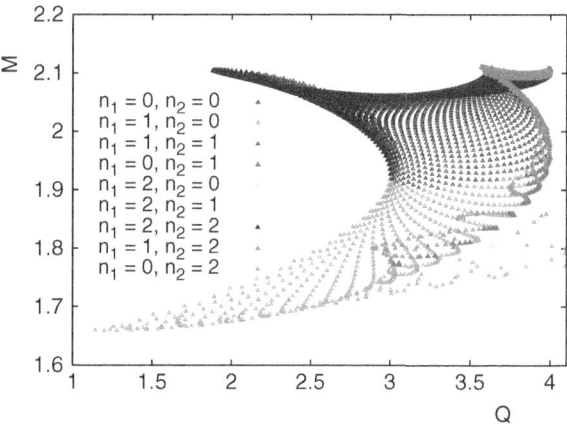

Fig. 2.11 Black hole mass M versus magnetic charge Q for $\mathfrak{su}(3)$ EYM black holes with $r_h = 1$ and $\Lambda = -0.1$. There are many different combinations of number of zeros of the gauge field functions (see Fig. 2.8), which are indicated by different colors. Here we have performed a scan over a grid of possible values of the gauge field functions at the event horizon, $\omega_1(r_h)$, $\omega_2(r_h)$, leading to discrete points in the spectrum. This is to enable the complicated structure of the spectrum to be seen

the gauge field functions (cf. Fig. 2.8). We have used a discrete grid of initial values of the gauge field functions at the event horizon $(\omega_1(r_h), \omega_2(r_h))$ and plotted discrete points so that at least some of the structure can be seen. In this case, because we have a four-parameter $(\Lambda, r_h, \omega_1(r_h), \omega_2(r_h))$ space of solutions of the field equations, even when Λ and r_h are fixed, we obtain two-dimensional regions in the (M, Q) plane, rather than curves as in the $\mathfrak{su}(2)$ case. It can be seen from Fig. 2.11 that the spectrum is very complicated, with the regions corresponding to different numbers of zeros of the gauge field functions overlapping. It is certainly the case that the black holes cannot be uniquely characterized by the four parameters (Λ, r_h, M, Q).

2.5.2 Analytic Work

For any fixed value of N, it is possible to examine the space of solutions numerically. However, we would like to know whether there are solutions for *all N*, and, in particular, whether for all N there are some solutions for which all the gauge field functions have no zeros, which we expect to be the case for sufficiently large $|\Lambda|$. Answering this question for general N requires analytic rather than numerical work.

In [175], the existence of black hole solutions for which the gauge function $\omega(r)$ had no zeros was proven analytically in the $\mathfrak{su}(2)$ case. Since $\mathfrak{su}(2)$ solutions can be embedded as $\mathfrak{su}(N)$ solutions via (2.30), we have automatically an analytic proof of the existence of nodeless $\mathfrak{su}(N)$ EYM black holes in adS. However, these embedded solutions are "trivial" in the sense that they are described by just three parameters: r_h, Λ and $\omega(r_h)$. The question is therefore whether the existence of "non-trivial" (that is, genuinely $\mathfrak{su}(N)$) solutions in which all the gauge field functions $\omega_j(r)$ have no zeros can be proven analytically. The answer to this question is affirmative and involves a generalization to $\mathfrak{su}(N)$ of the continuity-type argument used in [175]. The details are lengthy and will be presented elsewhere [11]. Here we simply outline the key steps in the proof.

The main idea of the proof is sketched in Fig. 2.12. We wish to find black hole solutions which are regular on the event horizon, regular everywhere outside the event horizon and regular at infinity. The proof proceeds via the following steps:

1. We first prove (generalizing the analysis of [100] to include Λ) that the field equations (2.12) and (2.15) and initial conditions at the event horizon (2.20) possess, locally in a neighborhood of the horizon, solutions which are analytic in r, r_h, Λ and the parameters $\omega_j(r_h)$. As might be expected, the analysis of [100] requires only minor modifications to include a negative cosmological constant.
2. This enables us to prove that, in a sufficiently small neighborhood of any embedded $\mathfrak{su}(2)$ solution in which $\omega(r)$ has no nodes, there exists (at least in a neighborhood of the event horizon) an $\mathfrak{su}(N)$ solution in which all the $\omega_j(r)$ have no nodes.
3. Using the analyticity properties of the solutions of the field equations, we then show that these $\mathfrak{su}(N)$ solutions can be extended out to large $r_L \gg r_h$, provided the initial parameters $\omega_j(r_h)$ are sufficiently close to those of an embedded $\mathfrak{su}(2)$ solution in which $\omega(r)$ has no zeros. Furthermore, by analyticity, none of the $\omega_j(r)$ will have any zeros between the event horizon r_h and r_L.
4. The key part of the proof lies in then showing that these $\mathfrak{su}(N)$ solutions can be further extended out to $r \to \infty$ and that they satisfy the boundary conditions (2.23) at infinity. This part of the analysis uses the properties of the Yang–Mills field equations (2.12) in the asymptotically adS regime. As in the $\mathfrak{su}(2)$ case [175], these have very different properties from the asymptotically flat case, and this makes it much easier to prove the existence of solutions. Furthermore, it can be shown that the gauge field functions $\omega_j(r)$ will have no zeros for $r \geq r_L$.

Fig. 2.12 Sketch of the main steps in the proof of the existence of non-trivial $\mathfrak{su}(N)$ EYM black holes in adS for which all the gauge field functions have no zeros. We wish to find black hole solutions which are regular on the event horizon regular everywhere outside the event horizon and regular at infinity. We thank J. E. Baxter for providing this sketch

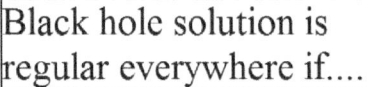

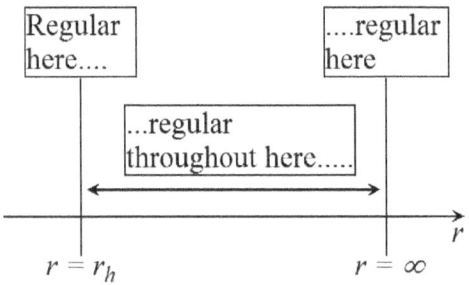

In summary, this process gives genuinely $\mathfrak{su}(N)$ black hole solutions in which all the gauge field functions have no zeros and which are characterized by the $N + 1$ parameters r_h, Λ and $\omega_j(r_h)$.

2.5.3 Stability Analysis of the Spherically Symmetric Solutions

The remaining outstanding question is whether these new black holes, with potentially unbounded amounts of gauge field hair, are stable. We consider linear, spherically symmetric perturbations only for simplicity. The analysis of [146, 178] in the $\mathfrak{su}(2)$ case revealed that, for sufficiently large $|\Lambda|$, stability under spherically symmetric perturbations continued to hold also for non-spherically symmetric perturbations, and one might hope that a similar result will hold in the more complex $\mathfrak{su}(N)$ case. However, we leave this for future work. Even for spherically symmetric perturbations, the analysis is highly involved in the $\mathfrak{su}(N)$ case and the details will be presented elsewhere [8, 11]. Here we briefly outline just the key features. The perturbation equations themselves can be found in Sect. 2.2.4.

2.5.3.1 Sphaleronic Sector

The sphaleronic sector consists of the perturbation equations (2.45) and (2.46) together with the Gauss constraint (2.47). The analysis of this sector essentially follows that of [47] in the asymptotically flat case. We begin by defining yet more new variables, $\delta\epsilon_j$, for $j = 1,\ldots,N$ by

$$\delta\epsilon_j = r\sqrt{\mu}\,\delta\beta_j, \tag{2.73}$$

then, after much algebra, the sphaleronic sector perturbation equations can be cast in the form

$$-\underline{\ddot{\Psi}} = \mathscr{M}_S\underline{\Psi}, \tag{2.74}$$

where the $(2N-1)$-dimensional vector $\underline{\Psi}$ is defined by

$$\underline{\Psi} = (\delta\epsilon_1,\ldots,\delta\epsilon_N,\delta\Phi_1,\ldots,\delta\Phi_{N-1})\,. \tag{2.75}$$

and $\mathscr{M}_S$ is a self-adjoint, second order, differential operator (involving derivatives with respect to r but not t), depending on the equilibrium functions $\omega_j(r)$, $m(r)$ and $S(r)$. The operator $\mathscr{M}_S$ can be written as the sum of three parts. The first is of the form $\chi^{\dagger}\chi$ for a particular first-order differential operator χ (whose precise form can be found in [8, 11]) and is therefore manifestly positive and is regular if the gauge field functions ω_j have no zeros. The second part vanishes when applied to a physical perturbation due to the Gauss constraint (2.47). The third part is a matrix $\mathscr{V}$ which does not contain any differential operators. It can be shown that the matrix $\mathscr{V}$ is regular and positive definite provided the unperturbed gauge functions $\omega_j(r)$ have no zeros and satisfy the $N - 1$ inequalities

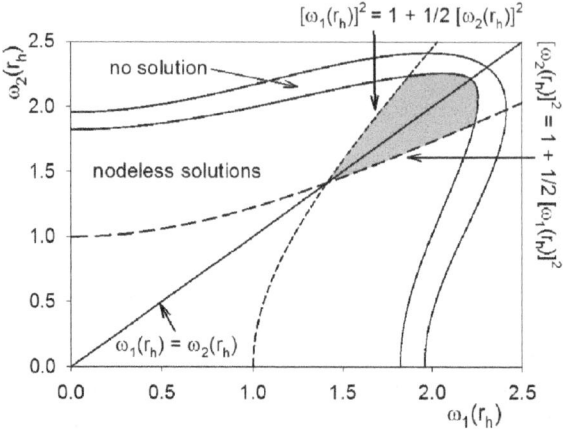

Fig. 2.13 Phase space of black hole solutions in $\mathfrak{su}(3)$ EYM with $\Lambda = -10$ and $r_h = 1$. The *shaded region* shows where solutions exist which satisfy the inequalities (2.76) at the event horizon. Taken from [10]

$$\omega_j^2 > 1 + \frac{1}{2}\left(\omega_{j+1}^2 + \omega_{j-1}^2\right) \tag{2.76}$$

for all $j = 1, \ldots N - 1$ and all $r \geq r_h$. The inequalities (2.76) define a non-empty subset of the parameter space. For example, we show in Fig. 2.13 where the inequalities (2.76) are satisfied for the gauge field functions at the event horizon, for the particular case of $\Lambda = -10$ and $r_h = 1$. From Fig. 2.13 we can see that there are some nodeless solutions which satisfy the inequalities (2.76) at the event horizon. For any N, it can also be proved analytically that, for sufficiently large $|\Lambda|$, there are non-trivial $\mathfrak{su}(N)$ solutions, in a neighborhood of some embedded $\mathfrak{su}(2)$ solutions, such that the inequalities (2.76) are satisfied at the event horizon.

However, the requirements of (2.76) are considerably stronger, as the inequalities have to be satisfied for *all* $r \geq r_h$. Our analytic work shows that, in fact, for any N and sufficiently large $|\Lambda|$, there do exist solutions to the field equations for which the inequalities (2.76) are indeed satisfied for all r. This involves proving that for at least some solutions for which the gauge field function values at the event horizon lie within the region where the inequalities (2.76) are satisfied, the gauge field functions remain within this open region. In Fig. 2.14 we show an example of such a solution for $\mathfrak{su}(3)$ EYM.

2.5.3.2 Gravitational Sector

As might be expected, the gravitational sector perturbation equations (2.48) are more difficult to analyze than the sphaleronic sector perturbation equations. For stable solutions, we require the matrix $\mathcal{M}_G$ (2.49) to be negative definite. For sufficiently large $|\Lambda|$, it can be shown that $\mathcal{M}_G$ is indeed negative definite for embedded

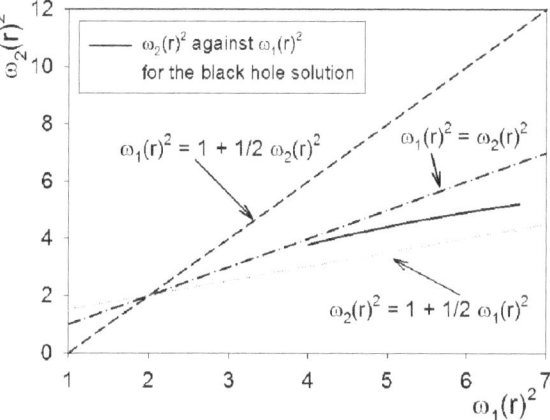

Fig. 2.14 An example of an $\mathfrak{su}(3)$ solution for which the inequalities (2.76) are satisfied for all $r \geq r_h$. In this example, $\Lambda = -10$, $r_h = 1$ and the values of the gauge field functions at the event horizon are $\omega_1(r_h) = 2$, $\omega_2(r_h) = 1.95$. Taken from [10]

$\mathfrak{su}(2)$ solutions, provided that $\omega^2(r) > 1$ for all $r \geq r_h$ (the existence of such $\mathfrak{su}(2)$ solutions is proved, for sufficiently large $|\Lambda|$, in [175]). As described in Sect. 2.5.2 above, our analytic work ensures the existence of genuinely $\mathfrak{su}(N)$ solutions in a sufficiently small neighborhood of these embedded $\mathfrak{su}(2)$ solutions. These $\mathfrak{su}(N)$ solutions are such that the inequalities (2.76) are satisfied for all $r \geq r_h$ (and therefore the solutions are stable under sphaleronic perturbations). The negativity of $\mathcal{M}_G$ can then be extended to these genuinely $\mathfrak{su}(N)$ solutions using an analyticity argument, based on the nodal theorem of [2] (see also [178] for a similar argument for the non-spherically symmetric perturbations of the $\mathfrak{su}(2)$ EYM black holes). The technical details of this argument will be presented elsewhere [11].

The conclusion of the work in this section is that there are at least some genuinely $\mathfrak{su}(N)$ EYM black holes in adS, for sufficiently large $|\Lambda|$, for which all the gauge field functions ω_j have no zeros, and which are stable under spherically symmetric perturbations in both the sphaleronic and the gravitational sectors.

2.6 Summary and Outlook

In this review we have studied classical, hairy black hole solutions of $\mathfrak{su}(N)$ EYM theory, particularly spherically symmetric spacetimes and black holes in adS. We very briefly discussed some of the key aspects of the solutions in asymptotically flat space, which have been extensively reviewed in [171]. Hairy black hole solutions exist for all N, with $N-1$ gauge field degrees of freedom [116], however, all these solutions are unstable [47]. Therefore, while these hairy black holes violate the "letter" of the no-hair conjecture (that is, their geometry is not completely fixed by global charges measurable at infinity), its "spirit" is maintained. In particular, stable equilibrium black holes are comparatively simple objects, described completely by just a few parameters.

The main conclusion of this article is that this is not true in adS. The existence of stable hairy black holes in $\mathfrak{su}(2)$ EYM [175] did not really contradict the "spirit" of the no-hair conjecture, as only a single additional parameter was required to fix the geometry outside the event horizon. However, the recent work [10] which shows that there are stable hairy black holes in $\mathfrak{su}(N)$ EYM in adS for arbitrarily large N changes the picture completely. For sufficiently large $|\Lambda|$, an infinite number of parameters are required in order to describe stable black holes. We might flip-pantly describe these as "furry" black holes, since they possess copious amounts of hair.

What are the consequences for black hole physics in adS of these "furry" black holes? These need to be explored. Given the huge amount of interest in the adS/CFT correspondence in string theory [111, 180, 181], a natural question is how black hole hair in the bulk asymptotically adS spacetime relates to the dual CFT. In par-ticular, it has been suggested [76] that there should be observables in the dual (de-formed) CFT which are sensitive to the presence of black hole hair. Another ex-ample of this approach can be found in [70], where an adS/CFT interpretation is given of some stable seven-dimensional black holes with $\mathfrak{so}(5)$ gauge fields. We would expect that, in analogy with the $\mathfrak{su}(2)$ case [49, 50, 74, 84, 113, 130, 132], there are solutions in some super-gravity theories with a gauge group containing an $\mathfrak{su}(N)$ factor, which will need to be studied in the context of adS/CFT. There is evidence [117] that there are non-trivial black hole solutions of $\mathfrak{su}(\infty)$ EYM in adS, giving black holes not just with unbounded amounts of hair, but infinite amounts of hair, at least in the limit $|\Lambda| \rightarrow \infty$. It remains to be seen whether exact solutions of the $\mathfrak{su}(\infty)$ field equations can be found for finite $\Lambda < 0$ and whether any of these black holes are stable. If so, then their role in adS/CFT would be puzzling indeed.

Due to space restrictions, there are many aspects of black holes in EYM which we have not been able to discuss. In particular, we have not mentioned the vast number of solutions which involve modifications of the EYM action (2.1), including higher curvature terms (see, for example, [88, 89]) or the inclusion of dilaton (see, for example, [134]), Higgs (see, for example, [15, 108, 109]) or other modifications of the EYM action (see, for example, [120, 147, 148]). Here we have also only studied four-dimensional spacetimes, while recent work has considered EYM in higher-dimensional spacetimes (see, for example, [32, 33, 35–40, 75, 123, 133, 135] and [166] for a review).

The black hole solutions of EYM and its variants certainly exhibit an abundantly rich structure, and no doubt will have more surprises in store for us in the future.

Acknowledgments We thank the organizers of the 4th Aegean Summer School for a most enjoy-able and informative conference. The work in Sect. 2.5 was done in collaboration with J. E. Baxter and Marc Helbling. We thank B. Bear, R. F. W. Jackson and the Chester Chronicle for suggesting the terminology "furry" black holes. We also thank Eugen Radu for numerous enlightening dis-cussions. This work was supported by STFC (UK), grant reference numbers PPA/G/S/2003/00082 and PP/D000351/1.

References

1. P. C. Aichelburg and P. Bizon, Magnetically charged black holes and their stability. Phys. Rev. **D 48**, 607–615 (1993).
2. H. Amann and P. Quittner, A nodal theorem for coupled systems of Schrödinger equations and the number of bound states. J. Math. Phys. **36**, 4553–4560 (1995).
3. A. Ashtekar, A. Corichi and D. Sudarsky, Hairy black holes, horizon mass and solitons. Class. Quant. Grav. **18**, 919–940 (2001).
4. J. Baacke and H. Lange, Stability analysis of the electroweak sphaleron. Mod. Phys. Lett. **A 7**, 1455–1470 (1992).
5. L. Barack and C. Cutler, Using LISA EMRI sources to test off-Kerr deviations in the geometry of massive black holes. Phys. Rev. **D 75**, 042003 (2007).
6. R. Bartnik, The structure of spherically symmetric $\mathfrak{su}(n)$ Yang-Mills fields. J. Math. Phys. **38**, 3623–3638 (1997).
7. R. Bartnik and J. McKinnon, Particle-like solutions of the Einstein-Yang-Mills equations. Phys. Rev. Lett. **61**, 141–144 (1988).
8. J. E. Baxter, Existence and stability of solitons and black holes in $\mathfrak{su}(N)$ Einstein-Yang-Mills theory with a negative cosmological constant. PhD thesis, University of Sheffield (2006).
9. J. E. Baxter, M. Helbling and E. Winstanley, Soliton and black hole solutions of $\mathfrak{su}(N)$ Einstein-Yang-Mills theory in anti-de Sitter space. Phys. Rev. **D 76**, 104017 (2007).
10. J. E. Baxter, M. Helbling and E. Winstanley, Abundant stable gauge field hair for black holes in anti-de Sitter space. Phys. Rev. Lett. (2008).
11. J. E. Baxter and E. Winstanley, On the existence of soliton and hairy black hole of su(N) Einstein-Yang-Mills theory with a negative cosmological constant, `arXiv:0808.2977`.
12. J. D. Bekenstein, Exact solutions of Einstein-conformal scalar equations. Ann. Phys. (NY) **82**, 535–547 (1974).
13. J. D. Bekenstein, Black holes with scalar charge. Ann. Phys. (NY) **91**, 75–82 (1975).
14. J. D. Bekenstein, Black hole hair: twenty-five years after, in *Proceedings of the Second International Sakharov Conference on Physics, Moscow, Russia, 20-23 May 1996*, ed. by I. M. Dremin and A. M. Semikhatov (World Scientific, Singapore, 1997), pp. 216–219.
15. J. van der Bij and E. Radu, Gravitating sphalerons and sphaleron black holes in asymptotically anti-de Sitter space-time. Phys. Rev. **D 64**, 064020 (2001).
16. J. van der Bij and E. Radu, New hairy black holes with negative cosmological constant. Phys. Lett. **B 536**, 107–113 (2002).
17. J. van der Bij and E. Radu, On rotating regular non-Abelian solutions. Int. J. Mod. Phys. **A 17**, 1477–1490 (2002).
18. D. Birmingham, Topological black holes in anti-de Sitter space. Class. Quant. Grav. **16**, 1197–1205 (1999).
19. P. Bizon, Colored black holes. Phys. Rev. Lett. **64**, 2844–2847 (1990).
20. P. Bizon, Stability of Einstein-Yang-Mills black holes. Phys. Lett. **B 259**, 53–57 (1991).
21. P. Bizon, Gravitating solitons and hairy black holes. Acta Phys. Polon. **B 25**, 877–898 (1994).
22. P. Bizon and T. Chmaj, Gravitating Skyrmions. Phys. Lett. **B 297**, 55–62 (1992).
23. P. Bizon and O. T. Popp, No hair theorem for spherical monopoles and dyons in $\mathfrak{su}(2)$ Einstein-Yang-Mills theory. Class. Quant. Grav. **9**, 193–205 (1992).
24. P. Bizon and R. M. Wald, The $n = 1$ colored black hole is unstable. Phys. Lett. **B 267**, 173–174 (1991).
25. J. Bjoraker and Y. Hosotani, Stable monopole and dyon solutions in the Einstein-Yang-Mills theory in asymptotically anti-de Sitter space. Phys. Rev. Lett. **84**, 1853–1856 (2000).
26. J. Bjoraker and Y. Hosotani, Monopoles, dyons and black holes in the four-dimensional Einstein-Yang-Mills theory. Phys. Rev. **D 62**, 043513 (2000).
27. N. M. Bocharova, K. A. Bronnikov and V. N. Mel'nikov, An exact solution of the system of Einstein equations and mass-free scalar field. Vestnik Moskov. Univ. Fizika **25**, 706–709 (1970).

28. H. Boutaleb-Joutei, A. Chakrabarti and A. Comtet, Gauge field configurations in curved space-times. Phys. Rev. **D 20**, 1884–1897 (1979).
29. P. Breitenlohner, P. Forgacs and D. Maison, On static spherically symmetric solutions of the Einstein-Yang-Mills equations. Commun. Math. Phys. **163**, 141–172 (1994).
30. P. Breitenlohner, P. Forgacs and D. Maison, Classification of static, spherically symmetric solutions of the Einstein-Yang-Mills theory with positive cosmological constant. Commun. Math. Phys. **261**, 569–611 (2006).
31. P. Breitenlohner, D. Maison and G. V. Lavrelashvili, Non-Abelian gravitating solitons with negative cosmological constant. Class. Quant. Grav. **21**, 1667–1684 (2004).
32. P. Breitenlohner, D. Maison and D. H. Tchrakian, Regular solutions to higher order curvature Einstein-Yang-Mills systems in higher dimensions. Class. Quant. Grav. **22**, 5201–5222 (2005).
33. Y. Brihaye, A. Chakrabarti, B. Hartmann and D. H. Tchrakian, Higher order curvature generalizations of Bartnik-McKinnon and colored black hole solutions in $D = 5$. Phys. Lett. **B 561**, 161–173 (2003).
34. Y. Brihaye, A. Chakrabarti and D. H. Tchrakian, Finite energy-action solutions of p_1 Yang-Mills equations on p_2 Schwarzschild and de Sitter backgrounds in dimensions $d \geq 4$. J. Math. Phys. **41**, 5490–5509 (2000).
35. Y. Brihaye, A. Chakrabarti and D. H. Tchrakian, Particle-like solutions to higher order curvature Einstein-Yang-Mills systems in d-dimensions. Class. Quant. Grav. **20**, 2765–2784 (2003).
36. Y. Brihaye, F. Clement and B. Hartmann, Spherically symmetric Yang-Mills solutions in a $(4 + n)$-dimensional space-time. Phys. Rev. **D 70**, 084003 (2004).
37. Y. Brihaye and T. Delsate, Black strings and solitons in five dimensional space-time with positive cosmological constant. Phys. Rev. **D 75**, 044013 (2007).
38. Y. Brihaye and B. Hartmann, Spherically symmetric solutions of a $(4 + n)$-dimensional Einstein-Yang-Mills model with cosmological constant. Class. Quant. Grav. **22**, 183–194 (2005).
39. Y. Brihaye, E. Radu and D. H. Tchrakian, Einstein-Yang-Mills solutions in higher dimensional de Sitter space-time. Phys. Rev. **D 75**, 024022 (2007).
40. Y. Brihaye, E. Radu and D. H. Tchrakian, AdS(5) rotating non-Abelian black holes. Phys. Rev. **D 76**, 105005 (2007).
41. Y. Brihaye, E. Radu and D. H. Tchrakian, Einstein-Yang-Mills solutions in higher dimensional de Sitter space-time. Phys. Rev. **D 75**, 024022 (2007).
42. O. Brodbeck, M. Heusler, G. V. Lavrelashvili, N. Straumann and M. S. Volkov, Stability analysis of new solutions of the EYM system with cosmological constant. Phys. Rev. **D 54**, 7338–7352 (1996).
43. O. Brodbeck, M. Heusler and O. Sarbach, The generalization of the Regge-Wheeler equation for self-gravitating matter fields. Phys. Rev. Lett. **84**, 3033–3036 (2000).
44. O. Brodbeck, M. Heusler, N. Straumann and M. S. Volkov, Rotating solitons and nonrotating, nonstatic black holes. Phys. Rev. Lett. **79**, 4310–4313 (1997).
45. O. Brodbeck and N. Straumann, A generalized Birkhoff theorem for the Einstein-Yang-Mills system. J. Math. Phys. **34**, 2412–2423 (1993).
46. O. Brodbeck and N. Straumann, Instability of Einstein-Yang-Mills solitons for arbitrary gauge groups. Phys. Lett. **B 324**, 309–314 (1994).
47. O. Brodbeck and N. Straumann, Instability proof for Einstein-Yang-Mills solitons and black holes with arbitrary gauge groups. J. Math. Phys. **37**, 1414–1433 (1996).
48. K. A. Bronnikov and Y. N. Kireyev, Instability of black holes with scalar charge. Phys. Lett. **A 67**, 95–96 (1978).
49. A. H. Chamseddine and M. S. Volkov, Non-Abelian BPS monopoles in $N = 4$ gauged supergravity. Phys. Rev. Lett. **79**, 3343–3346 (1997).
50. A. H. Chamseddine and M. S. Volkov, Non-Abelian solitons in $N = 4$ gauged supergravity and leading order string theory. Phys. Rev. **D 57**, 6242–6254 (1998).
51. P. T. Chruściel, "No-hair" theorems – folklore, conjectures, results. Contemp. Math. **170**, 23–49 (1994).

52. P. T. Chruściel, Uniqueness of stationary, electro-vacuum black holes revisited. Helv. Phys. Acta **69**, 529–552 (1996).
53. S. Coleman, Classical lumps and their quantum descendents. In *New Phenomena in Subnuclear Physics*, ed. by A. Zichichi (Plenum, New York, 1976).
54. A. Corichi, U. Nucamendi and D. Sudarsky, Einstein-Yang-Mills isolated horizons: phase space, mechanics, hair, and conjectures. Phys. Rev. **D 62**, 044046 (2000).
55. A. Corichi, U. Nucamendi and D. Sudarsky, A mass formula for EYM solitons. Phys. Rev. **D 64**, 107501 (2001).
56. S. Deser, Absence of static solutions in source-free Yang-Mills theory. Phys. Lett. **B 64**, 463–465 (1976).
57. S. Deser, Absence of static Einstein-Yang-Mills excitations in three dimensions. Class. Quant. Grav. **1**, L1–L4 (1984).
58. S. Droz, M. Heusler and N. Straumann, New black hole solutions with hair. Phys. Lett. **B 268**, 371–376 (1991).
59. H. Elvang and P. Figueras, Black Saturn. J. High Energy Phys. **0705**, 050 (2007).
60. R. Emparan and H. Reall, A rotating black ring in five dimensions. Phys. Rev. Lett. **88**, 101101 (2002).
61. R. Emparan and H. Reall, Black rings. Class. Quant. Grav. **23**, R169–R197 (2006).
62. A. A. Ershov and D. V. Gal'tsov, Non-existence of regular monopoles and dyons in the $\mathfrak{su}(2)$ Einstein-Yang-Mills theory. Phys. Lett. **A 150**, 159–162 (1990).
63. P. Forgacs and S. Reuillon, On the number of instabilities of cosmological solutions in an Einstein-Yang-Mills system. Phys. Lett. **B 568**, 291–297 (2003).
64. G. J. Galloway, K. Schleich, D. M. Witt and E. Woolgar, Topological censorship and higher genus black holes. Phys. Rev. **D 60**, 104039 (1999).
65. D. V. Galt'sov, Einstein-Yang-Mills solitons: towards new degrees of freedom. Preprint arXiv:gr-qc/9808002.
66. D. V. Gal'tsov, Gravitating lumps, in *Proceedings of the 16th International Conference on General Relativity and Gravitation (GR16)*, ed. N. T. Bishop and S. D. Maharaj (World Scientific, Singapore 2002).
67. D. V. Gal'tsov and A. A. Ershov, Non-Abelian baldness of colored black holes. Phys. Lett. **A 138**, 160–164 (1989).
68. D. V. Galtsov and M. S. Volkov, Instability of Einstein-Yang-Mills black holes. Phys. Lett. **A 162**, 144–148 (1992).
69. D. V. Gal'tsov and M. S. Volkov, Charged non-Abelian $\mathfrak{su}(3)$ Einstein-Yang-Mills black holes. Phys. Lett. **B 274**, 173–178 (1992).
70. J. P. Gauntlett, N. Kim and D. Waldram, M-fivebranes wrapped on supersymmetric cycles. Phys. Rev. **D 63** 126001 (2001).
71. G. W. Gibbons, Anti-gravitating black hole solutions with scalar hair in $N = 4$ supergravity. Nucl. Phys. **B 207**, 337–349 (1982).
72. Gibbons, G.W.: *Self-gravitating Magnetic Monopoles, Global Monopoles and Black Holes*. Lect. Notes Phys. **383**, 110–133 (1991).
73. B. R. Greene, S. D. Mathur and C. M. O'Neill, Eluding the no hair conjecture: black holes in spontaneously broken gauge theories. Phys. Rev. **D 47**, 2242–2259 (1993).
74. S. S. Gubser, A. A. Tseytlin and M. S. Volkov, Non-Abelian $4 - d$ black holes, wrapped 5-branes, and their dual descriptions. JHEP **0109**, 017 (2001).
75. B. Hartmann, Y. Brihaye and B. Bertrand, Spherically symmetric Yang-Mills solutions in a five-dimensional anti-de Sitter space-time. Phys. Lett. **B 570**, 137–144 (2003).
76. T. Hertog and K. Maeda, Black holes with scalar hair and asymptotics in $N = 8$ supergravity. JHEP **0407**, 051 (2004).
77. M. Heusler, *Black Hole Uniqueness Theorems* (Cambridge University Press, Cambridge 1996).
78. M. Heusler, No hair theorems and black holes with hair. Helv. Phys. Acta **69** 501–528 (1996).
79. M. Heusler, Stationary black holes: uniqueness and beyond. Living Rev. Relativity **1**, 6 (1998).

80. M. Heusler, S. Droz and N. Straumann, Stability analysis of self-gravitating skyrmions. Phys. Lett. **B 271**, 61–67 (1991).

81. M. Heusler, S. Droz and N. Straumann, Linear stability of Einstein-Skyrme black holes. Phys. Lett. **B 285**, 21–26 (1992).

82. S. Hod, Einstein-Yang-Mills solitons: the role of gravity. Phys. Lett. **B 657**, 255–256 (2007).

83. Y. Hosotani, Scaling behavior in the Einstein-Yang-Mills monopoles and dyons. J. Math. Phys. **43**, 597–603 (2002).

84. M. Hubscher, P. Meessen, T. Ortin and S. Vaula, Supersymmetric $N = 2$ Einstein-Yang-Mills monopoles and covariant attractors. Preprint arXiv:0712.1530 [hep-th].

85. R. Ibadov, B. Kleihaus, J. Kunz and Y. Shnir, New regular solutions with axial symmetry in Einstein-Yang-Mills theory. Phys. Lett. **B 609**, 150–156 (2005).

86. W. Israel, Event horizons in static vacuum space-times. Phys. Rev. **164**, 1776–1779 (1967).

87. W. Israel, Event horizons in static electrovac space-times. Commun. Math. Phys. **8**, 245–260 (1968).

88. P. Kanti and K. Tamvakis, Colored black holes in higher curvature string gravity. Phys. Lett. **B 392**, 30–38 (1997).

89. P. Kanti and E. Winstanley, Do stringy corrections stabilize colored black holes? Phys. Rev. **D 61**, 084032 (2000).

90. B. Kleihaus and J. Kunz, Static black hole solutions with axial symmetry. Phys. Rev. Lett. **79**, 1595–1598 (1997).

91. B. Kleihaus and J. Kunz, Static axially symmetric Einstein-Yang-Mills dilaton solutions: 1. Regular solutions. Phys. Rev. **D 57**, 834–856 (1998).

92. B. Kleihaus and J. Kunz, Rotating hairy black holes. Phys. Rev. Lett. **86**, 3704–3707 (2001).

93. B. Kleihaus, J. Kunz and F. Navarro-Lerida, Rotating Einstein-Yang-Mills black holes. Phys. Rev. **D 66**, 104001 (2002).

94. B. Kleihaus, J. Kunz and A. Sood, $\mathfrak{su}(3)$ Einstein-Yang-Mills sphalerons and black holes. Phys. Lett. **B 354**, 240–246 (1995).

95. B. Kleihaus, J. Kunz and A. Sood, Charged $\mathfrak{su}(N)$ Einstein-Yang-Mills black holes. Phys. Lett. **B 418**, 284–293 (1998).

96. B. Kleihaus, J. Kunz, A. Sood and M. Wirschins, Sequences of globally regular and black hole solutions in $\mathfrak{su}(4)$ Einstein-Yang-Mills theory. Phys. Rev. **D 58**, 084006 (1998).

97. D. Klemm, V. Moretti and L. Vanzo, Rotating topological black holes. Phys. Rev. **D 57**, 6127–6137 (1998)

98. D. Klemm, V. Moretti and L. Vanzo, Rotating topological black holes. Phys. Rev. **D 60**, 109902 (1999).

99. H. P. Kunzle, $\mathfrak{su}(n)$-Einstein-Yang-Mills fields with spherical symmetry. Class. Quant. Grav. **8**, 2283–2297 (1991).

100. H. P. Kunzle, Analysis of the static spherically symmetric $\mathfrak{su}(n)$ Einstein-Yang-Mills equations. Comm. Math. Phys. **162**, 371–397 (1994).

101. H. P. Kunzle and A. K. M. Masood-ul-Alam, Spherically symmetric static $\mathfrak{su}(2)$ Einstein-Yang-Mills fields. J. Math. Phys. **31**, 928–935 (1990).

102. G. V. Lavrelashvili and D. Maison, A remark on the instability of the Bartnik-McKinnon solutions. Phys. Lett. **B 343**, 214–217 (1995).

103. J. P. S. Lemos, Cylindrical black hole in general relativity. Phys. Lett. **B 353**, 46–51 (1995).

104. A. Lichnerowicz, *Théories Relativistes de la Gravitation et de l'Électromagnétisme* (Masson, Paris, 1955).

105. A. N. Linden, Horizons in spherically symmetric static Einstein-$\mathfrak{su}(2)$-Yang-Mills space-times. Class. Quant. Grav. **18**, 695–708 (2001).

106. A. N. Linden, Far field behavior of noncompact static spherically symmetric solutions of Einstein-$\mathfrak{su}(2)$-Yang-Mills equations. J. Math. Phys. **42**, 1196–1201 (2001).

107. A. N. Linden, Existence of noncompact static spherically symmetric solutions of Einstein-$\mathfrak{su}(2)$-Yang-Mills equations with small cosmological constant. Commun. Math. Phys. **221**, 525–547 (2001).

108. A. R. Lugo and F. A. Schaposnik, Monopole and dyon solutions in AdS space. Phys. Lett. **B 467**, 43–53 (1999).

109. A. R. Lugo, E. F. Moreno and F. A. Schaposnik, Monopole solutions in AdS space. Phys. Lett. **B 473**, 35–42 (2000).
110. K. I. Maeda, T. Tachizawa, T. Torii and T. Maki, Stability of non-Abelian black holes and catastrophe theory. Phys. Rev. Lett. **72**, 450–453 (1994).
111. J. M. Maldacena, The large N limit of superconformal field theories and supergravity. Adv. Theor. Math. Phys. **2**, 231–252 (1998).
112. R. B. Mann, Topological black holes: outside looking in. In *Internal Structure of Black Holes and Space-time Singularities*, ed. by L. M. Burko and A. Ori, Annals of the Israel Physical Society, vol. 13 (Israel Physical Society, Jerusalem, and Institute of Physics Publishing, Bristol, 1997).
113. R. B. Mann, E. Radu and D. H. Tchrakian, Non-Abelian solutions in AdS(4) and $d = 11$ supergravity. Phys. Rev. **D 74**, 064015 (2006).
114. S. G. Matinyan, Chaos in non-Abelian gauge fields, gravity, and cosmology. in *Proceedings of the Ninth Marcel Grossman Meeting*, ed. by V. G. Gurzadyan, R. T. Jantzen and R. Ruffini (World Scientific, 2002).
115. N. E. Mavromatos and E. Winstanley, Aspects of hairy black holes in spontaneously broken Einstein-Yang-Mills systems: Stability analysis and entropy considerations. Phys. Rev. **D 53**, 3190–3214 (1996).
116. N. E. Mavromatos and E. Winstanley, Existence theorems for hairy black holes in $\mathfrak{su}(N)$ Einstein-Yang-Mills theories. J. Math. Phys. **39**, 4849–4873 (1998).
117. N. E. Mavromatos and E. Winstanley, Infinitely coloured black holes. Class. Quant. Grav. **17**, 1595–1611 (2000).
118. P. O. Mazur, Black hole uniqueness theorems. in *Proceedings of the 11th International Conference on General Relativity and Gravitation*, ed. by M. A. H. MacCallum (Cambridge University Press, Cambridge 1987), pp. 130–157.
119. P. G. Molnar, Numerical solutions of the Einstein-Yang-Mills system with cosmological constant. Preprint `arXiv:gr-qc/9503036`.
120. I. G. Moss, N. Shiiki and E. Winstanley, Monopole black hole skyrmions. Class. Quant. Grav. **17**, 4161–4174 (2000).
121. R. C. Myers and M. J. Perry, Black holes in higher dimensions. Annals Phys. **172**, 304–347 (1986).
122. D. Nunez, H. Quevedo and D. Sudarsky, Black holes have no short hair. Phys. Rev. Lett. **76**, 571–574 (1996).
123. N. Okuyama and K.-I. Maeda, Five-dimensional black hole and particle solution with non-Abelian gauge field. Phys. Rev. **D 67**, 104012 (2003).
124. T. A. Oliynyk and H. P. Kunzle, On all possible static spherically symmetric EYM solitons and black holes. Class. Quant. Grav. **19**, 457–482 (2002).
125. T. A. Oliynyk and H. P. Kunzle, Local existence proofs for the boundary value problem for static spherically symmetric Einstein-Yang-Mills fields with compact gauge groups. J. Math. Phys. **43**, 2363–2393 (2002).
126. T. A. Oliynyk and H. P. Kunzle, Global behavior of solutions to the static spherically symmetric EYM equations. Class. Quant. Grav. **20** 4653–4682 (2003).
127. V. Paturyan, E. Radu and D. H. Tchrakian, Rotating regular solutions in Einstein-Yang-Mills-Higgs theory. Phys. Lett. **B 609**, 360–366 (2005).
128. I. Racz, On further generalization of the rigidity theorem for space-times with a stationary event horizon or a compact Cauchy horizon. Class. Quant. Grav. **17**, 153–178 (2000).
129. E. Radu, Static axially symmetric solutions of Einstein-Yang-Mills equations with a negative cosmological constant: the regular case. Phys. Rev. **D 65**, 044005 (2002).
130. E. Radu, New non-Abelian solutions in $D = 4$, $N = 4$ gauged supergravity. Phys. Lett. **B 542**, 275–281 (2002).
131. E. Radu, Rotating Yang-Mills dyons in anti-de Sitter space-time. Phys. Lett. **B 548**, 224–230 (2002).
132. E. Radu, Non-Abelian solutions in $N = 4$, $D = 5$ gauged supergravity. Class. Quant. Grav. **23**, 4369–4386 (2006).

133. E. Radu, C. Stelea and D. H. Tchrakian, Features of gravity-Yang-Mills hierarchies in d-dimensions. Phys. Rev. **D 73**, 084015 (2006).

134. E. Radu and D. H, Tchrakian, New hairy black hole solutions with a dilaton potential. Class. Quant. Grav. **22**, 879–892 (2005).

135. E. Radu and D. H. Tchrakian, No hair conjecture, non-Abelian hierarchies and anti-de Sitter space-time. Phys. Rev. **D 73**, 024006 (2006).

136. E. Radu and E. Winstanley, Static axially symmetric solutions of Einstein-Yang-Mills equations with a negative cosmological constant: black hole solutions. Phys. Rev. **D 70**, 084023 (2004).

137. C. Rebbi and P. Rossi, Multimonopole solutions in the Prasad-Sommerfield limit. Phys. Rev. **D 22**, 2010–2017 (1980).

138. S. A. Ridgway and E. J. Weinberg, Static black hole solutions without rotational symmetry. Phys. Rev. **D 52**, 3440–3456 (1995).

139. S. A. Ridgway and E. J. Weinberg, Are all static black hole solutions spherically symmetric? Gen. Rel. Grav. **27**, 1017–1021 (1995).

140. W. H. Ruan, Existence of infinitely many black holes in $\mathfrak{su}(3)$ Einstein-Yang-Mills theory. Nonlin. Anal. **47**, 6109–6119 (2001).

141. W. H. Ruan, Hairy black hole solutions to $\mathfrak{su}(3)$ Einstein-Yang-Mills equations. Commun. Math. Phys. **224**, 373–397 (2001).

142. R. Ruffini and J. A. Wheeler, Introducing the black hole. Phys. Today **24**, 30–41 (1971).

143. O. Sarbach, On the generalization of the Regge-Wheeler equation for self-gravitating matter fields. PhD Thesis, University of Zurich (2000).

144. O. Sarbach, M. Heusler and O. Brodbeck, Perturbation theory for self-gravitating gauge fields. 1. The odd parity sector. Phys. Rev. **D 62**, 084001 (2000).

145. O. Sarbach, M. Heusler and O. Brodbeck, Self-adjoint wave equations for dynamical perturbations of self-gravitating fields. Phys. Rev. **D 63**, 104015 (2001).

146. O. Sarbach and E. Winstanley, On the linear stability of solitons and hairy black holes with a negative cosmological constant: the odd parity sector. Class. Quant. Grav. **18**, 2125–2146 (2001).

147. N. Shiiki and N. Sawado, Black hole skyrmions with negative cosmological constant. Phys. Rev. **D 71** 104031 (2005).

148. N. Shiiki and N. Sawado, Regular and black hole solutions in the Einstein-Skyrme theory with negative cosmological constant. Class. Quant. Grav. **22** 3561–3574 (2005).

149. J. A. Smoller and A. G. Wasserman, Existence of infinitely many smooth, static, global solutions of the Einstein-Yang-Mills equations. Commun. Math. Phys. **151**, 303–325 (1993).

150. J. A. Smoller, A. G. Wasserman and S.-T. Yau, Existence of black hole solutions for the Einstein-Yang-Mills equations. Commun. Math. Phys. **154**, 377–401 (1993).

151. J. A. Smoller, A. G. Wasserman, S.-T. Yau and J. B. McLeod, Smooth static solutions of the Einstein-Yang-Mills equations. Commun. Math. Phys. **143**, 115–147 (1991).

152. Straumann, N.: *Black Holes with Hair*. Lect. Notes Phys. **410**, 294–304 (1992).

153. N. Straumann, Black holes with hair. Class. Quant. Grav. **10**, S155–S165 (1993).

154. N. Straumann and Z.-H. Zhou, Instability of the Bartnik-McKinnon solution of the Einstein-Yang-Mills equations. Phys. Lett. **B 237**, 353–356 (1990).

155. N. Straumann and Z.-H. Zhou, Instability of a colored black hole solution. Phys. Lett. **B 243**, 33–35 (1990).

156. D. Sudarsky and R. M. Wald, Extrema of mass, stationarity, and staticity, and solutions to the Einstein-Yang-Mills equations. Phys. Rev. **D 46**, 1453–1474 (1992).

157. D. Sudarsky and R. M. Wald, Mass formulas for stationary Einstein-Yang-Mills black holes and a simple proof of two staticity theorems. Phys. Rev. **D 47**, R5209–R5213 (1993).

158. D. Sudarsky and T. Zannias, Spherical black holes cannot support scalar hair. Phys. Rev. **D 58**, 087502 (1998).

159. T. Tachizawa, K. I. Maeda and T. Torii, Non-Abelian black holes and catastrophe theory. 2. Charged type. Phys. Rev. **D 51**, 4054–4066 (1995).

160. T. Tamaki, T. Torii and K. I. Maeda, Stability analysis of black holes via catastrophe theory and black hole thermodynamics in generalized theories of gravity. Phys. Rev. **D 68**, 024028 (2003).

161. T. Torii, K. Maeda and M. Narita, Toward the no-scalar hair conjecture in asymptotically de-Sitter space-time. Phys. Rev. **D 59**, 064027 (1999).

162. T. Torii, K. Maeda and M. Narita, Scalar hair on the black hole in asymptotically anti-de Sitter space-time. Phys. Rev. **D 64**, 044007 (2001).

163. T. Torii, K.-I. Maeda and T. Tachizawa, Non-Abelian black holes and catastrophe theory. 1. Neutral type. Phys. Rev. **D 51**, 1510–1524 (1995).

164. T. Torii, K.-I. Maeda and T. Tachizawa, Cosmic colored black holes. Phys. Rev. **D 52**, 4272–4276 (1995).

165. L. Vanzo, Black holes with unusual topology. Phys. Rev. **D 56**, 6475–6483 (1997).

166. M. S. Volkov, Gravitating non-Abelian solitons and hairy black holes in higher dimensions. in *Proceedings of the 11th Marcel Grossmann Meeting on Recent Developments in Theoretical and Experimental General Relativity, Gravitation, and Relativistic Field Theories, Berlin, Germany, 23-29 Jul 2006*.

167. M. S. Volkov, O. Brodbeck, G. V. Lavrelashvili and N. Straumann, The number of sphaleron instabilities of the Bartnik-McKinnon solitons and non-Abelian black holes. Phys. Lett. **B 349**, 438–442 (1995).

168. M. S. Volkov and D. V. Gal'tsov, Non-Abelian Einstein-Yang-Mills black holes. JETP Lett. **50**, 346–350 (1989).

169. M. S. Volkov and D. V. Gal'tsov, Black holes in Einstein-Yang-Mills theory. Sov. J. Nucl. Phys. **51**, 747–753 (1990).

170. M. S. Volkov and D. V. Gal'tsov, Odd parity negative modes of Einstein-Yang-Mills black holes and sphalerons. Phys. Lett. **B 341**, 279–285 (1995).

171. M. S. Volkov and D. V. Gal'tsov, Gravitating non-Abelian solitons and black holes with Yang-Mills fields. Phys. Rept. **319**, 1–83 (1999).

172. M. S. Volkov, N. Straumann, G. V. Lavrelashvili, M. Heusler and O. Brodbeck, Cosmological analogs of the Bartnik-McKinnon solutions. Phys. Rev. **D 54**, 7243–7251 (1996).

173. R. M. Wald, On the instability of the $n = 1$ Einstein-Yang-Mills black holes and mathematically related systems. J. Math. Phys. **33**, 248–255 (1992).

174. C. M. Will, Testing the general relativistic "no-hair" theorems using the galactic centre black hole SgrA*. Preprint arXiv:0711.1677 [astro-ph].

175. E. Winstanley, Existence of stable hairy black holes in $\mathfrak{su}(2)$ Einstein-Yang-Mills theory with a negative cosmological constant. Class. Quant. Grav. **16**, 1963–1978 (1999).

176. E. Winstanley, On the existence of conformally coupled scalar field hair for black holes in (anti-)de Sitter space. Found. Phys. **33**, 111–143 (2003).

177. E. Winstanley, Dressing a black hole with non-minimally coupled scalar field hair. Class. Quant. Grav. **22**, 2233–2248 (2005).

178. E. Winstanley and O. Sarbach, On the linear stability of solitons and hairy black holes with a negative cosmological constant: the even parity sector. Class. Quant. Grav. **19**, 689–724 (2002).

179. E. Witten, Some exact multipseudoparticle solutions of classical Yang-Mills theory. Phys. Rev. Lett. **38**, 121–124 (1977).

180. E. Witten, Anti-de Sitter space and holography. Adv. Theor. Math. Phys. **2**, 253–291 (1998).

181. E. Witten, Anti-de Sitter space, thermal phase transition, and confinement in gauge theories. Adv. Theor. Math. Phys. **2**, 505–532 (1998).

182. P. B. Yasskin, Solutions for gravity coupled to massless gauge fields. Phys. Rev. **D 12**, 2212–2217 (1975).

183. Z. H. Zhou, Instability of $\mathfrak{su}(2)$ Einstein-Yang-Mills solitons and non-Abelian black holes. Helv. Phys. Acta **65**, 767–819 (1992).

184. Z. H. Zhou and N. Straumann, Non-linear perturbations of Einstein-Yang-Mills solitons and non-Abelian black holes. Nucl. Phys. **B 360**, 180–196 (1991).

Chapter 3
Black Hole Thermodynamics and Statistical Mechanics

S. Carlip

Abstract We have known for more than 30 years that black holes behave as thermo-dynamic systems, radiating as black bodies with characteristic temperatures and entropies. This behavior is not only interesting in its own right; it could also, through a statistical mechanical description, cast light on some of the deep problems of quantizing gravity. In these lectures, I review what we currently know about black hole thermodynamics and statistical mechanics, suggest a rather speculative "universal" characterization of the underlying states, and describe some key open questions.

3.1 Introduction

Black holes are black bodies.

Since the seminal work of Hawking [1] and Bekenstein [2], we have understood that black holes behave as thermodynamic objects, with characteristic temperatures and entropies. Hawking radiation has not yet been directly observed, of course; a typical stellar mass black hole has a Hawking temperature of well under a microkelvin, far lower than that of the cosmic microwave background. But the thermodynamic properties of black holes are well understood, having been confirmed by a great many independent methods that all yield the same quantitative results: a temperature

$$kT_{Hawking} = \frac{\hbar\kappa}{2\pi} \tag{3.1}$$

and an entropy

$$S_{BH} = \frac{A_{horizon}}{4\hbar G}, \tag{3.2}$$

where $A_{horizon}$ is the horizon area and κ is the surface gravity.

S. Carlip (✉)
Physics Department, University of California at Davis, Davis, CA 95616, USA, and ITF, Utrecht University, 3584 CE Utrecht, The Netherlands
carlip@physics.ucdavis.edu

Carlip, S.: *Black Hole Thermodynamics and Statistical Mechanics*. Lect. Notes Phys. **769**, 89–123 (2009)
DOI 10.1007/978-3-540-88460-6_3   © Springer-Verlag Berlin Heidelberg 2009

In a typical thermodynamic system, thermal properties are the macroscopic echoes of microscopic physics. Temperature is a measure of the average energy of microscopic constituents; entropy counts the number of microstates. It is natural to ask whether the same is true for the black hole. This is an important question: the Bekenstein–Hawking entropy depends on both Planck's and Newton's constants, and a statistical mechanical description of black hole thermodynamics might tell us something profound about quantum gravity. Until about 10 years ago, virtually nothing was known about black hole statistical mechanics. Today, in contrast, we suffer an embarrassment of riches: we have many competing microscopic pictures, describing different states and different dynamics but all predicting the same thermodynamic behavior.

In these lectures, I will review what is currently know—and not known—about black hole thermodynamics and statistical mechanics. This is a large subject, and I will have to skip many interesting aspects. In particular, I will not discuss stability analysis, the peculiarities of negative heat capacity, or the complicated question of black hole phase transitions, and I will only lightly touch upon the profound issues of information loss and holography.

Even so, my approach will necessarily be sketchy and idiosyncratic, though I will also try to suggest further references with different emphases and different degrees of detail. I will aim for a broad overview, rather than focusing on the fine points of any one particular approach. Some books and review articles that I have found helpful include [3–7]. In an appendix, I discuss basic black hole properties and explain my notation.

3.2 Black Hole Thermodynamics

I will begin with two somewhat intuitive routes to black hole thermodynamics. One of these is based on the second law of thermodynamics, the other on the four laws of black hole mechanics. Neither route is completely convincing, but together they provide a good foundation for some of the harder quantitative approaches that I shall discuss later.

3.2.1 Entropy and the Second Law

Imagine dropping a small box of hot gas into a black hole. The initial state includes the gas and the black hole; the final state consists solely of a slightly larger black hole. The initial state certainly has nonzero entropy, in the form of the entropy of the gas. If the second law of thermodynamics is to hold, the final state must have nonzero entropy as well: the larger black hole must gain enough entropy to compensate for the entropy lost when the gas disappears behind the horizon.

We can make this argument somewhat more quantitative [8]. Suppose the box of gas has linear size L, mass m, and temperature T, and that the black hole has mass M and horizon radius $R = 2GM$ (and thus horizon area $A = 16\pi G^2 M^2$). The box of gas will merge with the black hole when its proper distance ρ from the horizon is of order L, at which point the disappearance of the gas will lead to a loss of entropy

$$\Delta S \sim -m/T.$$

For a Schwarzschild black hole, the proper distance from the horizon is

$$\rho = \int_{2GM}^{2GM+\delta r} \frac{dr}{\sqrt{1 - 2GM/r}} \sim \sqrt{GM\delta r},$$

so $\rho \sim L$ when $\delta r \sim L^2/GM$. The gas initially has mass m, but its energy as seen from infinity is red shifted as the box falls toward the black hole; when the box reaches $r = 2GM + \delta r$, the black hole will gain a mass

$$\Delta M \sim m\sqrt{1 - \frac{2GM}{2GM + \delta r}} \sim \frac{mL}{GM}.$$

If we now suppose that the box must be as large as the thermal wavelength of the gas, $L \sim \hbar/T$, we see that

$$\Delta S \sim -\frac{mL}{\hbar} \sim -\frac{GM\Delta M}{\hbar} \sim -\frac{\Delta A}{\hbar G}.$$

To preserve the second law of thermodynamics, the black hole must gain an entropy of at least order $\Delta A/\hbar G$.

One can perform a similar analysis for a single particle falling into a Kerr black hole (assuming the particle contains at least one bit of entropy) [2], a box containing a simple harmonic oscillator [2], and, using a more sophisticated analysis, a much more general system falling through a horizon [3, 9–11]. In each case, a "generalized second law" holds, provided one includes a change of entropy of order $\Delta A/\hbar G$ for the black hole. Such reasoning led Bekenstein to suggest in 1972 that a black hole should itself be attributed an entropy of order $A/\hbar G$ [2].

At the time, there seemed to be a compelling argument against such a hypothesis. Classical black holes are, after all, black: when placed in contact with a heat bath they will absorb energy while emitting none, thus behaving as if they have a temperature of zero [12]. Two years later, Hawking showed that this problem was cured by quantum theory. I shall return to this result below, but let us first consider another classical argument for black hole thermodynamics.

3.2.2 The Four Laws of Black Hole Mechanics

In four spacetime dimensions, a stationary asymptotically flat black hole is uniquely characterized by its mass M, angular momentum J, and charge Q. (In the presence of nonabelian gauge fields or certain exotic scalar fields, other kinds of black hole "hair" can occur [13], but this does not change the basic argument.) In the early 1970s, a set of relations among neighboring solutions were found, culminating in Bardeen, Carter, and Hawking's "four laws of black hole mechanics" [7, 12]. These take a form strikingly similar to the four laws of thermodynamics:

1. The surface gravity κ is constant over the event horizon.
2. For any two stationary black holes differing only by small variations in the parameters M, J, and Q,

$$\delta M = \frac{\kappa}{8\pi G}\delta A + \Omega_H \delta J + \Phi_H \delta Q, \qquad (3.3)$$

where Ω_H is the angular velocity and Φ_H is the electric potential at the horizon.
3. The area of the event horizon of a black hole never decreases,

$$\delta A \geq 0.$$

4. It is impossible by any procedure to reduce the surface gravity κ to zero in a finite number of steps.

As in ordinary thermodynamics, there are a number of formulations of the third law, which are not strictly equivalent; for a proof of the version given here, which is analogous to the Nernst form of the third law of thermodynamics, see [14]. These laws can be generalized beyond the particular four-dimensional "electrovac" setting in which they were first formulated; the first law, in particular, holds for arbitrary isolated horizons [15], and for much more general gravitational actions, for which the entropy can be understood as a Noether charge [16].

Bardeen, Carter, and Hawking noted that these laws closely parallel the ordinary laws of thermodynamics, with the horizon area playing the role of entropy and the surface gravity playing the role of temperature. But they added, "It should however be emphasized that $\kappa/8\pi$ and A are distinct from the temperature and entropy of the black hole. In fact the effective temperature of a black hole is absolute zero. In this sense a black hole can be said to transcend the second law of thermodynamics."[1]

3.2.3 Black Holes Radiate

The first suggestion that black holes might emit radiation was made by Zel'dovich [17], but his argument was qualitative, and applied only to superradiant modes of rotating black holes. In 1974, though, Hawking demonstrated that all black holes emit blackbody radiation [1, 18]. The result was startling, and according to Page [19], Hawking himself did not initially believe it. In hindsight, though, one can give a somewhat intuitive description of the effect [20].

Such a description has two main ingredients. The first is that the quantum mechanical vacuum is filled with virtual particle–antiparticle pairs that fluctuate briefly

---

[1] See [12], p. 168.

into and out of existence. Energy is conserved, so one member of each pair must have negative energy. (To avoid a common confusion, note that either the particle or the antiparticle can be the negative-energy partner.) Normally, negative energy is forbidden—in a stable quantum field theory, the vacuum must be the lowest energy state—but energy has a quantum mechanical uncertainty of order $\hbar/t$, so a virtual pair of energy $\pm E$ can exist for a time of order $\hbar/E$. The existence of such virtual pairs is experimentally well-tested: for example, virtual pairs of charged particles make the vacuum a polarizable medium, and vacuum polarization is observed in such phenomena as the Lamb shift and in energy levels of muonic atoms [21].

The second ingredient is the observation that in general relativity, energy—and, in particular, the sign of energy—can be frame dependent. The easiest way to see this is to note that the Hamiltonian is the generator of time translations, and thus depends on one's choice of a time coordinate. One must therefore be careful about what one means by positive and negative energy for a virtual pair.

In particular, consider the Schwarzschild metric,

$$ds^2 = \left(1 - \frac{2GM}{r}\right) dt^2 - \left(1 - \frac{2GM}{r}\right)^{-1} dr^2 - r^2 d\Omega^2. \tag{3.4}$$

Outside the event horizon, t is the usual time coordinate, measuring the proper time of an observer at infinity. Inside the horizon, though, components of the metric change sign, and r becomes a time coordinate, while t becomes a spatial coordinate: an observer moving forward in time is one moving in the direction of decreasing r, and not necessarily increasing t.[2] Hence an ingoing virtual particle that has negative energy relative to an external observer may have positive energy relative to an observer inside the horizon. The uncertainty principle can thus be circumvented: if the negative-energy member of a virtual pair crosses the horizon, it need no longer vanish in a time $\hbar/E$, and its positive-energy partner may escape to infinity.

We can again make this argument a bit more quantitative. Consider a virtual pair momentarily at rest at a coordinate distance δr from the horizon. As in Sect. 3.2.1, the proper time for one member of the pair to reach the horizon will be

$$\tau \sim \sqrt{GM\delta r}.$$

Setting this equal to the lifetime $\hbar/E$ of the pair, we find that

$$|E| \sim \frac{\hbar}{\sqrt{GM\delta r}},$$

which should also be the energy of the escaping positive-energy partner. This is the energy at $2GM + \delta r$, though; the energy at infinity will be red shifted to

---

[2] Strictly speaking, the coordinates labeled r and t for $r > 2GM$ are different from those with the same labels for $r < 2GM$, since the Schwarzschild coordinate system is only defined in nonoverlapping patches inside and outside the horizon. But one can rephrase the argument in terms of proper time of infalling observers in a way that dodges this mathematical subtlety [20].

$$E_\infty \sim \frac{\hbar}{\sqrt{GM\delta r}} \sqrt{1 - \frac{2GM}{2GM + \delta r}} \sim \frac{\hbar}{GM}, \tag{3.5}$$

independent of the initial position δr. We might thus expect a black hole to radiate with a characteristic temperature $kT \sim \hbar/GM$. In fact, the precise computations I shall describe below yield a temperature $kT_{Hawking} = \hbar\kappa/2\pi$, which for a Schwarzschild black hole is $\hbar/8\pi GM$.

Inserting the Hawking temperature (3.1) into the first law of black hole mechanics (3.3), we see that black holes can indeed be viewed as thermal objects, with an entropy (3.2). This result is fundamentally quantum mechanical—the Hawking temperature depends explicitly on $\hbar$—and in some sense quantum gravitational, since the Bekenstein–Hawking entropy depends on G as well.

3.2.4 Can Hawking Radiation Be Observed?

I will return to the more precise and detailed derivations of Hawking radiation below. But let us first address the question of whether this effect can be observed.

For a black hole of mass M, the Hawking temperature (3.5) is

$$T_{Hawking} \sim 6 \times 10^{-8} \left(\frac{M_\odot}{M}\right) K,$$

some eight orders of magnitude smaller than the cosmic microwave background temperature for a stellar mass black hole and far smaller for a supermassive black hole. While there is a chance that we could see Hawking radiation from the final stages of evaporation of primordial black holes [22, 23], such events are expected to be rare and difficult to identify.

Another highly speculative possibility for the detection of Hawking radiation comes from models of TeV-scale gravity. In such models—which typically arise from "brane world" scenarios in which our four-dimensional universe is a submanifold of a higher-dimensional spacetime—gravity may become strong at energies far below the Planck scale. If this is the case, black holes might be produced copiously at accelerators such as the LHC, and their quantum properties could be studied in detail [24, 25].

A third, less direct, route is to look for analogs of Hawking radiation in condensed matter systems. As Unruh first pointed out [26], one can create a sonic event horizon in a fluid flow by allowing the flow to become supersonic beyond some boundary. The same analysis that predicts Hawking radiation from a black hole leads to a prediction of phonon radiation from the sonic horizon of such a "dumb hole." Similar phenomena can occur in a variety of condensed matter systems, from Bose–Einstein condensates to "slow light" to superfluid quasiparticles, and a number of experimental efforts are underway; for reviews, see [27, 28]. It is worth emphasizing that while such experiments could provide strong evidence for Hawking radiation, which is essentially a kinematical property, they would not test the Bekenstein–Hawking entropy, which depends critically on the dynamics of general relativity [29].

3.2.5 The Many Derivations of Hawking Radiation

In the absence of direct experimental evidence, how confident should we be about
Hawking radiation and black hole thermodynamics? Although Hawking's derivation involves only standard quantum field theory, we can see from the arguments
of Sect. 3.2.3 that the radiation involves modes with arbitrarily high energies: while
the asymptotic energies (3.5) may be small, they come from red-shifted quanta with
much higher energies near the horizon. This has led some to suggest that the derivations might involve an extrapolation of quantum field theory beyond the range it can
be trusted [26, 30, 31].

I will return this issue below, but for now let me suggest a partial answer. If only
one derivation of Hawking radiation existed, we would clearly need to look very
carefully for hidden assumptions and unjustified extrapolations. In fact, though, we
have a rather large number of different derivations, which involve very different assumptions and extrapolations and nevertheless all agree. Some of these derivations
look at eternal black holes, others at black holes formed from collapse; some involve
explicit, detailed computations in particular field theories, others use general properties of axiomatic quantum field theory; some involve Planck-scale fluctuations,
others cut off energies well below the Planck scale; some some predict only the
Hawking temperature, others also allow a computation of the Bekenstein–Hawking
entropy. While it is still possible that these derivations all share a common flawed
assumption, it seems unlikely that so many methods would converge on the same
answer if that answer were wrong. None of this vitiates the need for observational
tests—after all, the entire general relativistic description of black holes could be
wrong—but it suggests that a failure of black hole thermodynamics would have to
be either very subtle or very radical.

I will describe some of these derivations below. Given the nature of these lectures, I will not attempt a full description of any one method; my aim is to give a
broad overview, with references that will allow the reader to delve into individual
approaches in more detail.

3.2.5.1 Bogoliubov Transformations and Inequivalent Vacua

As noted above, a crucial ingredient in understanding Hawking radiation is the fact
that energy—and, in particular, "positive" and "negative" energy—is frame dependent. Consider, for simplicity, a free real scalar field φ. Recall that in ordinary quantum field theory in flat spacetime, we quantize φ by first decomposing the field into
Fourier modes,

$$\varphi = \sum_{\mathbf{k}} \left(a_{\mathbf{k}} u_{\mathbf{k}}(t, \mathbf{x}) + a_{\mathbf{k}}^{\dagger} u_{\mathbf{k}}^{*}(t, \mathbf{x}) \right) \quad \text{with} \quad u_{\mathbf{k}} = e^{i\mathbf{k}\cdot\mathbf{x} - i\omega_{\mathbf{k}}t}, \quad \omega_{\mathbf{k}}$$

$$= \left(|\mathbf{k}|^2 + m^2 \right)^{1/2}, \tag{3.6}$$

and then interpret the $a_\mathbf{k}$ as annihilation operators and the $a_\mathbf{k}^\dagger$ as creation operators. The Fourier modes $u_\mathbf{k}$ can be understood as a set of orthonormal functions satisfying

$$(\Box + m^2)u_\mathbf{k}(t,\mathbf{x}) = 0, \qquad \partial_t u_\mathbf{k}(t,\mathbf{x}) = -i\omega_\mathbf{k} u_\mathbf{k}(t,\mathbf{x}), \qquad (3.7)$$

where the second condition determines what we mean by positive and negative frequency, and thus allows us to distinguish creation and annihilation operators. The vacuum is then defined as the state annihilated by all the $a_\mathbf{k}$,

$$a_\mathbf{k}|0\rangle = 0.$$

In a curved spacetime, or a noninertial coordinate system in flat spacetime, standard Fourier modes are no longer available. With a choice of time coordinate t, though, one can still find modes of the form (3.7) and perform a decomposition (3.6) to obtain creation and annihilation operators. Given two different reference frames with time coordinates t and $\bar{t}$, two such decompositions exist:

$$\varphi = \sum_i \left(a_i u_i + a_i^\dagger u_i^* \right) = \sum_i \left(\bar{a}_i \bar{u}_i + \bar{a}_i^\dagger \bar{u}_i^* \right), \qquad (3.8)$$

and since the (u_i, u_i^*) are a complete set of functions, we can write

$$\bar{u}_j = \sum_i (\alpha_{ji} u_i + \beta_{ji} u_i^*). \qquad (3.9)$$

This relation is known as a Bogoliubov transformation, and the coefficients α_{ji} and β_{ji} are Bogoliubov coefficients [32].

We now have two vacuum states, one annihilated by the a_i and one by the $\bar{a}_i$, and two number operators $N_i = a_i^\dagger a_i$ and $\bar{N}_i = \bar{a}_i^\dagger \bar{a}_i$. Using the orthonormality of the mode functions, it is straightforward to show that

$$\langle \bar{0}|N_i|\bar{0}\rangle = \sum_j |\beta_{ji}|^2. \qquad (3.10)$$

Thus if the coefficients β_{ji} are not all zero, the "barred" vacuum will have a nonvanishing "unbarred" particle content.

In [1] and [18], Hawking considered a mass collapsing to form a black hole, and computed the Bogoliubov coefficients connecting an initial vacuum far outside the collapsing matter to a final vacuum after the black hole formed. He found that the "barred" observer at future null infinity will observe a thermal distribution of particles, with a temperature (3.1).[3] I will not go into details here; three very nice reviews can be found in [4, 29, 33]. The essential physical feature is that ingoing vacuum modes "pile up" at the horizon, giving an exponential relationship between ingoing and outgoing surfaces of constant phase; the integrals that determine the Bogoliubov coefficients β_{ji} take the form

---

[3] The final distribution is actually not quite thermal, but contains a "graybody factor" that reflects the backscattering of some of the emitted radiation into the black hole.

$$\int dv\, e^{i\omega v} e^{-i\frac{\omega}{\kappa}\ln v},$$

yielding gamma functions of complex arguments whose absolute squares give the exponential behavior of a thermal distribution.

Hawking's derivation was based on a particular choice of vacuum state, but generalizations are possible. For example, one may compare the vacuum of a freely falling observer near the horizon to the vacuum of an observer at future null infinity [34]. One can also look beyond the expectation value of the number operator, and express the full final state in terms of initial modes; one finds that it is exactly thermal [35, 36]. Generalizations to spinor and gauge fields are straightforward, and yield the correct fermionic and bosonic distribution functions.

It is also possible to simplify the problem, by looking at the easier model of an accelerated observer in flat spacetime. Such an observer is naturally described in Rindler coordinates [37]

$$ds^2 = e^{2a\xi}\left(d\eta^2 - d\xi^2\right),$$

in which the exponential relationship between the unaccelerated and the accelerated modes is easy to verify. A straightforward calculation of Bogoliubov coefficients shows that the accelerated observer will see a thermal bath of "Unruh radiation" with a temperature $kT = \hbar a/2\pi$, where a is proper acceleration [34]. By the principle of equivalence, an observer at rest near the horizon of a black hole should experience the same effect, with the acceleration a replaced by the appropriately blue shifted surface gravity κ, the acceleration necessary to hold the observer at rest.

As I noted in the preceding section, the exponential relationship between "barred" and "unbarred" modes may be a cause of concern. The modes observed as Hawking radiation by an observer far from the black hole are red shifted from Planck-scale modes near the horizon, and it seems that one has extrapolated quantum field theory far beyond the range in which it is known to be valid. To address this question, a number of authors have looked at the effect of modifying the dispersion relations in a way that removes very high-energy modes (see, for example, [26, 38–40]). For example [41], one can replace the standard expression for the energy of a massless field, $\omega_{\mathbf{k}} = |\mathbf{k}|$, with

$$\omega_{\mathbf{k}}^2 = |\mathbf{k}|^2 - \frac{|\mathbf{k}|^4}{k_0{}^2},$$

eliminating modes with trans-Planckian energies. Both numerical and analytical computations show that despite these drastic changes in the high-frequency behavior, thermal Hawking radiation persists. We now have strong evidence that a few simple assumptions—a vacuum near the horizon as seen in a freely falling frame, fluctuations that start in the ground state, and adiabatic evolution of the modes—are sufficient to guarantee thermal radiation [42].

3.2.5.2 Particle Detectors in a Black Hole Background

The definitions of vacuum and particle number in the preceding section were taken from ordinary quantum field theory. But finding observables in quantum gravity is notoriously difficult, and one might worry about the applicability of these definitions in a highly curved spacetime. To address this issue, Unruh [34] and DeWitt [43] considered the response of a particle detector in a black hole background and showed that such a detector sees thermal radiation at the Hawking temperature. Similarly, a static atom outside a black hole will be excited as one would expect in a thermal bath [44].

3.2.5.3 The Stress–Energy Tensor

One can obtain further invariant information about black hole radiation by evaluating the expectation value of the stress–energy tensor of a quantum field in a black hole background. This is a large subject; good introductions can be found in the books [6] and [45]. For these lectures, the most relevant result is that an ingoing negative energy flux at the horizon balances the outgoing flux of Hawking radiation observed at infinity, leading to a back-reaction in which the black hole's mass decreases (as expected from the intuitive argument of Sect. 3.2.3) and ensuring energy conservation.

The computation of $\langle T_{\mu\nu} \rangle$ in a black hole background is generally very difficult (see, for example, [46] or Chap. 11 of [6]). In the special case of a massless scalar field—or more generally, a conformally invariant field—in two dimensions, the calculation drastically simplifies [47]. The key difference is that in two dimensions, conservation of the stress–energy tensor is sufficient to determine the full expectation value in terms of the trace anomaly $\langle T^{\mu}{}_{\mu} \rangle$, which, in turn, depends only on characteristics of the field in a flat background. The resulting expectation values are thermal, and the total flux can be used to determine the temperature, which matches the Hawking temperature (3.1).

Quite recently, Robinson and Wilczek have shown how to extend this result to more than two dimensions, by dimensionally reducing an arbitrary field to two dimensions (or equivalently looking at a partial wave expansion) and trading the trace anomaly for a chiral anomaly [48]. Their method, with some variations (for example, [49]), has been quickly extended to a wide variety of black holes. In a beautiful piece of work, Iso, Morita, and Umetsu have further shown that by looking at higher-order correlators, one can use similar techniques to obtain not just the total flux, but the full blackbody spectrum of Hawking radiation [50, 51].

3.2.5.4 Tunneling Through the Horizon

For many physical systems, we know that classically forbidden processes can occur through quantum tunneling. This is the case for Hawking radiation. The idea of a tunneling description dates back to at least 1975 [52], but the nicest form is more recent, coming from Parikh and Wilczek's insight that one can think of the horizon tunneling past the emitted radiation rather than vice versa [53–55].

Consider a spherically symmetric system of mass M consisting of a Schwarzschild black hole of mass $M - \omega$ emitting a shell of radiation of mass $\omega \ll M$. In Painlevé–Gullstrand coordinates, chosen because they are stationary and nonsingular at the horizon, the shell moves in a spacetime with metric

$$ds^2 = \left(1 - \frac{2G(M - \omega)}{r}\right) dt^2 - 2\sqrt{\frac{2G(M - \omega)}{r}} dt\, dr - dr^2 - r^2 d\Omega^2,$$

and outgoing radial null geodesics satisfy

$$\dot{r} = 1 - \sqrt{\frac{2G(M - \omega)}{r}}.$$

Now consider the imaginary part of the action for an outgoing positive energy shell—to be interpreted as an s-wave particle—crossing the horizon from r_{in} to r_{out}:

$$ImI = Im \int_{r_{in}}^{r_{out}} p_r dr = Im \int_{r_{in}}^{r_{out}} \int_0^{p_r} dp'_r dr = Im \int_M^{M-\omega} \int_{r_{in}}^{r_{out}} \frac{dr}{\dot{r}} dH, \qquad (3.11)$$

where I have used Hamilton's equations of motion to write $dp_r = dH/\dot{r}$ and noted that the horizon moves inward from GM to $G(M - \omega)$ as the particle is emitted. Setting $H = M - \omega$ and inserting the value of $\dot{r}$ obtained from the null geodesic equation, one can perform the integral easily through a contour deformation, obtaining

$$ImI = 4\pi \omega G \left(M - \frac{\omega}{2}\right) \qquad (3.12)$$

with $r_{in} > r_{out}$. Again, the physical picture is that the horizon tunnels inward as the black hole's mass decreases.

By standard quantum mechanics, the tunneling rate in the WKB approximation is then

$$\Gamma = e^{-2 ImI/\hbar} = e^{-8\pi \omega G\left(M - \frac{\omega}{2}\right)/\hbar} = e^{\Delta S_{BH}} \qquad (3.13)$$

where ΔS_{BH} is the change in the Bekenstein–Hawking entropy (3.2). By the first law of black hole mechanics, this is $\hbar \omega / T_H$, and we recover thermal Hawking radiation.

The tunneling derivation may be easily extended to other classes of black holes, and consistently reproduces the standard results. Its relationship to Hawking's original derivation is not obvious, but Parikh and Wilczek note that the same analysis can describe a negative-energy particle tunneling into the black hole, thus offering a similar physical picture.

3.2.5.5 Periodic Greens Functions

Consider the two-point function of a scalar field φ in a thermal ensemble of inverse temperature β:

$$
\begin{aligned}
G_\beta(x,0;x',t) &= Tr\left(e^{-\beta H}\varphi(x,0)\varphi(x',t)\right) \\
&= Tr\left(\varphi(x,0)e^{-\beta H}e^{\beta H}\varphi(x',t)e^{-\beta H}\right) \\
&= Tr\left(\varphi(x,0)e^{-\beta H}\varphi(x',t+i\beta)\right) = G_\beta(x',t+i\beta;x,0), \qquad (3.14)
\end{aligned}
$$

where I have used cyclicity of the trace and the fact that the Hamiltonian generates time translations, so $e^{\beta H}\varphi(x',t)e^{-\beta H} = \varphi(x',t+i\beta)$. In particular, (3.14) implies that if a thermal Greens function is symmetric in its arguments, it must be periodic in time with period $i\beta$. This argument may be run backwards, and such periodicity in imaginary time may be taken as the *definition* of a thermal Greens function; in axiomatic quantum field theory, this is formalized as the KMS condition [56–58].

As early as 1976, Bisognano and Wichmann showed that the Greens function for a uniformly accelerated observer obeys the KMS condition [59]. By the equivalence principle, the same should hold for an observer at rest near the horizon of a black hole. This is indeed the case, as shown by Gibbons and Perry [60, 61], who further demonstrated that the periodicity corresponds exactly to the expected Hawking temperature (3.1).

3.2.5.6 Gravitational Instantons

The periodicity of Greens functions described above suggests that it might be worthwhile to consider the analytic continuation of black hole spacetimes to "imaginary time." Near the horizon $r = r_+$, a stationary black hole metric takes the approximate form

$$
ds^2 = 2\kappa(r-r_+)dt^2 - \frac{1}{2\kappa(r-r_+)}dr^2 - r_+^2 d\Omega^2.
$$

Continuing to imaginary time $t = i\tau$ and replacing r by the proper distance

$$
\rho = \frac{1}{\kappa}\sqrt{2\kappa(r-r_+)}
$$

to the horizon, we obtain the "Euclidean black hole" metric

$$
ds^2 = d\rho^2 + \kappa^2\rho^2 d\tau^2 + r_+^2 d\Omega^2. \qquad (3.15)
$$

The ρ–τ portion of this metric may be recognized as that of a flat two-plane in polar coordinates, with imaginary time τ serving as the angular coordinate. The horizon $\rho = 0$ has shrunk to a point. To avoid a conical singularity at the origin, we must require that $\kappa\tau$ have period 2π, i.e., that τ have period $2\pi/\kappa = 1/kT_{Hawking}$.

This result provides a simple way to understand the periodicity of the Lorentzian Greens functions in imaginary time. But it does more: it allows a steepest descent ("instanton") approximation to the gravitational path integral and a semiclassical derivation of the Bekenstein–Hawking entropy [62]. The key ingredient is the observation that on a manifold with boundary, the ordinary Einstein–Hilbert action must be supplemented by a boundary term, without which it may have no extrema [62, 63]. At an extremum, the "bulk" contribution to the action,

$$\frac{1}{16\pi G} \int d^4x \sqrt{|g|} R,$$

vanishes, but the boundary term can give a nonzero contribution. In the original work in this field, the boundary term was taken at infinity [62, 64], but it may more intuitively be placed at the origin of the Euclidean black hole, that is, at the horizon [65–67]. This boundary term may be evaluated in a number of ways—a particularly elegant approach involves dimensional reduction to a disk in the ρ–τ plane [65]—and yields an extremal action

$$\bar{I}_{Euc} = \frac{A_{horizon}}{4\hbar G} - \beta (M + \Omega J + \Phi Q). \tag{3.16}$$

This Euclidean saddle point contributes $e^{\bar{I}_{Euc}}$ to the partition function, and from (3.16), we can recognize the result as the grand canonical partition function for a system with entropy $S_{BH} = A_{horizon}/4\hbar G$.

These results can be extended to much more general stationary configurations containing horizons [68]. The essential ingredient is a Killing vector with zeros, which become boundaries upon continuation to Euclidean signature. One can also obtain an equivalent result by canonically quantizing the system while including the boundary terms; the boundary term at the horizon gives rise to a new term in the Wheeler–DeWitt equation, from which one can again recover the Bekenstein–Hawking entropy [69].

3.2.5.7 Black Hole Pair Creation

A further path integral derivation of black hole entropy comes from studying the spontaneous pair creation rate for black holes in a background magnetic field [70], electric field [71], de Sitter space [72], or more complicated combinations of external fields [73]. One consistently finds that the production rate is enhanced by a factor of $e^{S_{BH}}$, exactly the phase space factor one would expect for a system in which the Bekenstein–Hawking entropy gives the logarithm of the number of states.

3.2.5.8 Quantum Field Theory and the Eternal Black Hole

Yet another derivation of Hawking radiation comes from considering quantum field theory on an eternal black hole background. Recall that in Kruskal coordinates, a black hole spacetime splits into four regions, as shown in Fig. 3.1. Consider a state

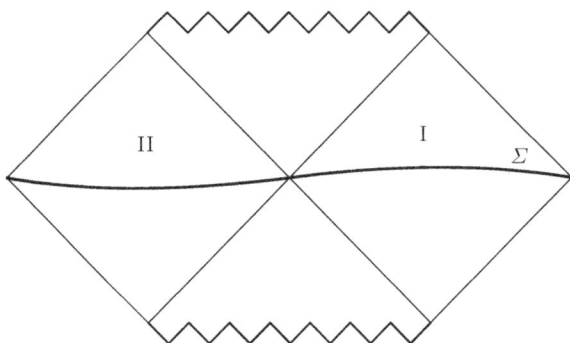

Fig. 3.1 A Carter-Penrose diagram for an eternal black hole

defined on a Cauchy surface Σ that passes through the bifurcation sphere. Region *II* is invisible to an observer living in region *I*, so such an observer should trace over the degrees of freedom in that region. Even if the initial state is pure, such a trace will lead to a density matrix describing the physics in region *I*. This makes it plausible that the region *I* observer will see thermal behavior, and detailed calculations show that this is indeed the case (Fig. 3.1).

In particular, for a free quantum field there is at most one quantum state, the Hartle–Hawking vacuum state, that is regular everywhere on the horizon [5, 74]. For a scalar field, a direct computation shows that the density matrix obtained by tracing over region *II* is thermal, with a temperature $T_{Hawking}$ [75]. For more general fields, the same can be shown by means of fairly sophisticated quantum field theory [5, 74] or by general path integral arguments [76].

3.2.5.9 Quantum Gravity in 2+1 Dimensions

Most standard derivations of black hole thermodynamics hold in an arbitrary number of dimensions, with changes only in the graybody factors for Hawking radiation. In three spacetime dimensions, though, many approaches become much simpler. The BTZ solution [77, 78] is a vacuum solution of the Einstein field equations in 2+1 dimensions with a negative cosmological constant. It has all the standard features of a rotating black hole—an event horizon, an inner Cauchy horizon, the same causal

structure as that of a (3+1)-dimensional asymptotically anti-de Sitter black hole—
but is, at the same time, a space of constant negative curvature. This latter feature
greatly simplifies many derivations: for example, Greens functions can be computed
exactly and their periodicity in imaginary time exhibited explicitly (see [78] for a
review). As was first suggested in [79], it might even be possible to use the relationship between three-dimensional general relativity and two-dimensional conformal
field theory [80] to find an exact description of the quantum states of the BTZ black
hole; the present status of this conjecture is discussed in [81].

The simplicity of the (2+1)-dimensional setting also permits an approach that
is not readily available in higher dimensions. The methods I have described so far
are based on properties of quantum fields in a classical, or at best semiclassical,
black hole background. In three dimensions, one can work in the opposite direction,
starting with a *quantum* black hole coupled to a *classical* source. As I shall discuss
further in Sect. 3.4.3, three-dimensional gravity with a negative cosmological constant is closely related to a two-dimensional field theory living at the "boundary" of
asymptotically anti-de Sitter space. Emparan and Sachs have shown how to couple
this two-dimensional field theory to a classical scalar field, allowing the computation of transition rates among black hole states due to emission and absorption of
the classical field [82]. By using detailed balance arguments, they recover standard
Hawking radiation, including the correct graybody factors, from this fundamentally
quantum gravitational picture.

3.2.5.10   Other Microscopic Approaches

The derivations I have described so far are essentially "thermodynamic," based on
macroscopic properties of black holes. As I shall discuss in the following sections,
we now also have a large number of "statistical mechanical" derivations, based on
analyses of the microscopic states of the black hole. These microscopic approaches
are not complete—string theory derivations, for example, are most reliable for extremal and near-extremal black holes, while loop quantum gravity derivations contain an order one parameter that, so far, must be adjusted by hand—but they seem
to work well within their ranges of validity. When combined with the macroscopic
approaches above, they provide strong evidence for the reality of black hole thermodynamics.

3.3 Black Hole Statistical Mechanics

In ordinary thermodynamic systems, thermal properties are macroscopic reflections
of the underlying microscopic physics. Temperature is a measure of the average
energy of the constituents of a system, for instance, while entropy is essentially
the logarithm of the number of states with specified macroscopic properties. The

connection between the microscopic and the macroscopic properties, given by statistical mechanics, has been remarkably successful across physics.

Given the thermodynamic properties of black holes, it is natural to ask whether these, too, have a statistical mechanical interpretation. Such an explanation would almost certainly involve quantum gravity—the Bekenstein–Hawking entropy (3.2) involves both Planck's constant $\hbar$ and Newton's constant G—and we might hope to learn something about the deep mysteries of quantum gravity.

To find such a statistical mechanical description, one should, in principle, carry out a number of steps:

1. Find a candidate quantum theory of gravity (not an easy task)
2. Identify black holes in the theory (also not easy)
3. Identify observables such as horizon area (surprisingly hard—finding physical observables in a quantum theory of gravity is notoriously difficult [83])
4. Count the microstates for a black hole configuration (perhaps easier, but still not trivial)
5. Compare to the Bekenstein–Hawking entropy (perhaps relatively easy)
6. Compute interactions with external fields, evaluate Hawking radiation, etc. (not at all easy)
7. Try to identify new quantum gravitational effects (the horizon area spectrum? evaporation remnants? higher order corrections to the Bekenstein–Hawking entropy? correlations across the horizon?).

Until recently, these steps seemed far beyond reach. In 1996, though, Strominger and Vafa published a remarkable paper in which they explicitly computed the entropy of a class of extremal black holes in string theory from the microscopic quantum theory [84]. Since then, a flood of new microscopic derivations of black hole thermodynamics has appeared. The new puzzle—the "problem of universality"—is that although these derivations seem to be using very different methods to count very different states, they all obtain the same thermodynamic properties.

3.3.1 The Many Faces of Black Hole Statistical Mechanics

In this section, I will briefly review some of the statistical mechanical approaches to black hole thermodynamics, and in particular the Bekenstein–Hawking entropy. As in Sect. 3.2.5, I will not go into detail, but will instead try to provide an overall flavor of the work, along with references for further study.

3.3.1.1 String Theory: Weakly Coupled Strings and Branes

The first breakthrough in the counting of black hole microstates came with the work of Strominger and Vafa on extremal black holes in string theory [84]. Their approach can be summarized as follows.

The effective low-energy field theory coming from string theory contains a number of gauge fields, each of which can give a charge to a black hole. An extremal supersymmetric (BPS) black hole is uniquely characterized by its charges; in particular, its horizon area can be expressed in terms of these charges. Given such a black hole, one can imagine tuning down the couplings, weakening gravity until the black hole "dissolves" into a gas of weakly coupled strings and branes. In this weakly coupled system, the charges can be expressed in terms of the number of strings and branes and the quantized momentum carried by strings. Furthermore, the states—the excitations of the string–brane system—can be explicitly counted [85]. We can therefore write the number of states in terms of the numbers of strings and branes, and thus the charges. Comparing this number to the horizon area, we recover the standard Bekenstein–Hawking entropy as the logarithm of the number of states.

One might worry that the number of states might not be the same in the weakly coupled system as in the strongly coupled black hole. For the supersymmetric case, though, this number is protected by nonrenormalization theorems. For black holes far from extremality, on the other hand, the computations are much more difficult; there are qualitative arguments that give an entropy proportional to the horizon area, but the exact proportionality factor of $1/4$ is difficult to obtain [86, 87].

It was quickly realized that the Strominger–Vafa results could be extended to a wide variety of extremal and near-extremal black holes, and through duality relations to a number of nonextremal black holes as well. Nice reviews can be found in [88] and [89]; for recent progress on the four-dimensional Kerr black hole, see [90].

This string theory approach has been remarkably successful, determining not only the Bekenstein–Hawking entropy for extremal and near-extremal black holes but also describing their interactions with other fields and their emission of Hawking radiation. The method has one peculiarity, though, to which I will return below. Suppose you ask me for the entropy of a three-charge black hole in five dimensions. I will compute the horizon area in the strongly coupled theory in terms of the charges, compute the number of states in the weakly coupled theory in terms of the charges, compare the two, and reply that the entropy is one-fourth of the horizon area. If you now ask me for the entropy of a four-charge black hole, or a black hole in six dimension, I cannot simply tell you that it is one-fourth of the horizon area; I must recompute the horizon area and the number of states in terms of the new parameters and compare the answers again. Each new black hole requires a new calculation: the theory tells us that the number of microstates of a black hole matches the Bekenstein–Hawking entropy (3.2), but it tells us so one black hole at a time.

3.3.1.2 String Theory: "Fuzzballs"

One can run the argument in the preceding section backward: given a particular excitation of the weakly coupled string and brane system, one can ask exactly what geometry results at strong coupling. The result is a "fuzzball" picture, in which particular black hole states correspond to complicated geometries that have *no* horizon or singularity, but that look very much like black hole geometries outside the would-be horizon [91, 92]. In special cases, one can count the number of such "fuzzball" geometries and reproduce the Bekenstein–Hawking entropy, and it seems likely that this result can be extended to more general black holes, although it is an open question whether simple geometric descriptions will always suffice [93]. Samir Mathur has discussed this approach extensively in his lectures, to which I refer the reader [85].

3.3.1.3 String Theory: The AdS/CFT Correspondence

Yet another string theory approach to black hole statistical mechanics is based on Maldacena's celebrated AdS/CFT correspondence [94–96]. This very well-supported conjecture asserts a duality between string theory in d-dimensional asymptotically anti-de Sitter spacetime and a conformal field theory in a flat $(d-1)$-dimensional space that can, in a sense, be viewed as the boundary of the AdS space-time. This correspondence is naturally "holographic" (see Sect. 3.5.2), describing the black hole in terms of a lower-dimensional theory and thus offering a frame-work for understanding the dependence of entropy on area rather than volume.

For asymptotically anti-de Sitter black holes, this correspondence makes it pos-sible to compute entropy by counting states in a (nongravitational) dual conformal field theory. The simplest case is the (2+1)-dimensional BTZ black hole discussed in Sect. 3.2.5, whose dual is a two-dimensional conformal field theory. As I shall dis-cuss in Sect. 3.4.1, the density of states in such a theory has an asymptotic behavior controlled by a single parameter, the central charge c. For asymptotically anti-de Sitter gravity in 2+1 dimensions, this central charge is dominated by a classical contribution, which was discovered some time ago by Brown and Henneaux [97]. Strominger [98] and Birmingham et al. [99] independently realized that this result could be used to compute the BTZ black hole entropy, reproducing the Bekenstein–Hawking expression.

While this result applies directly only to the special case of three-dimensional spacetime, it has an important generalization. Many of the higher dimensional near-extremal black holes of string theory—including black holes that are not themselves asymptotically anti-de Sitter—have a near-horizon geometry of the general form $BTZ \times trivial$, where the "trivial" part merely renormalizes constants in the calcula-tion of entropy. As a consequence, the BTZ results can be used to find the entropy of a large class of stringy black holes, including most of the black holes whose states can be counted in the weak coupling approach of Sect. 3.3.1 [100].

3.3.1.4 Loop Quantum Gravity

In the quest for quantum gravity, the leading alternative to string theory is loop quantum gravity [101]. The fundamental "position" variable in this theory is a three-dimensional $SU(2)$ connection; a state is a complex-valued function of (generalized) connections. A useful basis of states consists of spin networks, graphs with edges labeled by SU(2) representations ("spins") and vertices labeled by intertwiners. A spin network state can be evaluated on a given connection to give a complex number by computing the holonomies along the edges in the specified representations and combining them with the intertwiners at the vertices.

Given a surface Σ, one can define an area operator $\hat{A}_\Sigma$ that acts on loop quantum gravity states. It may be shown that spin networks are eigenfunctions of these operators, with eigenvalues of the form

$$A_\Sigma = 8\pi\gamma G \sum_j \sqrt{j(j+1)},$$

where the sum is over the spins j of edges of the spin network that cross Σ. The parameter γ, the Barbero–Immirzi parameter, represents a quantization ambiguity, and its physical significance is poorly understood; theories with different values of γ may be inequivalent, but it has been suggested that γ may not appear in properly renormalized observables [102] or in a slightly different approach to quantization [103].

Given this structure, a natural first attempt to count black hole states is to enumerate inequivalent spin networks crossing the horizon that yield a specified area [104, 105]. The more careful variation of this idea [106, 107] takes into account the fact that when one restricts to a black hole spacetime, one must place "boundary conditions" on the horizon to ensure that it is, in fact, a horizon. These conditions, in turn, require the addition of boundary terms to the Einstein–Hilbert action, which induce a three-dimensional Chern–Simons action on the horizon. The number of states of this Chern–Simons theory is closely related to the number of spin networks that induce the correct horizon area, but with slightly more subtle combinatorics. The ultimate result is that the black hole entropy takes the form [108, 109]

$$S = \frac{\gamma_M}{\gamma} \frac{A_{horizon}}{4\hbar G}, \tag{3.17}$$

where γ is the Barbero–Immirzi parameter and

$$\gamma_M \approx .23753$$

is a numerical constant determined as the solution of a particular combinatoric problem. If one chooses $\gamma = \gamma_M$, one thus recovers the standard Bekenstein–Hawking entropy.

The physical significance of this rather peculiar value of the Barbero–Immirzi parameter is not understood, and it may reflect an inadequacy in the quantization procedure or the definition of the area operator [103]. Note, though, that γ only appears in the combination $G\gamma$, so this choice may be viewed as a finite renormalization of Newton's constant. If the same shift occurs in the attraction between two masses, its interpretation becomes straightforward. Unfortunately, the Newtonian limit of loop quantum gravity is not yet well enough understood to see whether this is the case.

In any case, though, once γ is fixed for one type of black hole—the static Schwarzschild solution, say—the loop quantum gravity computations give the correct entropy for a wide variety of others, including charged black holes, rotating black holes, black holes with dilaton couplings, black holes with higher genus horizons, and black holes with arbitrarily distorted horizons [110, 111]. In particular, there is no need to restrict oneself to near-extremal black holes. Hawking radiation, on the other hand, is not yet very well understood in this approach, although there has been some progress [112, 113].

An alternate approach to black hole entropy also exists within the framework of loop quantum gravity [114]. Here, one again looks at a horizon area determined by edges of a spin network, but instead of counting states in an induced boundary theory, one merely asks the number of ways the spins can be joined to a single interior vertex. This amounts, in essence, to completely coarse graining the interior state of the black hole and is comparable in spirit to the thermodynamic derivation of Sect. 3.2.5. One again obtains an entropy proportional to the horizon area, although with a different value of the Barbero–Immirzi parameter.

3.3.1.5 Induced Gravity

In 1967, Sakharov suggested that the Einstein–Hilbert action for gravity might not be fundamental [115]. If one starts with a theory of fields propagating in a curved spacetime, counterterms from renormalization will automatically induce a gravitational action, which will almost always include an Einstein–Hilbert term at lowest order [116, 117]. Gravitational dynamics would then be, in Sakharov's terms, a sort of "metric elasticity" induced by quantum fluctuations.

One can write down an explicit set of "heavy" fields that can be integrated out in the path integral to induce the Einstein–Hilbert action. By including nonminimally coupled scalar fields, one can obtain finite values of Newton's constant and the cosmological constant. It is then possible to go back and count states of the heavy fields in a black hole background [118]. The nonminimal couplings lead to some subtleties in the definition of entropy, but in the end the computation reproduces the standard Bekenstein–Hawking value. Furthermore, the reduction to a two-dimensional conformally invariant system near the horizon, in the spirit of the thermodynamic approach of Sect. 3.2.5, allows a counting of states by standard methods of conformal field theory [119]. We thus obtain a new, and apparently quite different, view of the microstates of a black hole as those of the ordinary quantum fields responsible for inducing the gravitational action.

3.3.1.6 Entanglement Entropy

As discussed in Sect. 3.2.5, one way to obtain the thermodynamic properties of a black hole is to trace out the degrees of freedom behind the horizon, generating a density matrix for the external observer from a globally pure state. This process also produces a quantum mechanical "entanglement entropy," which can be thought of as a measure of the loss of information about correlations across the horizon. The suggestion that this entanglement entropy might account for the Bekenstein–Hawking entropy is an old one [120, 121], and it is not hard to show that for many (although not all [122]) states, the entanglement entropy is proportional to the horizon area: the main contribution comes from correlations among degrees of freedom very close to the horizon and does not involve "bulk" degrees of freedom. The *coefficient* of this entropy, on the other hand, is infinite, and must be cut off [123], leading to an expression that depends strongly on both the nongravitational content of the theory (the number and species of "entangled" fields contributing to the entropy) and the value of the cutoff.

The same modes that cause the entanglement entropy to diverge also give divergent contributions to the renormalization of Newton's constant, and it has been suggested that the two divergences may compensate [124]. This notion has recently gained new life with a proposal by Ryu and Takayanagi for a "holographic" description of entanglement entropy [125, 126], in which the d-dimensional spacetime containing a black hole is embedded at the asymptotic boundary of $(d+1)$-dimensional anti-de Sitter space. The idea is inspired by the string theory AdS/CFT correspondence and can be largely proven to work in situations in which such a correspondence exists [127]; the bulk anti-de Sitter metric provides a natural cutoff, yielding finite contributions to both S and G. When applied to a black hole, the proposal correctly reproduces the standard Bekenstein–Hawking entropy [128], providing yet another physical picture of the relevant microstates.

3.3.1.7 Other Approaches

A variety of other microscopic descriptions of black hole thermodynamics have also been proposed. In the causal set formulation of quantum gravity—in which a continuous spacetime is replaced by a discrete set of points with prescribed causal relations—there is evidence that the Bekenstein–Hawking entropy is given by the number of points in the future domain of dependence of a spatial cross-section of the horizon [129]. York has estimated the entropy obtained by quantizing the quasinormal modes [130] of the Schwarzschild black hole, finding a result that lies within a few percent of the Bekenstein–Hawking value [131]. Black hole entropy can be related to the Kolmogorov-Sinai entropy of a string spreading out on the black hole horizon [132]. A number of mini- and midisuperspace models—models in which most of the degrees of freedom of the gravitational field are frozen out—have also been proposed to explain black hole statistical mechanics [133–135], though none is yet very convincing.

One can also build "phenomenological" models of black hole microstates, in which the horizon area is simply assumed to be quantized [136–139]. Such models do not, of course, tell us *why* area is quantized, and thus do not address the fundamental physical questions of black hole statistical mechanics, but they can suggest useful directions for further research.

Suppose, for example, that the black hole area spectrum is discrete and equally spaced and that the exponential of the entropy (3.2) gives an exact count of the number of states at a given horizon area. Then the difference between two adjacent values must be an integer; that is,

$$\Delta A = 4\hbar G \ln k \tag{3.18}$$

for some integer k. Hod has pointed out [140] that for the Schwarzschild black hole, the most highly damped quasinormal modes [130]—the damped "ringing modes" of an excited black hole—have frequencies whose real part approaches

$$Re\,\omega = \ln 3/8\pi GM$$

(a numerical result later verified analytically [141]). If one applies the Bohr correspondence principle and argues that area eigenstates of the black hole should change by emission of quanta of energy $\hbar\omega$, one obtains

$$\Delta A = 32\pi G^2 M \Delta M = 4\hbar G \ln 3,$$

matching (3.18) with $k = 3$. It is not yet clear whether this result has deep significance. It seems to extend to general single-horizon black holes [142] and in a more complicated way to many "stringy" black holes [143], but results for charged and rotating black holes are unclear (for an optimistic view, see [144]).

One can also describe the Bekenstein-Hawking entropy as a count of the number of distinct ways that a black hole with specified macroscopic properties can be made from collapsing matter [10]. Like the phenomenological models of area quantization, this result does not really describe the microscopic degrees of freedom of the black hole itself (except perhaps in the "membrane paradigm" [145]), but it strongly suggests that if the formation of a black hole is a unitary process, such degrees of freedom must exist.

3.4 The Problem of Universality

One of the main lessons of the preceding section is that a great many different models of black hole microphysics yield the same thermodynamic properties. Some of these models are clearly ad hoc, but others are carefully worked out consequences of serious approaches to quantum gravity. So the new question is why everyone is getting the same answer.

To some extent, this "problem of universality" is a selection issue: there are undoubtedly computations that gave the "wrong" answer for black hole entropy and were discarded without being published. But as noted in Sect. 3.3.1, even within a particular well-motivated and successful string theory model we do not yet understand the universality of the entropy–area relationship. And regardless of what one may think about any one particular approach, one must still explain why *any* microscopic model reproduces the results of Hawking's original thermodynamic computation, a computation that seems to require no information about quantum gravity at all.

There are other situations, of course, in which thermodynamic properties do not depend too delicately on an underlying quantum theory. For example, for a large range of parameters the entropy of a box of gas depends only very weakly on whether the molecules are fermions or bosons. But in cases like this, we have a *classical* microscopic description, and the correspondence principle guarantees that the quantum theory will give a good approximation for the classical results. For a black hole, things are different: the only classical description we have is one in which black holes have no hair—no phase space volume—and thus no entropy. We need something new, some new principle that determines the quantum mechanical density of states in terms of the classical characteristics of a black hole.

I do not know the ultimate explanation for this universal behavior, but in the remainder of this section, I will make a tentative suggestion and offer some evidence that it may be correct.

3.4.1 The Cardy formula

I only know of one well-understood case in which universality of the sort we see in black hole statistical mechanics appears elsewhere in physics. Consider a two-dimensional conformal field theory, that is, a theory in two spacetime dimensions that is invariant under diffeomorphisms ("generally covariant") and Weyl transformations ("locally scale invariant"). If we choose complex coordinates z and $\bar{z}$, the basic symmetries of such a theory are the holomorphic and antiholomorphic diffeomorphisms $z \to f(z), \bar{z} \to \bar{f}(\bar{z})$. These are canonically generated by "Virasoro generators" $L[\xi]$ and $\bar{L}[\bar{\xi}]$ [146]. Such a theory has two conserved charges, $L_0 = L[\xi_0]$ and $\bar{L}_0 = \bar{L}[\bar{\xi}_0]$, which can be thought of as "energies" with respect to constant holomorphic and antiholomorphic transformations, or alternatively as linear combinations of energy and angular momentum.

As generators of diffeomorphisms, the Virasoro generators have an algebra that is almost unique [147]:

$$\{L[\xi], L[\eta]\} = L[\eta\xi' - \xi\eta'] + \frac{c}{48\pi} \int dz \left(\eta'\xi'' - \xi'\eta''\right)$$

$$\{L[\xi], \bar{L}[\bar{\eta}]\} = 0 \qquad\qquad (3.19)$$

$$\{\bar{L}[\bar{\xi}], \bar{L}[\bar{\eta}]\} = \bar{L}[\bar{\eta}\bar{\xi}' - \bar{\xi}\bar{\eta}'] + \frac{\bar{c}}{48\pi} \int d\bar{z} \left(\bar{\eta}'\bar{\xi}'' - \bar{\xi}'\bar{\eta}''\right).$$

The central charges c and $\bar{c}$ determine the unique central extension of the ordinary algebra of diffeomorphisms. These constants can occur classically, coming, for instance, from boundary terms in the generators [97], or can appear upon quantization.

Now consider a conformal field theory for which the lowest eigenvalues of L_0 and $\bar{L}_0$ are nonnegative numbers Δ_0 and $\bar{\Delta}_0$. In 1986, Cardy discovered a remarkable result [148, 149]: the density of states $\rho(\Delta, \bar{\Delta})$ at eigenvalues $(\Delta, \bar{\Delta})$ of L_0 and $\bar{L}_0$ has the simple asymptotic behavior

$$\ln \rho(\Delta, \bar{\Delta}) \sim 2\pi \left\{ \sqrt{\frac{c_{eff}\Delta}{6}} + \sqrt{\frac{\bar{c}_{eff}\bar{\Delta}}{6}} \right\}, \text{ with } c_{eff} = c - 24\Delta_0, \bar{c}_{eff} = \bar{c} - 24\bar{\Delta}_0.$$

(3.20)

The entropy is thus determined by the symmetry, independent of any other details— exactly the sort of universality we are looking for.

A typical black hole is neither two-dimensional nor conformally invariant, of course, so this result may at first seem irrelevant. But there is a sense in which black holes become *approximately* two-dimensional and conformal near the horizon. For fields in a black hole background, for instance, excitations in the r–t plane become so blue shifted relative to transverse excitations and dimensionful quantities that an effective two-dimensional conformal description becomes possible [150–152]. Indeed, as noted in Sect. 3.2.5, the full Hawking radiation spectrum can be derived from such an effective description [50, 51]. Martin, Medved, and Visser have further shown that a generic near-horizon region has a conformal symmetry, in the form of an approximate conformal Killing vector [153, 154].

3.4.2 Horizons and Constraints

For the special case of the (2+1)-dimensional BTZ black hole, the Cardy formula can be used directly to count states. For this solution, the boundary at infinity is geometrically a two-dimensional flat cylinder, and the asymptotic diffeomorphisms that respect boundary conditions satisfy a Virasoro algebra with a classical central charge [97], which can be used in the Cardy formula [98, 99]. As described in Sect. 3.3.1, this calculation can be extended to a number of near-extremal black holes whose near-horizon geometry contains a *BTZ* factor. For more general black holes, though, something new is needed.

One key question, I believe, is how to specify that one is talking about a black hole in quantum gravity. One cannot simply require a fixed metric: the components of the metric do not all commute and cannot be simultaneously specified in a quantum theory. For the BTZ case, the key element is a set of boundary conditions at infinity, but in general it seems more natural to consider conditions at the horizon. Two approaches to this question are currently under investigation, each leading to an effective two-dimensional conformal description in which the Cardy formula might be applicable.

3.4.2.1 The Horizon as a Boundary

The first approach [155, 156] is to introduce "boundary conditions" at the horizon. The horizon is not, of course, a genuine boundary, but it is a place at which we must restrict the value of the metric, precisely to ensure that it is a horizon. As in the BTZ case, such a restriction forces us to add new boundary terms to the canonical generators of diffeomorphisms, changing their algebra. One finds a conformal symmetry in the r–t plane with a classical central charge. For a large variety of black holes, it has been shown that the Cardy formula then yields the correct entropy.[4]

On the other hand, the diffeomorphisms whose algebra yields that central charge, essentially those that leave the lapse function invariant, are generated by vector fields that blow up at the horizon. This is not necessarily a bad thing—from the perspective of an external observer, many physical quantities diverge at the horizon— but the status of these transformations is not clear. In addition, the "horizon as boundary" method has trouble with the two-dimensional black hole, and some normalization issues are not completely sorted out. A related approach is to look for approximate conformal symmetry near the horizon [158, 159]; one again finds a Virasoro algebra with a central charge that seems to lead to the correct entropy, but there are again some normalization ambiguities.

3.4.2.2 Horizon Constraints

A more recent approach [160, 161] is to impose the presence of a horizon by adding "horizon constraints" in the canonical formulation of gravity, that is, introducing new constraints that restrict data on a specified surface to be that of a black hole horizon. In outline, the procedure is this:

1. Dimensionally reduce to the two-dimensional r–t plane near the horizon
2. Continue to Euclidean signature, shrinking the horizon to a point as in Sect. 3.2.5, and evolve radially
3. Impose constraints on a small circle around the horizon that force the initial data be that of a "stretched horizon"
4. Adjust the diffeomorphism constraints on the stretched horizon a la Bergmann and Komar [162–164] to make them commute with the new horizon constraints
5. Find the resulting algebra and central charge

The Cardy formula again reproduces the correct Bekenstein–Hawking entropy.

3.4.2.3 Universality Again

If either of these approaches is to be an answer to the "problem of universality," it must be that the horizon conformal symmetries are secretly present in the various

---

[4] For this section, see [157] for further references.

other computations of black hole entropy. I do not know whether this is the case; it is a subject of continuing research.

One fairly simple test is to compare the near-horizon Virasoro algebra of Sect. 3.4.2 with the asymptotic Virasoro algebra of the BTZ black hole, which is the key element in the AdS/CFT computations of Sect. 3.3.1. It is shown in [161] that after a suitable matching of coordinate choices, the central charges and conformal weights exactly coincide, providing one piece of evidence for the proposed explanation of universality. There is also an intriguing link to the loop quantum gravity approach of Sect. 3.3.1: the induced horizon Chern-Simons theory in loop quantum gravity is naturally associated with a two-dimensional conformal field theory [165], whose central charge matches the horizon central charge of Sect. 3.4.2. Searches for hidden conformal symmetry in loop quantum gravity, the fuzzball approach, and induced gravity are currently underway.

3.4.3 What Are the States?

In light of the problem of universality, is there anything general we can say about the states responsible for black hole thermodynamics? At first sight, the answer must be "no": if a universal underlying structure controls the density of states, there should be many different models with different degrees of freedom but with the same thermodynamic properties. Nevertheless, it may still be possible to find an *effective* description that is valid across models.

To see this, let us first return to the BTZ black hole. In three spacetime dimensions, general relativity has a peculiar feature: it is a topological theory, with no propagating degrees of freedom [166]. Where, then, do the black hole degrees of freedom come from?

The answer to this paradox is at least partially understood [81]. For the (2+1)-dimensional Einstein–Hilbert action to have any black hole extrema, one must impose anti-de Sitter boundary conditions at infinity. Diffeomorphisms that do not respect these boundary conditions are no longer true invariances of the theory, and states one might naively take to be physically equivalent—states that differ only by a diffeomorphism—must be considered distinct if the diffeomorphism connecting them is incompatible with the boundary conditions. New physical degrees of freedom thus appear, which can be labeled by diffeomorphisms that fail to respect the anti-de Sitter boundary conditions. The action for these new degrees of freedom can be extracted explicitly from the Einstein–Hilbert action [167], and the resulting dynamics is that of a Liouville theory, a two-dimensional conformal field theory whose central charge matches the classical value obtained by Brown and Henneaux [97]. Whether one can actually count the states in this theory to reproduce the Bekenstein–Hawking entropy remains an open question [81, 168].

For higher-dimensional black holes, the problem is quite a bit more difficult. One possible approach is to start with the Virasoro algebra (3.19) for the near-horizon conformal algebra of Sect. 3.4.2.2. In Dirac quantization, the existence of a constraint ordinarily restricts the physical states: we should require that

$$L[\xi]|phys\rangle = \bar{L}[\bar{\xi}]|phys\rangle = 0. \tag{3.21}$$

But if the central charge c is nonzero, these conditions are incompatible with the algebra (3.4.2). The solution is known in conformal field theory—one can, for instance, require only that the positive frequency parts of the Virasoro generators annihilate physical states [146]—but the result is much the same as for the BTZ black hole: certain states that were originally counted as nonphysical have now become physical. While it is not exactly the same, this phenomenon is reminiscent of the Goldstone mechanism [169], in which a spontaneously broken symmetry leads to massless excitations in the "broken" directions. And like the Goldstone mechanism, it can provide an effective description of degrees of freedom that is independent of their fundamental physical makeup.

One way to see whether this picture makes sense is to examine the path integral measure. The effect of adding a central charge to the Virasoro algebra is to make certain constraints second class [163, 164]. The presence of such second class constraints leads to a new term in the measure, similar to the Faddeev–Popov determinant in quantum field theory [170]. Such a determinant can be interpreted as a contribution to the phase space volume, or the density of states, and might explain the counting of black hole states. For the present case, the relevant determinant is of the form

$$\det\left|-\frac{c}{12}\frac{d^3}{dx^3} + \frac{d}{dx}L + L\frac{d}{dx}\right|^{1/2} \qquad \text{with } L = L_0 + L_1 e^{2ix} + L_{-1} e^{-2ix}.$$

Work on evaluating and understanding this expression is in progress.

Perhaps the most important test of this idea would be to couple the effective horizon degrees of freedom to external matter and see if one could reproduce Hawking radiation. In 2+1 dimensions, this can be done [82]. In higher dimensions, it may be possible to take advantage of the conformal description of Hawking radiation discussed in Sect. 3.2.5, but this remains to be seen.

3.5 Open Questions

Some 35 years after the seminal papers of Hawking and Bekenstein, black hole equilibrium thermodynamics is a mature subject. The role of trans-Planckian excitations near the horizon, discussed in Sect. 3.2.5, is not yet fully understood, and questions of possible observational tests remain of great interest, but I will risk the claim that the macroscopic thermodynamic properties of black holes are largely under control.

The microscopic, statistical mechanical picture of the black hole, in contrast, is poorly understood and is the subject of a great deal of research. This is hardly surprising—black hole microstates are almost certainly quantum gravitational, and we are still far from a complete, compelling theory of quantum gravity.

Much of the current research focuses on particular microscopic models of black holes, from string theory, loop quantum gravity, and a number of other perspectives. But there are also some broader open questions. In these lectures, I have emphasized one of these, the problem of universality, mainly because it is a focus of my own research. But I will close by briefly mentioning two other deep questions.

3.5.1 The Information Loss Paradox

Consider a configuration of matter in a pure state—a spherically symmetric state of a scalar field, for instance—that collapses to form a black hole, which then evaporates by Hawking radiation. If Hawking radiation is exactly thermal, and if the black hole evaporates completely, the ultimate result will be a transition from an initial pure state to a final mixed (thermal) state [171]. Such an evolution is not unitary and seems to violate the basic principles of quantum mechanics. Similarly, we can imagine a black hole held at equilibrium by the continual ingestion of mass to balance its Hawking radiation; this would seem to allow us to convert an arbitrarily large amount of matter from a pure to a mixed state.

The solution to this paradox is heavily debated [172–174]. If the black hole horizon is fundamental (as it is not in, for instance, the "fuzzball" proposal discussed in Mathur's lectures [85]), there is wide agreement that any answer must involve a breakdown of locality; see, for example, [175–178]. But there is certainly no consensus as to how such a breakdown might occur. The answer is likely to involve deep problems of quantum gravity, a setting in which nonlocality is both inevitable and very poorly understood [83].

3.5.2 Holography

As a count of microscopic degrees of freedom, the Bekenstein–Hawking entropy (3.2) has a peculiar feature: the number of degrees of freedom is determined by the area of a surface rather than the volume it encloses. This is very different from conventional thermodynamics, in which entropy is an extensive quantity, and it implies that the number of degrees of freedom grows much more slowly with size than one would expect in an ordinary thermodynamic system.

This "holographic" behavior [179, 180] seems fundamental to black hole statistical mechanics, and it has been conjectured that it is a general property of quantum gravity. It may be that the generalized second law of thermodynamics requires a similar bound for any matter that can be dropped into a black hole; a nice review of such entropy bounds can be found in [181]. The AdS/CFT correspondence discussed in Sect. 3.3.1 is perhaps the cleanest realization of holography in quantum gravity, but it requires specific boundary conditions. A more general formulation proposed by Bousso [182] is supported by classical computations [183], and is currently a very active subject of research, extending far beyond its birthplace in black hole physics to cosmology, string theory, and quantum gravity.

Acknowledgments These lectures were given during an appointment to the Kramers Chair at Utrecht University, for whose hospitality I am very grateful. This work was supported in part by U.S. Department of Energy grant DE-FG02-91ER40674.

Appendix: Black Hole Basics

Intuitively, a black hole is a "region of no return," an area of spacetime from which not even light can escape. For a spacetime that looks asymptotically close enough to Minkowski space, this intuitive picture is formalized by the notion of an event horizon, the boundary of the past of future null infinity, that is, the boundary beyond which no light ray can reach infinity [184]. The event horizon has been extensively studied and has many interesting global properties: for example, it cannot bifurcate and cannot decrease in area.

Unfortunately, while the event horizon has nice properties, it does not seem to be quite the right object to capture local physics. The problem is that the event horizon is teleological: that is, its definition requires knowledge of the indefinite future. To illustrate this with a thought experiment, imagine that we are at the center of a highly energetic ingoing spherical shell of light, currently two light years from Earth. Suppose this shell is so energetic that it has a Schwarzschild radius of one light year.[5] If I now shine a flashlight into the sky, 1 year from now the light will have traveled one light year, where it will meet the incoming shell just as the shell reaches its own Schwarzschild radius. At that point, the pulse of light from the flashlight will be trapped at the horizon of an ordinary Schwarzschild black hole and will be unable to travel any farther outward. In other words, in this scenario we are *now* at the event horizon of a black hole, even though we will detect no change in our local observations until we are abruptly crushed out of existence 2 years from now.

Since it seems implausible that Hawking radiation "now" can depend on such future events, the event horizon is probably not quite the right object for the study of black hole thermodynamics. Over the past few years, a number of attempts have been made to suitably "localize" the horizon; a nice review can be found in [185].

In these lectures, I will mainly use the concept of an "isolated horizon" [186], a locally defined surface that seems appropriate for equilibrium black hole thermodynamics. An isolated horizon is essentially a null surface whose area remains constant in time, as the horizon of a stationary black hole does. A thought experiment may again be helpful. Imagine a spherical lattice studded with equally spaced flashbulbs,

---

[5] This is admittedly not very likely, but note that it cannot be ruled out observationally: no signal could propagate faster than such a shell, so we would not know of its existence until it reached us.

set to all go off at the same time (as measured in the lattice rest frame). When the bulbs flash, they will emit two spherical shells of light, one ingoing and one outgoing. In ordinary nearly flat spacetime, the area of the outgoing sphere increases with time. At the horizon of a Schwarzschild black hole, on the other hand, it is not hard to check that the area of the outgoing sphere remains constant, while inside the horizon, both spheres decrease in area.[6]

To generalize this example, we first define a nonexpanding horizon $\mathscr{H}$ in a d-dimensional spacetime to be a $(d-1)$-dimensional submanifold such that [15, 186]

1. $\mathscr{H}$ is null, with null normal ℓ_a;
2. the expansion of $\mathscr{H}$ vanishes: $\vartheta_{(\ell)} = q^{ab}\nabla_a\ell_b = 0$, where q_{ab} is the induced metric on $\mathscr{H}$;
3. $-T^a{}_b\ell^b$ is future directed and causal.

These conditions imply the existence of a one-form ω_a such that

$$\nabla_a\ell^b = \omega_a\ell^b \quad \text{on } \mathscr{H}.$$

The surface gravity for the normal ℓ^a is then defined as

$$\kappa_{(\ell)} = \ell^a\omega_a. \tag{3.22}$$

Note, though, that the normal ℓ^a is not unique: a null vector has no canonical normalization, so if ℓ^a is a null normal to $\mathscr{H}$ and φ is an arbitrary function, $e^\varphi\ell^a$ is also a null normal to $\mathscr{H}$. We can partially fix this scaling ambiguity by demanding further time independence: we define a weakly isolated horizon by adding the requirement

4. $\mathscr{L}_\ell\omega = 0$ on $\mathscr{H}$,

where $\mathscr{L}$ denotes the Lie derivative. This constraint implies the zeroth law of black hole mechanics that the surface gravity is constant on the horizon.

Even with this last condition, the null normal ℓ^a may be rescaled by an arbitrary constant. Such a rescaling also scales the surface gravity, so the numerical value of $\kappa_{(\ell)}$ remains undetermined. This reflects a genuine physical ambiguity in the choice of time at the horizon. Note that the first law of black hole mechanics (3.3) requires such an ambiguity: mass is only defined relative to a choice of time, so for consistency, rescaling time must also rescale the surface gravity.

For a stationary black hole, ℓ^a can be chosen to coincide with the Killing vector that generates the horizon, whose normalization is fixed at infinity—that is, we can use the global properties of the solution to adjust clocks at the horizon by comparing them to clocks at infinity. If, on the other hand, we wish to focus on physics only at or very near the horizon, the normalization becomes more problematic. One can use the known properties of exact solutions to write an expression for the surface gravity in terms of other quantities at the horizon, thereby fixing ℓ^a [15], but so far the procedure seems somewhat artificial.

---

[6] The outgoing sphere remains outgoing with respect to the lattice, of course; as the lattice collapses, its area decreases even faster than that of the outgoing light sphere.

As noted in Sect. 3.2.2, weakly isolated horizons obey the four laws of black hole mechanics, the second law in the strong form that the area, by definition, remains constant. Generalization to dynamical, evolving horizons are also possible, and could provide a setting for nonequilibrium black hole thermodynamics; for a recent review, see [187].

References

1. S. W. Hawking, Nature **248**, 30 (1974).
2. J. D. Bekenstein, Phys. Rev. **D7**, 2333 (1973).
3. R. M. Wald, Living Rev. Relativity **4**, 6 (2001), URL: http://www.livingreviews.org/lrr-2001-6, eprint gr-qc/9912119.
4. T. Jacobson, in *Valdivia 2002, Lectures on Quantum Gravity*, edited by A. Gomberoff and D. Marolf (Springer, New York 2005), eprint gr-qc/0308048.
5. R. M. Wald, *Quantum Field Theory in Curved Spacetime and Black Hole Thermodynamics* (University of Chicago Press, Chicago 1994).
6. V. P. Frolov and I. D. Novikov, *Black Hole Physics* (Springer, New York 1998).
7. C. DeWitt and B. S. DeWitt (eds.) *Black Holes*, Proceedings of the 1972 Les Houches Summer school (Gordon and Breach, Newark 1973).
8. C. Kiefer, in *Classical and Quantum Black Holes*, edited by P. Fré, V. Gorini, G. Magli and U. Moschella (IOP Publishing, Bristol 1999).
9. J. D. Bekenstein, Phys. Rev. **D9**, 3292 (1974).
10. W. H. Zurek and K. S. Thorne, Phys. Rev. Lett. **54**, 2171 (1985).
11. V. P. Frolov and D. N. Page, Phys. Rev. Lett. **71**, 3902 (1993), eprint gr-qc/9302017.
12. J. M. Bardeen, B. Carter and S. W. Hawking, Commun. Math. Phys. **31**, 161 (1973).
13. Winstanley, E.: *Physics of Black Holesthis*. Lect. Notes Phys. **769**, Springer, Berlin (2009), eprint arXiv:0801.0527.
14. W. Israel, Phys. Rev. Lett. **57**, 397 (1986).
15. A. Ashtekar, S. Fairhurst and B. Krishnan, Phys. Rev. **D62**, 104025 (2000), eprint gr-qc/0005083.
16. R. M. Wald, Phys. Rev. **D48**, 3427 (1993), eprint gr-qc/9307038.
17. Ya. B. Zel'dovich, Sov. Phys. JETP Lett. **14**, 180 (1970).
18. S. W. Hawking, Commun. Math. Phys. **43**, 199 (1975).
19. D. N. Page, New J. Phys. **7**, 203 (2005), eprint hep-th/0409024.
20. B. F. Schutz, *A First Course in General Relativity* (Cambridge University Press, Cambridge 1990), Sect. 11.4.
21. S. Weinberg, *The Quantum Theory of Fields* (Cambridge University Press, Cambridge 1995), Chap. 11.2.
22. J. H. MacGibbon and B. J. Carr, Astrophys. J. **371**, 447 (1991).
23. D. B. Cline, Phys. Rept. **307**, 173 (1998).
24. S. B. Giddings, AIP Conf. Proc. **957**, 69 (2007), eprint arXiv:0709.1107.
25. Kanti, P.: *Physics of Black Holesthis*. Lect. Notes Phys. **769**, Springer, Berlin (2009).
26. W. G. Unruh, Phys. Rev. Lett. **46**, 1351 (1981).
27. C. Barceló, S. Liberati and M. Visser, Living Rev. Relativity **8**, 12 (2005), URL: http://www.livingreviews.org/lrr-2005-12, eprint gr-qc/0505065.
28. M. Novello, M. Visser and G. E. Volovik, *Artificial Black Holes* (World Scientific, Singapore 2002).
29. M. Visser, Int. J. Mod. Phys. **D12**, 649 (2003), eprint hep-th/0106111.
30. T. Jacobson, Phys. Rev. **D44**, 1731 (1991).
31. A. D. Helfer, Rept. Prog. Phys. **66**, 943 (2003), eprint gr-qc/0304042.
32. N. N. Bogoliubov, Sov. Phys. JETP **7**, 51 (1958).

33. J. H. Traschen, in *Mathematical Methods in Physics*, Proceedings of the 1999 Londrona Winter School, edited by A. A. Bytsenko and F. L. Williams (World Scientific, Singapore 2000), eprint gr-qc/0010055.
34. W. G. Unruh, Phys. Rev. **D14**, 870 (1976).
35. R. M. Wald, Commun. Math. Phys. **45**, 9 (1975).
36. L. Parker, Phys. Rev. **D12**, 1519 (1975).
37. W. Rindler, Am. J. Phys. **34**, 1174 (1966).
38. R. Brout, S. Massar, R. Parentani and Ph. Spindel, Phys. Rev. **D52**, 4559 (1995), eprint hep-th/9506121.
39. S. Corley, Phys. Rev. **D57**, 6280 (1998), eprint hep-th/9710075.
40. T. Jacobson, Phys. Rev. **D53**, 7082 (1996), eprint hep-th/9601064.
41. S. Corley and T. Jacobson, Phys. Rev. **D54**, 1568 (1996), eprint hep-th/9601073.
42. W. G. Unruh and R. Schutzhold, Phys. Rev. **D71**, 024028 (2005), eprint gr-qc/0408009.
43. B. S. DeWitt, in *General Relativity: An Einstein Centenary Survey*, edited by S. W. Hawking and W. Israel (Cambridge University Press, Cambridge 1979).
44. H. Yu and W. Zhou, Phys. Rev. **D76**, 044023 (2007), eprint arXiv:0707.2613.
45. N. D. Birrell and P. C. W. Davies, *Quantum Fields in Curved Space* (Cambridge University Press, Cambridge 1982).
46. D. N. Page, Phys. Rev. **D25**, 1499 (1982).
47. S. M. Christensen and S. A. Fulling, Phys. Rev. **D15**, 2088 (1977).
48. S. P. Robinson and F. Wilczek, Phys. Rev. Lett. **95**, 011303 (2005), eprint gr-qc/0502074.
49. R. Banerjee and S. Kulkarni, Phys. Rev. **D77**, 024018 (2008), eprint arXiv:0707.2449.
50. S. Iso, T. Morita and H. Umetsu, Phys. Rev. **D76**, 064015 (2007), eprint arXiv:0705.3494.
51. S. Iso, T. Morita and H. Umetsu, eprint arXiv:0710.0456.
52. T. Damour and R. Ruffini, Phys. Rev. **D14**, 332 (1976).
53. M. K. Parikh and F. Wilczek, Phys. Rev. Lett. **85**, 5042 (2000), eprint hep-th/9907001.
54. M. K. Parikh, Int. J. Mod. Phys. **D13**, 2351 (2004).
55. M. K. Parikh, Gen. Rel. Grav. **36**, 2419 (2004), eprint hep-th/0405160.
56. R. Kubo, J. Phys. Soc. Japan **12**, 570 (1957).
57. P. C. Martin and J. Schwinger, Phys. Rev. **115**, 1342 (1959).
58. R. Haag, *Local Quantum Physics* (Springer, New York 1993).
59. J. J. Bisognano and E. H. Wichmann, J. Math. Phys. **17**, 303 (1976).
60. G. W. Gibbons and M. J. Perry, Phys. Rev. Lett. **36**, 985 (1976).
61. G. W. Gibbons and M. J. Perry, Proc. Roy. Soc. Lond. **A358**, 467 (1978).
62. G. W. Gibbons and S. W. Hawking, Phys. Rev. **D15**, 2752 (1977).
63. T. Regge and C. Teitelboim, Annals Phys. **88**, 286 (1974).
64. S. W. Hawking, in *General Relativity: An Einstein Centenary Survey*, edited by S. W. Hawking and W. Israel (Cambridge University Press, Cambridge 1979).
65. M. Banados, C. Teitelboim and J. Zanelli, Phys. Rev. Lett. **72**, 957 (1994), eprint gr-qc/9309026.
66. C. Teitelboim, Phys. Rev. **D51**, 4315 (1995), eprint hep-th/9410103.
67. S. W. Hawking and G. T. Horowitz, Class. Quant. Grav. **13**, 1487 (1996), eprint gr-qc/9501014.
68. S. W. Hawking and C. J. Hunter, Phys. Rev. **D59**, 044025 (1999), eprint hep-th/9808085.
69. S. Carlip and C. Teitelboim, Class. Quant. Grav. **12**, 1699 (1995), eprint gr-qc/9312002.
70. D. Garfinkle, S. B. Giddings and A. Strominger, Phys. Rev. **D49** 958 (1994), eprint gr-qc/9306023.
71. J. D. Brown, Phys. Rev. **D51**, 5725 (1995), eprint gr-qc/9412018.
72. R. B. Mann and S. F. Ross, Phys. Rev. **D52**, 2254 (1995), eprint gr-qc/9504015.
73. I. S. Booth and R. B. Mann, Phys. Rev. Lett. **81**, 5052 (1998), eprint gr-qc/9806015
74. B. S. Kay and R. M. Wald, Phys. Rept. **207**, 49 (1991).
75. W. Israel, Phys. Lett. **A 57**, 107 (1976).
76. T. Jacobson, Phys. Rev. **D50**, 6031 (1994), eprint gr-qc/9407022.
77. M. Banados, C. Teitelboim and J. Zanelli, Phys. Rev. Lett. **69**, 1849 (1992), eprint hep-th/9204099.

78. S. Carlip, Class. Quant. Grav. **12**, 2853 (1995), eprint gr-qc/9506079.
79. S. Carlip, Phys. Rev. **D51**, 632 (1995), eprint gr-qc/9409052.
80. E. Witten, Nucl. Phys. **B311**, 46 (1988).
81. S. Carlip, Class. Quant. Grav. **22**, R85 (2005), eprint gr-qc/0503022.
82. R. Emparan and I. Sachs, Phys. Rev. Lett. **81**, 2408 (1998), eprint hep-th/9806122.
83. S. Carlip, Rept. Prog. Phys. **64**, 885 (2001), eprint gr-qc/0108040.
84. A. Strominger and C. Vafa, Phys. Lett. **B379**, 99 (1996), eprint hep-th/9601029.
85. Mathur, S.: *Physics of Black Holesthis*. Lect. Notes Phys. **769**, Springer, Berlin (2009).
86. L. Susskind, in *The black Hole: 25 Years After*, edited by C. Teitelboim and J. Zanelli (World Scientific, Singapore 1988), eprint hep-th/9309145.
87. G. T. Horowitz and J. Polchinski, Phys. Rev. **D55**, 6189 (1997), eprint hep-th/9612146.
88. A. W. Peet, in *TASI 99: Strings, Branes, and Gravity*, edited by J. Harvey, S. Kachru and E. Silverstein (World Scientific, Singapore 2001), eprint hep-th/0008241.
89. S. R. Das and S. D. Mathur, Ann. Rev. Nucl. Part. Sci. **50**, 153 (2000), eprint gr-qc/0105063.
90. G. T. Horowitz and M. M. Roberts, Phys. Rev. Lett. **99**, 221601 (2007), eprint arXiv:0708.1346.
91. S. D. Mathur, Fortsch. Phys. **53**, 793 (2005), eprint hep-th/0502050.
92. S. D. Mathur, Class. Quant. Grav. **23**, R115 (2006), eprint hep-th/0510180.
93. I. Kanitscheider, K. Skenderis and M. Taylor, JHEP 0706, **056** (2007), eprint arXiv:0704.0690.
94. J. M. Maldacena, Adv. Theor. Math. Phys. **2**, 231 (1998).
95. J. M. Maldacena, Int. J. Theor. Phys. **38**, 1113 (1999), eprint hep-th/9711200.
96. O. Aharony, S. S. Gubser, J. M. Maldacena, H. Ooguri and Y. Oz, Phys. Rept. **323**, 183 (2000), eprint hep-th/9905111.
97. J. D. Brown and M. Henneaux, Commun. Math. Phys. **104**, 207 (1986).
98. A. Strominger, JHEP 9802, **009** (1998), eprint hep-th/9712251.
99. D. Birmingham, I. Sachs and S. Sen, Phys. Lett. **B424**, 275 (1998), eprint hep-th/9801019.
100. Skenderis, K.: Lect. Notes Phys. **541**, 325 (2000), eprint hep-th/9901050.
101. C. Rovelli, Living Rev. Relativity **1**, 1 (1998), URL: http://www.livingreviews.org/lrr-1998-1, eprint gr-qc/9710008.
102. T. Jacobson, Class. Quant. Grav. **24**, 4875 (2007), eprint arXiv:0707.4026.
103. S. Alexandrov and E. R. Livine, Phys. Rev. **D67**, 044009 (2003), eprint gr-qc/0209105.
104. K. V. Krasnov, Phys. Rev. **D55**, 3505 (1997), eprint gr-qc/9603025.
105. C. Rovelli, Phys. Rev. Lett. **77**, 3288 (1996), eprint gr-qc/9603063.
106. A. Ashtekar, J. Baez, A. Corichi and K. Krasnov, Phys. Rev. Lett. **80**, 904 (1998), eprint gr-qc/9710007.
107. A. Ashtekar, J. C. Baez and K. Krasnov, Adv. Theor. Math. Phys. **4**, 1 (2000), eprint gr-qc/0005126.
108. M. Domagala and J. Lewandowski, Class. Quant. Grav. **21**, 5233 (2004), eprint gr-qc/0407051.
109. K. A. Meissner, Class. Quant. Grav. **21**, 5245 (2004), eprint gr-qc/0407052.
110. A. Ashtekar and J. Lewandowski, Class. Quant. Grav. **21**, R53 (2004), eprint gr-qc/0404018.
111. A. Ashtekar, J. Engle and C. Van Den Broeck, Class. Quant. Grav. **22**, L27 (2005), eprint gr-qc/0412003.
112. M. Barreira, M. Carfora and C. Rovelli, Gen. Rel. Grav. **28**, 1293 (1996), eprint gr-qc/9603064.
113. K. V. Krasnov, Class. Quant. Grav. **16**, 563 (1999), eprint gr-qc/9710006.
114. E. R. Livine and D. R. Terno, Nucl. Phys. **B741**, 131 (2006), eprint gr-qc/0508085.
115. A. D. Sakharov, Sov. Phys. Dokl. **12**, 1040 (1968), reprinted in Gen. Rel. Grav. **32**, 365 (2000).
116. S. L. Adler, Rev. Mod. Phys. **54**, 729 (1982).
117. S. L. Adler, Rev. Mod. Phys. **55**, 837 (1983) (Erratum).
118. V. P. Frolov and D. V. Fursaev, Phys. Rev. **D56**, 2212 (1997), eprint hep-th/9703178.
119. V. P. Frolov, D. Fursaev and A. Zelnikov, JHEP 0303, **038** (2003), eprint hep-th/0302207.
120. L. Bombelli, R. K. Koul, J. Lee and R. D. Sorkin, Phys. Rev. **D 34**, 373 (1986).

121. M. Srednicki, Phys. Rev. Lett. **71**, 666 (1993), eprint hep-th/9303048.
122. M. Requardt, eprint arXiv:0708.0901.
123. G. 't Hooft, Nucl. Phys. **B256**, 727 (1985).
124. L. Susskind and J. Uglum, Phys. Rev. **D50**, 2700 (1994), eprint hep-th/9401070.
125. S. Ryu and T. Takayanagi, Phys. Rev. Lett. **96**, 181602 (2006), eprint arXiv:hep-th/0603001.
126. V. E. Hubeny, M. Rangamani and T. Takayanagi, JHEP 0707, **062** (2007), eprint arXiv:0705.0016.
127. D. V. Fursaev, JHEP 0609, **018** (2006), eprint hep-th/0606184.
128. R. Emparan, JHEP 012 **0606** (2006), eprint hep-th/0603081.
129. D. Rideout and S. Zohren, Class. Quant. Grav. **23**, 6195 (2006), eprint gr-qc/0606065.
130. Siopsis, G.: *Physics of Black Holesthis*. Lect. Notes Phys. **769**, Springer, Berlin (2009)
131. J. W. York, Phys. Rev. **D28**, 2929 (1983).
132. K. Ropotenko, eprint arXiv:0711.3131.
133. C. Vaz, Phys. Rev. **D61**, 064017 (2000), eprint gr-qc/9903051.
134. J. Makela and A. Peltola, Phys. Rev. **D69**, 124008 (2004), eprint gr-qc/0307025.
135. C. Kiefer, J. Mueller-Hill, T. P. Singh and C. Vaz, Phys. Rev. **D75**, 124010 (2007), eprint gr-qc/0703008.
136. J. D. Bekenstein, Lett. Nuovo Cim. **11**, 467 (1974).
137. H. A. Kastrup, Phys. Lett. **B413**, 267 (1997), eprint gr-qc/9707009.
138. A. Barvinsky, S. Das and G. Kunstatter, Phys. Lett. **B517**, 415 (2001), eprint hep-th/0102061.
139. J. D. Bekenstein and G. Gour, Phys. Rev. **D66**, 024005 (2002), eprint gr-qc/0202034.
140. S. Hod, Phys. Rev. Lett. **81**, 4293 (1998), eprint gr-qc/9812002.
141. L. Motl and A. Neitzke, Adv. Theor. Math. Phys. **7**, 307 (2003), eprint hep-th/0301173.
142. R. G. Daghigh and G. Kunstatter, Class. Quant. Grav. **22**, 4113 (2005), eprint gr-qc/0505044.
143. D. Birmingham and S. Carlip, Phys. Rev. Lett. **92**, 111302 (2004), eprint hep-th/0311090.
144. S. Hod, Class. Quant. Grav. **24**, 4871 (2007), eprint arXiv:0709.2041.
145. K. S. Thorne, R. H. Price and D. A. Macdonald, *Black Holes: The Membrane Paradigm* (Yale University Press, New York 1986).
146. P. Di Francesco, P. Mathieu and D. Sénéchal, *Conformal Field Theory* (Springer, New York 1997).
147. C. Teitelboim, in *Quantum Theory of Gravity*, edited by S. M. Christensen (Adam Hilger, Bristol 1984).
148. J. A. Cardy, Nucl. Phys. **B270**, 186 (1986).
149. H. W. J. Blöte, J. A. Cardy and M. P. Nightingale, Phys. Rev. Lett. **56**, 72 (1986).
150. D. Birmingham, K. S. Gupta and S. Sen, Phys. Lett. **B505**, 191 (2001), eprint hep-th/0102051.
151. K. S. Gupta and S. Sen, Phys. Lett. **B526**, 121 (2002), eprint hep-th/0112041.
152. H. E. Camblong and C. R. Ordóñez, Phys. Rev. **D71**, 104029 (2005), eprint hep-th/0411008.
153. A. J. M. Medved, D. Martin and M. Visser, Class. Quant. Grav. **21**, 3111 (2004), eprint gr-qc/0402069.
154. A. J. M. Medved, D. Martin and M. Visser, Phys. Rev. **D70**, 024009 (2004), eprint gr-qc/0403026.
155. S. Carlip, Phys. Rev. Lett. **82**, 2828 (1999), eprint hep-th/9812013.
156. S. Carlip, Class. Quant. Grav. **16**, 3327 (1999), eprint gr-qc/9906126.
157. S. Carlip, Int. J. Theor. Phys. **46**, 2192 (2007), eprint gr-qc/0601041.
158. S. N. Solodukhin, Phys. Lett. **B454**, 213 (1999), eprint hep-th/9812056.
159. S. Carlip, Phys. Rev. Lett. **88**, 241301 (2002), eprint gr-qc/0203001.
160. S. Carlip, Class. Quant. Grav. **22**, 1303 (2005), eprint hep-th/0408123.
161. S. Carlip, Phys. Rev. Lett. **99**, 021301 (2007), eprint gr-qc/0702107.
162. P. G. Bergmann and A. B. Komar, Phys. Rev. Lett. **4**, 432 (1960).
163. P. A. M. Dirac, Can. J. Math. **2**, 129 (1950).
164. P. A. M. Dirac, Can. J. Math. **3**, 1 (1951).
165. E. Witten, Commun. Math. Phys. **121**, 351 (1989).
166. S. Carlip, Living Rev. Relativity **8**, 1 (2005), URL: http://www.livingreviews.org/lrr-2005-1, eprint gr-qc/0409039.

167. S. Carlip, Class. Quant. Grav. **22**, 3055 (2005), eprint gr-qc/0501033.
168. Y.-J. Chen, Class. Quant. Grav. **21**, 1153 (2004), eprint hep-th/0310234.
169. S. Weinberg, *The Quantum Theory of Fields* (Cambridge University Press, Cambridge 1995), Chap. 19.2.
170. M. Henneaux and C. Teitelboim, *Quantization of Gauge Systems* (Princeton University Press, Princeton 1992).
171. S. W. Hawking, Phys. Rev. **D14**, 2460 (1976).
172. J. Preskill, in *Black Holes, Membranes, Wormholes and Superstrings*, edited by S. Kalara and D. V. Nanopoulos (World Scientific, Singapore 1993), eprint hep-th/9209058.
173. T. Banks, Nucl. Phys. Proc. Suppl. **41**, 21 (1995), eprint hep-th/9412131.
174. C. R. Stephens, G. 't Hooft and B. F. Whiting, Class. Quant. Grav. **11**, 621 (1994.), eprint gr-qc/9310006.
175. S. B. Giddings, Phys. Rev. **D74**, 106005 (2006), eprint hep-th/0605196.
176. V. Balasubramanian, D. Marolf and M. Rozali, Gen. Rel. Grav. **38**, 1529 (2006).
177. V. Balasubramanian, D. Marolf and M. Rozali, Int. J. Mod. Phys. **D15**, 2285 (2006), eprint hep-th/0604045.
178. A. Ashtekar, V. Taveras and M. Varadarajan, eprint arXiv:0801.1811.
179. G. 't Hooft, in *Salamfestschrift: A Collection of Talks*, edited by A. Ali, J. Ellis and S.R̄andjbar-Daemi (World Scientific, Singapore 1993), eprint gr-qc/9310026.
180. L. Susskind, J. Math. Phys. **36**, 6377 (1995), eprint hep-th/9409089.
181. R. Bousso, Rev. Mod. Phys. **74**, 825 (2002), eprint hep-th/0203101.
182. R. Bousso, JHEP 9907, **004** (1999), eprint hep-th/9905177.
183. E. E. Flanagan, D. Marolf and R. M. Wald, Phys. Rev. **D62**, 084035 (2000), eprint hep-th/9908070.
184. S. W. Hawking and G. F. R. Ellis, *The Large Scale Structure of Space-Time* (Cambridge University Press, Cambridge 1973).
185. I. Booth, Can. J. Phys. **83**, 1073 (2005), eprint gr-qc/0508107.
186. A. Ashtekar, C. Beetle and S. Fairhurst, Class. Quant. Grav. **16**, L1 (1999), eprint gr-qc/9812065.
187. A. Ashtekar and B. Krishnan, Living Rev. Relativity **7**, 10 (2004), URL: http://www.livingreviews.org/lrr-2004-10, eprint gr-qc/0407042.

Chapter 4
Colliding Black Holes and Gravitational Waves

U. Sperhake

Abstract This article presents a summary of numerical simulations of black-hole spacetimes in the framework of general relativity. The first part deals with the 3+1 decomposition of generic spacetimes as well as the Einstein equations which forms the basis of most work in numerical relativity. Technical aspects of the resulting numerical evolutions and the diagnostics of the resulting spacetimes are discussed. The second part presents an overview of the history of numerical simulations of black-hole spacetimes. Finally, we summarize results derived from numerical black-hole simulations obtained after the breakthrough in 2005. The relevance of these results in the context of astrophysics, gravitational wave physics, and fundamental physics is discussed.

4.1 Introduction

In Einstein's theory of general relativity gravitation is a manifestation of the curvature of the spacetime rather than a force in the traditional sense. The fundamental quantity which encapsulates all information about the spacetime curvature is the spacetime metric, a set of ten functions of space and time. This metric obeys the Einstein equations which equates the Einstein tensor, a complex combination of the metric and its first and second derivatives, with the mass–energy tensor describing the matter distribution. The Einstein equations thus represent a system of ten second-order partial differential equations, one of the most complicated systems of equations in all of physics. Einstein himself did not expect physically meaningful solutions to be found analytically and it came as a surprise when Karl Schwarzschild found his famous solution of a static, spherically symmetric vacuum spacetime just a few months after the publication of general relativity in 1916. This solution is now known as a "Schwarzschild black hole", but the term black hole was not coined until much later by John Wheeler. The Schwarzschild solution has lead to invaluable insight into general relativity and was soon generalized to include electric charge

U. Sperhake (✉)
Theoretisch Physikaslisches Institut, Universität Jena, Max-Wien-Platz 1, 07743 Jena, Germany
U.Sperhake@uni-jena.de

Sperhake, U.: *Colliding Black Holes and Gravitational Waves*. Lect. Notes Phys. **769**, 125–175 (2009)
DOI 10.1007/978-3-540-88460-6_4

in the form of the Reissner-Nordström solution. The key simplification leading to these analytic solutions is the high degree of symmetry of the spacetime which reduces the Einstein equations to a 1D problem with no time dependence. Relaxing the assumption of spherical symmetry to allow for a spacetime with non-vanishing angular momentum led to a much more complex system of equations even in the limit of stationarity. It took more than four decades until Roy Kerr found the analytic expressions for the metric of an axisymmetric spacetime containing a rotating black hole [168]. Again, the inclusion of electric charge resulted in a generalization, the so-called Kerr–Newman solution.

For a long time, these black-hole solutions were considered a mathematical curiosity rather than objects of physical relevance. This picture has changed dramatically in the course of recent decades, however. Not only are black holes now accepted as a common end product of the evolution of very massive stars, they are also recognized as almost ubiquitously present in the form on supermassive black holes (SMBH) at the centers of at least more massive galaxies [176]. The formation history of these SMBHs is subject of ongoing research in astrophysics and is likely to be closely interrelated with structure formation in the universe in general (see, e.g., [118, 150, 151, 181, 190, 191, 268–270]). Observations of the central regions of galaxies have also revealed significant correlation of the masses of black holes with the structure of the galaxy cores, specifically the velocity dispersions and the density profiles [65, 123, 135, 197, 199, 202]. Given the all absorbing nature of black holes, it is quite remarkable, that they also form the engine for the strongest sources of electromagnetic radiation observed in the universe. Active galactic nuclei are now commonly believed to be driven by accretion around black holes. Their observation at cosmological redshifts provides valuable constraints on the formation history of SMBHs. Most notably, the discovery of the most luminous quasar at $z \approx 6$ in the Sloan Digital Sky Survey [121] implies that black holes of masses around $10^9 \, M_{\odot}$ were already in existence less than 10^9 years after the big bang.

Black holes also play a fundamental role in the ongoing effort to detect gravitational waves. This type of radiation is general relativity's analogue of electromagnetic waves and is a direct consequence of the Einstein equations. In fact, such radiative solutions were recognized by Einstein himself, but were subject of a long-lasting debate on whether they represent gauge effects or truly physical phenomena. There is no doubt left on the physical nature of gravitational waves (GW) now, but their direct detection is made enormously difficult by their extremely weak interaction with matter. To date, therefore, the only evidence for the existence of gravitational waves is indirect and based on observations of binary pulsar systems. Most notably, decade long observations of the Hulse–Taylor pulsar 1913+16 show a gradual decrease in the orbital period which is in excellent agreement with the energy loss of the system expected from emission of gravitational waves according to the theory of general relativity [164, 260]. This indirect evidence has led to the award of the 1993 Nobel Prize to Hulse and Taylor and also provided motivation for the construction of laser interferometric detectors in multinational collaborations such as the American LIGO [1, 76], the European GEO600 [188] and VIRGO [4, 47], and the Japanese TAMA [15, 259]. A space-based interferometer, LISA [155], is targeted for launch in 2018 and will facilitate high signal-to-noise ratio measurements of low-frequency gravitational wave sources. The strongest source for all these detectors is the inspiral and coalescence of black-hole binaries. Obtaining a detailed theoretical understanding of these binary systems is crucial to support the effort to directly detect GWs.

The enormous complexity of the Einstein equations in the absence of strong symmetry and/or time independence makes it impossible to study binary-black-hole systems analytically in the framework of full general relativity. In consequence, the theoretical modeling has pursued two alternative approaches. The first replaces general relativity by an approximative description of the physics which allows for analytic studies. In particular, binaries can be described with good accuracy in the framework of post-Newtonian (PN) theory as long as they orbit each other with sufficient separation (see [56] for a review). In the late stages after the merger of the binary, in contrast, the system closely resembles a single Kerr hole and is described well by perturbation theory, i.e., the linearization of Einstein's equations around a Kerr background. We will return to both of these approximation theories below. The second approach to studying binary systems is to use numerical methods to solve the full Einstein equations. The research field concerned with this approach is called numerical relativity and is the main subject of this report.

It is a remarkable coincidence that major breakthroughs in numerical relativity have been achieved at almost exactly the same time that the above-mentioned ground-based laser interferometers have advanced to the stage that they are capable of performing observation runs at or close to the design sensitivity. These are therefore very exciting times for black-hole and gravitational wave physics and the community is about to open an entirely new window to the universe with unprecedented opportunities to gain fundamentally new insight into the structure and evolution of the universe.

This article is structured as follows. In Sect. 4.2 we summarize the 3+1 decomposition of spacetime. A list of ingredients for a numerical simulation is given in Sect. 4.3. Methods to extract physical information from a simulation are discussed in Sect. 4.4. In Sects. 4.5 and 4.6, we present a brief overview of the history of black-hole simulations and summarize results obtained in the last few years following the breakthrough in binary simulations in 2005.

Notation: We use geometric units, that is we set the gravitational constant G and the speed of light c to unity. We use Einstein summation and let Latin indices run from 1 to 3 and Greek indices from 0 to 3. In sign convention we follow [140] and use the convention of Misner, Thorne, and Wheeler (MTW) [205].

4.2 The 3+1 Decomposition of General Relativity

The numerical solution of the Einstein equations faces a multitude of conceptual difficulties commonly not present in other areas of computational physics. We will discuss various of these problems further below, but we cannot even get started without addressing the most fundamental problem; the Einstein equations are expressed in terms of geometrical objects, so-called tensors. Computers, in contrast, exclusively operate on numbers. Furthermore, general relativity is based on the unification of space and time in the 4D spacetime, whereas physical systems are commonly described in terms of a time evolution of the state of the system.

The most common approach to tackle these problems in numerical relativity is the so-called 3+1 decomposition based on the canonical work of Arnowitt, Deser, and Misner (ADM) [25] and later formulated by York [277, 279]. The key idea here is to decompose the 4D spacetime into a one-parameter family of 3D spatial slices. Each of these slices describes a snapshot of the system under consideration and the Einstein equations tell us how the system evolves from one snapshot to the next. Each slice is described in terms of the components of two fundamental tensors or "forms", the 3D metric and the extrinsic curvature. These components can be represented in the computer as arrays of numbers. In order to assign unique meaning to these fields of numbers, however, we need a basis expansion of the tensors. Commonly this basis is a coordinate basis, so that there remains the task of determining or evolving the coordinates. The physics of the system can only be interpreted as a combination of the tensor components (their numerical values) in combination with the meaning of the coordinates. While the choice of coordinates is in principle arbitrary because of the invariance of general relativity under coordinate or gauge transformations, the actual choice of coordinates turns out to be crucial for obtaining a stable numerical scheme. We will discuss this issue in more detail below.

Most of this article will focus on numerical work based on the 3+1 decomposition. We emphasize, however, that alternative approaches have been investigated. The most important alternative is based on the characteristics of the Einstein equations, the light or null cones. These characteristic or null foliations of spacetime have been pioneered in the seminal work of Bondi and Sachs [66, 236] and lead to a remarkably simple hierarchy of the Einstein equations. The main difficulty of the characteristic approach is the breakdown of the characteristic coordinate systems in regions of strong curvature due to the formation of caustics. The characteristic approach is still subject of considerable research and has also inspired the combined use with 3+1 or Cauchy formulations of general relativity in the form of Cauchy-characteristic matching. For further details the reader is referred to Winicour's review article [272] and references therein.

An alternative combination of the benefits of the 3+1 and characteristic decomposition can be obtained using the conformal field equations based on the studies by Friedrich [129]. Here one evolves hyperboloidal surfaces which are spatial everywhere but asymptote toward null infinity. For more information on this direction we recommend Frauendiener's review article [128] as well as references therein.

We conclude this introduction by pointing to further review articles on numerical relativity and black-hole simulations. It is beyond the scope of this work to give a comprehensive account of all the foundations of numerical relativity and we will primarily adopt a more practical point of view here. Further details and mathematical rigor can be found in York's articles. We also recommend Gourgoulhon's excellent review for a detailed introduction to the 3+1 decomposition in lecture-style format [140]. A review focusing more on the mathematical aspects of numerical relativity and the relationship between numerical and mathematical relativity is given in [167]. Reference [45] presents more details on the numerical techniques used for modeling compact binaries and also discusses neutron star simulations in more depth. The most recent review article by Pretorius [231] summarizes black-hole simulations performed in the last 2 years with particular emphasis on the final stages of the coalescence.

4.2.1 The Einstein Equations

The fundamental quantity which we need to determine is the 4D spacetime metric $g_{\alpha\beta}$. As the metric is a symmetric tensor, this corresponds to ten independent components. Once we know the metric, we can calculate the Christoffel connection

$$\Gamma^{\alpha}_{\beta\gamma} = \frac{1}{2} g^{\mu\alpha} \left(\partial_{\beta} g_{\gamma\mu} + \partial_{\gamma} g_{\mu\beta} - \partial_{\mu} g_{\beta\gamma} \right), \tag{4.1}$$

where $g^{\alpha\beta}$ is the inverse of the metric. From the connection we obtain the covariant derivative and the Riemann tensor

$$R^{\alpha}{}_{\beta\gamma\delta} = \partial_{\gamma} \Gamma^{\alpha}_{\beta\delta} - \partial_{\delta} \Gamma^{\alpha}_{\beta\gamma} + \Gamma^{\alpha}_{\mu\gamma} \Gamma^{\mu}_{\beta\delta} - \Gamma^{\alpha}_{\mu\delta} \Gamma^{\mu}_{\beta\gamma}. \tag{4.2}$$

We thus have all the information to compute geodesics in this spacetime, geodesic deviation and, as we will see below, the total mass and the gravitational radiation generated in the spacetime.

In order to determine the metric, we need to solve the Einstein equations

$$G_{\alpha\beta} \equiv R_{\alpha\beta} - \frac{1}{2} R g_{\alpha\beta} = T_{\alpha\beta}, \tag{4.3}$$

where the Ricci tensor and scalar are defined as contractions of the Riemann tensor: $R_{\beta\delta} = R^{\mu}{}_{\beta\mu\delta}$ and $R = R^{\mu}{}_{\mu}$. The matter energy tensor $T_{\alpha\beta}$ describes the matter distribution of the spacetime.

Finding solutions to the Einstein equations is actually a simple task. Just take any metric, compute the Riemann tensor according to Eq. (4.2), calculate the Ricci tensor and scalar and finally the matter tensor $T_{\alpha\beta}$ from Eq. (4.3). This provides a solution to the Einstein equations with the matter sources $T_{\alpha\beta}$. The problem with this approach is that matter tensors calculated in this way will in general not correspond to any physically meaningful or realistic matter distribution. The difficult part is therefore not finding solutions to the Einstein equations, but rather finding physically meaningful solutions. This is also the remarkable feature of the Schwarzschild and Kerr solutions. They are believed to closely resemble real, existing, physical objects. We therefore need to first prescribe the energy matter tensor $T_{\alpha\beta}$ and then determine the metric from the system of partial differential equations (4.3).

Black holes are vacuum solutions to the Einstein equations and as such obey the vacuum Einstein equations which can be written as

$$R_{\alpha\beta} = 0. \tag{4.4}$$

In the remainder of this work we will exclusively study systems with vanishing energy matter tensor $T_{\alpha\beta} = 0$. We do not abandon the energy matter tensor, however, without emphasizing that the simulation of compact binaries involving neutron stars has been subject to comparable numerical efforts as has been the study of black-hole binaries. Indeed, the first orbital simulations of compact binaries to have been achieved in numerical relativity were neutron star binary systems [195, 201, 247]. For more details and recent developments of neutron star, mixed black-hole neutron star as well as boson-star binaries, the reader is referred to these papers as well as [14, 120, 217, 248] and references therein.

As a starting point for the 3+1 decomposition we consider a 4D manifold $\mathcal{M}$ with coordinates x^{α} and a metric of signature $-+++$. We next require a foliation. That is, we assume that there exists a function $t(x^{\alpha})$ of the spacetime coordinates x^{α} with non-vanishing gradient everywhere. Without loss of generality we assume that the gradient satisfies $g_{\mu\nu}\nabla^{\mu}t\nabla^{\nu}t < 0$. In consequence, the slices $t = \text{const}$ are spacelike in the sense that the norm of any vector tangent to the slices is positive, i.e., has the opposite sign of the norm of ∇t. The foliation is graphically illustrated in Fig. 4.1 where we show two hypersurfaces corresponding to $t = 0$ and $t = dt$. We next consider vectors v tangent to a hypersurface Σ_t with fixed t. By definition these vectors have vanishing inner product with the gradient of t: $v^{\mu}\nabla_{\mu}t = 0$. The timelike normal field of the hypersurfaces is therefore given by

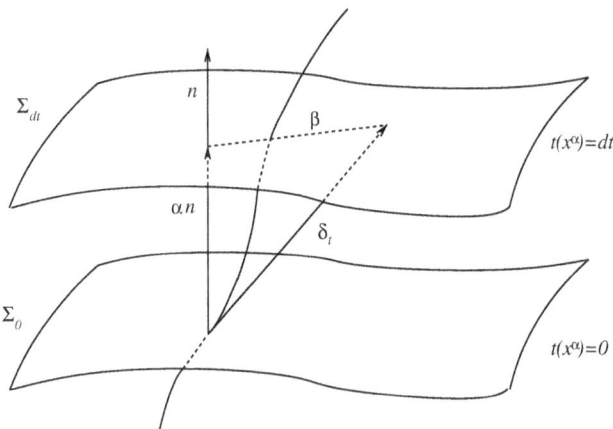

Fig. 4.1 Illustration of a time-like foliation of spacetime. One spatial dimension is suppressed for presentation purposes

$$n_\alpha = \frac{\nabla_\alpha t}{\sqrt{-\nabla^\mu t \nabla_\mu t}}, \tag{4.5}$$

and its dual vector field is $n^\alpha = g^{\alpha\mu} n_\mu$.

It turns out to be convenient to use coordinates adapted to the 3+1 foliation. These are given by t and three further coordinates labeling points inside each hypersurface Σ_t. The spatial coordinates x^i define a three-parameter family of curves $x^i = \text{const}$ which thread the foliation, that is, any such curve intersects each hypersurface Σ_t exactly once. In Fig. 4.1 we have illustrated such a curve together with its tangent vector ∂_t. We emphasize that ∂_t is in general not orthogonal to the hypersurfaces Σ_t.

We have now split the coordinate freedom into two different parts. First, we can choose the foliation via the function t, second we have the freedom to label the points inside any hypersurface by choosing the spatial coordinates x^i. In the majority of formulations of the Einstein equations, this freedom is encapsulated in the following two functions. First, the lapse function is defined as

$$\alpha = \sqrt{-\nabla^\mu t \nabla_\mu t}. \tag{4.6}$$

Loosely speaking, it represents a measure for the separation in proper time between two neighboring hypersurfaces Σ_t and Σ_{t+dt}. Translated into a more numerical language, the lapse function enables us to control the advance in proper time corresponding to an advance in coordinate time dt. Often, one wants to slowdown the advance in proper time in regions where the code encounters a singularity by locally decreasing the lapse toward zero.

The second gauge function is the shift vector defined by

$$\beta^i = (\partial_t)^i - \alpha n^i, \tag{4.7}$$

as illustrated in Fig. 4.1. The shift vector determines how points with identical spatial labels x^i are identified on neighboring slices.

Given the decomposition of spacetime into a time-like foliation of space-like slices, it will be helpful to apply a similar decomposition to the geometric objects. For this purpose we define the projection operator

$$\perp^\mu{}_\alpha = \delta^\mu{}_\alpha + n^\mu n_\alpha. \tag{4.8}$$

For any given tensor this enables us to define its spatial projection. For example, for a tensor $T^\alpha{}_\beta$ we have

$$\perp T^\alpha{}_\beta = \perp^\alpha{}_\mu \perp^\nu{}_\beta T^\mu{}_\nu, \tag{4.9}$$

and likewise for tensors with different arrangements of indices. Projections onto the time direction are directly obtained from contraction with the unit normal field n^α. For our example we obtain the time projection $T^\mu{}_\nu n_\mu n^\nu$. We can also define mixed projections, as for example $\perp^\alpha{}_\mu T^\mu{}_\nu n^\nu$.

In particular, we can apply the projection operator to the metric itself and obtain

$$\gamma_{\alpha\beta} \equiv \perp g_{\alpha\beta} = g_{\alpha\beta} + n_{\alpha\beta} = \perp_{\alpha\beta}. \tag{4.10}$$

This projection of the metric defines an induced 3D metric on the hypersurface in the sense that its effect on all geometric objects tangent to the hypersurface is the same as if the spacetime metric were acting on them.

We now recall the definitions of the connection and the Riemann tensor in Eqs. (4.1) and (4.2). These definitions are valid for an arbitrary dimension and thus also apply to the induced metric. We merely replace Greek with Latin indices in these definitions and obtain the 3D Christoffel connection and Riemann tensor. From the connection we derive the 3D covariant derivative D_a. For example, for a 3D tensor with one upper and one lower index, the covariant derivative is

$$D_a T^b{}_c = \partial_a T^b{}_c + \Gamma^b_{ia} T^i{}_c - \Gamma^i_{ca} T^b{}_i. \tag{4.11}$$

If we use coordinates adapted to the 3+1 decomposition, this can be shown to be identical to the spatial components of

$$D_\alpha T^\beta{}_\gamma = \perp^\alpha{}_\mu \perp^\beta{}_\nu \perp^\rho{}_\gamma \nabla_\mu T^\nu{}_\rho. \tag{4.12}$$

In summary, we can apply the entire machinery of differential geometry to the induced three-metric $\gamma_{\alpha\beta}$ just as we applied it to the four-metric $g_{\alpha\beta}$. We still need to work out, however, how these 3D objects are related to their 4D counterparts.

Before we address this question, though, we need to introduce the extrinsic curvature which is defined as

$$K_{\alpha\beta} = -\perp^\mu{}_\alpha \perp^\nu{}_\beta \nabla_\nu n_\mu. \tag{4.13}$$

As illustrated in Figs. 2.2, 2.3, and 2.4 of [140], the extrinsic curvature can be interpreted as the variation of the time-like unit normal field on the hypersurface. We emphasize that $K_{\alpha\beta}$ is by definition a purely spatial quantity. A straightforward calculation leads to the important equivalent relation

$$K_{\alpha\beta} = -\frac{1}{2} \mathscr{L}_n \gamma_{\alpha\beta}, \tag{4.14}$$

where $\mathscr{L}_n$ is the Lie derivative along the unit normal field n^α.

We next address the question of how the 4D Riemann tensor is related to the 3D quantities. This is best done by considering the projections of the 4D Riemann tensor. The calculations are lengthy but straightforward and the interested reader is referred to [140]. Here we merely list the resulting relations

$$\perp^\mu{}_\alpha \perp^\nu{}_\beta \perp^\gamma{}_\rho \perp^\sigma{}_\delta R^\rho{}_{\sigma\mu\nu} = \mathscr{R}^\gamma{}_{\delta\alpha\beta} + K^\gamma{}_\alpha K_{\delta\beta} - K^\gamma{}_\beta K_{\alpha\delta}, \tag{4.15}$$

$$\perp^\mu{}_\alpha \perp^\nu{}_\beta \perp^\gamma{}_\rho n^\sigma R^\gamma{}_{\sigma\alpha\beta} = D_\beta K^\gamma{}_\alpha - D_\alpha K^\gamma{}_\beta, \tag{4.16}$$

$$\perp_{\rho\alpha} \perp^\mu{}_\beta n^\sigma n^\nu R^\rho{}_{\sigma\mu\nu} = \mathscr{L}_n K_{\alpha\beta} + \frac{1}{\alpha} D_\alpha D_\beta \alpha + K_{\alpha\mu} K^\mu{}_\beta, \tag{4.17}$$

where we use the symbol $\mathscr{R}$ to distinguish the 3D Riemann tensor from its 4D counterpart R. These equations are often referred to as the Gauss–Codacci or Gauss–Codacci–Mainardi equations. We note that all further projections vanish due to the symmetry of the Riemann tensor. Contracted versions of these equations are straightforwardly obtained by multiplication with the metric $g^{\alpha\beta}$.

If we look at the right-hand sides of these relations, all terms except for the Lie derivative of $K_{\alpha\beta}$ are purely spatial expressions. In adapted coordinates (t, x^i) we are therefore allowed to replace Greek by Latin indices which run from 1 to 3 only. The Lie derivative of the extrinsic curvature, on the other hand, can be rewritten as

$$\mathscr{L}_n K_{\alpha\beta} = \mathscr{L}_{\frac{1}{\alpha}(\partial_t - \beta)} K_{\alpha\beta} = \frac{1}{\alpha}\left(\mathscr{L}_{\partial_t} - \mathscr{L}_\beta\right) K_{\alpha\beta} = \frac{1}{\alpha}\left(\partial_t K_{\alpha\beta} - \mathscr{L}_\beta\right) K_{\alpha\beta}, \quad (4.18)$$

where the Lie derivative of the extrinsic curvature along the shift vector β is again a purely spatial quantity.

Finally, we are in the position to decompose the Einstein equations $R_{\alpha\beta} = 0$ themselves. Again we refer the reader for details of the calculations to [140] and summarize the results. As with the Riemann tensor, there are three projections. First, we can project both indices onto the time direction. Inserting the above projections of the Riemann tensor into $R_{\mu\nu}n^\mu n^\nu$ leads to

$$\mathscr{R} + K^2 - K_{mn}K^{mn} = 0, \quad (4.19)$$

where $K = \gamma^{mn}K_{mn}$ is the trace of the extrinsic curvature. This equation is known as the Hamiltonian constraint. It does not contain any time derivatives but instead is a relation which must be obeyed by the three-metric γ_{ij} and the extrinsic curvature K_{ij} on each hypersurface. Similarly, we obtain the momentum constraint from the mixed projection $\perp R_{\alpha\mu}n^\mu$

$$D^i K - D_m K^{im} = 0. \quad (4.20)$$

All the information about the time evolution is contained in the spatial projection $\perp R_{\alpha\beta} = 0$ which leads to

$$(\partial_t - \mathscr{L}_\beta)K_{ij} = -D_i D_j \alpha + \alpha\left(\mathscr{R}_{ij} - 2K_{im}K^m{}_j + K_{ij}K\right). \quad (4.21)$$

Together with Eq. (4.14), this equation forms a second order in time evolution system for the induced metric γ_{ij}. This system together with the Hamiltonian and momentum constraints are often referred to as the "ADM" equations. This term is not strictly correct because Arnowitt, Deser, and Misner used the canonical momenta in place of the extrinsic curvature in their original work [25]. We will follow common notation here, however, and will talk of the ADM equations in the remainder of this work.

It is this set of equations which is at the heart of the majority of work in numerical relativity. It is highly instructive to discuss these equations in more detail. First, we note that the equations do not provide any information on the gauge functions α and β^i. This is expected as these functions incorporate the coordinate freedom of general relativity and therefore can be specified arbitrarily. Second, we count the degrees of freedom. We have a second-order system in time for the six independent components of the symmetric three-metric γ_{ij}. Four of these are determined by the constraints, so that there remain two dynamic degrees of freedom, the two degrees of freedom of gravitation. Finally, the Bianchi identities

$$\nabla_{[\nu}R_{\alpha\beta]\lambda\mu} = 0, \qquad \nabla_\nu G^{\mu\nu} = 0 \tag{4.22}$$

can be shown to propagate the constraint equations through the evolution. That is, if the constraints are satisfied on some initial hypersurface and the evolution equations hold, then the constraints are automatically satisfied on all other hypersurfaces. This greatly simplifies the task of numerically evolving data; it is sufficient to enforce the constraints on the initial data and evolve these using Eqs. (4.14) and (4.21).

In summary, we have reformulated the Einstein equations as an initial value problem. Given an initial snapshot of the three-metric γ_{ij} and the extrinsic curvature K_{ij}, we merely need to specify gauge functions α and β^i and subsequently can evolve the data and reconstruct the entire spacetime. It is this conceptual simplicity of numerical relativity that has inspired the community with a great deal of optimism following the early work in the 1970s. In the next section, we will discuss the difficulties which have prevented the community from successfully implementing the above recipe for several decades and also the solutions which finally have resulted in the breakthroughs of 2005.

4.3 The Ingredients of Numerical Relativity

4.3.1 The Formulation of the Einstein Equations

We have discussed in detail how the ADM equations provide a conceptually simple recipe for evolving a given set of initial data using numerical methods in general relativity. Unfortunately, all attempts of implementing these equations have resulted in numerical instabilities after timescales much shorter than the dynamical timescale of the systems under consideration. Because of their enormous complexity, the evolution equations defy all attempts of applying standard stability analysis. Most likely, the instabilities observed in numerical relativity for such a long time are a consequence of various causes. It is now commonly believed, however, that the structure of the ADM equations makes them an unlikely candidate for providing long-term stable numerical evolutions.

The key difficulty here is that the Einstein equations are a constraint system. We have seen above, how the Einstein equations can be decomposed into evolution equations and constraints. This decomposition is not unique, however. For example, we can add any combination of the constraints to the right-hand side of the evolution equations and thus obtain a different system. All such decompositions describe the same physics and will have identical physical (constraint satisfying) solutions. But the evolution equations also admit non-physical (constraint violating) solutions and this unphysical solution space depends on the decomposition. In particular, some decompositions will allow for unphysical solutions which rapidly grow beyond control. We need to bear in mind in this context that any numerical solution will inevitably satisfy the constraints only within some accuracy, so that such rapidly growing solutions, if present, are likely to be excited by numerical noise. It is desirable, for this purpose, to have a smooth dependence of the spacetime solution on the initial data. This quality is encapsulated in the well-posedness of the system of equations. While a well-posed system does not guarantee stable numerical evolutions, it is generally accepted that a well-posed evolution system is a much more likely candidate for successful numerical simulations.

The common approach to obtain well-posedness and thus some bounds on the deviation in the time evolution of neighboring initial data sets is based on using strongly or symmetric hyperbolic systems (see, e.g., [145] and references therein for definitions). Indeed, it was shown in [172] that a first-order reduction of the ADM equations is weakly hyperbolic and that the standard finite differencing applied to weakly hyperbolic systems results in ill-posed systems [90].

As a result of the continued problems encountered in evolutions using the ADM equations, a wealth of alternative formulations of the Einstein equations has been suggested in the literature [13, 43, 59, 62, 130, 134, 172, 207, 239, 246]. To date, however, only two of these have been demonstrated to facilitate long-term stable evolutions of black-hole binary spacetimes. These are the Baumgarte–Shapiro–Shibata–Nakamura (BSSN) system [43, 246] and the generalized harmonic gauge (GHG) formulation [134, 227]. We will now discuss these two systems in some more detail.

4.3.1.1 The BSSN System

The BSSN system results from the ADM equations by applying the following modifications. First, the extrinsic curvature is split into its trace and a tracefree part. Second, a conformal transformation is applied to the three-metric and the extrinsic curvature. Finally, a contracted version of the Christoffel symbols of the conformal metric is introduced as an additional variable. The BSSN variables are then given by

$$\phi = \frac{1}{12} \ln(\det \gamma_{ij}), \qquad \tilde{\gamma}_{ij} = e^{-4\phi} \gamma_{ij},$$

$$K = \gamma_{mn} K^{mn}, \qquad \tilde{A}_{ij} = e^{-4\phi} \left(K_{ij} - \frac{1}{3} \gamma_{ij} K \right),$$

$$\tilde{\Gamma}^i = \tilde{\gamma}^{mn} \Gamma^i_{mn} = -\partial_m \tilde{\gamma}^{im}. \tag{4.23}$$

This corresponds to a rearrangement of the degrees of freedom which is similar to the York–Lichnerowicz split underlying most of the initial data calculation which we will discuss below in Sect. 4.3.3. Expressing the ADM equations in terms of these variables leads to the BSSN system

$$\partial_t \tilde{\gamma}_{ij} = \beta^m \partial_m \tilde{\gamma}_{ij} + 2\tilde{\gamma}_{m(i}\partial_{j)}\beta^m - \frac{2}{3}\tilde{\gamma}_{ij}\partial_m\beta^m - 2\alpha\tilde{A}_{ij}, \tag{4.24}$$

$$\partial_t \phi = \beta^m \partial_m \phi + \frac{1}{6}(\partial_m\beta^m - \alpha K), \tag{4.25}$$

$$\partial_t \tilde{A}_{ij} = \beta^m \partial_m \tilde{A}_{ij} + 2\tilde{A}_{m(i}\partial_{j)}\beta^m - \frac{2}{3}\tilde{A}_{ij}\partial_m\beta^m + e^{-4\phi}\left(\alpha\mathscr{R}_{ij} - D_iD_j\alpha\right)^{\text{TF}}$$
$$+ \alpha\left(K\tilde{A}_{ij} - 2\tilde{A}_i{}^m\tilde{A}_{mj}\right), \tag{4.26}$$

$$\partial_t K = \beta^m \partial_m K - D^m D_m \alpha + \alpha\left(\tilde{A}^{mn}\tilde{A}_{mn} + \frac{1}{3}K^2\right), \tag{4.27}$$

$$\partial_t \tilde{\Gamma}^i = \beta^m \partial_m \tilde{\Gamma}^i - \tilde{\Gamma}^m \partial_m \beta^i + \frac{2}{3}\tilde{\Gamma}^i\partial_m\beta^m + 2\alpha\tilde{\Gamma}^i_{mn}\tilde{A}^{mn}$$
$$+ \frac{1}{3}\tilde{\gamma}^{im}\partial_m\partial_n\beta^n + \tilde{\gamma}^{mn}\partial_m\partial_n\beta^i$$
$$- \frac{4}{3}\alpha\tilde{\gamma}^{im}\partial_m K + 2\tilde{A}^{im}\left(6\alpha\partial_m\phi - \partial_m\alpha\right)$$
$$- \left(\sigma + \frac{2}{3}\right)\left(\tilde{\Gamma}^i - \tilde{\gamma}^{mn}\tilde{\Gamma}^i_{mn}\right)\partial_k\beta^k, \tag{4.28}$$

where the superscript ^{TF} means that we take the tracefree part of the preceding expression. The last term on the right-hand side of Eq. (4.28) vanishes in the continuum limit by virtue of the definition of $\tilde{\Gamma}^i$ in Eq. (4.23). It has been shown in [273], however, to cure instability problems observed in simulations which do not employ octant symmetry [7]. In practice, setting the free parameter $\sigma = 0$ proves satisfactory. Alternatively to using this term, Alcubierre et al. [9] achieve stable evolutions by recalculating $\tilde{\Gamma}^i$ from the metric $\tilde{\gamma}_{ij}$ whenever it appears on the right-hand side of Eqs. (4.24), (4.25), (4.26), (4.27), and (4.28) in undifferentiated form. So far, all successful implementations of the BSSN equations also require us to enforce the vanishing of the trace of $\tilde{A}_{ij}$. This is realized numerically by replacing $\tilde{A}_{ij}$ with $\tilde{A}_{ij} - \tilde{\gamma}_{ij}\tilde{\gamma}^{mn}\tilde{A}_{mn}$ after each timestep. Some codes also enforce in a similar way the constraint $\det\tilde{\gamma}_{ij} = 1$.

A further modification of the BSSN system has been introduced in [94] who evolve the conformal factor in terms of the variable $\chi = e^{4\phi}$. Using this "χ-version" of the BSSN system has in some instances been found to result in better convergence properties [81].

The hyperbolicity of the BSSN system was studied in [238] and provided first insight into how well-posedness of the BSSN system is actually achieved. The sensitivity of the hyperbolicity properties of the system under minor changes in the equations may also explain why certain modifications, such as the enforcement of $\text{tr}\tilde{A}_{ij} = 0$, appear to be necessary to obtain long-term stability. Notwithstanding the various open questions underlying the stability properties of the different formulations of the Einstein equations, the BSSN system has become the most popular choice in practice for writing the Einstein equations in simulations of black-hole and/or neutron star binaries.

4.3.1.2 The Generalized Harmonic Formulation

In contrast to the BSSN system, the generalized harmonic formulation is not derived from the ADM equations. Instead, it is based on the 4D version of the Einstein equations in harmonic gauge. The harmonic gauge condition is

$$\Box x^{\alpha} \equiv \nabla^{\mu}\nabla_{\mu}x^{\alpha} = 0 \tag{4.29}$$

and casts the Einstein equations in a particularly convenient form. Specifically, the Ricci tensor can be written as

$$R_{\alpha\beta} = -\frac{1}{2}g^{\mu\nu}\partial_{\mu}\partial_{\nu}g_{\alpha\beta} + \dots, \tag{4.30}$$

where the dots denote further terms containing the metric and its first derivatives, but no second derivatives. The principal part of the Einstein equations $R_{\alpha\beta} = 0$ is therefore identical to that of the wave equation which has made this gauge very popular in analytic studies of the Einstein equations (see e.g., [83]).

Even though this structure is also very appealing from a numerical point of view, it has not been used successfully in black-hole simulations. It has been shown in [134] how one can generalize this system to accommodate arbitrary gauge choices while still preserving the wave-like character of the principal part. This is realized by introducing the source functions

$$H_{\alpha} = \Box x_{\alpha}, \tag{4.31}$$

which vanish for the special choice of harmonic gauge. With these functions, the Einstein equations in vacuum can be written as

$$R_{\alpha\beta} = -\frac{1}{2}g^{\mu\nu}\partial_{\mu}\partial_{\nu}g_{\alpha\beta} + \dots - \frac{1}{2}\left(\partial_{\alpha}H_{\beta} + \partial_{\beta}H_{\alpha}\right), \tag{4.32}$$

where again the dots denote terms only involving the metric and its first derivative. The introduction of the auxiliary gauge functions H_{α} thus preserves the wave-like principal part of the Einstein equations for arbitrary gauge choices.

As yet, no simple geometric interpretation of the H_{α} analogous to that of lapse α and shift β^{i} has been found, but the two sets of gauge functions are connected via the differential relations [226]

$$H_{\mu}n^{\mu} = -K - \frac{1}{\alpha^{2}}\left(\partial_{t}\alpha - \beta^{i}\partial_{i}\alpha\right), \tag{4.33}$$

$$\perp^{i}{}_{\mu}H^{\mu} = \frac{1}{\alpha}\gamma^{ik}\partial_{k}\alpha + \frac{1}{\alpha^{2}}\left(\partial_{t}\beta^{i} - \beta^{k}\partial_{k}\beta^{i}\right) - \gamma^{mn}\Gamma^{i}_{mn}. \tag{4.34}$$

Just as lapse and shift need to be specified in addition to the evolution of the BSSN equations, the functions H_{α} need to be specified by the user in the GHG system. In analogy to the introduction of the variables $\tilde{\Gamma}^{i}$ in the BSSN system in Eq. (4.23), the definition (4.31) of H_{α} takes on the role of an auxiliary constraint

$$C_\alpha = H_\alpha - \Gamma^\mu_{\mu\alpha} + g^{\mu\nu}\partial_\mu g_{\nu\alpha}. \qquad (4.35)$$

While this constraint is propagated in the continuum limit by the evolution equations in the same way as the Hamiltonian and momentum constraints, this can become problematic in numerical simulations, where constraints will always be violated due to numerical inaccuracies. If these constraint violations grow without control, they may give rise to numerical instabilities.

We have already seen, how the addition of the constraint to the right-hand sides of the evolution equations can cure numerical instabilities in the case of the BSSN equation (4.28) for the variable $\tilde{\Gamma}^i$. A similar cure using the constraint C_α as suggested by Gundlach et al. [144] turned out to be an important ingredient in Pretorius' first simulation of a black-hole binary through inspiral and merger [227]. These cases represent good examples of the intricacies involved in numerically evolving the Einstein equations.

4.3.2 Gauge Conditions

We have seen in the previous sections that the Einstein equations do not predict the evolution of the gauge variables α and β^i in the BSSN system or H_α in the generalized harmonic formulation. Instead, these functions are specified by the user and represent the coordinate freedom of general relativity. Indeed, any choice for these functions is guaranteed not to affect the physical properties of the system under investigation. If the choice of gauge has no impact on the physics of the system, one may wonder why it is necessary to discuss gauge conditions at all. The reason is that the choice of gauge does have a strong impact on the performance and stability of a numerical code. A simple example to illustrate this problem arises in evolutions of a single Schwarzschild black hole [253]: A simulation starting on a time-symmetric hypersurface using geodesic slicing, i.e., $\alpha = 1$ everywhere, in combination with vanishing shift will hit the singularity after a short coordinate time of $\Delta t = \pi M$, where M is the mass of the Schwarzschild hole. Because of the divergent nature of the metric components at the singularity, a computer is not capable of representing the singularity using numbers and instead produces "non-assigned-numbers" at some grid-points. These quickly swamp the entire computational grid and render the entire simulation useless.

A common strategy to avoid this problem is to reduce the lapse function α as the hypersurfaces get closer to a singularity [9, 182, 253]. The corresponding slowdown in the advance of proper time "bends" the hypersurfaces around the singularity. Such singularity avoiding slicings are frequently used in numerical codes. A potential danger arising out of this procedure, however, is the so-called slice stretching (see, for example, Sect. V B in [18]). Whereas the advance of proper time at points x^i close to the black-hole singularity is slowed down, points further away from the black hole advance almost normally, i.e., with $\alpha \approx 1$. As the evolution proceeds, these differences accumulate and eventually neighboring points on the numerical grid represent spacetime events far away from each other. Unless the shift vector is carefully chosen to counteract this effect, this leads to resolution problems near the black hole and gives rise to numerical instabilities.

The detailed study of the impact of gauge conditions and the reasons why some conditions work so much better than others are still subject to ongoing research and there remain many questions, in particular in connection with the shift vector. To date, the choice of gauge conditions in numerical codes has been motivated by the avoidance of singularities and slice stretching, but fail-safe recipes for their derivation are currently not known. Instead, the selection of gauge conditions is based on a combination of educated guessing and empirical testing in black-hole simulations.

Gauge conditions used in early numerical simulations were inspired by geometrical ideas. The maximal slicing condition $K = 0$ derives its name from the fact that the 3D volume of spatial hypersurfaces obeying this condition is maximal [253]. This condition leads to an elliptic equation for the lapse function and is therefore computationally expensive and non-trivial to implement. Similarly, the minimization of the strain (cf. Eq. (4.5) of [253]) leads to an elliptic condition for the shift vector. To the authors knowledge, these gauge conditions have not yet been implemented in more recent simulations of black-hole mergers, so that it remains unclear, to what extent the instabilities encountered in early simulations are based on this choice of gauge. In any case, however, maximal slicing and the minimal distortion shift form the basis of many modified gauge conditions employed in the course of the following decades.

A remarkable simplification of the implementation of gauge conditions like maximal slicing is the idea of driver conditions [7, 9, 41]. Here, the elliptic equation is replaced by a parabolic or hyperbolic equation which drives the gauge ever closer to an equilibrium state similar to the equation of heat conduction. The key numerical advantage is that such evolution equations are substantially easier to implement than the solving of elliptic equations.

The idea of driver conditions was particularly appealing for stationary or quasi-stationary spacetimes and has commonly been used for puncture-type initial data (see Sect. 4.3.3 below). By using co-moving or co-rotating coordinates, most of the black-hole dynamics can be absorbed in the coordinates and the spacetime variables show little actual change in coordinate time [9, 11, 80, 116, 282]. The most prominent conditions are the "1+log" slicing

$$\partial_t \alpha = -2\alpha K \tag{4.36}$$

and a second order in time Γ-driver condition for the shift. Different groups use slightly different Γ-driver conditions. For example, the version reported in [11] is

$$\partial_t \beta^i = \frac{3}{4} \alpha^p \psi_{\mathrm{BL}}^{-n} B^i, \tag{4.37}$$

$$\partial_t B^i = \partial_t \tilde{\Gamma}^i - \eta \beta^i, \tag{4.38}$$

with parameter choices $p = 1$ or 2, $n = 2$ or 4, and $\eta \in [2,5]$. In these simulations, the conformal factor is split into an analytic part ψ_{BL} of the Brill–Lindquist solution and a regular remainder (see, for example, Sect. IV C of [9]).

Gauge conditions were studied in a more general way by Bona and Massó [60] (for further discussion see also [8] and references therein). The above-mentioned harmonic gauge as well as the driver conditions are special cases of the Bona-Massó family of gauge conditions [60]. This general class of gauge conditions has been used in various analytic studies to investigate singularity avoidance and the formation of gauge shocks [10, 12, 232].

While these ingredients still form a major part of the current generation of numerical codes, the simulation of black-hole binaries through merger has so far only been successfully accomplished after abandoning the idea of co-rotating coordinates and instead allowing the black holes to move throughout the computational domain. The first inspiral and merger was obtained by Pretorius [227] who used the generalization of the harmonic formulation described in the previous section. Specifically, he constructed his gauge by evolving the gauge functions according to

$$\Box H_t = -\xi_1 \frac{\alpha - 1}{\alpha^\eta} + \xi_2 n^\nu \partial_\nu H_t, \tag{4.39}$$

$$H_i = 0. \tag{4.40}$$

A few months after Pretorius' breakthrough, the relativity groups of the University of Brownsville and NASA Goddard independently discovered an evolution method now commonly referred to as the moving-puncture approach [34, 94]. In contrast to previous puncture simulations, the conformal factor is not decomposed into an analytically known part plus a regular piece and is instead evolved as a single quantity. In combination with modifications of the "1+log" slicing and the Γ-driver condition which allow the black holes to move across the computational domain they thus obtained a remarkably straightforward technique for evolving black-hole binaries. Several groups have now developed codes using this moving-puncture method. All codes use the modified "1+log" slicing condition

$$\partial_t \alpha = \beta^i \partial_i \alpha - 2\alpha K, \tag{4.41}$$

but they differ in the modifications applied to the $\tilde{\Gamma}$-driver condition for the shift vector. A sample of the exact gauge conditions reported by various groups is given as follows

Code	Reference		
UTB	[94]	$\partial_t \beta^i = B^i,$	$\partial_t B^i = \frac{3}{4}\partial_t \tilde{\Gamma}^i - \eta B^i,$
Goddard	[34]	$\partial_t \beta^i = \frac{3}{4}\alpha B^i,$	$\partial_t B^i = \partial_0 \tilde{\Gamma}^i - \eta B^i,$
PSU	[156]	$\partial_t \beta^i = \frac{3}{4}\alpha B^i,$	$\partial_t B^i = \partial_0 \tilde{\Gamma}^i - \eta B^i,$
LEAN	[257]	$\partial_t \beta^i = B^i,$	$\partial_t B^i = \partial_t \tilde{\Gamma}^i - \eta B^i,$
BAM	[81]	$\partial_0 \beta^i = \frac{3}{4}B^i,$	$\partial_0 B^i = \partial_0 \tilde{\Gamma}^i - \eta B^i,$
AEI	[34]	$\partial_t \beta^i = \frac{3}{4}\alpha B^i,$	$\partial_t B^i = \partial_0 \tilde{\Gamma}^i - \eta B^i,$

where $\partial_0 = \partial_t - \beta^i \partial_i$. The free parameter η has an influence on the eventual co-ordinate radius of the black holes [81] and typical choices for this parameter are in the range $0.5 \leq \eta \leq 2$ with no significant impact on the quality of the simu-lations except for instabilities arising at outer refinement boundaries observed in some cases for large values of η (cf. [258]). A more detailed analysis of different gauge conditions in moving-puncture evolutions of black-hole binary spacetimes is given in [267], but the general picture appears to be that all of the above conditions provide binary simulations of comparable quality.

4.3.3 Initial Data

We have so far discussed the differential equations determining the time evolution of the spacetime hypersurfaces. In order to start an evolution, however, we first need to construct an initial data set. This task confronts us with two problems. First, the initial data need to satisfy the Hamiltonian and momentum constraints (4.19), (4.20). The second problem is that the initial data set must represent a snapshot of an astrophysically realistic system. The construction of initial data is an entire branch of research in numerical relativity and we cannot cover all aspects of this work in this report. For a more comprehensive summary of the initial data calculation we refer the reader to Cook's review article [107].

Most of the work on solving the constraints is based on the York–Lichnerowicz split [182, 274–277], which rearranges the degrees of freedom via a conformal rescaling and the split of the extrinsic curvature into its trace and a tracefree part according to

$$\gamma_{ij} = \psi^4 \tilde{\gamma}_{ij}, \tag{4.42}$$

$$K_{ij} = A_{ij} + \frac{1}{3}\gamma_{ij}K. \tag{4.43}$$

It turns out to be convenient to further decompose the tracefree part of the ex-trinsic curvature into a longitudinal and a transverse part. Two approaches to this decomposition have been used. In the physical traceless decomposition [213–215], this procedure is applied directly to the traceless part of the extrinsic curvature A_{ij}, in the conformal traceless decomposition [277, 278], it is applied to a conformally rescaled version

$$A_{ij} = \psi^{-10}\tilde{A}_{ij} \qquad \text{or} \qquad A_{ij} = \psi^{-2}\tilde{A}_{ij}. \tag{4.44}$$

Both approaches eventually require us to specify the conformal metric $\tilde{\gamma}_{ij}$, the trace of the extrinsic curvature K and the symmetric transverse tracefree part of the extrinsic curvature. The four constraint equations are solved with these freely spec-ified functions and provide solutions for the conformal factor ψ and the potential of the longitudinal part of the extrinsic curvature. The detailed equations can be found in Sect. 2.2 of [107]. A particularly useful property of these splits is that the momentum constraint decouples from the Hamiltonian constraint if K is a constant. This simplification is frequently used in the practical calculation of initial data sets.

More recently, an alternative approach called the thin-sandwich decomposition [280] has become a very popular alternative to this approach. Loosely speaking, the key idea here is to replace the extrinsic curvature in terms of the time derivative of the metric using the evolution equation (4.14) for the metric. Eventually, one freely specifies the conformal metric $\tilde{\gamma}_{ij}$, its time derivative, the trace of the extrinsic curvature, and a conformally rescaled version of the lapse function. Solving the constraint provides us not only with the extrinsic curvature and the three-metric on the initial slice, but also with a lapse function and shift vector. The advantage of this approach is that we can directly impose a condition on the time derivative of the three-metric and obtain lapse and shift corresponding to this condition. This is particularly useful in the construction of quasi-equilibrium data, as, for example, a circularized binary in co-rotating coordinates, where the time derivative of the metric is assumed to vanish. A more detailed description of the thin-sandwich approach is presented in Sect. 2.3 of [107].

Having obtained the framework which facilitates an efficient solving of the constraint equations, there remains the second difficulty we mentioned at the beginning of this section. How do we obtain realistic black-hole initial data? There are two main approaches to this problem. First we discuss the generalization of analytically known single black-hole solutions.

As one might expect, the approaches discussed above provide relatively simple methods to derive the Schwarzschild solution. If, for example, we assume a time-symmetric initial data set, i.e., $K_{ij} = 0$, the momentum constraints can be shown to be trivially satisfied and the Hamiltonian constraint becomes

$$\bar{\nabla}^2 \psi = 0, \tag{4.45}$$

where $\bar{\nabla}$ is the flat space Laplace operator. The simplest solution to this equation is

$$\psi = 1 + \frac{M}{2r}, \tag{4.46}$$

which gives us the Schwarzschild solution in isotropic coordinates. This solution can be generalized straightforwardly to any number of black holes. Indeed, the linearity of the Hamiltonian constraint (4.45) immediately allows us to superpose solutions to obtain [75, 203]

$$\psi = \sum_{i=1}^{N} \frac{M_i}{|r - r_i|}. \tag{4.47}$$

These are known as Brill–Lindquist initial data and represent N holes at positions r_i. It can be shown that each of the poles in these solutions corresponds to spatial infinity in an asymptotically flat hypersurface, that is, each hole provides a connection to a different universe, so that we have in total $N + 1$ universes. A similar solution where all holes provide a connection between the same two asymptotically flat universes has been found by Misner [204].

Both, the Brill–Lindquist and the Misner data, represent N black holes at the moment of time symmetry, that is, black holes with vanishing linear and angular momentum. It is a remarkable property that analytic solutions for the momentum constraints can even be found in the generalized case of Misner data with non-vanishing momenta [69]. These data are commonly referred to as Bowen–York data and start again with the simplifying assumption of conformal and asymptotic flatness as well as maximal slicing $K = 0$. With the analytic solution of the momentum constraints, there merely remains the task of solving numerically the Hamiltonian constraint for the conformal factor. Even more remarkable, the total linear momenta $\mathbf{P}_i$ and spins $\mathbf{S}_i$ associated with the individual holes in the limit of isolated holes appear as explicit parameters in the analytic Bowen–York extrinsic curvature and thus provide us with a straightforward physical interpretation of the initial data. The total energy of the spacetime is also obtained relatively straightforwardly from the $1/r$ falloff term of the conformal factor as $r \to \infty$. The corresponding generalization to Brill–Lindquist data was developed by Brandt and Brügmann [73]. These data are known as puncture data and form the starting point for most of the so-called "moving puncture simulations" mentioned above.

In spite of the great popularity of these initial data, there are some concerns associated with the underlying simplifying assumptions. First, it has been shown that there are no spatial hypersurfaces of the Kerr spacetime with non-zero spin parameter for which the three-metric can be written in a conformally flat way [133]. It turns out that the initial data thus calculated represent the snapshot of a rotating black hole plus a non-vanishing gravitational wave content. We will return to this spurious gravitational radiation further below. At this point, we merely note that all binary-black-hole data successfully evolved to date contain such spurious initial radiation. In comparison with the merger waveform, however, this spurious or junk radiation is rather low in amplitude and appears to represent a smaller problem than anticipated, at least in the case of non-or slowly rotating black holes. Alternative non-conformally flat black hole initial data based on generalizations of the single hole Kerr–Schild solution [168, 169] have been investigated in initial data studies as well as numerical evolutions [67, 74, 193, 194, 196, 256, 257].

A popular alternative to puncture-type initial data is often referred to as "excision data". The idea here is to incorporate black holes in the form of horizon boundary conditions into the initial data. A black hole is defined by the presence of an event horizon, that is, a boundary which defines a region of spacetime from which null geodesics cannot extend all the way to null infinity. A more convenient framework encapsulating horizons in numerical relativity is that of apparent and isolated horizons ([28, 64, 117, 139] and references therein) which provides boundary conditions for the metric and extrinsic curvature components at the horizon. These conditions are particularly convenient to apply in combination with the quasi-equilibrium assumption and, thus, the thin-sandwich approach. Black-hole data have been constructed along these lines in [23, 108–110, 166, 223] and form the starting point for most of the simulations performed with the generalized harmonic formulation [39, 70, 86, 224].

We conclude this discussion with a counting of the physical parameters of a general black-hole binary. First, we need to fix the total scale of our problem which corresponds to fixing the total ADM mass of the system. Once we have fixed the scale, six parameters are required to determine the spins S_1 and S_2 of the two holes and one parameter for the mass ratio $q = M_1/M_2$. In general we also need to take into account the eccentricity of the orbits. The emission of gravitational waves has the effect of circularizing the orbit [221], however, so that for many purposes, it is sufficient to consider quasi-circular orbits.[1] In Most cases, we therefore have seven physical parameters (see, however, [158, 258] and references therein for investigations of eccentric binaries and their relevance in astrophysics). In practice, current numerical codes are able to evolve a binary for at most a few tens of orbits at acceptable computational cost, so that we need to specify an initial separation of the binary. This can be done, for example, in the form of a coordinate separation or an initial orbital frequency. Constructing a quasi-circular orbit then requires the accurate specification of the orbital angular momentum corresponding to a circular orbit. Three methods have been used in the literature to minimize the eccentricity of the initial configuration. The effective potential method [44, 106] is inspired by Newtonian physics and starts with a fixed value of the orbital angular momentum. It then varies the separation of the orbit and defines the quasi-circular configuration as that which minimizes the binding energy of the binary. The second method is based on the approximate stationarity of a circular binary in co-rotating coordinates. Mathematically, this corresponds to the existence of an approximate helical Killing vector which is used to impose the approximate symmetry of the binary under rotations [138, 141, 264] (see also [265] for a sequence of parameters for quasi-circular puncture initial data sets). Finally, the post-Newtonian formalism predicts the angular momentum of a binary with given separation on a quasi-circular orbit (see, formula (64) in [81] based on the 3PN accurate calculations in ADM-transverse-traceless gauge of [111]).

A comparison of the three methods applied to a non-spinning, equal-mass binary starting about two orbits prior two merger is given in [81] and finds excellent agreement between the resulting momentum parameters. The phase of the resulting waveforms turns out to be rather sensitive to the initial parameters, however, so that the three methods lead to notably different merger times. We will return to the issue of residual eccentricity in the initial data below in Sect. 4.6.1 when we discuss methods to further improve the initial momentum parameters.

4.3.4 Mesh Refinement and Outer Boundary Conditions

We now turn our attention to the more technical aspects of numerical simulations of black-hole spacetimes. A major difficulty arises out of the presence of different length scales in the spacetimes under consideration. The black-hole size is approxi-

---

[1] The term 'quasi' refers to the fact that the orbit is continuously shrinking because of the energy loss of the system.

mately given by the mass of the hole M. Gravitational waves, however, have wavelengths about one or two orders of magnitude larger and need to be extracted sufficiently far away from the strong-field region, ideally in the wave zone. This wave zone starts approximately at distances of $10^2 M$. In order to avoid contaminations from the outer boundary, the computational domain needs to be several times as large as that value. With current computational resources, it is impossible to evolve such large domains with the resolution required to resolve the steep gradients near the black-hole horizons. The only solution to this problem is the use of mesh refinement, that is, the use of different resolutions in different parts of the computational domain. This applies both to the finite differencing codes and to the Caltech–Cornell spectral code.

Mesh refinement has been made popular in numerical relativity by Choptuik who thus obtained the required accuracy in his discovery of critical phenomena [104]. Because of the movement of black holes, it is not sufficient to use fixed mesh refinement, where the regions of increased resolutions remain stationary in time. Mesh refinement where the zones of refinement change in time is called adapted and generally measures the steepness of the gradients to determine what resolution is needed in a particular region of the domain. The implementation of mesh-refinement in black-hole evolutions does not require the full machinery of adapted mesh refinement because it is relatively straightforward to locate black holes via their apparent horizons and black holes are rigid objects and preserve their shape to a remarkable degree. A common approach in the current generation of black-hole codes is the so-called moving boxes method. That is, the computational grid consists of a nested set of rectangular boxes with decreasing size and increasing resolution. A subset of these boxes follows the black-hole motion and thus guarantees that sufficient resolution is maintained near the black holes. This is illustrated in Fig. 4.2 where the black holes are represented by their apparent horizons (white hemispheres).

While mesh refinement is conceptually rather straightforward, it represents a formidable book keeping exercise in general relativistic simulations and also a potential source of instabilities. Indeed, it is often hard to generalize stability studies to numerical techniques with mesh refinement and commonly the success of a method is only established in practice by evolving black-hole data.

An alternative to mesh refinement is the use of coordinates which are "stretched" further away from the black holes and thus result in an effectively lower resolution. These so-called "fish-eye" coordinates allow one to push the outer boundary to larger radii at acceptable computational costs [32].

Fixed mesh refinement was first used in black-hole simulations by Brügmann [78] in fixed form for a dynamically sliced Schwarzschild hole. Pretorius' first simulations of a black-hole inspiral and merger used mesh refinement based on a modified Berger–Oliger [50] scheme (see [228, 229] for details). Further refinement packages include CARPET [101, 242] which provides mesh refinement for several codes [99, 156, 175, 257] using the CACTUS computational toolkit [89], PARAMESH [189] used by the Goddard group [35], SAMRAI [237] which is used by OPENGR [216], and Steven Liebling's HAD [147, 183] used for mixed binary evolutions in [14].

Fig. 4.2 Illustration of mesh refinement in black hole simulations

A closely related topic concerns the specification of conditions at the outer boundaries. A potential danger arising from outer boundary conditions is the violation of the constraints. Furthermore, it is not immediately clear, to what extent a given set of boundary conditions preserves the hyperbolicity of the set of evolution equations. Several conditions ensuring the constraints and/or the hyperbolicity have been suggested in the literature [91, 131, 132, 143, 173, 184, 240]. To our knowledge, however, only that of [184] has been successfully used for black-hole binary evolutions in [70, 224]. Pretorius [227, 229] instead uses a compactification of the spacetime. In contrast to characteristic formulations where compactification is natural and common, it inevitably implies a loss of resolution at sufficiently large distances from the binary when applied to slices approaching spatial infinity. The reason is simply that the characteristics, i.e., null geodesics are curves of constant phase of gravitational waves, whereas space-like curves are not. Pretorius solves this difficulty by using numerical dissipation in the outer parts of the computational domain and thus avoids high-frequency noise.

All other codes use the relatively simple outgoing Sommerfeld condition; see, for example, Sect. VI A of [9]. For the studies performed so far, this choice appears to provide sufficient accuracy, provided the outer boundary is located at sufficiently large distances from the strong-field sources. It remains to be seen to what extent improvements will be needed in future studies.

4.3.5 Singularity Treatment

A further complication in black-hole simulations normally not encountered in other areas of computational physics is the presence of singular points in the spacetimes. Two types of singularities can arise in general relativity; coordinate singularities, as, for example, the famous $r = 2M$ in the Schwarzschild metric, and physical singularities. If the code encounters either of these, it will crash because metric components will diverge at the singularities. We have already discussed this point in the context of singularity avoiding slicings which slows down the evolution in the vicinity of singular points.

An alternative to this approach has been suggested by Unruh as cited in [261] and is based on the cosmic censorship conjecture which stipulates that there exist no naked singularities. Instead, a singularity will always be surrounded by a horizon, that is, a causal boundary that disconnects a region of spacetime from the exterior in the sense that no information, not even light, can travel from the inner region to the exterior. In consequence, the external spacetime is completely independent from what is happening in the interior. It is possible therefore, to remove this interior part containing the singularity from the computational domain and evolve exclusively the exterior spacetime. This is graphically illustrated in Fig. 4.3, where the dots represent grid points and the circle the horizon. Points outside the horizon are evolved normally (black dots). Inside the horizon there is a layer of boundary points (grey) where data are commonly obtained from extrapolation from exterior points or using one-sided derivatives. The inner points (white) are simply ignored, that is "excised" from the numerical simulation.

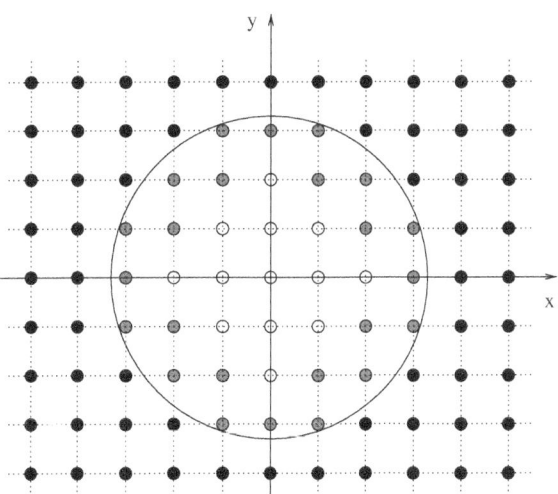

Fig. 4.3 Illustration of black-hole excision

Black-hole excision has been used in the 1990s using a technique called causal differencing [17, 245]. More recent implementations in finite differencing codes have been based on straightforward extrapolation. The so-called "simple excision" method of Alcubierre and Brügmann [7] provided a remarkably straightforward method to obtain long-term stable evolutions of single black holes and has also been used in simulations of orbiting binaries in co-rotating coordinates [80, 116]. More general techniques accommodating moving black holes via dynamic excision have been used in [74, 224, 226, 227, 249, 256]

An alternative method to handle the coordinate singularity inherent to puncture data was the decomposition of the conformal factor. In the simulation, only the regular piece was evolved. The recent "moving puncture" simulations differ from that approach in that they evolve the entire conformal factor. One consequence is that the nature of the coordinate singularity also changes its nature [77, 152] and looses contact with the asymptotic spatial infinity. Given the finite numerical resolution, however, these features inside the black-hole horizon are not resolved in a numerical simulation, and the moving-puncture method appears to provide a kind of automatic and natural excision.

Among the current generation of black-hole codes, explicit excision is implemented in the generalized harmonic codes of Pretorius and the Caltech–Cornell effort. The implementation in the spectral Caltech–Cornell code is special in the sense that they use a so-called dual-coordinate frame to accommodate the motion of the holes. Using two coordinate systems and transforming variables between these systems avoid the necessity to move the excision region across the computational domain; see [241] for details.

4.4 Diagnostics

Once we have successfully evolved a spacetime containing black holes, there still remains the task of extracting physical information from the simulation. This procedure faces two major difficulties. First, a computer simulation only produces a large set of numbers which represent coordinate-dependent quantities. We need to construct physical, that is gauge invariant, combinations from these quantities. A second problem is that not all physical concepts familiar from Newtonian physics are well defined in general relativity. In particular, this applies to local quantities as, for example, the energy contained in a particular region of spacetime. In the following we will discuss all important quantities currently used in numerical relativity to extract physical information from the simulations. For this purpose we assume that all ADM variables are known in some part of the spacetime. These variables are the lapse α, the shift β, the three-metric γ_{ij}, and the extrinsic curvature K_{ij}. These variables can always be computed straightforwardly from the evolution variables, even if we use a formulation not based on the ADM equations.

4.4.1 Global Quantities

Global quantities provide us with information characterizing the entire spacetime. They are usually defined by evaluating variables at spatial or null infinity. In most contemporary numerical evolutions, the computational domain does not extend all the way to infinity, so that we need to approximate global quantities by calculating them at large but finite distances from the strong-field region near the black holes.

The total mass or energy of the spacetime is given by the so-called ADM mass [25] which is obtained from the three-metric by the surface integral

$$M_{\mathrm{ADM}} = \frac{1}{16\pi} \lim_{r \to \infty} \int_{S_r} \sqrt{\gamma} \gamma^{ij} \gamma^{kl} (\partial_j \gamma_{ik} - \partial_k \gamma_{ij}) dS_l. \tag{4.48}$$

Here γ is the determinant of the three-metric, S_r the coordinate sphere $r = \mathrm{const}$, $\hat{n}_m$ the unit normal field on that sphere, and $dS_l = \hat{n}_l d\theta d\phi$ with standard angular coordinates θ and ϕ.

Similarly, the total linear and angular momentum of the spacetime can be calculated from (see e.g., [277])

$$P_i = \frac{1}{8\pi} \lim_{r \to \infty} \int_{S_r} \sqrt{\gamma} (K^m{}_i - K\delta^m{}_i) dS_m, \tag{4.49}$$

$$J_i = \frac{1}{8\pi} \epsilon_{il}{}^m \lim_{r \to \infty} \int_{S_r} \sqrt{\gamma} x^l (K^n{}_m - K\delta^n{}_m) dS_n. \tag{4.50}$$

We emphasize that all these quantities are by construction time independent. In contrast, the Bondi-mass [66] is evaluated at null infinity, thus takes into account the radiation of energy to null infinity in the form of gravitational waves and varies with retarded time. It is a natural diagnostic tool in characteristic formulations but not directly available in 3+1 evolutions.

4.4.2 Local Quantities

We have already mentioned that it is often impossible to define local concepts of energy and momenta. In the case of black holes, however, it is possible to use the concept of horizons [28] to define mass and spin associated with the horizon and thus with the black hole. Imagine for that purpose a 3D hypersurface Σ and a closed 2D surface S embedded in Σ (see for example Fig. 1 in [117]). On each point of S one can define in and outgoing null vectors $\hat{n}_\alpha$ and ℓ_α. The expansion of in and outgoing light cones is given in terms of these null vectors by

$$\theta_{(\ell)} = q^{\alpha\beta} \nabla_\alpha \ell_\beta, \qquad \theta_{(\hat{n})} = q^{\alpha\beta} \nabla_\alpha \hat{n}_\beta, \tag{4.51}$$

where $q_{\alpha\beta}$ is the induced two-metric on the surface S. A marginally trapped surface is defined by the condition that the outgoing expansion vanishes $\theta_{(\ell)} = 0$ and the ingoing expansion satisfies $\theta_{(\hat{n})} < 0$. Loosely speaking, this means that all light cones on the trapped surface are tilted inwards to such an extent that light rays cannot escape outwards. In general, a black-hole spacetime has more than one marginally trapped surface and the apparent horizon is defined as the outermost marginally trapped surface.

The main task in a numerical simulation is to locate surfaces S with vanishing expansion $\theta_{(\ell)} = 0$. Various apparent horizon finders have been developed by the numerical relativity community. For more details on the numerical methods to locate the apparent horizon and the physical interpretation of the horizon properties the reader is referred to [22, 28, 42, 117, 163, 177, 262] and references therein.

For the discussion in the remainder of the work, the most important quantity is the irreducible mass of the horizon which is defined in terms of the horizon area by

$$M_{\text{irr}} = \sqrt{\frac{A_{\text{AH}}}{16\pi}}. \tag{4.52}$$

In the limit of an isolated hole, i.e. a black hole whose interaction with other holes or matter sources is negligible, one can use the world tube of apparent horizons, the so-called isolated horizon, to define the angular momentum associated with the horizon

$$J_{(i)} = \frac{1}{8\pi} \oint_S \phi_{(i)}^m R^n K_{mn} dS, \tag{4.53}$$

where R^n is the outgoing unit norm field on S and $\phi_{(i)}^m$ is the Killing vector associated with the rotational symmetry and the index (i) labels the axis of the rotation, e.g., the x, y, or z component of the spin (see [28] for more details). Finally, we can use the spin of the black-hole to calculate the total black-hole mass M according to Christodoulou's formula [105]

$$M^2 = M_{\text{irr}}^2 + \frac{J^2}{4M_{\text{irr}}^2}. \tag{4.54}$$

In the limit of a stationary spacetime with a single black hole, this mass corresponds to the ADM mass. In spacetimes with a black-hole binary we can use the individual black hole masses and the ADM mass to define the binding energy

$$E_b = M_{\text{ADM}} - M_1 - M_2. \tag{4.55}$$

This definition assumes, however, that there are no other forms of energy present in the spacetime. In numerical practice, this condition is normally violated because initial data sets contain some spurious gravitational radiation in addition to the black holes. In many cases, however, this spurious energy content turns out to be small compared with the right-hand side of Eq. (4.55) and the resulting error in the binding energy is small.

4.4.3 Gravitational Waves

Arguably the most important information resulting from a simulation of black holes is the amount and structure of the gravitational waves emitted in the course of the inspiral and merger. The gravitational wave signal enables us to calculate the loss of energy and linear and angular momentum of the system and also predicts the strain $h_{+,\times}$ exerted upon a distant gravitational wave detector.

The most common method to extract gravitational waves from a numerical simulation is based on the Newman Penrose formalism [211]. Specifically, one defines a tetrad ℓ^α, $\hat{n}^\alpha$, m^α and $\bar{m}^\alpha$ where $\hat{n}$ and ℓ are ingoing and outgoing null vectors, and m is a complex linear combination constructed out of two spatial unit vectors such that

$$-\ell \cdot \hat{n} = 1 = m \cdot \bar{m} \tag{4.56}$$

and all other inner products vanish.

The Newman–Penrose scalar Ψ_4 is defined in terms of this tetrad and the Weyl tensor as

$$\Psi_4 = C_{\alpha\beta\gamma\delta}\hat{n}^\alpha \bar{m}^\beta \hat{n}^\gamma \bar{m}^\delta. \tag{4.57}$$

In 3+1 simulations, the Weyl tensor is obtained from the fundamental forms according to the Gauss–Codacci equations (4.15), (4.16), and (4.17). In practice, Ψ_4 is calculated on a sphere of constant coordinate radius r_{ex} and is therefore a function of the angular coordinates θ, ϕ, and the time t.

It can be shown that under a tetrad rotation which leaves ℓ and n unchanged but rotates m, $\bar{m}$ through an angle ϑ, the Newman–Penrose scalar Ψ_4 transforms into $e^{-2i\vartheta}\Psi_4$, that is as a spin-weight -2 field. It is therefore convenient to decompose Ψ_4 in a series of spin-weight -2 spherical harmonics $Y_{\ell m}^{-2}$, where $\ell = 2,\dots$ and $m = -\ell,\dots,\ell$ denote the multipole indices [263]. At extraction radius r_{ex} we can describe the gravitational wave signal in the form of mode coefficients $\psi_{\ell m}(t)$ of the series expansion

$$\Psi_4 = \sum_{\ell,m} \psi_{\ell m}(t) Y_{\ell m}^{-2}(\theta,\phi). \tag{4.58}$$

It turns out that the complete signal is often dominated by a small number of modes, normally including the quadrupole moments $\ell = 2$. It is for this reason that gravitational waveforms are often presented in the form of 1D plots showing some $\psi_{\ell m}(t)$.

In order to ensure that Ψ_4 is a measure for the outgoing gravitational waves, the tetrad has to be chosen with care. In the case of spacetimes perturbatively close to the Kerr-solution, the appropriate choice is the Kinnersley tetrad [174]. In general numerical simulations, however, it is not clear how one can unambiguously identify the Kinnersley tetrad. Instead, one commonly constructs the tetrad from the time-like unit normal field n and three spatial triad vectors u, v, and w according to

$$\ell^\alpha = \frac{1}{\sqrt{2}} \left(n^\alpha + u^\alpha \right), \tag{4.59}$$

$$\hat{n}^\alpha = \frac{1}{\sqrt{2}} \left(n^\alpha - u^\alpha \right), \tag{4.60}$$

$$m^\alpha = \frac{1}{\sqrt{2}} \left(v^\alpha + i w^\alpha \right). \tag{4.61}$$

The triad vectors, in turn, are constructed by applying a Gram–Schmidt orthogonalization to the coordinate triad

$$u^i = [x, y, z], \tag{4.62}$$

$$v^i = [xz, yz, -x^2 - y^2], \tag{4.63}$$

$$w^i = \epsilon^i{}_{mn} v^m w^n, \tag{4.64}$$

with the 3D Levi-Civita tensor $\epsilon^i{}_{mn}$. There remains some freedom in starting the orthogonalization with u, v, or w and different implementations have been used by the community. So far, the choice does not seem to have a notable impact on the resulting waveforms.

The approximative character of the tetrad makes it necessary to extract gravitational waves at a sufficiently large distance from the strong-field region near the black holes. In practice, extraction radii of the order of $100\, M_{ADM}$ are used in most current simulations. The uncertainties arising from the use of finite extraction radii have been estimated in [70]. A more general discussion of various issues in the standard wave extraction procedure is given in [180]. Methods for approximating the Kinnersley tetrad more efficiently have been studied in [48, 208–210].

The energy and linear and angular momentum radiated in the form of gravitational waves are given in terms of the Newman–Penrose scalar Ψ_4 via the integrals (see, e.g., [92]) are

$$\frac{dE}{dt} = \lim_{r \to \infty} \left(\frac{r^2}{16\pi} \int_\Omega \left| \int_{-\infty}^t \Psi_4 d\tilde{t} \right|^2 d\Omega \right), \tag{4.65}$$

$$\frac{dP_i}{dt} = -\lim_{r \to \infty} \left(\frac{r^2}{16\pi} \int_\Omega \ell_i \left| \int_{-\infty}^t \Psi_4 d\tilde{t} \right|^2 d\Omega \right), \tag{4.66}$$

$$\frac{dJ_z}{dt} = -\lim_{r \to \infty} \left\{ \frac{r^2}{16\pi} \mathrm{Re} \left[\int_\Omega \left(\partial_\phi \int_{-\infty}^t \Psi_4 d\tilde{t} \right) \left(\int_{-\infty}^t \int_{-\infty}^{\hat{t}} \bar{\Psi}_4 d\hat{t} d\tilde{t} \right) \right] d\Omega \right\}, \tag{4.67}$$

where $\ell_i = [-\sin\theta\cos\phi, \ -\sin\theta\cos\sin\phi, \ -\cos\theta]$. In practice, the integrals are evaluated on coordinate spheres with radius r_{ex} where one also calculates Ψ_4. The errors arising from the use of finite radii can be estimated by calculating the quantities at different extraction radii and studying the variation of Ψ_4 and the momenta analogous to a convergence study of the code's performance at different grid resolutions. The uncertainties depend on the details of the simulation, but in general are of the order of a few percent or less for extraction radii of the order of $10^2 \, M_{ADM}$ (see, e.g., [257]).

An alternative method for extracting gravitational waves is based on the Zerilli–Moncrief formalism [206, 281] and provides the GW signal in the form of two gauge invariant perturbation functions. More details about this method and applications can be found in [2, 3, 175, 256] and references therein.

4.5 A Brief History of Black-Hole Simulations

Attempts at solving the Einstein equations numerically date back to the 1960s and 1970s and the pioneering work by Hahn, Lindquist, Eppley, Smarr, and coworkers [119, 149, 250–252]. These early attempts used the ADM formulation of the Einstein equations and focused on axisymmetric spacetimes and were therefore restricted to head-on collisions of black holes. The resulting simulations turned out to be relatively short-lived, however, compared with the dynamic timescale of the problem. Considering that modern supercomputers are just about powerful enough to facilitate numerical simulations of black-hole binaries, it is clear, that the early numerical studies were inhibited by the computational resources available at the time.

It was therefore more than a decade later, before the significant increase in computer power led to a systematic reinvestigation of the problem in the framework of the "Grand Challenge" (see e.g., [16, 19–21, 29]). These studies predicted a total radiation of the order of $10^{-3} M_{ADM}$ emitted in the head-on collision of two black holes [19]. Simulations of unequal-mass binaries revealed a gravitational recoil or kick of up to 10–20 km/s [21]. Simulations were also performed for the first time in three dimensions [18]. In spite of this progress, however, the fundamental difficulties with instabilities in the numerical simulations were not overcome. After the end of the Grand Challenge, a joined effort by the universities of Pittsburgh, Penn State, and Texas investigated grazing collisions using the black-hole excision method [74].

The 1990s also saw the first investigation of alternative ways to write the Einstein equations. Bona and Massó wrote the evolution equations in the form of balance laws [59–61], not dissimilar to the way the equations of hydrodynamics are commonly implemented numerically. Even though their efforts did not overcome the stability problems, the idea of using alternative formulations was gradually adopted by other groups and eventually provided a major ingredient in solving the binary-black-hole problem (cf. Sect. 4.3.1). Most importantly, test simulations using the BSSN formulation [43, 246] demonstrated improved stability properties.

The BSSN system also played an important role in the studies of the numerical relativity group of the Albert Einstein Institute in Potsdam starting in the late 1990s. These efforts used initial data of puncture type, factored out the Brill–Lindquist conformal factor during the evolution, and employed coordinate conditions which keep the black hole centers fixed on the numerical grid. These studies resulted in the first grazing collisions of black holes [79], the first use of mesh refinement in black-hole simulations [78] and the first long-term stable simulations of black-hole head-on collisions [9] as well as single black-hole spacetimes [7]. A guiding principle for many of these studies was to absorb as much as possible the dynamics of the system in the coordinates and use gauge conditions which drive the system into quasi-stationarity. Eventually, this approach lead to simulations of orbiting binaries on timescales similar to the orbital period [80, 116].

In view of the persistent stability problems, the Lazarus project attempted to use fully non-linear evolutions until shortly before the merger of the binary, but then match the evolution to a perturbative treatment (see [30, 31] and references therein). This approach facilitated the evolution of relatively short, plunging configurations and provided estimates on the gravitational recoil [93] as well as the first results on spinning binaries [33].

By early 2005, the combined methods of the BSSN formulation, improved gauge conditions, and/or black-hole excision allowed the community to study head-on collisions of black holes using mesh refinement, more accurate fourth-order numerical schemes, and/or non-conformally flat initial data of Kerr–Schild type [125, 256, 282].

The year 2005 also saw the eventual breakthrough, when Pretorius used the remarkable combination of the generalized harmonic formulation of the Einstein equations, implicit numerical schemes, and spatial compactification to provide the first simulation of a binary through merger with accurate gravitational waveforms [226]. About half a year later, the groups at Brownsville and Goddard independently discovered a method to evolve and merge black-hole binaries of puncture type using a relatively straightforward to implement generalization of previous puncture evolutions with the BSSN system [34, 94]. Retrospectively, it is quite remarkable that these two notably different methods have provided within a few months a success-ful path to the "holy grail of numerical relativity". As of 2008, there exist about ten independent numerical codes of one or the other method which have been demon-strated to produce stable and convergent simulations of at least some types of black-hole binary spacetimes [35, 81, 95, 120, 156, 175, 229, 241, 257].

In the next section we will summarize the results obtained with these codes in the course of the last 2 1/2 years.

4.6 Properties of Black-Hole Binaries

Following the breakthroughs of 2005, the numerical relativity community has gen-erated a wealth of results on black-hole binary spacetimes. Before we discuss these results in more detail, we need to comment on the scale invariance of black-hole spacetimes. In Sect. 4.3.3 we have summarized the physical parameters of generic black-hole initial configurations. In particular, we noted that the total ADM mass of the system merely represents a scaling factor in a numerical simulation. That is, expressed in units of the ADM mass, all quantities of a simulation have the same numerical value irrespective of the magnitude of the ADM mass itself. A single numerical simulation thus represents a one-parameter family of solutions. We emphasize, however, that the total mass of the system is very important from a gravitational wave detector's point of view. Suppose, a characteristic frequency is given by $\omega = C/M_{ADM}$, with C some constant. The maximal sensitivity of the LIGO detector, for example, is located in a window around 150 Hz. The ADM mass of the system will determine the systems characteristic frequency and thus where this frequency is located in the LIGO sensitivity range. For the case of a binary of two non-spinning holes of mass 5 $M_\odot$ each, for example, it is the earlier inspiral phase which falls into the maximum sensitivity range of LIGO, whereas for a system of two black holes of mass 50 $M_\odot$, it is the merger and ringdown signal. This is illustrated in Fig. 4 of [218]. From the gravitational wave data analysis point of view, there are also parameters which describe the relative location of the black-hole binary relative to the earth: the source's position on the sky and its inclination relative to the plane of the detector. These parameters are not related to the physical properties of the binary, however, and need not concern us in this work.

4.6.1 Non-spinning, Equal-Mass Binaries

The inspiral of two non-spinning black holes of equal mass represents the simplest binary configuration and has been the first to be evolved successfully through inspiral, merger, and ringdown [34, 94, 227]. This scenario is currently the best understood type of binary systems and we will use it here to also illustrate the fundamental characteristics of a black-hole inspiral and merger and the resulting waveform patterns.

In realistic astrophysical scenarios, the binary will complete thousands of orbits or more before coalescence. Unless there is significant interaction with third party objects, the binary will loose all orbital eccentricity due to the circularizing effect of gravitational wave emission [221]. Numerical simulations are currently able to simulate only up to about 15 orbits by which time most binaries are expected to be in quasi-circular configuration. In order to accurately model such systems, numerical simulations need to start from initial data which represent as closely as possible a snapshot of a binary in quasi-circular inspiral. In practice, this has commonly been approximated in one of the three methods we discussed in Sect. 4.3.3. All of these methods, however, result in measurable eccentricity in the orbits (see, for example, [37, 86, 165, 224]). This small residual eccentricity is a major source of uncertainty in the comparison of numerical with post-Newtonian results [70] and improved methods to further reduce the eccentricity have been designed using iterative procedures [70, 224] or the integration of post-Newtonian equations over a larger number of orbits [165].

For illustration of the binary inspiral, we show in the upper panel of Fig. 4.4 the puncture trajectories of the holes as obtained for the simulation of a relatively short inspiral starting from the so-called R1-configuration (see Table I of [35]). For

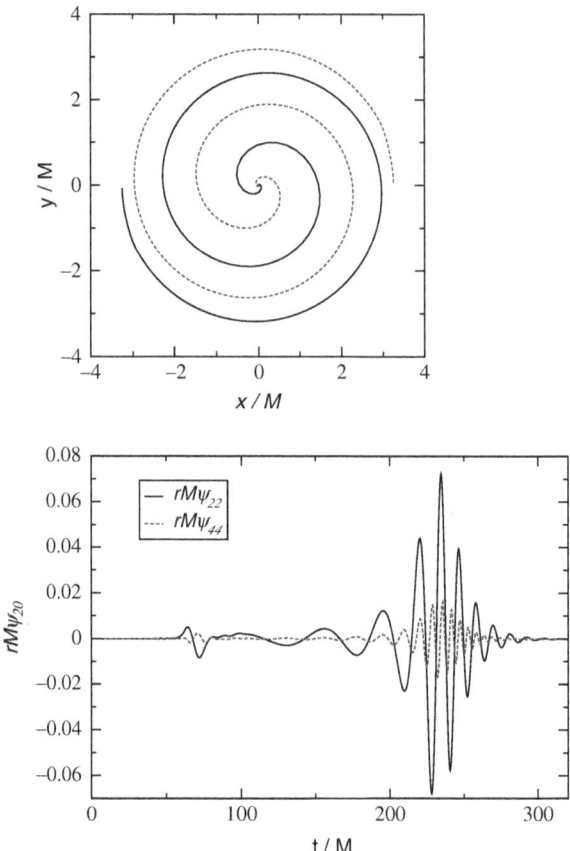

Fig. 4.4 *Upper panel*: Trajectories of the two holes in the inspiral of two equal mass, non-spinning holes starting from a coordinate separation $d/M = 6.514\,M$ (see entry R1 in Table I of [35]). *Lower panel*: The $\ell = 2$, $m = 2$ mode of the Newman–Penrose scalar Ψ_4 extracted from the same simulation at radius $r_{\text{ex}} = 60\,M$

trajectories containing more orbits, see, for example, Fig. III of [70]. One of the most remarkable features of both moving punctures and evolutions using the generalized harmonic formulation is the similarity of the black-hole coordinate trajectories with the intuitively expected picture. Bearing in mind the gauge dependence of the trajectories, this result was by no means to be expected; indeed, the frequencies derived from the trajectories agree remarkably well with those derived from the gravitational waveforms [86]. Figure 4.4 also illustrates the relatively smooth transition from the inspiral to the ringdown. A more careful investigation reveals that the merger only lasts for about 0.5–0.75 orbits [86] and does not exhibit strange features in the waveforms.

A more fundamental question concerns the dependence of the results on the choice of initial data. Orbital simulations starting from Cook–Pfeiffer excision and puncture data have been compared in [39] and showed qualitatively good agreement. A more detailed comparison is inhibited by residual spin in the excision data, though. Such problems are not present in comparisons of head-on collisions of black-hole binaries. The comparison of collisions starting from Brill–Lindquist, Misner, and superposed Kerr–Schild data exhibits good quantitative agreement [257]. Minor differences in the evolutions of Kerr–Schild data might be attributed to spurious radiation present in these data [57]. In summary, these results are reassuring, though more detailed comparisons will be needed in the future.

All simulations of equal mass, non-spinning binaries agree rather well on the total radiated energy and angular momentum in the course of the inspiral, merger, and ringdown. About 3.5% of the total mass and about 21% of the total angular momentum of the binary system are carried away in the form of gravitational waves. Furthermore, the radiation is dominated by the $\ell = 2, m = \pm 2$ quadrupole contribution which carries $> 98\%$ of the total radiated energy [35, 52, 81, 95, 257]. This is illustrated in the upper panel of Fig. 4.4 where we show the $\ell = 2, m = 2$ mode as well as the next strongest mode $\ell = 4, m = 4$. From Eq. (4.65) we see that the energy scales with the square of the wave amplitude and it becomes clear that only a small fraction of the energy is contained in $\ell = 4, m = 4$. All other modes are negligible compared with these two. Finally, most of the energy and angular momentum is radiated in the final plunge and merger of the binary, whereas contributions from the early inspiral phase are rather small (see, for example, Fig. 11 in [81]).

The dominant role of the quadrupole radiation is no surprise and directly follows from post-Newtonian studies (see [56] and references therein). Indeed, most of the inspiral phase up to about the last few orbits is rather well described by the post-Newtonian approximation and one of the most important question facing the community right now is to determine, how close to the merger the PN approximation breaks down. For this comparison it is often convenient to split the complex Newman–Penrose scalar Ψ_4 into phase and amplitude

$$\Psi_4(t) = A(t)e^{i\phi(t)}. \tag{4.68}$$

Sometimes, the wave signal is also expressed in terms of the gravitational wave polarizations $+$ and $\times$ related to Ψ_4 according to

$$\Psi_4 = \partial_t \partial_t (h_+ - ih_\times) \tag{4.69}$$

(see, e.g., [52] for a discussion of the constants of integration) and the amplitude phase decomposition is applied to h_+ and $h_\times$. The Newman–Penrose scalar is the standard choice of describing gravitational waves in numerical relativity, whereas h_+ and $h_\times$ are more directly related to the displacements in gravitational wave detectors and thus more popular in GW data analysis. For the comparison between numerical and post-Newtonian results, the choice of variables is not important, however.

First comparisons between numerical and PN results demonstrated that the PN-adiabatic model agrees better with the numerical results at larger BH-separations, but gives reasonable results even when extrapolated to the formation of a common apparent horizon [86]. Their study also found the highest 3PN and 3.5PN order to result in the best agreement with the numerical data. A study using numerical waveforms covering the last 14 cycles of the inspiral was presented in [37, 38] and revealed an accumulated phase discrepancy of about 1 rad until shortly before the merger. Subsequent numerical simulations using improved initial data and higher order finite differencing or spectral methods resulted in even better agreement [70, 153]. The most comprehensive comparison performed by the Caltech group [70] showed a phase difference between numerical and various PN waveforms of about 0.1 rad. The particularly good agreement observed for the Taylor T4 approximant result appears to be more coincidental, as it is significantly smaller than the discrepancies among the different PN results. Comparisons with post-Newtonian results uniformly found higher order amplitude corrections to the PN waveforms to improve the agreement with numerical results [70, 86, 153].

So far we have focused on quasi-circular binaries. While the majority of systems are indeed expected to have vanishing eccentricity, some astrophysical scenarios, as for example third body interactions, may induce eccentric orbits. The effect of significant eccentricities on the dynamics of the binary and the gravitational wave signal has been studied in [158, 258]. Relatively small eccentricities cause a small increase in the radiated energy and angular momentum, whereas binaries with large eccentricities plunge rather than inspiral which significantly reduces the energy and momentum emission. Binaries with larger eccentricity also emit an increasing fraction of their energy in the $\ell = 2$, $m = 0$ mode as opposed to the dominating $\ell = 2$, $m = \pm 2$ modes. The results obtained so far indicate that there exists a relatively sharp distinction between orbiting and plunging configurations: simulations with orbital angular momentum $L \lesssim 0.8\, M^2$ plunge, those with $L \gtrsim 0.8\, M^2$ inspiral. The study in [158] also demonstrated that the GW merger signal shows universality for angular momenta above the critical value.

A remarkable behavior of black-hole binaries has been found in [230] when fine tuning the linear momentum parameter of the holes in the initial data. Such fine tuning leads to binaries which exhibit "zoom-whirl" behavior, that is, they inspiral initially, but then may stall at some finite separation for a while and eventually merge or separate. A similar behavior is known in the structure of geodesics of single black hole spacetimes. While it is tempting to think of critical phenomena in the context of the fine tuning of initial parameters (see [146] for a review), a clear relation between the two effects has as yet not been established.

4.6.2 Unequal Mass Binaries

Spacetimes containing black-hole binaries of unequal mass are no longer symmetric under rotations by 180° around the axis defined by the orbital angular momentum. This loss of symmetry has important consequences for the gravitational wave emission. In particular, the radiation of linear momentum is no longer isotropic and results in a net-loss of linear momentum of the binary system. By conservation of

linear momentum, this imparts a recoil or kick on the final merged hole. At the
leading order, this effect arises from the overlap of the mass-quadrupole with the
octupole and flux-quadrupole moments [49, 68, 220]. This kick is a genuinely relativistic effect and has significant repercussions on astrophysical systems containing
black holes [63, 65, 142, 151, 160, 181, 185, 190, 199, 212, 269, 271]. It might also
manifest itself directly in astrophysical observations of quasi-stellar objects without
host galaxies [148, 159, 186, 192, 200] or the distorted morphology of x-shaped
radio sources [190, 198, 199].

The kick generated by the inspiral and merger of unequal-mass binaries has been
the subject of various approximative studies [55, 114, 122, 126, 127, 254, 255],
but highly accurate results require the solution in the framework of fully non-linear
general relativity and, thus, numerical relativity. First numerical studies of certain
mass ratios revealed kick velocities of the order of 100 km/s [36, 156]. In order to
find the maximum kick resulting from unequal-mass binary inspiral, [136] calculated the kick for mass ratios ranging from $q = 1$ to $q = 4$ and found a maximum
kick of 175 ± 11 km/s for the mass ratio $\eta = 0.195 \pm 0.005$. This is illustrated in
Fig. 4.5. This velocity is larger than the escape velocities of about 30 km/s for glob-
ular clusters and falls into the range of escape velocities predicted for dwarf galaxies, but is significantly smaller than that from giant elliptic galaxies of the order of
1000 km/s [199]. The resulting ejection or displacement of the black hole following
a merger has important repercussions on models for the formation history of black
holes as well as the structure of host galaxies and the population of intergalactic
black-hole populations (see e.g., [65, 151, 199, 212, 269]).

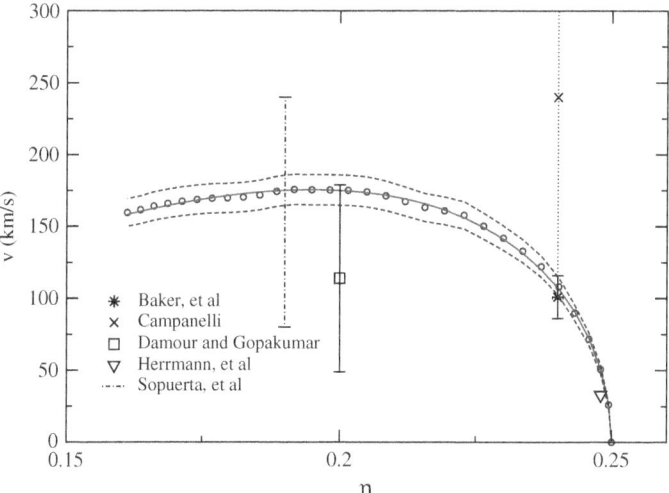

Fig. 4.5 The recoil velocity resulting from the inspiral and merger of a non-spinning binary with
mass ratio $\eta = M_1 M_2 / (M_1 + M_2)^2$ as calculated in [136]. For comparison the figure also includes
values from [36, 93, 114, 156, 254]

In contrast to the emission of linear momentum, the radiated energy and angular momentum are maximal in the equal-mass case. Radiated energy and the final spin parameter of the single hole are well approximated by fitting formulas [52, 136]

$$E_{\text{rad}} = 0.0363\, M \left[\frac{4q}{(1+q)^2}\right], \qquad (4.70)$$

$$j_{\text{fin}} = 0.089 + 2.4\frac{q}{(1+q)^2}. \qquad (4.71)$$

A further consequence of the reduced symmetry of unequal-mass binaries is the more complex structure of higher order multipoles. In the equal-mass case, all radiation modes with odd m vanish by symmetry. Additionally, it turns out that the percentage of energy radiated in higher ($\ell > 2$) modes increases from less than 2% for $q = 1$ to more than 10% for $q = 4$. This is illustrated in Fig. 12 of [52]. This sensitivity of higher order modes to the mass ratio is significant for gravitational wave data analysis because the inclusion of higher order modes in the analysis is likely to improve the accuracy of parameter estimates and the detection range of gravitational wave observations (see e.g., [26, 27]).

The comparison of numerical with post-Newtonian results for unequal-mass binaries represents a more challenging task, because of the increased computational cost of numerical simulations as the mass ratio q deviates more strongly from 1. There are currently not as accurate and long numerical waveforms available for the comparison. The sequence of unequal-mass binaries generated for the kick calculations in [136] was used in [52] for a comparison with post-Newtonian results in the inspiral and black-hole quasi-normal mode studies (see [51, 102, 179] and references therein) in the ringdown phase. Similar to the equal-mass study in [86], post-Newtonian results were found to predict remarkably well the relation between wave frequency and amplitude. The convergence of the PN series is non-monotonic but the inclusion of higher order terms improved the agreement with the numerical results. Spin and mass parameter estimates obtained from the black-hole ring down were in excellent agreement with the values derived from the measured gravitational radiation and balance arguments. Intriguing oscillations observed in the quality factor estimates obtained in the ringdown phase could indicate non-linear effects but might also be artifacts of numerical noise. Simulations of higher accuracy are required to conclusively address this issue.

The first study on the use of numerically generated waveforms in gravitational wave data analysis was performed in [46]. They discuss sources of uncertainties in using numerical waveforms and estimate that first detection efforts will require about 100 templates to cover the zero spin part of the parameter space. Reference [218] used a set of numerical waveforms of equal-and unequal-mass binaries and studied the agreement of the numerical waveforms with a variety of PN template families. For this study they used the fitting factor (FF) [24] which takes into account the instrumental sensitivity and is a standard tool in matched filtering data analysis. They thus found good agreement with FF ≥ 0.96 for total masses of 10–20 $M_\odot$ and ground-based detectors. For larger masses of the binary, the detectors become increasingly sensitive to the merger and ringdown part of the waveform, but they found that the addition of a phenomenological 4PN term extends the range of high fitting factors to about 120 $M_\odot$. The effective-one-body (EOB) method (see [88] and references therein) as well as the phenomenological Buonanno–Chen–Valisneri [85] family of waveforms similarly lead to high-fitting factors in the mass range 10–120 $M_\odot$. The EOB approach is compared in further detail with numerical simulations in [88], where the addition of a 4PN term is shown to result in phase agreement within 8% of a GW cycle at the end of the ringdown phase. The EOB method was also used in [115] to compare the predictions for the spin of the final merged hole. Agreement of about 2% with the numerical results was found.

The generation of phenomenological waveforms is the subject of [5, 6]. Hybrid waveforms obtained from matching numerical with PN waveforms are used to create a parameterization of unequal-mass inspiral waveforms and study their use in GW data analysis. The results indicate that the detection range of ground-based interferometers might be enhanced significantly by using such waveform families.

A particular type of binaries of relevance for gravitational wave physics are the so-called extreme mass ratio inspirals (EMRI) consisting of a stellar size compact object orbiting around a supermassive black hole. EMRIs are considered one of the most important sources of the space interferometer LISA (see, e.g., [161, 162]). Mass ratios of $q \sim 10^{-6}$ characteristic of such systems are currently beyond the range of capabilities of numerical relativity, and the modeling of these scenarios is commonly done in the framework of perturbation theory and self-force calculations (see [225] for a review). Numerical results might still be of interest for less extreme mass ratios, as simulations with $q = 10$ appear to be feasible and their comparison with approximative studies might allow for some calibration of the methods analogous to the comparison between numerical and PN results.

4.6.3 Spinning Binaries

Spinning binaries are by far the most complex black-hole binaries. Bearing in mind, that six out of the seven free physical parameters determine the spin, this is not surprising. Indeed, the resulting parameter space is so large, that only a subset has been studied in any detail so far. The majority of work has gone into studying binaries where the spins are aligned or anti-aligned with the orbital angular momentum. The case of the spin being aligned with the orbital angular momentum may also be the astrophysically most likely scenario as accretion processes have been argued to result in alignment of spin and orbital angular momentum [58].

A particularly intriguing question concerns the formation of naked singularities as would be the case for Kerr holes with spin parameter $a/M \geq 1$. In particular, spins aligned with the orbital angular momentum might be suspected to lead to a very large spin of the final merged hole. The simulations presented in [96], however, demonstrate the difficulties in creating a maximally spinning black hole in this way. The larger the spin magnitude, the longer the inspiral lasts and the more angular momentum and energy is radiated from the system before merger. For the interpretation of this delayed inspiral it is helpful to consider the innermost stable circular orbit (ISCO) [44, 54, 84, 106, 112, 113, 141, 170, 222]. In particular, it can be shown that the ISCO separation decreases for binaries with aligned spins and increases for spins anti-aligned with the orbital angular momentum [222]. Assuming that the ISCO gives a measure for the merger separation, this result agrees with the delayed and accelerated inspiral observed for aligned and anti-aligned inspirals, respectively.

Binaries with spins which are not parallel to the orbital angular momentum exhibit spin-precession. References [98, 99] studied the precession using configurations where the spins are either in the orbital plane or oriented at 45° relative to the plane. They use a simplified method to determine the spin of the individual holes where they integrate the flat space Killing vectors over the horizon surface. Their simulations demonstrate the precession of the individual spins as well as the realignment of the spin of the final black hole, the so-called spin-flip, which may explain the reorientation of jets observed in radio galaxies [178, 219]. The spin–orbit interaction was studied in special configurations starting either without spin but with orbital angular momentum or the other way round in [97]. In both cases the result is a transfer of momentum from spin to orbit or vice versa. This coupling also contributes to the generally more complex structure of spinning binaries.

An effect we have already discussed in the context of unequal-mass binaries, is the recoil or rocket effect in binary-black-hole mergers. Post-Newtonian studies predicted contributions to the recoil arising from the spin–spin and spin–orbit coupling in black-hole binaries [171]. One of the most surprising results as yet obtained from numerical simulations of black hole binaries is the magnitude of the recoil in spinning binaries. The first studies focused on spins parallel to the orbital angular momentum and anti-aligned with each other. These scenarios generate kicks of up to 500 km/s [99, 157, 175] for inspirals and some tens of km/s for head-on collisions [103]. Even larger kicks of up to $1,300\,\text{km/s}$ were predicted by [99] for configurations with spins in the orbital plane, but pointing in opposite directions. Subsequent numerical studies of this scenario revealed unexpected kick magnitudes of about 2,500 km/s for spin amplitudes $a/M \approx 0.7$ which implies maximum values of 4,000 km/s extrapolated for $a/M \rightarrow 1$ [99, 100, 137]. Kicks above 1000 km/s are also predicted by the EOB model [243]. Such large recoil velocities would in fact be sufficient to eject black holes even from giant elliptic galaxies. Given that galaxies with bulges appear to ubiquitously harbor supermassive black holes [124], it appears that these "superkicks", while theoretically possible, are not realized very often in actual galactic mergers. This is also indicated by the Monte Carlo study employing the EOB model in [243] who predict that only a few percent of mergers with mass ratios $1 \le q \le 10$ and spin magnitude $a_1 = a_2 = 0.9$ with random spin orientation results in kicks above 1000 km/s.

The surprising magnitude of the recoil for spinning configurations has sparked a wealth of more detailed investigations and attempts to generate fitting formulas valid for general types of initial configurations. A multipolar analysis of the recoil was presented in [244] for unequal masses and non-zero and non-precessing spins.

Including specific multipoles with $\ell \leq 4$ was found to determine the kick within a few percent, higher order multipoles being almost negligible. They further found these multipoles to describe well, how the kick is built-up during the inspiral and merger, including breaking effects in the late stages. The numerical results were found to be well reproduced by an "effective Newtonian" formula. A heuristic formula suggested in [99] for the kick magnitude was tested in [187] using numerical simulations of three families of unequal-mass, spinning binaries. They observe good agreement between the model and the numerical simulations and find most of the kick to be generated close to the merger of the holes. The most recent investigation by [40], however, called into question the kick magnitude for unequal masses. In particular, they observe a dependence on η^3 instead of η^2 which implies fewer kicks above 1000 km/s, though still more than the number predicted by the EOB study in [243]. The dependence of the kick on the orientation angle of the spin in the orbital plane was systematically analyzed in [82]. The sinusoidal dependence is in agreement with the heuristic model of [99]. It was also shown that the recoil is with good approximation proportional to the difference between the $\ell = 2$, $m = +2$, and $m = -2$ modes. 2.5 Post-Newtonian order predictions were found to accurately model the spin evolution up to about 60 M before the merger, but not beyond that, illustrating the need for more sophisticated models, such as that of [243]. The asymmetry in the quadrupole radiation of these superkick also implies that such GW sources appear brighter in some directions than others. Implications of spinning binaries for GW detection were also the subject of [266]. They calculate the match between waveforms resulting from different spinning binary configurations and find the inclusion of higher order multipoles necessary to break the degeneracy between the waveforms in the context of matched filtering analysis.

Comparisons of PN predictions with numerical results for the emitted gravitational waveforms from spinning binaries are currently restricted to the case of spins aligned with the orbital angular momentum. First results indicate that these scenarios might be modeled by PN theory with comparable accuracy as in the non-spinning case [154]. There still remains a lot of work to do before more comprehensive statements can be made.

A question of significant astrophysical interest concerns the spin distribution arising from black-hole mergers. This effect was investigated in a series of papers [233–235] which provided semi-analytic fits. An analytic study based on conservation of momentum was presented in [71, 72] and suggests a series of numerical simulations to nail down remaining free parameters in their predictions. The analytic study by [87] pointed out a particularly intriguing scenario: the generation of a non-spinning hole in a merger of a binary with spins anti-aligned with the orbital angular momentum. According to their model, this special case can only be realized in the case of unequal masses. The fitting formulas of [233] as well as numerical simulations presented in [53] agree remarkably well with the study of [87].

4.7 Conclusions

In summary, numerical relativity has achieved what has for a long time been called its "holy grail": The simulation of a black-hole binary through inspiral and merger. The methods used for this breakthrough have turned out to be remarkably robust and have so far been applied with great success to a wider class of black-hole binaries. In the course of the last few years, numerical relativity has thus produced important results for astrophysics, including kicks and diagrams for the spin distribution of black holes. At the same time, the field has established a connection to approximative theories. The good agreement with PN results is encouraging from the point of view of generating hybrid waveforms for use in gravitational wave detection and parameter estimates. The use of numerical waveforms in the data analysis pipeline is currently being started and is widely expected to improve the detection range even of the current generation of GW detectors. Numerical relativity has also opened the door to studying a variety of fundamental questions such as the existence of zoom-whirl orbits and the presence of non-linear phenomena in black-hole ring down.

In spite of the dramatic progress of the field, many open questions remain. Most outstanding among these are a more systematic investigation of the spin parameter space including calibration of the results versus approximative theories. The accuracy of the simulations performed to date has probably been higher than anticipated, but it remains to be seen, whether it will prove sufficient for the daunting task to generate complete waveform template banks for the ongoing effort to detect and observe gravitational waves. It will also be interesting to probe a larger range of parameters, as for example the mass ratio or the kinetic energy of binary spacetimes and compare results with analytic or perturbative predictions. Questions such as these will keep the community busy for years to come and it remains to be seen, how many surprises are still to be discovered in the dynamics of spacetimes involving black holes.

Acknowledgments This work was supported by the DFG grant SFB/Transregio 7 "Gravitational Wave Astronomy" and the ILIAS Sixth Framework programme.

References

1. B. Abbott and the LIGO Scientific Collaboration, LIGO: The Laser Interferometer Gravitational-Wave Observatory. (2007) arXiv:0711.3041 [gr-qc].
2. A. M. Abrahams and C. R. Evans, Gauge-invariant treatment of gravitational radiation near the source: Analysis and numerical simulations. Phys. Rev. D **42**, 2585 (1990).
3. A. M. Abrahams, D. Bernstein, D. Hobill, E. Seidel and L. Smarr, Numerically generated black-hole spacetimes: Interaction with gravitational waves. Phys. Rev. D **45**, 3544 (1992).
4. F. Acernese and the VIRGO Collaboration, Status of VIRGO. Class. Quant. Grav. **22**, S869 (2005).
5. P. Ajith et al., Phenomenological template family for black-hole coalescence waveforms. Class. Quant. Grav. **24** S689 (2007).

6. P. Ajith et al., A template bank for gravitational waveforms from coalescing binary black holes: I. non-spinning binaries. arXiv:0710.2335 (2007).
7. M. Alcubierre and B. Brügmann, Simple excision of a black hole in 3+1 numerical relativity. Phys. Rev. D. **63**, 104006 (2001).
8. M. Alcubierre et al., The 3D grazing collision of two black holes. Phys. Rev. Lett. **87**, 271103 (2001).
9. M. Alcubierre et al., Gauge conditions for long-term numerical black hole evolutions without excision. Phys. Rev. D. **67**, 084023 (2003).
10. M. Alcubierre, Hyperbolic slicings of spacetime: Singularity avoidance and gauge shocks. Class. Quant. Grav. **20**, 607 (2003).
11. M. Alcubierre et al., Dynamical evolution of quasi-circular binary black hole data. Phys. Rev. D. **72**, 044004 (2005).
12. M. Alcubierre, Are gauge shocks really shocks? Class. Quant. Grav. **22**, 4071 (2005).
13. A. Anderson and J. W. York Jr., Fixing Einstein's Equations. Phys. Rev. Lett. **82**, 4384 (1999).
14. M. Anderson et al., Simulating binary neutron stars: Dynamics and gravitational waves. arXiv:0708.2720 [gr-qc]. (2007).
15. M. Ando and the TAMA Collaboration, Current status of the TAMA300 gravitational wave detector. Class. Quant. Grav. **22**, S881 (2005).
16. P. Anninos, D. Hobill, E. Seidel, L. Smarr and W.-M. Suen, Collision of two black holes. Phys. Rev. Lett. **71**, 2851 (1993).
17. P. Anninos, G. Daues, J. Massó, E. Seidel and W.-M. Suen, Horizon boundary condition for black hole spacetimes. Phys. Rev. D. **51**, 5562 (1995).
18. P. Anninos, K. Camarda, J. Massó, E. Seidel and W.-M. Suen, Three-dimensional numerical relativity: The evolution of black holes. Phys. Rev. D. **52**, 2059 (1995).
19. P. Anninos, D. Hobill, E. Seidel, L. Smarr and W.-M. Suen, Head-on collision of two equal mass black holes. Phys. Rev. D. **52**, 2044 (1995).
20. P. Anninos, R. H. Price, J. Pullin, E. Seidel and W.-M. Suen, Head-on collision of two black holes: Comparison of different approaches. Phys. Rev. D. **52**, 4462 (1995).
21. P. Anninos and S. Brandt, Head-on collision of two unequal mass black holes. Phys. Rev. Lett. **81**, 508 (1998).
22. P. Anninos, K. Camarda, J. Libson, J. Massó, E. Seidel and W.-M. Suen, Finding apparent horizons in dynamic 3D numerical spacetimes. Phys. Rev. D. **58**, 024003 (1998).
23. M. Ansorg, Multi-domain spectral method for initial data of arbitrary binaries in general relativity. Class. Quant. Grav. **24**, S1 (2007).
24. T. Apostolatos, Search templates for gravitational waves from precessing, inspiralling binaries. Phys. Rev. D **52** 605 (1995).
25. R. Arnowitt, S. Deser and C. W. Misner, (1962). The dynamics of general relativity. In L. Witten (Ed.), *Gravitation an Introduction to Current Research* (pp. 227–265). New York: John Wiley. gr-qc/0405109.
26. K. G. Arun, B. R. Iyer, B. S. Sathyaprakash and S. Sinha, Higher harmonics increase LISA's mass reach for supermassive black holes. Phys. Rev. D **75** 124002 (2007).
27. K. G. Arun, B. R. Iyer, B. S. Sathyaprakash, S. Sinha and C. Van Den Broek, Higher signal harmonics, LISA's angular resolution and dark energy. Phys. Rev. D **76** 104016 (2007).
28. A. Ashtekar and B. Brishnan, Isolated and dynamical horizons and their applications. Living Rev. Relativity **2004-10** url: http://relativity.livingreviews.org/Articles/lrr-2004-10/download/index.html. Cited 29 Jan 2008.
29. J. G. Baker et al., Collision of boosted black holes. Phys. Rev. D **55**, 829 (1997).
30. J. G. Baker, B. Brügmann and M. Campanelli, Gravitational waves from black hole collisions via an eclectic approach. Class. Quant. Grav. **17**, L149 (2000).
31. J. G. Baker, M. Campanelli, C. O. Lousto and R. Takahashi, Modeling gravitational radiation from coalescing binary black holes. Phys. Rev. D **65**, 124012 (2002).
32. J. Baker, M. Campanelli and C. O. Lousto, The Lazarus project: A pragmatic approach to binary black hole evolutions. Phys. Rev. D. **65** 044001 (2002).

33. J. G. Baker, M. Campanelli, C. O. Lousto and R. Takahashi, The coalescence remnant of spinning binaries. Phys. Rev. D **69**, 027505 (2004).
34. J. G. Baker, J. Centrella, D.-I. Choi, M. Koppitz, and J. van Meter, Gravitational-wave extraction from an inspiraling configuration+ of merging black holes. Phys. Rev. Lett. **96**, 111102 (2006).
35. J. G. Baker, J. Centrella, D.-I. Choi, M. Koppitz and J. van Meter, Binary black hole merger dynamics and waveforms. Phys. Rev. D **73**, 104002 (2006).
36. J. G. Baker et al., Getting a kick out of numerical relativity. Astrophys. J. **653** L93 (2006).
37. J. G. Baker, J. R. van Meter, S. T. McWilliams, J. Centrella and B. J. Kelly, Consistency of post-Newtonian waveforms with numerical relativity. Phys. Rev. Lett. **99**, 181101 (2007).
38. J. G. Baker et al., Binary black hole late inspiral: Simulations for gravitational wave observations. Phys. Rev. D **75**, 124024 (2007).
39. J. G. Baker, M. Campanelli, F. Pretorius and Y. Zlochower, Comparisons of binary black hole merger waveforms. Class. Quant. Grav. **24**, S25 (2007).
40. J. G. Baker et al., Modeling kicks from the merger of generic black-hole binaries. (2005), arXiv:0802.0416 [astro-ph].
41. J. Balakrishna, G. Daues, E. Seidel, W.-M. Suen, M. Tobias and E. Wang, Coordinate conditions in three-dimensional numerical relativity. Class. Quant. Grav. **13** L135 (1996).
42. T. W. Baumgarte, G. B. Cook, M. A. Scheel, S. L. Shapiro and S. A. Teukolsky, Implementing an apparent-horizon finder in three dimensions. Phys. Rev. D **54** 4849 (1996).
43. T. W. Baumgarte and S. L. Shapiro, On the numerical integration of Einstein's field equations. Phys. Rev. D **59** 024007 (1998).
44. T. W. Baumgarte, Innermost stable circular orbit of binary black holes. Phys. Rev. D **62** 024018 (2000).
45. T. W. Baumgarte and S. L. Shapiro, Numerical relativity and compact binaries. Phys. Rept. **376** 41 (2003).
46. T. Baumgarte, P. Brady, J. D. E. Creighton, L. Lehner, F. Pretorius and R. De Voe, Learning about compact binary merger: The interplay between numerical relativity and gravitational-wave astronomy (2006), gr-qc/0612100.
47. F. Acernese et al., A comparison of methods for gravitational wave burst searches from LIGO and Virgo. (2007) gr-qc/0701026.
48. C. Beetle, M. Bruni, L. M. Burko and A. Nerozzi, Towards a novel wave-extraction method for numerical relativity. I. Foundations and initial-value formulation. Phys. Rev. D. **72** 024013 (2005).
49. J. D. Bekenstein, Gravitational-radiation recoil and runaway black holes. Astrophys. J. **183** 657 (1973).
50. M. J. Berger and J. Oliger, Adaptive mesh refinement for hyperbolic partial differential equations. J. Comput. Phys. **53** 484 (1984).
51. E. Berti and V. Cardoso, Quasinormal ringing of Kerr black holes. I: The excitation factors. Phys. Rev. D **74** 104020 (2006).
52. E. Berti et al., Inspiral, merger and ringdown of unequal mass black hole binaries: A multipolar analysis. Phys. Rev. D **76** 064034 (2007).
53. E. Berti, V. Cardoso, J. A. González, U. Sperhake and B. Brügmann, Multipolar analysis of spinning binaries. (2007) arXiv:0711.1097 [gr-qc].
54. L. Blanchet, Innermost circular orbit of binary black holes at the third post-Newtonian approximation. Phys. Rev. D **65** 124009 (2002).
55. L. Blanchet, M. S. S. Qusailah and C. M. Will, Gravitational recoil of sinpiralling black hole binaries to second post-Newtonian order. Astrophys. J. **635** 508 (2005).
56. L. Blanchet, Gravitational radiation from post-newtonian sources and inspiralling compact binaries. Living Rev. Relativity **2006-4** url: http://www.livingreviews.org/Articles/lrr-2006-4/download/index.html. Cited 29 Jan 2008.
57. T. Bode, D. Shoemaker, F. Herrmann and I. Hinder, Delicacy of Binary Black Hole Mergers in the Presence of Spurious Radiation. (2007) arXiv:0711.0669 [gr-qc].
58. T. Bogdanovic, C. S. Reynolds and M. C. Miller, Alignment of the spins of supermassive black holes prior to coalescence (2007), astro-ph/0703054.

59. C. Bona and J. Massó, Hyperbolic evolution systems for numerical relativity. Phys. Rev. Lett. **68**, 1097 (1992).
60. C. Bona, J. Massó, E. Seidel and J. Stela, A new formalism for numerical relativity. Phys. Rev. Lett. **75**, 600 (1995).
61. C. Bona, J. Massó, E. Seidel and J. Stela, First order hyperbolic formalism for numerical relativity. Phys. Rev. D **56**, 3405 (1997).
62. C. Bona, T. Ledvinka and C. Palenzuela, General-covariant evolution formalism for numerical relativity. Phys. Rev. D. **67** 104005 (2003).
63. E. W. Bonning, G. A. Shields and S. Salviander, Recoiling Black Holes in Quasars (2005) arXiv:0705.4263 [astro-ph].
64. I. Booth, Black hole boundaries. Can. J. Phys. **83**, 1073 (2005).
65. M. Boylan-Kolchin, C.-P. Ma and E. Quataert, Core formation in Galactic nuclei due to recoiling black holes. Astrophys. J. **613**, L37 (2004).
66. H. Bondi, M. G. J. van der Burg and R. A. Metzner, Gravitational waves in general relativity VII. Waves from axi-symmetric isolated systems. Proc. Roy. Soc. A. **269**, 21 (1962).
67. E. Bonning, P. Marronetti, D. Neilsen and R. A. Matzner, Physics and initial data for multiple black hole spacetimes. Phys. Rev. D. **68**, 044019 (2003).
68. W. B. Bonnor and M. A. Rotenberg, Transport of momentum by gravitational waves: The linear approximation. Proc. Roy. Soc. A **265**, 109 (1961).
69. J. M. Bowen and J. W. York Jr., Time-asymmetric initial data for black holes and black-hole collisions. Phys. Rev. D. **21**, 2047 (1980).
70. M. Boyle et al., High-accuracy comparison of numerical relativity simulations with post-Newtonian expansions. (2007) arXiv:0710.0158 [gr-qc].
71. L. Boyle, M. Kesden and S. Nissanke, Binary black hole merger: Symmetry and the spin expansion. (2007) arXiv:0709.0299 [gr-qc].
72. L. Boyle and M. Kesden, The spin expansion for binary black hole merger: New predictions and future directions. (2007) arXiv:0712.2819 [astro-ph].
73. S. Brandt and B. Brügmann, A simple construction of initial data for multiple black holes. Phys. Rev. Lett. **78**, 3606 (1997).
74. S. Brandt et al., Grazing collisions of black holes via the excision of singularities. Phys. Rev. Lett. **85**, 5496 (2000).
75. D. R. Brill and R. W. Lindquist, Interaction energy in geometrostatics. Phys. Rev. **131**, 471 (1963).
76. D. Brown et al., Searching for gravitational waves from binary inspiral with LIGO. Class. Quant. Grav. **21**, S1625 (2004).
77. J. D. Brown, Puncture Evolution of Schwarzschild Black Holes. (2007) arXiv:0705.1359 [gr-qc].
78. B. Brügmann, Adaptive mesh and geodesically sliced Schwarzschild spacetime in 3+1 dimensions. Phys. Rev. D **54**, 7361 (1996).
79. B. Brügmann, Binary black hole mergers in 3D numerical relativity. Int. J. Mod. Phys. **8**, 85 (1999).
80. B. Brügmann, W. Tichy and N. Jansen, Numerical simulation of orbiting black holes. Phys. Rev. Lett. **92**, 211101 (2004).
81. B. Brügmann et al., Calibration of moving puncture simulations. Phys. Rev. D. **77**, 024027 (2008).
82. B. Brügmann, J. A. González, M. D. Hannam, S. Husa and U. Sperhake, Exploring black hole superkicks. (2007) arXiv:0707.0135 [gr-qc].
83. Y. Bruhat, The Cauchy problem. In L. Witten (Ed.), *Gravitation: An Introduction to Current Research*. (Cambirdge University Press, Cambridge, 1962).
84. A. Buonanno and T. Damour, Effective one-body approach to general relativistic two-body dynamics. Phys. Rev. D **59** 084006 (1999).
85. A. Buonanno, Y. Chen and M. Valisneri, Detection template families for gravitational waves from the final stages of binary-black-hole inspirals: Nonspinning case. Phys. Rev. D **67** 024016 (2003).

86. A. Buonanno, G. B. Cook and F. Pretorius, Inspiral, merger and ring-down of equal-mass black-hole binaries. Phys. Rev. D. **75**, 124018 (2007).

87. A. Buonanno, L. Kidder and L. Lehner, Estimating the final spin of a binary black hole coalescence. Phys. Rev. D. **77** 026004 (2008).

88. A. Buonanno et al., Toward faithful templates for non-spinning binary black holes using the effective-one-body approach. Phys. Rev. D **76** 104049 (2007).

89. Cactus Computational Toolkit homepage. url: http://www.cactuscode.org/. Cited 29 Jan 2008.

90. G. Calabrese, J. Pullin, O. Sarbach and M. Tiglio, Convergence and stability in numerical relativity. Phys. Rev. D. **66**, 041501 (2002).

91. G. Calabrese, J. Pullin, O. Sarbach, M. Tiglio and O. Reula, Well posed constraint-preserving boundary conditions for the linearized Einstein equations. Commun. Math. Phys. **240**, 377 (2003).

92. M. Campanelli and C. O. Lousto, Second order gauge invariant gravitational perturbations of a Kerr black hole. Phys. Rev. D **59**, 124022 (1999).

93. M. Campanelli, Understanding the fate of merging supermassive black holes. Class. Quant. Grav. **22**, S387 (2005).

94. M. Campanelli, C. O. Lousto, P. Marronetti and Y. Zlochower, Accurate evolutions of orbiting black-hole binaries without excision. Phys. Rev. Lett. **96**, 111101 (2006).

95. M. Campanelli, C. O. Lousto and Y. Zlochower, Last orbit of binary black holes. Phys. Rev. D **73**, 061501 (2006).

96. M. Campanelli, C. O. Lousto and Y. Zlochower, Gravitational radiation from spinning-black-hole binaries: The orbital hang up. Phys. Rev. D **74**, 041501 (2006).

97. M. Campanelli, C. O. Lousto and Y. Zlochower, Spin-orbit interactions in black-hole binaries. Phys. Rev. D **74**, 084023 (2006).

98. M. Campanelli, C. O. Lousto, Y. Zlochower, B. Krishnan and D. Merritt, Spin flips and precession in black-hole-binary mergers. Phys. Rev. D **75** 064030 (2007).

99. M. Campanelli, C. O. Lousto and Y. Zlochower, Large merger recoils and spin flips from generic black-hole binaries. Astrophys. J. **659**, L5 (2007).

100. M. Campanelli, C. O. Lousto, Y. Zlochower and D. Merritt, Maximum gravitational recoil. Phys. Rev. Lett. **98**, 231102 (2007).

101. Carpet Code homepage. url: http://www.carpetcode.org/. Cited 29 Jan 2008.

102. S. Chadrasekhar and S. Detweiler, The quasi-normal modes of the Schwarzschild black hole. Proc. Roy. Soc. A **344** 441 (1975).

103. D.-I. iChoi et al., Recoiling from a kick in the head-on collision of spinning black holes. Phys. Rev. D **76** 104026 (2007).

104. M. W. Choptuik, Universality and scaling in gravitational collapse of a massless scalar field. Phys. Rev. Lett. **70**, 9 (1993).

105. D. Christodoulou, Reversible and irreversible transformations in black hole physics. Phys. Rev. Lett. **25**, 1596 (1970).

106. G. B. Cook, Three-dimensional initial data for the collision of two black holes. II. Quasicircular orbits for equal-mass black holes. Phys. Rev. D **50**, 5025-5032 (1994).

107. G. B. Cook, Initial data for numerical relativity. Living Rev. Relativity **2000-5** url: http://relativity.livingreviews.org/Articles/lrr-2000-5/download/index.html. Cited 29 Jan 2008.

108. G. B. Cook, Corotating and irrotational binary black holes in quasicircular orbits. Phys. Rev. D **65**, 084003 (2002).

109. G. B. Cook and H. Pfeiffer, Excision boundary conditions for black hole initial data. Phys. Rev. D **70**, 104016 (2004).

110. S. Dain, J. L. Jaramillo and B. Krishnan, On the existence of initial data containing isolated black holes. Phys. Rev. D **71**, 064003 (2004).

111. T. Damour, P. Jaranowski and G. Schäfer, Determination of the last stable orbit for circular general relativistic binaries at the third post-Newtonian approximation. Phys. Rev. D **62** 084011 (2000).

112. T. Damour, P. Jaranowski and G. Schäfer, Dynamical invariants for general relativistic two-body systems at the third post-Newtonian approximation. Phys. Rev. D **62**, 044024 (2000).

113. T. Damour, E. Gourgoulhon and P. Grandclément, Circular orbits of corotating binary black holes: Comparison between analytical and numerical results. Phys. Rev. D **66** 024007 (2002).

114. T. Damour and A. Gopakumar, Gravitational recoil during binary black hole coalescence using the effective one body approach. Phys. Rev. D **73** 124006 (2006).

115. T. Damour and A. Nagar, Final spin of a coalescing black-hole binary: An effective-one-body approach. Phys. Rev. D **76** 044003 (2007).

116. P. Diener et al., Accurate evolution of orbiting binary black holes. Phys. Rev. Lett. **96**, 121101 (2006).

117. O. Dreyer, B. Krishnan, E. Schnetter and D. Shoemaker, Introduction to isolated horizons in numerical relativity. Phys. Rev. D **67**, 024018 (2003).

118. G. Efstathiou and M. Rees, High-redshift quasars in the Cold Dark Matter cosmogony. MN-RAS **230**, 5 (1988).

119. K. R. Eppley (1975). *The numerical evolution of the collision of two black holes* Phd Thesis, Princeton University.

120. Z. Etienne, J. A. Faber, Y. T. Liu, S. L. Shapiro, K. Taniguchi and T. Baumgarte, Fully general relativistic simulations of black-hole-neutron star mergers. (2007) arXiv:0712.2460 [astro-ph].

121. X. Fan et al., A survey of $z > 5.7$ Quasars in the sloan digital sky survey. II. Discovery of three additional quasars at $z > 6$. Astron. J. **125**, 1649 (2003).

122. M. Favata, S. A. Hughes and D. E. Holz, How black holes get their kicks: Gravitational radiation recoil revisited. Astrophys. J. **607**, L5 (2004).

123. L. Ferrares and D. Merritt, A fundamental relation between supermassive black holes and their host galaxies. Astrophys. J. **539**, L9 (2000).

124. L. Ferrarese and H. Ford, Supermassive black holes in galactic nuclei: Past, present and future research. Sp. Sci. Rev. **116** 523 (2005).

125. D. R. Fiske, Wave zone extraction of gravitational radiation in three-dimensional numerical relativity. Phys. Rev. D. **71**, 104036 (2005).

126. M. J. Fitchett, The influence of gravitational wave momentum losses on the centre of mass motion of a Newtonian binary system. MNRAS. **203** 1049 (1983).

127. M. J. Fitchett and S. Detweiler, Linear momentum and gravitational waves - Circular orbits around a Schwarzschild black hole. MNRAS. **211** 933 (1984).

128. J. Frauendiener, Conformal Infinity. Living Rev. Relativity **2004-1** url: http://relativity.livingreviews.org/Articles/lrr-2004-1/download/index.html. Cited 19 Feb 2008.

129. H. Friedrich, Cauchy problems for the conformal vacuum field equations in general relativity. Comm. Math. Phys. **91**, 445 (1983).

130. H. Friedrich, Hyperbolic reductions for Einstein's equations. Class. Quant. Grav. **13**, 1451 (1996).

131. H. Friedrich and G. Nagy, The initial boundary value problem for Einstein's vacuum field equations. Commun. Math. Phys. **201**, 619 (1999).

132. S. Frittelli and R. Gomez, Einstein boundary conditions in relation to constraint propagation for the initial-boundary value problem of the Einstein equations. Phys. Rev. D **69**, 124020 (2004).

133. A. Garat and R. H. Price, Nonexistence of conformally flat slices of the Kerr spacetime. Phys. Rev. D **61**, 124011 (2000).

134. D. Garfinkle, Harmonic coordinate method for simulating generic singularities. Phys. Rev. D. **65**, 044029 (2002).

135. K. Gebhardt et al., A relationship between nuclear black hole mass and galaxy velocity dispersion. Astrophys. J. **539**, L13 (2000).

136. J. A. González, U. Sperhake, B. Brügmann, M. D. Hannam and S. Husa, The maximum kick from nonspinning black-hole binary inspiral. Phys. Rev. Lett. **98** 091101 (2007).

137. J. A. González, M. D. Hannam, U. Sperhake, B. Brügmann and S. Husa, Supermassive kicks for spinning black holes. Phys. Rev. Lett. **98** 231101 (2007).

138. E. Gourgoulhon, P. Grandclément and S. Bonazzola, Binary black holes in circular orbits. I. A global spacetime approach. Phys. Rev. D **65**, 044020 (2002).
139. E. Gourgoulhon and J. L. Jaramillo, A 3+1 perspective on null hypersurfaces and isolated horizons. Phys. Rept. **423**, 159 (2006).
140. E. Gourgoulhon, 3+1 Formalism and bases of numerical relativity. (2000) gr-qc/0703035.
141. P. Grandclément, E. Gourgoulhon and S. Bonazzola, Binary black holes in circular orbits. II. Numerical methods and first results. Phys. Rev. D **65**, 044021 (2002).
142. A. Gualandris and D. Merritt, Ejection of supermassive black holes from galaxy cores. (2007) arXiv:0708.0771 [astro-ph].
143. C. Gundlach and J. M. Martín-García, Symmetric hyperbolicity and consistent boundary conditions for second-order Einstein equations. Phys. Rev. D **70**, 044032 (2004).
144. C. Gundlach, G. Calabrese, I. Hinder and J. M. Martín-García, Constraint damping in the Z4 formulation and harmonic gauge. Class. Quant. Grav. **22**, 3767 (2005).
145. C. Gundlach and J. M. Martín-García, Symmetric hyperbolic form of systems of second-order evolution equations subject to constraints. Phys. Rev. D. **70**, 044031 (2004).
146. C. Gundlach and J. M. Martín-García, Critical phenomena in gravitational collapse. Living Rev. Relativity **2007-5** url: http://relativity.livingreviews.org/Articles/lrr-2007-5/download/index.html. Cited 29 Jan 2008.
147. HAD homepage. url: http://had.liu.edu/. Cited 29 Jan 2008.
148. M. G. Haehnelt, M. B. Davies and M. J. Rees, Possible evidence for the ejection of a supermassive black hole from an ongoing merger of galaxies. MNRAS **366**, L22 (2005).
149. S. G. Hah and R. W. Lindquist, The two body problem in geometrodynamics. Ann. Phys. **29**, 304 (1964).
150. Z. Haiman and A. Loeb, What is the highest plausible redshift of luminous quasars? Astrophys. J. **552**, 459 (2001).
151. Z. Haiman, Constraints from gravitational recoil on the growth of supermassive black holes at high redshift. Astrophys. J. **613**, 36 (2004).
152. M. D. Hannam, S. Husa, D. Pollney, B. Brügmann and N. Ó Murchadha, Geometry and regularity of moving punctures. Phys. Rev. Lett. **99**, 241102 (2007).
153. M. D. Hannam et al., Where post-Newtonian and numerical-relativity waveforms meet. (2007) arXiv:0706.1305 [gr-qc].
154. M. D. Hannam, S. Husa, B. Brügmann and A. Gopakumar, Comparison between numerical-relativity and post-Newtonian waveforms from spinning binaries: The orbital hang-up case. (2007) arXiv:0712.3787 [gr-qc].
155. G. Heinzel et al., LISA interferometry: Recent developments. Class. Quant. Grav. **23**, S119 (2006).
156. F. Herrmann, I. Hinder, D. Shoemaker and P. Laguna, Unequal-mass binary black hole plunges and gravitational recoil. Class. Quant. Grav. **24**, S33 (2007).
157. F. Herrmann, I. Hinder, D. Shoemaker, P. Laguna and R. A. Matzner, Gravitational recoil from spinning binary black hole mergers. (2007) gr-qc/0701143.
158. I. Hinder, B. Vaishnav, F. Herrmann, D. Shoemaker and P. Laguna, Universality and final spin in eccentric binary black hole inspirals. (2007) arXiv:0710.5167 [gr-qc].
159. L. Hoffman and A. Loeb, Three-body kick to a bright quasar out of its galaxy during a merger. Astrophys. J. **638**, L75 (2006)
160. K. Holley-Bockelmann, K. Gultekin, D. Shoemaker and N. Yunes, Gravitational wave recoil and the retention of intermediate mass black holes. (2007) arXiv:0707.1334 [astro-ph].
161. S. A. Hughes, (Sort of) Testing relativity with extreme mass ratio inspirals. AIP Conf. Proc. **873** 233 (2006).
162. S. A. Hughes, LISA sources and science. (2007) arXiv:0711.0188 [gr-qc].
163. M. F. Huq, M. W. Choptuik and R. A. Matzner, Locating boosted Kerr and Schwarzschild apparent horizons. Phys. Rev. D **66**, 084024 (2002).
164. R. A. Hulse and J. H. Taylor, Discovery of a pulsar in a binary system. Astrophys. J. **195**, L51 (2004).

165. S. Husa, J. A. González, M. D. Hannam, B. Brügmann and U. Sperhake, Reducing phase error in long numerical binary black hole evolutions with sixth order finite differencing. (2007) arXiv:0706.0740 [gr-qc].
166. J. L. Jaramillo, E. Gourgoulhon and G. A. Mena Marugan, Inner boundary conditions for black hole Initial Data derived from Isolated Horizons. Phys. Rev. D **70**, 124036 (2004).
167. J. L. Jaramillo, J. A. Valiente Kroon and E. Gourgoulhon, From geometry to numerics: Interdisciplinary aspects in mathematical and numerical relativity. (2007) arXiv:0712.2332.
168. R. P. Kerr, Gravitational field of a spinning mass as an example of algebraically special metrics. Phys. Rev. Lett. **11**, 237 (1963).
169. R. P. Kerr and A. Schild (1965), Some algebraically degenerate solutions of Einstein's gravitational field equations. Proc. Symp. Appl. Math. **XVII**, (pp. 199–209).
170. L. E. Kidder, C. M. Will and A. G. Wiseman, Innermost stable orbits for coalescing binary systems of compact objects. Class. Quant. Grav. **9** L125 (1992).
171. L. Kidder, Coalescing binary systems of compact objects to (post)^5/2-Newtonian order. V. Spin effects. Phys. Rev. D **52** 821 (1995).
172. L. E. Kidder, M. A. Scheel and S. A. Teukolsky, Extending the lifetime of 3D black hole computations with a new hyperbolic system of evolution equations. Phys. Rev. D. **64**, 064017 (2001).
173. L. E. Kidder, L. Lindblom, M. A. Scheel, L. T. Buchman and H. P. Pfeiffer, Boundary conditions for the Einstein evolution system. Phys. Rev. D. **71**, 064020 (2005).
174. W. Kinnersley, Type D vacuum metrics. J. Math. Phys. **10**, 1195 (1969).
175. M. Koppitz et al., Recoil Velocities from Equal-Mass Binary-Black-Hole Mergers. Phys. Rev. Lett. **99**, 041102 (2007).
176. J. Kormendy and D. Richstone, Inward bound – The search for supermassive black holes in galactic nuclei. ARA&A **33**, 581 (1995).
177. B. Krishnan, Fundamental properties and applications of quasi-local black hole horizons. (2007) arXiv:0712.1575 [gr-qc].
178. J. P. Leahy and P. Parma (1992). Multiple outbursts in radio galaxies. In J. Roland, H. Sol and G. Pelletier (Eds.), *7.IAP Meeting: Extragalactic radio sources - from beams to jets* (pp. 307–308).
179. E. W. Leaver, Spectral decomposition of the perturbation response of the Schwarzschild geometry. Phys. Rev. D **34** 384 (1986).
180. L. Lehner and O. M. Moreschi, Dealing with delicate issues in waveform calculations. Phys. Rev. D. **76** 124040 (2007).
181. N. I. Libeskind, S. Cole, C. S. Frenk and J. C. Helly, The effect of gravitational recoil on black holes forming in a hierarchical universe. MNRAS **368**, 1381 (2006).
182. A. Lichnerowicz, L'integration des équations de la gravitation relativiste et le problème des n corps. J. Math. Pures et Appl. **23**, 37 (1944).
183. S. Liebling, Singularity threshold of the nonlinear sigma model using 3D adaptive mesh refinement. Phys. Rev. D **66**, 041703(R) (2002).
184. L. Lindblom, M. A. Scheel, L. E. Kidder, R. Owen and O. Rinne, A new generalized harmonic evolution system. Class. Quant. Grav. **23**, S447 (2006).
185. Z. Lippai, Z. Frei and Z. Haiman, Prompt shocks in the gas disk around a recoiling supermassive black hole binary. (2006) arXiv:0801.0739 [astro-ph].
186. A. Loeb, Observable signatures of a black hole ejected by gravitational radiation recoil in a galaxy merger. Class. Quant. Grav. **23**, L71 (2006).
187. C. O. Lousto and Y. Zlochower, Further insight into gravitational recoil. (2007) arXiv:0708.4048.
188. H. Lück et al., Status of the GEO600 detector. Class. Quant. Grav. **23**, L71 (2006).
189. P. MacNeice et al., PARAMESH: A parallel adaptive mesh refinement community toolkit. Comput. Phys. Comm. **136**, 330 (2000).
190. P. Madau and E. Quataert, The effect of gravitational-wave recoil on the demography of massive black holes. Astrophys. J. **606**, L17 (2004).
191. P. Madau, M. J. Rees, M. Volonteri, F. Haardt and S. P. Oh, Early reionization by mini-quasars. Astrophys. J. **604**, 484 (2004).

192. P. Magain et al., Discovery of a bright quasar without a massive host galaxy. Nature. **437**, 381 (2005).
193. P. Marronetti and R. A. Matzner, Solving the initial value problem of two black holes. Phys. Rev. Lett. **85**, 5500 (2000).
194. P. Marronetti, M. F. Huq, P. Laguna, L. Lehner, R. A. Matzner and D. Shoemaker, Approximate analytical solutions to the initial data problem of black hole binary systems. Phys. Rev. D **62**, 024017 (2000).
195. P. Marronetti, M. D. Duez, S. L. Shapiro and T. Baumgarte, Dynamical determination of the innermost stable circular orbit of binary neutron stars. Phys. Rev. Lett. **92**, 141101 (2004).
196. R. A. Matzner, M. F. Huq and D. Shoemaker, Initial data and coordinates for multiple black hole systems. Phys. Rev. D **59**, 024015 (1998).
197. D. Merritt and L. Ferrarese, Black hole demographics from the $M_\bullet$-σ relation. MNRAS **320**, L30 (2001).
198. D. Merritt and R. D. Ekers, Tracing black hole mergers through radio lobe morphology. Science **297**, 1310 (2002).
199. D. Merritt, M. Milosavljević, M. Favata, S. Hughes and D. Holz, Consequences of gravitational radiation recoil. Astrophys. J. **607**, L7 (2004).
200. D. Merritt et al., The nature of the HE0450-2958 system. MNRAS **367**, 1746 (2006).
201. M. Miller, P. Gressman and W.-M. Suen,, Towards a realistic neutron star binary inspiral: Initial data and multiple orbit evolution in full general relativity. Phys. Rev. D. **69**, 064026 (2004).
202. M. Milosavljević and D. Merrit, Formation of galactic nuclei. Astrophys. J. **563**, 34 (2001).
203. C. W. Misner and J. A. Wheeler, Classical physics as geometry. Ann. Phys. (N.Y.). **2**, 525 (1957).
204. C. W. Misner, Wormhole initial conditions. Phys. Rev. **118**, 1110 (1960).
205. C. W. Misner, K. S. Thorne and J. A. Wheeler, Gravitation. (W. H. Freeman, New York, 1973).
206. V. Moncrief, Gravitational perturbations of spherically symmetric systems. I. The exterior problem. Ann. Phys. **88**, 323 (1974).
207. G. Nagy, O. E. Ortiz and O. A. Reula, Strongly hyperbolic second order Einstein's evolution equations. Phys. Rev. D. **70**, 044012 (2004).
208. A. Nerozzi, C. Beetle, M. Bruni, L. M. Burko and D. Pollney, Towards wave extraction in numerical relativity: The quasi-Kinnersley frame. Phys. Rev. D. **72** 024014 (2005).
209. A. Nerozzi, M. Bruni, V. Re and L. M. Burko, Towards a wave-extraction method for numerical relativity: IV. Testing the quasi-Kinnersley method in the Bondi-Sachs framework. Phys. Rev. D. **73** 044020 (2006).
210. A. Nerozzi, Scalar functions for wave extraction in numerical relativity. Phys. Rev. D. **75** 104002 (2007).
211. E. T. Newman and R. Penrose, An approach to gravitational radiation by a method of spin coefficients. J. Math. Phys. **3** 566 (1962).
212. R. O'Leary, E. O'Shaughnessy and F. Rasio, Dynamical interactions and the black hole merger rate of the universe. Phys. Rev. D. **76**, 061504 (2007).
213. N. Ó Murchadha and J. W. York Jr., Initial-value problem of general relativity. I. General formulation and interpretation. Phys. Rev. D. **10**, 428 (1974).
214. N. Ó Murchadha and J. W. York Jr., Initial-value problem of general relativity. II. Stability of solution of the initial-value equations. Phys. Rev. D. **10**, 437 (1974).
215. N. Ó Murchadha and J. W. York Jr., Gravitational potentials: A constructive approach to genera l relativity. Gen. Relativ. Gravit. **7**, 257 (1976).
216. openGR homepage. url: http://wwwrel.ph.utexas.edu/openGR/. Cited 29 Jan 2008.
217. C. Palenzuela, L. Lehner and S. L. Liebling, Orbital dynamics of binary boson star systems. (2007) arXiv:0706.2435 [gr-qc].
218. Y. Pan et al., A data-analysis driven comparison of analytic and numerical coalescing binary waveforms: Nonspinning case. Phys. Rev. D **77**, 012014 (2008).
219. P. Parma, R. D. Ekers and R. Fanti, High resolution radio observations of low luminosity radio galaxies. Astron. Astrophys. Suppl. Ser. **59** 511 (1985).

220. A. Peres, Classical radiation recoil. Phys. Rev. **128**, 2471 (1962).
221. P. C. Peters, Gravitational radiation and the motion of two point masses. Phys. Rev. **136**, B1224 (1964).
222. H. P. Pfeiffer, S. A. Teukolsky and G. B. Cook, Quasicircular orbits for spinning binary black holes. Phys. Rev. D **62** 104018 (2000).
223. H. Pfeiffer (2003). *Initial data for black hole evolutions* Phd Thesis, Cornell University, gr-qc/0510016.
224. H. P. Pfeiffer et al., Reducing orbital eccentricity in binary black hole simulations. (2005) gr-qc/0702106.
225. E. Poisson, The motion of point particles in curved spacetime. Living Rev. Relativity **2004-6** url: http://relativity.livingreviews.org/Articles/lrr-2004-6/download/index.html. Cited 29 Jan 2008.
226. F. Pretorius, Numerical relativity using a generalized harmonic decomposition. Class. Quant. Grav. **22**, 425 (2005).
227. F. Pretorius, Evolution of binary black-hole spacetimes. Phys. Rev. Lett. **95**, 121101 (2005).
228. F. Pretorius and M. W. Choptuik, Adaptive mesh refinement for coupled elliptic-hyperbolic systems. J. Comput. Phys. **218**, 246 (2006).
229. F. Pretorius, Simulation of binary-black-hole spacetimes with a harmonic evolution scheme. Class. Quant. Grav. **23**, 529 (2006).
230. F. Pretorius and D. Khurana, Black hole mergers and unstable circular orbits. Class. Quant. Grav. **24**, S83 (2007).
231. F. Pretorius, Binary black hole coalescence. (2007) arXiv:0710.1338.
232. B. Reimann, Constraint and gauge shocks in one-dimensional numerical relativity. Phys. Rev. D. **71**, 064021 (2005).
233. L. Rezzolla et al., Spin diagrams for equal-mass black-hole binaries with aligned spins. (2007) arXiv:0708.3999.
234. L. Rezzolla et al., The final spin from the coalescence of aligned-spin black-hole binaries. (2007) arXiv:0710.3345 [gr-qc].
235. L. Rezzolla et al., On the final spin from the coalescence of two black holes. (2007) arXiv:0712.3541 [gr-qc].
236. R. K. Sachs, Gravitational waves in general relativity. Proc. Roy. Soc. A. **270**, 103 (1962).
237. Samrai homepage. url: https://computation.llnl.gov/casc/SAMRAI/. Cited 29 Jan 2008.
238. O. Sarbach, G. Calabrese, J. Pullin and M. Tiglio, Hyperbolicity of the BSSN system of Einstein evolution equations. Phys. Rev. D. **66**, 064022 (2002).
239. O. Sarbach and M. Tiglio, Exploiting gauge and constraint freedom in hyperbolic formulations of Einstein's equations. Phys. Rev. D. **66**, 064023 (2002).
240. O. Sarbach and M. Tiglio, Boundary conditions for Einstein's field equations: Analytical and numerical analysis. J. Hyperbol. Diff. Equat. **2**, 839 (2004).
241. M. A. Scheel et al., Solving Einstein's equations with dual coordinate frames. Phys. Rev. D **74**, 104006 (2006).
242. E. Schnetter, S. H. Hawley and I. Hawke, Evolutions in 3D numerical relativity using fixed mesh refinement. Class. Quant. Grav. **21**, 1465 (2004).
243. J. D. Schnittman and A. Buonanno, The distribution of recoil velocities from merging black holes. (2007) astro-ph/0702641.
244. J. D. Schnittman et al., Anatomy of the binary black hole recoil: A multipolar analysis. (2007) arXiv:0707.0301 [astro-ph].
245. E. Seidel, Towards a singularity-proof scheme in numerical relativity. Phys. Rev. Lett. **69**, 1845 (1992).
246. M. Shibata and T. Nakamura, Evolution of three-dimensional gravitational waves: Harmonic slicing case. Phys. Rev. D. **52**, 5428 (1995).
247. M. Shibata, K. Taniguchi and K. Uryū, Merger of binary neutron stars of unequal mass in full general relativity. Phys. Rev. D. **68**, 084020 (2003).
248. M. Shibata and K. Taniguchi, Merger of black hole and neutron star in general relativity: Tidal disruption, torus mass, and gravitational waves. (2003) arXiv:0711.1410 [astro-ph].

249. D. Shoemaker, K. Smith, U. Sperhake, P. Laguna, E. Schnetter and D. Fiske, Moving black holes via singularity excision. Class. Quant. Grav. **20**, 3729 (2003).
250. L. Smarr (1975). *The structure of general relativity with a numerical illustration: The collision of two black holes* Phd Thesis, University of Texas at Austin.
251. L. Smarr, A. Čadež, B. DeWitt and K. Eppley, Collision of two black holes: Theoretical framework. Phys. Rev. D **14**, 2443 (1976).
252. L. Smarr., Space-times generated by computers: Black holes with gravitational radiation. Ann. N. Y. Acad. Sciences. **302**, 569 (1977).
253. L. Smarr and J. W. York Jr., Kinematical conditions in the construction of spacetime. Phys. Rev. D. **17**, 2529 (1978).
254. C. F. Sopuerta, N. Yunes and P. Laguna, Gravitational recoil from binary black hole mergers: The close limit approximation. Phys. Rev. D **74** 124010 (2006).
255. C. F. Sopuerta, N. Yunes and P. Laguna, Gravitational recoil velocities from eccentric binary black hole mergers. Astrophys. J. **656** L9 (2007).
256. U. Sperhake, B. Kelly, P. Laguna, K. L. Smith and E. Schnetter, Black-hole head-on collisions and gravitational waves with fixed mesh-refinement and dynamic singularity excision. Phys. Rev. D. **71**, 124042 (2005).
257. U. Sperhake, Binary black-hole evolutions of excision and puncture data. Phys. Rev. D. **76**, 104015 (2007).
258. U. Sperhake, E. Berti, V. Cardoso, J. A. González and B. Brügmann, Eccentric binary black-hole mergers: The transition from inspiral to plunge in general relativity. (2007) arXiv:0710.3823 [gr-qc].
259. D. Tatsumi et al., Current status of Japanese detectors. (2007) arXiv:0704.2881 [gr-qc].
260. J. H. Taylor and J. M. Weisberg, Further experimental tests of relativistic gravity using the binary pulsar PSR 1913+16. Astrophys. J. **345**, 434 (1989).
261. J. Thornburg, Coordinates and boundary conditions for the general relativistic initial data problem. Class. Quant. Grav. **54**, 1119 (1987).
262. J. Thornburg, A Fast Apparent-Horizon Finder for 3-Dimensional Cartesian Grids in numerical relativity. Class. Quant. Grav. **21**, 743 (2004).
263. K. S. Thorne, Multipole expansions of gravitational radiation. Rev. Mod. Phys. **52** 299 (1980).
264. W. Tichy, B. Brügmann and P. Laguna, Gauge conditions for binary black hole puncture data based on an approximate helical Killing vector. Phys. Rev. D **68**, 064008 (2003).
265. W. Tichy and B. Brügmann, Quasi-equilibrium binary black hole sequences for puncture data derived from helical Killing vector conditions. Phys. Rev. D **69**, 024006 (2004).
266. B. Vaishnav, I. Hinder, F. Herrmann and D. Shoemaker, , Matched filtering of numerical relativity templates of spinning binary black holes. (2007) arXiv:0705.3829 [gr-qc].
267. J. R. van Meter, J. G. Baker, M. Koppitz and D.-I. Choi, How to move a black hole without excision: Gauge conditions for the numerical evolution of a moving puncture. Phys. Rev. D. **73**, 124011 (2006).
268. M. Volonteri, F. Haardt and P. Madau, The assembly and merging history of supermassive black holes in hierarchical models of galaxy formation. Astrophys. J. **582**, 559 (2003).
269. M. Volonteri and R. Perna, Dynamical evolutions of intermediate-mass black holes and their observational signatures in the nearby universe. MNRAS **358**, 913 (2005).
270. M. Volonteri, G. Lodato and P. Natarajan, The evolution of massive black hole seeds. (2007) arXiv:0709.0529 [astro-ph].
271. M. Volonteri, F. Haardt and K. Gultekin, Compact massive objects in Virgo galaxies: The black hole population. (2007) arXiv:0710.5770 [astro-ph].
272. J. Winicour, Characteristic evolution and matching. Living Rev. Relativity **2005-10** url: http://relativity.livingreviews.org/Articles/lrr-2005-10/download/index.html. Cited 29 Jan 2008.
273. H.-J. Yo, T. W. Baumgarte and S. L. Shapiro, Improved numerical stability of stationary black hole evolution calculations. Phys. Rev. D **66**, 084026 (2002).
274. J. W. York Jr., Gravitational degrees of freedom and the initial-value problem. Phys. Rev. Lett. **26**, 1656 (1971).

275. J. W. York Jr., Role of conformal three-geometry in the dynamics of gravitation. Phys. Rev. Lett. **28**, 1082 (1972).
276. J. W. York Jr., Covariant decompositions of symmetric tensors in the theory of gravitation. Ann. Inst. Henri Poincaré A. **21**, 319 (1974).
277. J. W. York Jr., Kinematics and dynamics of general relativity. In L. Smarr (Ed.), *Sources of Gravitational Radiation* (Cambirdge University Press, Cambridge, 1979), (pp. 82–126).
278. J. W. York Jr. and T. Piran, The initial value problem and beyond. In R. A. Matzner and L. C. Shepley (Eds.), *Spacetime and Geometry* (1982), (pp. 147–176).
279. J. W. York Jr., The initial value problem and dynamics. In N. Derielle & T. Piran (Eds.), *Gravitational Radiation* (North-Holland Publishing Company, 1979) (pp. 175–201).
280. J. W. York Jr., Conformal 'thin-sandwich' data for the initial-value problem of general relativity. Phys. Rev. Lett. **82**, 1350 (1999).
281. F. J. Zerilli, Tensor harmonics in canonical form for gravitational radiation and other applications. Phys. Rev. Lett. **82**, 1350 (1999).
282. Y. Zlochower, J. G. Baker, M. Campanelli and C. O. Lousto, Accurate black hole evolutions by fourth-order numerical relativity. Phys. Rev. D **72**, 024021 (2005).

Chapter 5
Numerical Simulations of Black Hole Formation

N. Stergioulas

Abstract Using recent advance in numerical relativity, three-dimensional simulations of the formation of black holes through gravitational collapse of rotating stars have been performed with unprecedented accuracy. In the case of rotating neutron stars, unstable to quasi-radial oscillations, the complete transition from one stationary solution of Einstein's equations to another, including the formation of horizons and gravitational wave emission has been demonstrated. In the case of differentially rotating supermassive stars, non-axisymmetric dynamical instabilities can lead to fragmentation and prompt collapse to supermassive black holes. Here, we present a summary of recent, detailed numerical simulations by Baiotti et al. [1] and Zink et al. [2, 3].

5.1 Introduction

The modeling of black hole spacetimes with collapsing matter-sources in multi-dimensions has proved to be a challenging task in numerical relativity. Recently, through the adoption of advanced numerical techniques, several obstacles have been overcome. This has allowed the detailed simulation of the collapse of dynamically unstable rotating stars to Kerr black holes, as well as the discovery of a new path to black hole formation in supermassive stars, through the onset of non-axisymmetric dynamical instabilities. These recent advances, once enriched with additional physical input, may ultimately allow for a better understanding of the process of black hole formation and the accompanying emission of gravitational waves and high-energy radiation.

The difficulties encountered in such simulations can be traced back to inherent difficulties and complexities of the system of equations which is to be integrated, the Einstein field equations coupled to the general-relativistic hydrodynamics equations, as well as to the immense computational resources needed for three-dimensional (3D) evolutions. In particular, the precise numerical computation

N. Stergioulas (✉)

Department of Physics, Aristotle University of Thessaloniki, 54124 Thessaloniki, Greece
niksterg@auth.gr

Stergioulas, N.: *Numerical Simulations of Black Hole Formation*. Lect. Notes Phys. **769**, 177–208 (2009)
DOI 10.1007/978-3-540-88460-6_5   © Springer-Verlag Berlin Heidelberg 2009

of the gravitational radiation emitted in the process is particularly challenging, as the energy released in gravitational waves is several orders of magnitude smaller than the total rest-mass energy of the system.

The covariant nature of Einstein's equations leads to difficulties in constructing an appropriate coordinate representation which would allow for stable and accurate simulations, especially when black holes form (characterized by unavoidable physical singularities). One of recent successful numerical approaches to this problem relies on reformulating the original ADM approach in order to achieve long-term stability. Building on the experience developed with lower-dimensional formulations, Nakamura, Oohara and Kojima [4] presented a conformal traceless reformulation of the ADM system which subsequent authors (see, e.g., [5–12]) showed to be robust for different classes of spacetimes. The most widespread version was given by [5, 6] and is commonly referred to as the BSSN formulation.

Successful long-term 3D evolutions of black holes in vacuum have been obtained using excision techniques (see, e.g., [13–22]). In this approach, part of the spacetime region within the black hole horizon (causally disconnected from the evolution outside the horizon) is not evolved. Instead, suitable boundary conditions are specified at an excision surface. In simulations where the black hole is not present initially, excision can be applied once the black hole apparent horizon is found. This technique extends the duration of the simulations past the time of black hole formation considerably, which allows for an accurate investigation of the dynamics of trapped surfaces formed during the collapse, from which important information on the mass and spin of the formed black hole can be extracted.

The numerical investigations of black hole formation (beyond spherical symmetry) started in the early 1980s with the pioneering work of Nakamura [23], who adopted the (2+1)+1 formulation of the Einstein equations in cylindrical coordinates and introduced regularity conditions to avoid divergences at coordinate singularities. Nakamura used a "hypergeometric" slicing condition which prevents the grid points from reaching the singularity when a black hole forms. The simulations could track the evolution of the collapse of a $10M_\odot$ "core" of a massive star with different amounts of rotational energy, up to the formation of a rotating black hole. However, the numerical scheme employed was not accurate enough to compute the emitted gravitational radiation. Later on, in a series of papers [24–27], Bardeen, Stark and Piran studied the collapse of rotating relativistic polytropes to black holes, also presenting a first estimate of the associated gravitational radiation. The gravitational field and hydrodynamics equations were formulated using the 3+1 formalism in two spatial dimensions, using the radial gauge and a mixture of singularity-avoiding polar and maximal slicings. The initial model was a spherically symmetric relativistic polytrope of mass M in equilibrium. The gravitational collapse was induced by lowering the pressure in the initial model by a prescribed (and often very large percentage). Simultaneously, an angular momentum distribution, approximating rigid-body rotation, was added to the initial data. With such a setup, the energy ΔE carried away through gravitational waves from the collapse to a Kerr black hole was found to be $\Delta E/Mc^2 < 7 \times 10^{-4}$.

Shibata [28] investigated the effects of rotation on the criterion for prompt adiabatic collapse of rigidly and differentially rotating polytropes to a black hole, finding that the criterion for black hole formation depends strongly on the amount of angular momentum, but only weakly on its (initial) distribution. The effects of shock heating when using a non-isentropic equation of state (EOS hereafter) are important in preventing prompt collapse to black holes in the case of large rotation rates. More recently, Shibata [8, 29] has performed axisymmetric simulations of the collapse of rotating supramassive neutron stars to black holes for a wide range of polytropic EOSs. Parameterizing the "stiffness" of the EOS through the polytropic index N, the final state of the collapse is a Kerr black hole without any noticeable disc formation, when the polytropic index N is in the range $2/3 \leq N \leq 2$. Based on the specific angular momentum distribution in the initial star, an upper limit to the mass of a possible disc was set at less than 10^{-3} of the initial stellar mass [29]. Furthermore, 3D, fully relativistic simulations of the collapse of supramassive, uniformly rotating neutron stars to rotating black holes were presented in [7]. The simulations focused on $N = 1$ polytropes and showed no evidence of massive disc formation or outflows. These results are in agreement with those obtained in axisymmetry [8, 29] and with the simulations presented in [30] (both in axisymmetry and in 3D) which show that for a rapidly rotating polytrope with $J/M^2 < 0.9$ (J being the angular momentum) all the mass falls promptly into the black hole, with no disc being formed.

Baiotti et al. [1] recently presented new, fully 3D simulations of gravitational collapse of uniformly rotating neutron stars, both secularly and dynamically unstable, which were modeled as relativistic polytropes, ranging from slowly rotating models to rapidly rotating models near the mass-shedding limit. For the first time in such 3D simulations, the *event* horizon of the forming black hole was detected, which allowed for a more accurate determination of the black hole mass and spin than it would be otherwise possible using the area of the *apparent* horizon. Several other approaches for measuring the properties of the newly formed Kerr black hole, including the recently proposed *isolated* and *dynamical-horizon* frameworks, were also considered. A comparison among the different methods indicated that the dynamical-horizon approach is simple to implement and yields estimates which are accurate and more robust than those of other methods. In all simulations presented in [1] no evidence for the formation of a sizable, stable disk outside the black hole was found. The gravitational waves extracted from such simulations [31, 32] confirmed the scaling of the emitted energy with the fourth power in the angular momentum of the initial model, but the largest emitted energy (for the most rapidly rotating model) was found to be several orders of magnitude smaller than what was found in the approximate simulations of Stark & Piran [25]. The difference is due to the approximations made when constructing initial data in the latter reference, while [1] constructed highly accurate initial data.

Due to the recent prospect of detecting gravitational radiation directly, the connection between the local dynamics of collapse and the gravitational wave emission is currently receiving increased attention (e.g., [1, 31–33]). In this context, a nonaxisymmetric instability in a star is expected to change the nature of the signal and to enhance the chances of detecting it [34].

Supermassive stars are possible progenitors of supermassive black holes, modeled as polytropes with $N \sim 3$. In spherical symmetry, every general-relativistic polytrope with index $N = 3$ is unstable to radial oscillations [35] – in turn, there exists a critical $N_c < 3$ for which the star is marginally stable. Rotation can increase this critical value again [36]. Collapse is initiated in unstable models when a radial or quasi-radial perturbation grows on a dynamical timescale (corresponding to a turning point along a sequence of constant angular momentum [37]). The cooling sequence of models of supermassive stars has such a turning point, confirmed by numerical simulations [38]. If the star is differentially rotating as it evolves along the cooling sequence, it might encounter a non-axisymmetric instability [39–41]. A recent investigation of the collapse of differentially rotating supermassive stars by Saijo [42] was based on a sequence of relativistic $N = 3$ polytropes.

Zink et al. [2] found that differentially rotating $N = 3$ polytropes exist that can have a quasi-toroidal shape (in addition to the usual quasi-spherical models). Moreover, it was found that such models may be unstable to non-axisymmetric instabilities, leading to fragmentation of the star. Since the initial models were marginally stable against collapse, the resulting fragments did, indeed, subsequently collapse. In one case, where a one-armed instability was dominant, the time evolution of the resulting fragment was followed (using adaptive mesh refinement) until the apparent horizon of a black hole was detected, demonstrating for the first time such a new path to black hole formation. The dynamical instability triggering black hole formation in the above setup may be related to low-$T/|W|$ instabilities, which seems to be associated with the existence of corotation points (points where the frequency of a mode equals the local angular velocity of the fluid), see [43] (T is the kinetic energy of the star, while $|W|$ is its binding energy), see [44].

In numerical simulations of dynamically unstable stars in three spatial dimensions, the collapse to a black hole proceeds in an almost axisymmetric manner, although the initial data is represented on discrete Cartesian grids. It has been found that even when non-axisymmetric perturbations are applied to the collapsing matter, no large deviations from axisymmetry are seen during the initial stages of collapse [42]. In such cases, either the amount of rotational over gravitational binding energy $T/|W|$ is insufficient or the collapse time is too short to admit growth of initial deviations from the symmetric state to significant levels. The situation can be quite different when the system is not unstable to axisymmetric perturbations, or if the collapse stabilizes around a new equilibrium with higher $T/|W|$. If a general-relativistic star encounters a non-axisymmetric instability, the nature of its subsequent evolution may be characterizable by certain properties of the equilibrium model, like the rotation law, $T/|W|$, compactness, and equation of state. For the limit of uniformly rotating, almost homogeneous models of low compactness, one expects, for $T/|W| > 0.27$, a dynamical transition to an ellipsoid by a principle of correspondence with Newtonian gravity. For polytropes with strong differential rotation, the initial model may be *quasi-toroidal*, i.e., it has at least one isodensity surface which is homeomorphic to a torus. If models of this kind, or purely toroidal ones, are subject to the development of a non-axisymmetric instability, they may exhibit fragmentation [45, 46]. It is this last kind of instability that is of particular interest in the case of supermassive stars. Watts et al. ([43], see also [47, 48]) have suggested a possible relation of low-$T/|W|$ instabilities to the location of oscillation-mode pattern speeds with respect to the corotation band.[1]

Since the parameter space of possible initial models is large, and given that three-dimensional simulations of this kind are still quite expensive in terms of computational resources, one is forced to restrict attention to several isolated sequences, where just one initial model parameter is varied, to gain evidence on its systematic effects, and to a plane in parameter space defined by a constant central rest-mass density and a fixed parameter $\Gamma = 4/3$ in the Γ-law equation of state $P = (\Gamma - 1)\rho\epsilon$. As long as one is concerned with the question of stability of modes along a sequence, the consideration of models of constant central density is not overly restrictive as far as the development of the instability is concerned, while the nature of the final remnant might be rather sensitive to this assumption. The choice $\Gamma = 4/3$ is well known to approximately correspond to the adiabatic coefficient of a degenerate, relativistic Fermi gas or to a radiation pressure-dominated gas [49], and is thus closely connected to iron cores and supermassive stars. For supermassive stars an event horizon can form before thermonuclear reactions become important, depending on the metallicity and mass of the progenitor [50]. Because of this, it is conceivable that the type of evolution in [2] can be used as an approximate model of supermassive black hole formation. Gravitational wave detection may uncover so-far unexpected processes involving black hole formation, and in that case it is useful to have a general understanding of possible dynamical scenarios.

Previous work on dynamical instabilities relevant for supermassive stars comes from three areas: the study of (i) fragmentation in Newtonian polytropes, (ii) non-axisymmetric instabilities in general-relativistic polytropes, and (iii) black hole formation by gravitational collapse. In the first area, of particular relevance is the work by Centrella, New et al. [46, 51], since the kind of initial model and subsequent evolution studied in these publications are similar to the ones presented in [2, 3], apart from the fact that Newtonian gravity and a softer equation of state ($\Gamma = 1.3$) were used. New and Shapiro [40] investigated equilibrium sequences of differentially rotating Newtonian polytropes with $\Gamma = 4/3$, in order to present an evolutionary scenario where supermassive stars develop a bar-mode instability, instead of collapsing axisymmetrically. This kind of scenario (see also [34, 39]) is also important when connecting the fragmentation due to non-axisymmetric instabilities presented here to the evolution of supermassive stars.

Non-axisymmetric dynamical instabilities in general relativistic, self-gravitating fluid stars have been studied by several authors [30, 52–56]. In addition, some evidence of fragmentation has been found in [30] in a ring resulting from a "supra-Kerr" collapse with $J/M^2 > 1$ (here J denotes the total angular momentum, and M the ADM mass), but no black hole was identified, although the pressure in the initial data was artificially reduced by a large factor in order to induce collapse. Black hole formation by gravitational collapse has been studied extensively, in recent years in

---

[1] The *corotation band* in a differentially rotating star is the set of frequencies associated with modes having at least one *corotation point*, i.e., a point where the local pattern speed of the instability matches the local angular velocity.

three spatial dimensions (see [1, 7, 30, 57–60] and references in the introduction). The collapse of differentially rotating supermassive stars in the approximation of spatial conformal flatness has been investigated by Saijo [42]. Finally, one should mention the work on low-$T/|W|$ instabilities by Watts et al. [43, 47, 48] and recent numerical studies of related interest can be found in [61–63].

5.2 Numerical Methods

The simulations in [1–3] were performed with a general-relativistic hydrodynamics code, the Whisky code [64], in which the Einstein and hydrodynamics equations are finite differenced on a Cartesian grid and solved using state-of-the-art numerical schemes. The code incorporates the expertise developed over the past years in the numerical solution of the Einstein equations and of the hydrodynamics equations in a curved spacetime (see [9, 10], but also [65] and references therein) and is the result of a collaboration among several European Institutes [66].

The Whisky code solves the general-relativistic hydrodynamics equations on a 3D numerical grid with Cartesian coordinates. The code has been constructed within the framework of the Cactus Computational Toolkit (see [67] for details), developed at the Albert Einstein Institute (Golm) and at the Louisiana State University (Baton Rouge). This public domain code provides high-level facilities such as parallelization, input/output, portability on different platforms, and several evolution schemes to solve general systems of partial differential equations. Special attention is dedicated to the solution of the Einstein equations, whose matter terms in non-vacuum spacetimes are handled by the Whisky code, which incorporates important recent developments regarding, in particular, new numerical methods for the solution of the hydrodynamics equations that have been described in detail in [68]. These include *(i)* the Piecewise Parabolic Method (PPM) [69] and the Essentially Non-Oscillatory (ENO) methods [70] for the cell-reconstruction procedure; *(ii)* the Harten-Lax-van Leer-Einfeldt (HLLE) [71] approximate Riemann solver, the Marquina flux formula [72]; *(iii)* the analytic expression for the left eigenvectors [73] and the compact flux formulae [74] for a Roe-type Riemann solver and a Marquina flux formula; *(iv)* the use of a "method of lines" (MoL) approach for the implementation of high-order time evolution schemes; *(v)* the possibility to couple the general-relativistic hydrodynamics equations with a conformally decomposed three metrics.

While the Cactus code provides at each time step a solution of the Einstein equations [9]

$$G_{\mu\nu} = 8\pi T_{\mu\nu} \,, \tag{5.1}$$

where $G_{\mu\nu}$ is the Einstein tensor and $T_{\mu\nu}$ is the stress–energy tensor, the Whisky code provides the time evolution of the hydrodynamics equations, expressed through the conservation equations for the stress–energy tensor $T^{\mu\nu}$ and for the matter current density J^μ

$$\nabla_\mu T^{\mu\nu} = 0 \,, \quad \nabla_\mu J^\mu = 0. \tag{5.2}$$

In what follows, a brief description is given, on how both the right-and the left-hand sides of (5.1) are computed within the coupled `Cactus/Whisky` codes.

5.2.1 Evolution of the Field Equations

In the ADM formulation [75, 76], the spacetime is foliated with a set of non-intersecting spacelike hypersurfaces. Two kinematic variables relate the hypersurfaces: the lapse function α, which describes the rate of advance of time along a timelike unit vector n^μ normal to a spacelike hypersurface, and the shift three-vector β^i that relates the coordinates of two spacelike hypersurfaces. In this construction the line element reads

$$ds^2 = -(\alpha^2 - \beta_i \beta^i) dt^2 + 2\beta_i dx^i dt + \gamma_{ij} dx^i dx^j . \tag{5.3}$$

The original ADM formulation casts the Einstein equations into a first-order (in time) quasi-linear [77] system of equations. The dependent variables are the three-metric γ_{ij} and the extrinsic curvature K_{ij}, with first-order evolution equations given by

$$\partial_t \gamma_{ij} = -2\alpha K_{ij} + \nabla_i \beta_j + \nabla_j \beta_i, \tag{5.4}$$

$$\partial_t K_{ij} = -\nabla_i \nabla_j \alpha + \alpha \left[R_{ij} + K\, K_{ij} - 2K_{im}K_j^m \right.$$

$$\left. -8\pi \left(S_{ij} - \frac{1}{2}\gamma_{ij}S \right) - 4\pi \rho_{\text{ADM}} \gamma_{ij} \right]$$

$$+ \beta^m \nabla_m K_{ij} + K_{im}\nabla_j \beta^m + K_{mj}\nabla_i \beta^m . \tag{5.5}$$

Here, ∇_i denotes the covariant derivative with respect to the three-metric γ_{ij}, R_{ij} is the Ricci curvature of the three-metric, $K \equiv \gamma^{ij}K_{ij}$ is the trace of the extrinsic curvature, S_{ij} is the projection of the stress–energy tensor onto the spacelike hypersurfaces and $S \equiv \gamma^{ij}S_{ij}$ (for a more detailed discussion, see [78]). In addition to the evolution equations, the Einstein equations also provide four constraint equations to be satisfied on each spacelike hypersurface. These are the Hamiltonian constraint equation

$$^{(3)}R + K^2 - K_{ij}K^{ij} - 16\pi \rho_{\text{ADM}} = 0 , \tag{5.6}$$

and the momentum constraint equations

$$\nabla_j K^{ij} - \gamma^{ij}\nabla_j K - 8\pi j^i = 0 . \tag{5.7}$$

In (5.4), (5.5), (5.6), and (5.7), ρ_{ADM} and j^i are the energy density and the momentum density as measured by an observer moving orthogonally to the spacelike hypersurfaces.

Details of the particular implementation of the conformal traceless reformulation of the ADM system (used in the `Cactus` code in place of the original ADM formulation), as proposed by [4–6] are extensively described in [9, 79]. This formulation makes use of a conformal decomposition of the three-metric, $\tilde{\gamma}_{ij} = e^{-4\phi}\gamma_{ij}$, and the trace-free part of the extrinsic curvature, $A_{ij} = K_{ij} - \gamma_{ij}K/3$, with the conformal factor ϕ chosen to satisfy $e^{4\phi} = \gamma^{1/3}$, where γ is the determinant of the spatial three-metric γ_{ij}. In addition to the evolution equations for the conformal three-metric $\tilde{\gamma}_{ij}$ and the conformal traceless extrinsic curvature $\tilde{A}_{ij}$, there are evolution equations for the conformal factor ϕ, for the trace of the extrinsic curvature K and for the "conformal connection functions" $\tilde{\Gamma}^i \equiv \tilde{\gamma}^{ij}{}_{,j}$. Although the final mixed, first and second-order, evolution system for $\{\phi, K, \tilde{\gamma}_{ij}, \tilde{A}_{ij}, \tilde{\Gamma}^i\}$ is not in any immediate sense hyperbolic, there is evidence showing that the formulation is at least equivalent to a hyperbolic system [80–82]. In the formulation of [5], the auxiliary variables $\tilde{F}_i = -\sum_j \tilde{\gamma}_{ij,j}$ were used instead of the $\tilde{\Gamma}^i$.

In [9, 83] the improved properties of this conformal traceless formulation of the Einstein equations were compared to the ADM system. In particular, in [9] a number of strongly gravitating systems were analyzed numerically with *convergent* high-resolution, shock-capturing (HRSC) methods with *total-variation-diminishing* (TVD) schemes using the equations described in [84]. These included weak and strong gravitational waves, black holes, boson stars and relativistic stars. The results showed that this treatment leads to numerical evolutions of the various strongly gravitating systems which did not show signs of numerical instabilities for sufficiently long times. However, it was also found that the conformal traceless formulation requires grid resolutions higher than the ones needed in the ADM formulation to achieve the same accuracy, when the foliation is made using the "K-driver" approach discussed in [85]. But, in long-term evolutions a small error-growth rate and the absence (or suppression) of numerical instabilities are the most desirable properties.

5.2.1.1 Gauge Choices

The `Cactus` code can handle arbitrary lapse and shift conditions, which can be chosen as appropriate for a given spacetime simulation. In particular, in [1] the hyperbolic K-driver slicing conditions of the form

$$(\partial_t - \beta^i \partial_i)\alpha = -f(\alpha)\,\alpha^2(K - K_0), \tag{5.8}$$

with $f(\alpha) > 0$ and $K_0 \equiv K(t=0)$ was used. This is a generalization of many well-known slicing conditions. For example, setting $f = 1$ one recovers the "harmonic" slicing condition, while setting $f = q/\alpha$, with q an integer, one recovers the generalized "1+log" slicing condition [86]. All of the simulations discussed in [1] used condition (5.8) with $f = 2/\alpha$. This choice was been made mostly because of its computational efficiency, even though "gauge pathologies" could develop with the "1+log" slicings [87, 88].

For the spatial gauge, in [1] the "Gamma-driver" shift conditions were used [18, 79], which essentially act so as to drive the $\tilde{\Gamma}^i$ to be constant. In this respect, the "Gamma-driver" shift conditions are similar to the "Gamma-freezing" condition $\partial_t \tilde{\Gamma}^k = 0$, which, in turn, is closely related to the well-known minimal distortion shift condition [89]. The differences between these two conditions involve the Christoffel symbols and are basically due to the fact that the minimal distortion condition is covariant, while the Gamma-freezing condition is not. In [1] the hyperbolic Gamma-driver condition,

$$\partial_t^2 \beta^i = F \, \partial_t \tilde{\Gamma}^i - \eta \, \partial_t \beta^i, \tag{5.9}$$

where F and η are, in general, positive functions of space and time were used. For the hyperbolic Gamma-driver conditions it is crucial to add a dissipation term with coefficient η to avoid strong oscillations in the shift. Experience has shown that by tuning the value of this dissipation coefficient it is possible to almost freeze the evolution of the system at late times. In [1] the values $F = \frac{3}{4}$ and $\eta = 3$ were chosen and were kept fixed in time.

5.2.2 Evolution of the Equations of Hydrodynamics

An important feature of the Whisky code is the implementation of a *conservative formulation* of the hydrodynamics equations [73, 90, 91], in which the set of (5.2) is written in a hyperbolic, first-order and flux-conservative form of the type

$$\partial_t \mathbf{q} + \partial_i \mathbf{f}^{(i)}(\mathbf{q}) = \mathbf{s}(\mathbf{q}) , \tag{5.10}$$

where $\mathbf{f}^{(i)}(\mathbf{q})$ and $\mathbf{s}(\mathbf{q})$ are the flux-vectors and source terms, respectively [65]. Note that the right-hand side (the source terms) depends only on the metric, and its first derivatives, and on the stress–energy tensor. Furthermore, while the system (5.10) is not strictly hyperbolic, strong hyperbolicity is recovered in a flat spacetime, where $\mathbf{s}(\mathbf{q}) = 0$. As shown by [91], in order to write system (5.2) in the form of system (5.10), the *primitive* hydrodynamical variables (i.e., the rest-mass density ρ and the pressure p (measured in the rest-frame of the fluid), the fluid three-velocity v^i (measured by a local zero-angular momentum observer), the specific internal energy ϵ and the Lorentz factor W are mapped to the so-called *conserved* variables $\mathbf{q} \equiv (D, S^i, \tau)$ via the relations

$$D \equiv \sqrt{\bar{\gamma}} W \rho ,$$

$$S^i \equiv \sqrt{\bar{\gamma}} \rho h W^2 v^i , \tag{5.11}$$

$$\tau \equiv \sqrt{\bar{\gamma}} \left(\rho h W^2 - p \right) - D ,$$

where $h \equiv 1 + \epsilon + p/\rho$ is the specific enthalpy and $W \equiv (1 - \gamma_{ij} v^i v^j)^{-1/2}$. Note that only five of the seven primitive variables are independent.

In order to close the system of equations for the hydrodynamics an EOS which relates the pressure to the rest-mass density and to the energy density must be specified. The (barotropic) polytropic EOS is

$$p = K\rho^{\Gamma} , \tag{5.12}$$

$$e = \rho + \frac{p}{\Gamma - 1} , \tag{5.13}$$

while the "ideal-fluid" EOS is

$$p = (\Gamma - 1)\rho\epsilon . \tag{5.14}$$

Here, e is the energy density in the rest frame of the fluid, K the polytropic constant (not to be confused with the trace of the extrinsic curvature defined earlier) and Γ the adiabatic exponent. In the case of the polytropic EOS (5.12), $\Gamma = 1 + 1/N$, where N is the polytropic index and the evolution equation for τ need not be solved. In the case of the ideal-fluid EOS (5.14), on the other hand, non-isentropic changes can take place in the fluid and the evolution equation for τ needs to be solved. Note that the polytropic EOS (5.12) does not allow any transfer of kinetic energy to thermal energy, a process which occurs in physical shocks (shock heating).

Additional details on numerical methods for general-relativistic hydrodynamics can be found in [65]. An important feature of the first-order hyperbolic form of the equations is that it has allowed to extend to a general-relativistic context the powerful numerical methods developed in classical hydrodynamics, in particular HRSC schemes based on linearized Riemann solvers. Such schemes are essential for a correct representation of shocks, whose presence is expected in several astrophysical scenarios. For an introduction to HRSC methods the reader is referred to [92–94].

5.2.3 Hydrodynamical Excision

Excision boundaries are usually based on the principle that a region of spacetime that is causally disconnected can be ignored without this affecting the solution in the remaining portion of the spacetime. Although this is true for signals and perturbations travelling at physical speeds, numerical calculations may violate this assumption and disturbances, such as gauge waves, may travel at larger speeds thus leaving the physically disconnected regions. Note that, in non-vacuum spacetimes, the excision boundaries for the hydrodynamical and the metric fields need not be the same. For the fluid quantities, in fact, all characteristics emanating from an event in spacetime will propagate within the sound-cone at that event, and, for physically realistic EOSs, this sound-cone will always be contained within the light-cone at that event. As a result, if a region of spacetime contains trapped surfaces, both the hydrodynamical and the metric fields are causally disconnected and both can be excised there. An effective implementation consists of applying the simplest outflow boundary condition (here, by outflow we mean flow into the excision region) at the excision boundary. In practice, one can apply a zeroth-order extrapolation to all variables at the boundary, i.e., a simple copy of the hydrodynamical variables across the excision boundary. Although the actual implementation of this excision technique may depend on the reconstruction method used, the working principle is always the same.

The location of the excision boundary itself is based on the determination of the apparent horizon which, within the Cactus code, is obtained using the fast finder of Thornburg [95]. In [1], the excision boundary is placed a few gridpoints (typically 4), within a surface which is 0.6 times the size of the apparent horizon. Similar or larger excision regions show no problems in vacuum evolutions and since the sound-cones are always contained within the light-cones, one expects no additional problems to arise from the hydrodynamics. More details on how the hydrodynamical excision is applied in practice, as well as tests showing that this method is stable, consistent and converges to the expected order can be found in [96].

5.2.4 General-Relativistic Hydrodynamics

The equations of general-relativistic hydrodynamics are derived from the conservation equations of the stress–energy tensor T^{ab} and the matter current density J^a:

$$\nabla_a T^{ab} = 0, \quad \nabla_a J^a = 0, \tag{5.15}$$

where $J^a = \rho u^a$, ρ is the rest-mass density and u^a the 4-velocity of the fluid. We use the perfect-fluid stress–energy tensor

$$T^{ab} = \rho h u^a u^b + P g^{ab}, \tag{5.16}$$

with P being the fluid pressure, $h = 1 + \epsilon + P/\rho$ the relativistic specific enthalpy and ϵ the specific internal energy of the fluid.

For evolving the hydrodynamical fields we employ the Whisky code [1, 64] which implements the general-relativistic hydrodynamics equations in the hyperbolic first-order flux-conservative form proposed and tested in [84, 91]. This code requires us to add an artificial atmosphere to the computational domain in regions of very low density. We typically choose an atmospheric density of 10^{-5} of the maximal density of the initial model. The evolved state vector $\mathcal{U} = (D, S_i, \tau)^T$ is defined in terms of the *primitive* hydrodynamical variables ρ, P and v^i, the 3-velocity, measured by an Eulerian observer:

$$\mathcal{U} = \begin{bmatrix} D \\ S_j \\ \tau \end{bmatrix} = \begin{bmatrix} \sqrt{\gamma} W \rho \\ \sqrt{\gamma} \rho h W^2 v_j \\ \sqrt{\gamma}(\rho h W^2 - P - W\rho) \end{bmatrix}, \tag{5.17}$$

where $\gamma = det\gamma_{ij}$ and $W = \alpha u^0 = 1/\sqrt{1 - \gamma_{ij}v^iv^j}$ is the Lorentz factor.

The set of equations then reads

$$\partial_t \mathscr{U} + \partial_i F^i = S, \qquad (5.18)$$

with the three flux vectors given by

$$F^i = \begin{bmatrix} \alpha \left(v^i - \frac{1}{\alpha}\beta^i\right) D \\ \alpha \left(\left(v^i - \frac{1}{\alpha}\beta^i\right) S_j + \sqrt{\gamma}P\delta^i_j\right) \\ \alpha \left(\left(v^i - \frac{1}{\alpha}\beta^i\right) \tau + \sqrt{\gamma}v^i P\right) \end{bmatrix}. \qquad (5.19)$$

The source vector S, which contains the curvature-related force and work terms, but no derivatives of the primitive variables, is given by

$$S = \begin{bmatrix} 0 \\ \alpha\sqrt{\gamma}T^{\mu\nu}g_{\nu\sigma}\Gamma^\sigma_{\mu j} \\ \alpha\sqrt{\gamma}(T^{\mu 0}\partial_\mu\alpha - \alpha T^{\mu\nu}\Gamma^0_{\mu\nu}) \end{bmatrix}, \qquad (5.20)$$

where $\Gamma^\alpha_{\mu\nu}$ are the standard 4-Christoffel symbols.

We choose the ideal-fluid Γ-law equation of state,

$$P(\rho,\epsilon) = (\Gamma - 1)\rho\epsilon \qquad (5.21)$$

to close the system of hydrodynamic equations.

5.2.5 Mesh Refinement

In order to ensure adequate spatial resolution while keeping the computational resource requirements of three-dimensional simulations to a minimum, one can use Berger-Oliger style [97] mesh refinement with subcycling in time as implemented by the open-source Carpet [98, 99] driver for the Cactus code. Carpet provides fixed, progressive [1] and adaptive mesh refinement. In [2, 3] a predefined refinement hierarchy with five levels of refinement, arranged in a box-in-box manner centered on the origin was used. The resolution factor between levels is two. Adaptive (or at least progressive) mesh refinement is necessary to track black hole formation in detail and has been performed on one model in [2].

5.2.6 Mode Extraction

To evaluate and quantify the stability or instability of a given model to non-axisymmetric perturbations, once can extract azimuthal modes $e^{im\varphi}$ by means of a Fourier analysis of the rest-mass density on a ring of fixed coordinate radius in the equatorial plane.[2] Following Tohline et al. [100], one can compute complex weighted averages

$$C_m = \frac{1}{2\pi} \int_0^{2\pi} \rho(\varpi, \varphi, z = 0)\, e^{im\varphi} d\varphi \qquad (5.22)$$

and define normalized real mode amplitudes

$$A_m = \frac{|C_m|}{C_0}. \qquad (5.23)$$

Here $\varpi = \sqrt{x^2 + y^2} = const.$ and is chosen to correspond to the initial equatorial radius of maximum density in the case of quasi-toroidal initial models, if not mentioned otherwise. The index m corresponds to the number of azimuthal density nodes and is used to characterize the modes.

5.3 Black Hole Formation in Rotating Neutron Star Collapse

Rotating relativistic stars with mass larger than the maximum mass of nonrotating models (i.e., *supramassive stars*) can become unstable to axisymmetric perturbations. Given equilibrium models of gravitational mass M and central energy density e_c along a sequence of fixed angular momentum or fixed rest mass, the Friedman, Ipser & Sorkin criterion $\partial M / \partial e_c = 0$ [37] can be used to locate the exact onset of the *secular* instability to axisymmetric collapse. The onset of the *dynamical* instability to collapse is located near that of the secular instability but at somewhat larger central energy densities. Unfortunately, no simple criterion exists to determine this location, but the expectation mentioned above has been confirmed by the simulations performed in [1, 7]. Note that in the absence of viscosity or strong magnetic fields, the star is not constrained to rotate uniformly after the onset of the secular instability and could develop differential rotation. In a realistic neutron star, however, viscosity or intense magnetic fields are likely to enforce a uniform rotation and cause the star to collapse soon after it passes the secular instability limit.

The initial data for the fully relativistic, 3D dynamical simulations are usually constructed with an independent, 2D numerical code, that computes accurate stationary equilibrium solutions for axisymmetric and rapidly rotating relativistic stars in polar coordinates, such as the rns code presented in [101]. The data are then transformed to Cartesian coordinates using standard coordinate transformations. Initial data constructed with the rns code have also been used in [9, 10, 102] and details on the accuracy of the code can be found in [44]. In [1] the focus was on initial models constructed with the polytropic EOS (5.12), choosing $\Gamma = 2$ and a

---

[2] These quantities are not gauge-invariant, but they provide a useful way of characterizing the representation of the instability within a given choice of coordinates.

polytropic constant $K_{ID} = 100$ to produce stellar models that are, at least qualitatively, representative of what is expected from observations of neutron stars. More specifically, four models along the line defining the onset of the secular instability and having polar-to-equatorial axes ratio of roughly 0.95, 0.85, 0.75 and 0.65 were selected (these models are indicated as S1–S4 in Fig. 5.1), respectively. Four additional models were defined by increasing the central energy density of the secularly unstable models by 5%, keeping the same axes ratio. These models (indicated as D1–D4 in Fig. 5.1) were found (as expected) to be dynamically unstable in the simulations that were performed.

Figure 5.1 shows the gravitational mass as a function of the central energy density for equilibrium models constructed with the chosen polytropic EOS. The solid, dashed and dotted lines correspond respectively to the sequence of non-rotating models, the sequence of models rotating at the mass-shedding limit and the sequence of models that are at the onset of the secular instability to axisymmetric perturbations. Furthermore, the secularly and dynamically unstable initial models used in the collapse simulations are shown as open and filled circles, respectively.

Table 5.1 summarizes the main equilibrium properties of the initial models. The circumferential equatorial radius is denoted as R_e, while Ω is the angular velocity with respect to an inertial observer at infinity, and r_p/r_e is the ratio of the polar to equatorial coordinate radii. The height of the corotating innermost stable circular

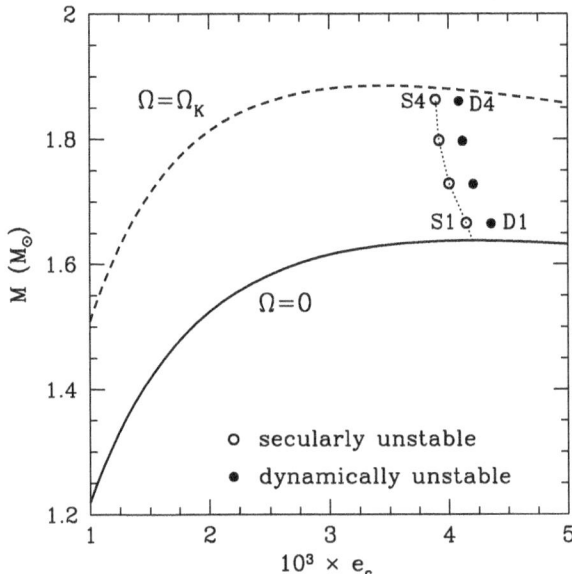

Fig. 5.1 Gravitational mass vs. central energy density for equilibrium models constructed with the polytropic EOS, for $\Gamma = 2$ and polytropic constant $K_{ID} = 100$. The solid, dashed and dotted lines correspond to the sequence of non-rotating models, the sequence of models rotating at the mass-shedding limit and the sequence of models that are at the onset of the secular instability to axisymmetric perturbations. Also shown are a number of secularly unstable (*open circles*) and dynamically unstable models (*filled circles*)

Table 5.1 Equilibrium properties of a number of initial stellar models. The different columns refer respectively to: the central rest-mass density ρ_c, the ratio of the polar to equatorial coordinate radii r_p/r_e, the gravitational mass M, the circumferential equatorial radius R_e, the angular velocity Ω, the ratio J/M^2 where J is the angular momentum, the ratio of rotational kinetic energy to gravitational binding energy $T/|W|$, and the "height" of the corotating ISCO h_+. For all models, $\Gamma = 2$ and $K_{\mathrm{ID}} = 100$

Model	$\rho_c^{\dagger}$	r_p/r_e	M	R_e	$\Omega^{\ddagger}$	J/M^2	$T/|W|^{\ddagger}$	h_+
S1	3.154	0.95	1.666	7.82	1.69	0.207	1.16	1.18
S2	3.066	0.85	1.729	8.30	2.83	0.363	3.53	0.51
S3	3.013	0.75	1.798	8.90	3.49	0.470	5.82	0.04
S4	2.995	0.65	1.863	9.76	3.88	0.545	7.72	–
D1	3.280	0.95	1.665	7.74	1.73	0.206	1.16	1.26
D2	3.189	0.85	1.728	8.21	2.88	0.362	3.52	0.58
D3	3.134	0.75	1.797	8.80	3.55	0.468	5.79	0.10
D4	3.116	0.65	1.861	9.65	3.95	0.543	7.67	–

$\dagger \times 10^{-3}$
$\ddagger \times 10^{-2}$

orbit (ISCO) is defined as $h_+ = R_+ - R_e$, where R_+ is the circumferential radius for a corotating ISCO observer. Note that in those models for which a value of h_+ is not reported, all circular geodesic orbits outside the stellar surface are stable. Other quantities shown are the central rest-mass density ρ_c, the angular momentum J and the ratio of rotational kinetic energy to gravitational binding energy $T/|W|$.

All of the simulations presented in [1] were computed using a uniformly spaced computational grid for which symmetry conditions were imposed across the equatorial plane. Different spatial resolutions were used to check convergence, up to a resolution of $288^2 \times 144$ cells. The outer boundary was set at ~ 2.0 times the initial stellar equatorial radius for D1 and at ~ 1.4 times the initial stellar equatorial radius for D4. The hydrodynamics equations were solved employing the Marquina flux formula and a third-order PPM reconstruction, which was shown in [103] to be superior to other methods in maintaining a highly accurate angular-velocity profile (see also [102] for recent 3D evolutions of rotating relativistic stars with a third-order order PPM reconstruction). The Einstein field equations, on the other hand, were evolved using the ICN evolution scheme, the "1+log" slicing condition and the "Gamma-driver" shift conditions [18]. Finally, both polytropic and ideal-fluid EOSs were used, although no significant difference was found in the dynamics between the two cases. This is most probably related to the small ratio of J/M^2 for uniformly rotating initial models. This implies a relatively rapid collapse and, as a result, no significant shock formation develops.

5.3.1 Dynamics of the Matter

Given an initial stellar model which is dynamically unstable, simple round-off errors would be sufficient to produce an evolution leading either to the gravitational collapse to a black hole or to the migration to the stable branch of the equilibrium configurations [10] (recall that both evolutions are equally probable mathematically, although astrophysically one only expects the case of collapse to occur). In general, however, leaving the onset of the dynamical instability to the cumulative effect of the numerical truncation error is not a good idea, since this produces instability growth times that are dependent on the grid resolution used. For this reason, the collapse is induced by slightly reducing the pressure in the initial configuration. This is done uniformly throughout the star by using a polytropic constant for the evolution, K, that is slightly smaller than the one used to compute the initial data, K_{ID}. Very small perturbations of order $(K_{\mathrm{ID}} - K)/K_{\mathrm{ID}} \lesssim 2\%$ are sufficient to induce collapse.

After imposing the pressure reduction, the Hamiltonian and momentum constraints (the initial-value problem, IVP) are solved to enforce that the constraint violation is at the truncation-error level of the Cartesian grid. This ensures that second order convergence holds from the start of the simulations. Strict second-order convergence is lost when excision is introduced, although the code remains convergent at a lower rate, while the norms of the Hamiltonian constraint start to grow exponentially. Details on how one can solve the IVP implementing the York–Lichnerowicz conformal transverse-traceless decomposition can be found in [104]. If, on the other hand, the IVP is not solved after the pressure change, the constraint violations increase twice as fast and converge to second order only after an initial period of about $20\,\mathrm{M} \sim 0.17$ ms.

The dynamics resulting from the collapse of models S1–S4 and D1–D4 are qualitatively similar. However, as one would expect, models D1–D4 collapse more rapidly to a black hole (the formation of the apparent horizon appears about 5% earlier in coordinate time) are computationally less expensive and therefore better suited for a detailed investigation. Specifically, model D4 which, being rapidly rotating, is already rather flattened initially (i.e., $r_p/r_e = 0.65$) and has the largest J/M^2 among the dynamically unstable models (cf., Fig. 5.1 and Table 5.3) represents the most interesting case. In Figs. 5.2 and 5.3 some representative snapshots of isocontour levels during the collapse of this rapidly rotating model are shown. As the collapse proceeds, the large angular velocity of the initial model produces significant deviations from a spherical infall. Indeed, the parts of the star around the rotation axis that are experiencing smaller centrifugal forces collapse more promptly and, as a result, the configuration increases its oblateness.

At about $t = 0.64$ ms the collapse of model D4 produces an apparent horizon. Soon after this, the central regions of the computational domain are excised, preventing the code from crashing and thus allowing for an extended time evolution. The dynamics of the matter at this stage is shown in the lower panel of Fig. 5.2, which refers to $t = 0.67$ ms and where both the location of the apparent horizon (thick dashed line) and of the excised region (area filled with squares) are shown. By this time, the star has flattened considerably and all of the matter near the rotation axis has fallen inside the apparent horizon, but a disc of low-density matter remains near the equatorial plane, orbiting at very high velocities $\gtrsim 0.2\,c$. The appearance of an effective barrier preventing a purely radial infall of matter far from the rotational axis is likely the consequence of the larger initial angular momentum

Fig. 5.2 Collapse sequence for the rapidly rotating model D4 [1]. Different panels refer to different snapshots during the collapse and show the isocontours of the rest-mass density and velocity field in the (x, z) plane. The isobaric surfaces are logarithmically spaced and a reference length for the vector field is shown in the lower-right panel, for a velocity of $0.2\,c$. The time of the different snapshots appears in the *top right corner* of each panel and is given in ms, while the units on both axes are expressed in km. Note that a region around the singularity that has formed in the *lower panel* is excised from the computational domain and is indicated as an area filled with *squares*. Also shown with a *thick dashed line* is the coordinate location of the apparent horizon

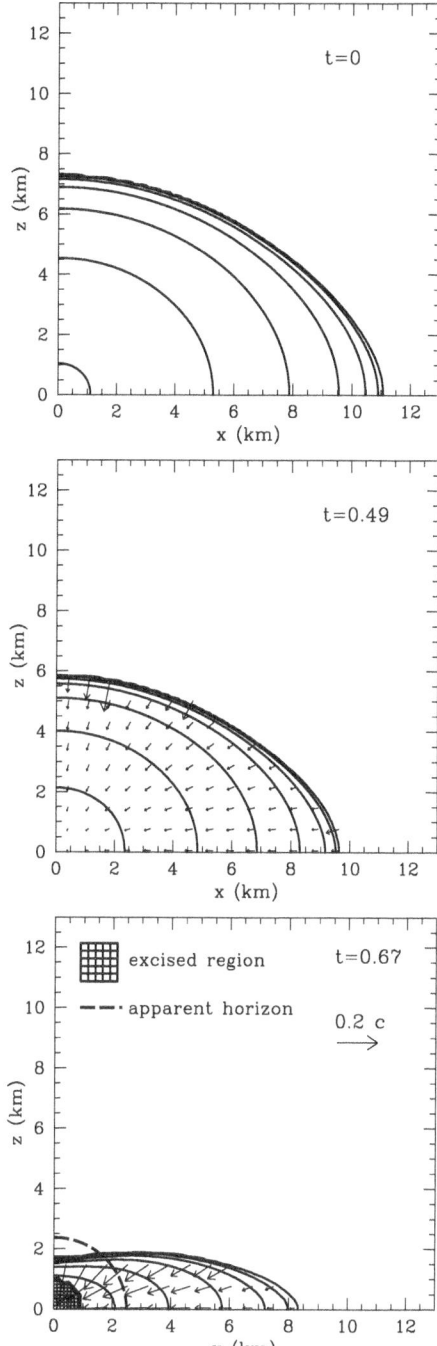

Fig. 5.3 Magnified view of the final stages of the collapse of model D4 [1]. Note that a region around the singularity that has formed is excised from the computational domain and this is indicated as an area filled with *squares*. Also shown with a *thick dashed line* is the coordinate location of the apparent horizon

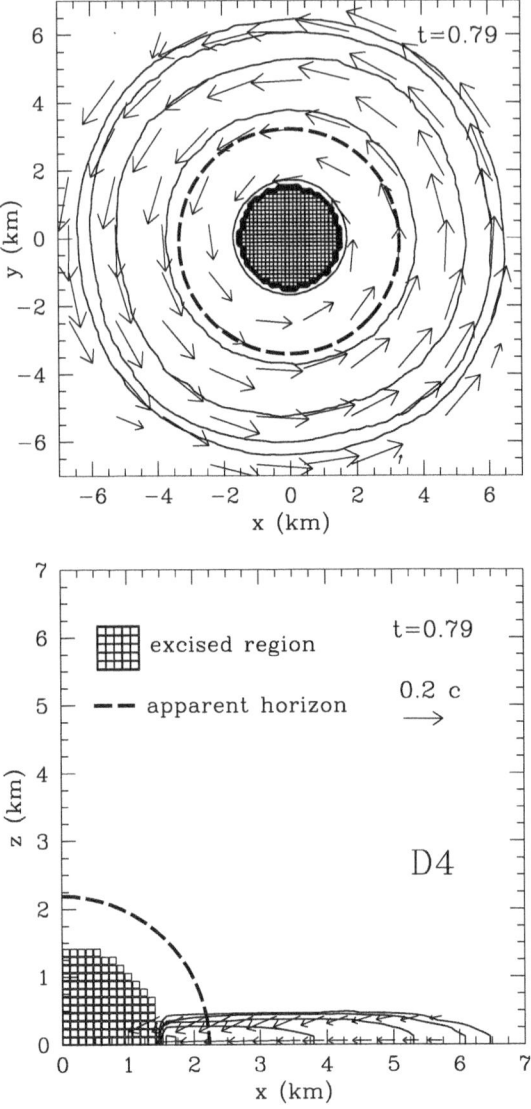

of the this collapsing matter and of the pressure wave originating from the faster collapse along the rotational axis. Note, in fact, that the radial velocity at the equator does not increase significantly at the stellar surface between $t \simeq 0.49$ and $t \simeq 0.67$ ms, but that it actually slightly decreases. This is the opposite of what happens for the radial velocity of the fluid elements in the stellar interior on the equatorial plane.

Note that the disc formed outside the apparent horizon is *not* dynamically stable and, in fact, it rapidly accretes onto the newly formed black hole. This is shown in

Fig. 5.3, which offers a magnified view at a later time of $t = 0.79$ ms. At this stage the disc is considerably flattened but also has large radial inward velocities which induce it to be accreted rapidly onto the black hole. Note that as the area of the apparent horizon increases, so does the excised region, which is allowed to grow accordingly. This can be appreciated by comparing the areas filled with squares in the lower panel of Fig. 5.2 (referring to $t = 0.67$ ms) with the corresponding ones in Fig. 5.3 (referring to $t = 0.79$ ms). By a time $t = 0.85$ ms, essentially all of the residual stellar matter has fallen within the trapped surface of the apparent horizon and the black hole thus formed approaches the Kerr solution. Note that a simple kinematic explanation can be given for the instability of the disc formed during this oblate collapse and comes from relating the position of the outer edge of the disc when this first forms, with the location of the ISCO of the newly formed Kerr black hole. Measuring accurately the mass and spin of the black hole reveals, in fact, that the ISCO is located at $x = 11.08$ km, which is larger than the outer edge of the disc (cf. lower panel of Fig. 5.2).

All simulations to-date agree that no massive and stable discs form for initial models of neutron stars that are uniformly rotating and when a polytropic EOS with $1 \leq N \leq 2$ is used. The results in [1] corroborate this view and in turn imply that the collapse of a rapidly rotating old and cold neutron star cannot lead to the formation of the central engine believed to operate in a gamma-ray burst, namely a rotating black hole surrounded by a centrifugally supported, self-gravitating torus. Relativistic simulations with more appropriate initial data, accounting in particular for the extended envelope of the massive progenitor star which is essential in the so-called collapsar model of gamma-ray bursts [105], will be necessary to shed light on the mechanism responsible for such events.

5.3.2 Dynamics of the Horizons

In order to investigate the formation of a black hole in a numerical simulation, one can use horizon finders, which compute the *apparent* horizon and the *event* horizon. The apparent horizon, which is defined as the outermost trapped surface (a closed surface on which all outgoing photons normal to it have zero expansion), is calculated at every time step and its location is used to set up the excised region inside the horizon.

In contrast, the event horizon, which is an expanding null surface composed of photons which will eventually find themselves trapped, is computed *a posteriori*, once the simulation is finished, by reconstructing the full spacetime from the 3D data each simulation produces. In stationary black hole systems, where no mass-energy falls into the black hole, the apparent and event horizons coincide, but generally (in dynamical spacetimes) the apparent horizon lies inside the event horizon. In the Cactus framework, one can use the fast solver of Thornburg [95] to locate the apparent horizon at every time step, and the level-set finder of Diener [106] to locate the event horizon after the simulation has been completed and the data produced is post-processed. In the collapse simulations presented in [1] the event horizon rapidly grows to its asymptotic value after formation. With a temporal gap of $\sim 10\,M$ after the formation of the event horizon, the apparent horizon is found and then it rapidly approaches the event horizon, always remaining within it.

A simple method for computing the black hole mass is to note that, for a stationary Kerr black hole, the mass can be found directly in terms of the event-horizon geometry as

$$M = \frac{C_{\text{eq}}}{4\pi} , \tag{5.24}$$

where $C_{\text{eq}} \equiv \int_0^{2\pi} \sqrt{g_{\phi\phi}}\,d\phi$ is the proper equatorial circumference. Provided there is a natural choice of equatorial plane, it is expected that, as the black hole settles down to the Kerr solution, C_{eq} will tend to the correct value. However, as numerical errors build up at late times, this asymptotic regime is only reached with a limited accuracy.

The computation of the black hole mass, M, coming from the use of (5.24), is shown in Fig. 5.4, which presents the evolution of the event-horizon mass $M = C_{eq}/4\pi$ for models D1 and D4. Clearly, as the equatorial circumference grows, the agreement with the expected ADM mass improves. The level-set approach of [106], in fact, needs initial guesses for the null surface, which converge exponentially to the correct event-horizon surface for decreasing times, hence introduce a systematic error in the calculation of the event horizon at late times. In Fig. 5.4, different

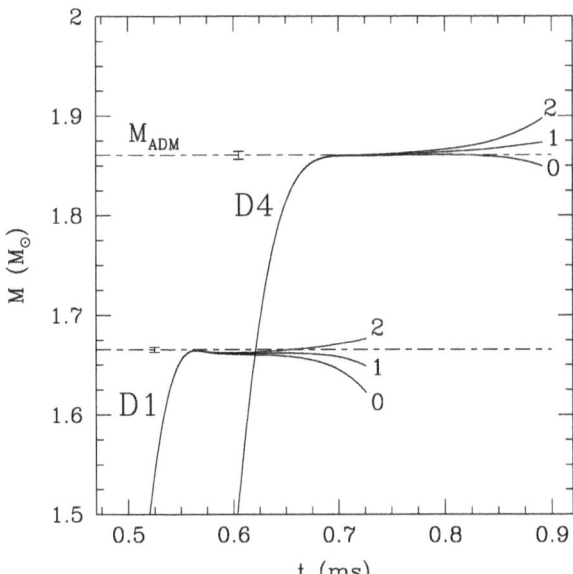

Fig. 5.4 Evolution of the event-horizon mass $M = C_{eq}/4\pi$ for models D1 and D4 [1]. Different lines refer to the different initial guesses for the null surface and are numbered 0, 1 and 2. Note that they all converge exponentially to the correct event-horizon surface for decreasing times

lines refer to the different initial guesses and are numbered "0", "1" and "2", respectively. All choices lead to the same mass determination. Additional methods for determining the mass and angular momentum of black holes formed during collapse are presented in detail in [1].

5.4 Black Hole Formation Through Non-axisymmetric Instabilities in Supermassive Stars

In a recent investigation of the collapse of differentially rotating supermassive stars, Saijo [42] focused on a sequence of relativistic $N = 3$ polytropes with a parameterized rotation law of the commonly used form $j(\Omega) = \tilde{A}^2(\Omega_c - \Omega)$, where Ω_c is the angular velocity at the center, and the parameter $\tilde{A}$ specifies the degree of differential rotation ($\tilde{A} \to \infty$ is uniform rotation). The selected sequence was specified by a constant central density of $\rho_c = 3.38 \times 10^{-6}$ (in units of $K = G = c = 1$) and by the choice $\tilde{A}/r_e = 1/3$, where r_e denotes the equatorial coordinate radius. To examine the indirect collapse by fragmentation of a polytrope with toroidal shape, Zink et al. [2] chose a model with the same central density as in Saijo's [42] models, but with a ratio of polar to equatorial coordinate radius $r_p/r_e = 0.24$ (hereafter called the *reference model*). The ratio of rotational kinetic energy to gravitational binding energy for this model was $T/|W| = 0.227$. While the critical limit for the dynamical f-mode instability in uniform density, uniformly rotating Maclaurin spheroids is $(T/|W|)_{\mathrm{dyn}} = 0.2738$ (e.g., [107]), recent investigations of the stability of soft ($N \sim 3$) differentially rotating polytropes in Newtonian gravity [46, 61, 62, 108] have shown that the Maclaurin approximation is inappropriate for such systems, and generally find the critical $(T/|W|)_{\mathrm{dyn}}$ to be below the Maclaurin value. Consequently, quasi-toroidal configurations with $(T/|W|)_{\mathrm{dyn}} \lesssim 0.27$ can be unstable to non-axisymmetric perturbations and this was the case for the model examined in [2]. In fact, non-axisymmetric instabilities in supermassive stars can lead to a new form of black hole formation.

5.4.1 Quasi-toroidal Polytropes

Equilibrium models of supermassive stars in which non-axisymmetric instabilities are present are general relativistic, differentially rotating polytropes, which are usually constructed as quasi-spherical objects (i.e., their density maximum is at the center). In [2] it was found that *quasi-toroidal* models also exist: Such models have at least one isodensity surface which is homeomorphic to a torus. To construct equilibrium polytropes of this kind, an extended version of the rns code was used [101]. To compute an equilibrium polytrope, the central rest-mass density ρ_c, the coordinate axes ratio r_p/r_e and a barotropic equation of state, $P(\rho)$, need to be specified. For differentially rotating models, the rotation law adopted in [2, 3] requires the additional parameter $A = \tilde{A}/r_e$. The rns code solves the equilibrium equations iteratively, starting from a nonrotating TOV system. When two different branches of models are present (i.e., both a quasi-spherical branch and a quasi-toroidal branch) it may be necessary to select a number of intermediate attractors as trial fields, in order to converge to the desired model. Some models are thus constructed by first obtaining a specific quasi-toroidal model, and then moving in parameter space along the quasi-toroidal branch to the target model. Equilibrium models thus exist in a four-dimensional parameter space $(\Gamma, \rho_c, A, r_p/r_e)$. While in [2] a *reference model* was studied, in [3] sequences in ρ_c, Γ and r_p/r_e containing this model were studied, focusing on the important case $\Gamma = 4/3$, since it approximately represents a radiation pressure-dominated star. Most polytropes were constructed with a meridional grid resolution of $n_r = 601$ radial zones and $n_{\cos\theta} = 301$ angular zones, a maximal harmonic index $\ell_{max} = 10$ for the angular expansion of the Green's functions and a solution accuracy of 10^{-7}. Selected models were tested for convergence with resolutions up to $n_r = 2401$, $n_{\cos\theta} = 1201$ and $\ell_{max} = 20$.

5.4.2 Numerical Setup

All simulations in [2, 3] were performed within the `Cactus/Whisky` framework in full general relativity. The only assumption on symmetry was a reflection invariance with respect to the equatorial plane of the star. The gauge freedom was fixed by the generalized "1+log" slicing condition for the lapse function [109] with $f(\alpha) = 2/\alpha$ and by the hyperbolic-like condition suggested in [110] for the shift vector. In addition, the *Carpet* driver [98] was used for mesh refinement in *Cactus*. The hydrodynamics part of the field equations was evolved using the high-resolution shock-capturing PPM-Marquina implementation in the *Whisky* module [1, 68], and an ideal-fluid equation of state.

The *reference model* has a quasi-toroidal structure, with an off-center density maximum, but a non-zero central density. After mapping the model to the hierarchy of Cartesian grids provided by *Carpet*, a small cylindrical density perturbation of the form

$$\rho(x) \rightarrow \rho(x) \left[1 + \frac{1}{r_e} \sum_{m=1}^{4} \lambda_m B f(\varpi) \sin(m\phi) \right] \qquad (5.25)$$

was added to the equilibrium polytrope. Here, $m \in \{1, 2, 3, 4\}$, λ_m is either 0 or 1, ϖ is the cylindrical radius and $f(\varpi)$ is a radial trial function. Experiments were made with $f(\varpi) = \varpi$ and $f(\varpi) = \varpi^m$, but the exact choice was found not to affect the results significantly. This is true quite generally, since one only requires the trial function to have some reasonable overlap with a set of quasi-normal modes. When a perturbation with $\lambda_i = \delta_{ij}$, $j \in \{1, 2, 3, 4\}$ leads to an instability with the associated number of node lines in the equatorial plane, this instability is denoted with the term $m = j$ *mode* (and the corresponding perturbation $m = j$ *perturbation*) (this is a simplification, since each m is expected to represent a discrete, infinite spectrum of modes [111], from which, however, only the fastest-growing unstable member is observed in numerical simulations).

After perturbing the model, the constraint equations were not solved again, since the amplitude B was chosen such that the violation of the constraints by the initial perturbation was about an order of magnitude smaller than that caused by the systematic error induced by the $m = 4$ symmetry of the Cartesian grid. A fixed *box-in-box* mesh refinement with five levels was used to accurately resolve the central high-density ring. The three innermost grids covered the star, while the two outermost ones pushed the outer boundaries to $6.4 r_e$. The typical resolution used was $65 \times 65 \times 33$ per grid patch, leading to a central resolution of $\sim 10^{-2} r_e$. However, runs with $89 \times 89 \times 45$, $97 \times 97 \times 52$ and $131 \times 131 \times 65$ points per grid patch were also performed to test the resolution dependence of the solution. In order to determine the amplitude of a specific mode in the equatorial plane, a projection onto Fourier modes at certain coordinate radii [112, 113] was performed (notice that, as soon as the system starts to deviate significantly from axisymmetry, care must be taken in interpreting the results since the interpretation of the projection curve as a circle assumes ∂_ϕ to be a Killing vector).

5.4.3 Dynamical Evolution of the Reference Model

Parameters and integral quantities for the reference model used in [2, 3] are shown in Table 5.2, while Table 5.3 lists the parameters for different simulation setups. For setup N, the evolution of the moduli of the equatorial Fourier components at the initial radius of highest density is shown in Fig. 5.5. It is evident that, initially, the $m = 4$ component (*thin dotted line*) induced by the Cartesian grid is dominant. However, the star is unstable to $m = 1$ (*thick solid line*) and $m = 2$ (*thick dashed line*), and these modes consequently grow into the nonlinear regime, their e-folding times being rather close. Setups M1 and M2 are variants of setup N, where only a specific unstable mode was imposed. An approximate measurement of the e-folding times

Table 5.2 Parameters and integral quantities of the reference quasi-toroidal polytrope [2]. The first four quantities are parameters, while Ω_e is the angular velocity at the equator and Ω_K is the associated Keplerian velocity of the same model

Polytropic index Γ	Γ	$4/3$
Central rest-mass density	ρ_c	3.38×10^{-6}
Degree of differential rotation	A	$1/3$
Coordinate axes ratio	r_p/r_e	0.24
Density ratio	ρ_{max}/ρ_c	16.71
ADM mass	M	7.003
Rest mass	M_0	7.052
Equatorial inverse compactness	R_e/M	11.71
Angular momentum	J	52.20
Normalized angular momentum	J/M^2	1.064
Kinetic over binding energy	$T/|W|$	0.227
(See caption)	Ω_e/Ω_K	0.467

Table 5.3 Different setups for the dynamical evolution of the reference model [2]. Adaptive mesh refinement (AMR) was used to investigate black hole formation

Setup	Patch resolution	Refinement levels	Perturbation Modes	Amplitude	AMR
$65 \times 65 \times 33$	5	$m = 1 - 4$	$B = 10^{-3}$	no	
H1	$89 \times 89 \times 45$	5	$m = 1 - 4$	$B = 10^{-3}$	no
H2	$97 \times 97 \times 52$	5	$m = 1 - 4$	$B = 10^{-3}$	no
H3	$131 \times 131 \times 65$	5	$m = 1 - 4$	$B = 10^{-3}$	no
I1	$65 \times 65 \times 33$	5	$m = 1 - 4$	$B = 10^{-4}$	no
I2	$65 \times 65 \times 33$	5	none	$B = 0$	no
M1	$65 \times 65 \times 33$	$5 - 12$	$m = 1$	$B = 10^{-3}$	yes
M2	$65 \times 65 \times 33$	$5 - 12$	$m = 2$	$B = 10^{-3}$	yes

and mode frequencies was obtained within an error of $5 - 10\%$, related to ambiguities in defining the interval of extraction. All setups showed consistent results within this uncertainty. In units of the dynamical timescale, (defined as $t_D = r_e \sqrt{r_e/M}$), the e-folding times were $\approx 0.93 t_D$ for $m = 1$ and $\approx 0.84 t_D$ for $m = 2$, respectively. Mode frequencies were $\approx 3.05/t_D$ for $m = 1$ and $\approx 3.31/t_D$ for $m = 2$, respectively.

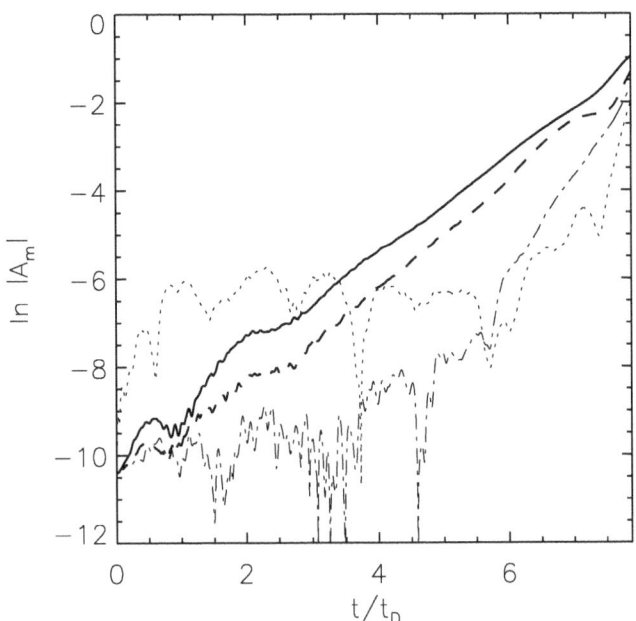

Fig. 5.5 Time evolution of the mode amplitudes in the standard grid setup N [2]. The amplitudes were obtained from a Fourier decomposition of the density profile on the equatorial plane circle at $\varpi = 0.25 r_e$, the initial radius of highest density. Shown are the $m = 1$ (*thick solid line*), $m = 2$ (*thick dashed line*), $m = 3$ (*thin dash-dotted line*) and $m = 4$ (*thin dotted line*) mode amplitudes

To establish whether a black hole is formed by a fragment it is necessary to cover the fragment with significantly more resolution than affordable by fixed mesh refinement. Hence, [2] implemented a simplified adaptive mesh refinement scheme, in order to follow the system to black hole formation: In this scheme, a tracking function (provided by the location of a density maximum) was used to construct a locally fixed hierarchy of grids moving with the fragment. Additional refinement levels were switched on during contraction, until an apparent horizon was found. Since the e-folding times for $m = 1$ and $m = 2$ turned out to be close, the number and interaction behavior of the fragments in the non-linear regime depended sensitively on the initial perturbation. Thus setups M1 and M2 were used to follow the formation and evolution of a specific number of fragments.

The time evolution of the density in the equatorial plane for setup M1 in [2] is shown in Fig. 5.6. While the initial model is axisymmetric, it has already developed a strong $m = 1$ type deviation from axisymmetry at $t = 6.43t_D$, which consequently evolves into a collapsing off-center fragment. At $t = 7.45t_D$, an apparent horizon was found, using the numerical code described in [114]. The horizon was centered on the collapsing fragment at a coordinate radius of $r_{AH} \approx 0.16r_e$ and had an irreducible mass of $M_{AH} \approx 0.24M_{star}$. Its coordinate representation was significantly deformed: its shape was close to ellipsoidal, with an axes ratio of $\sim 2 : 1.1 : 1$. The apparent horizon was covered by three refinement levels and 50–100 grid points along each axis.

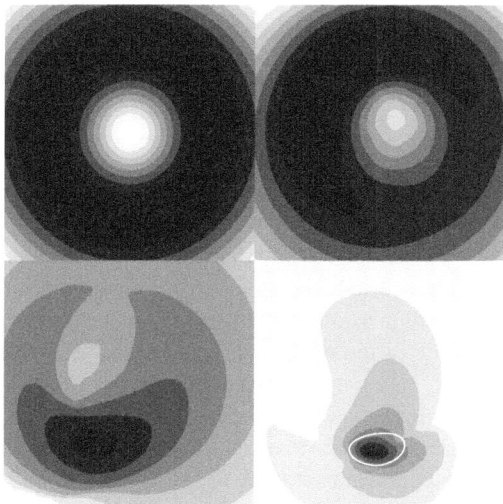

Fig. 5.6 Time evolution of the equatorial plane density for setup M1 [2]. Shown are isocontours of the logarithm of the rest-mass density. The four snapshots extend to $0.37r_e$ and are taken at $t/t_D = 0$, 6.43, 7.14, and 7.45, respectively. They show the formation and collapse of the fragment produced by the $m = 1$ instability. The last slice contains an apparent horizon demarked by the thick white line. Note that the shades of grey used for illustration are adapted to the current maximal density at each time, and that darker shades denote higher densities

Fig. 5.7 Time evolution of
the equatorial plane density
for setup M2 [2]. The snapshots are at the same times as
in Fig. 5.6. In this case, two
fragments formed. Constraint
violations forced a termina-
tion of the simulation before
apparent horizons could be
located

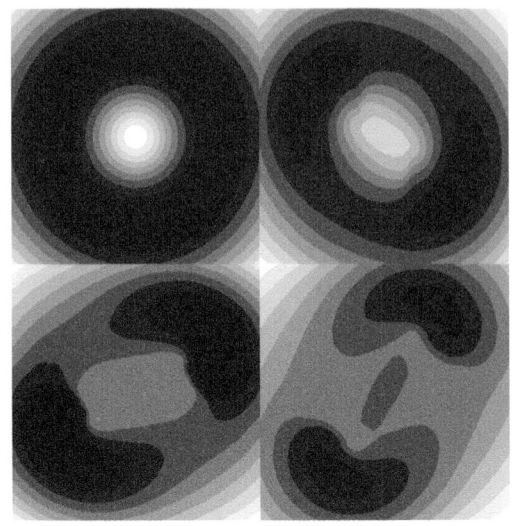

The evolution of the setup M2 in [2] is shown in Fig. 5.7. Two orbiting and collapsing fragments are formed. However, even with the adaptive mesh refinement method, constraint violations prevented the simulation to continue until formation of apparent horizons. Cell-based adaptive mesh refinement, a better choice of gauge, or methods based on numerical analysis (e.g., [115]) might be required in this case.

5.4.4 Models of $\Gamma = 4/3$ Polytropes with Fixed Central Rest-Mass Density

The influence of certain parameters on the stability properties of the relativistic quasi-toroidal polytropes can be studied by following sequences of models, which contain the reference models, along which specific parameters are varied. In [3] models in a two-dimensional parameter subspace, with fixed $\Gamma = 4/3$ and $\rho_c = 10^{-7}$, which differ in the rotation law parameter, A, and in the axes ratio, r_p/r_e, were discussed. The choice of the central density did not seem to affect the almost exponential development of a non-axisymmetric unstable mode in the linear regime considerably, even for very compact quasi-toroidal polytropes. Models with $\rho_c = 10^{-7}$ are already quite compact, with $R_e/M \approx 10\ldots100$ and $r_p/M \approx 2\ldots70$.

To investigate the stability of these models, the initial data have been evolved, imposing a perturbation of the form given by (5.25) with $\lambda_m = 1$, and with a resolution of $65 \times 65 \times 33$ grid points in the outer patches and $97 \times 97 \times 49$ in the innermost patch. Selected models were tested against individual $m = j$ perturbations with $\lambda_m = \delta_{mj}$, with different resolutions, and different densities of the artificial atmosphere, to test consistency and convergence. Also, central rest-mass densities different from 10^{-7} were investigated in a few models.

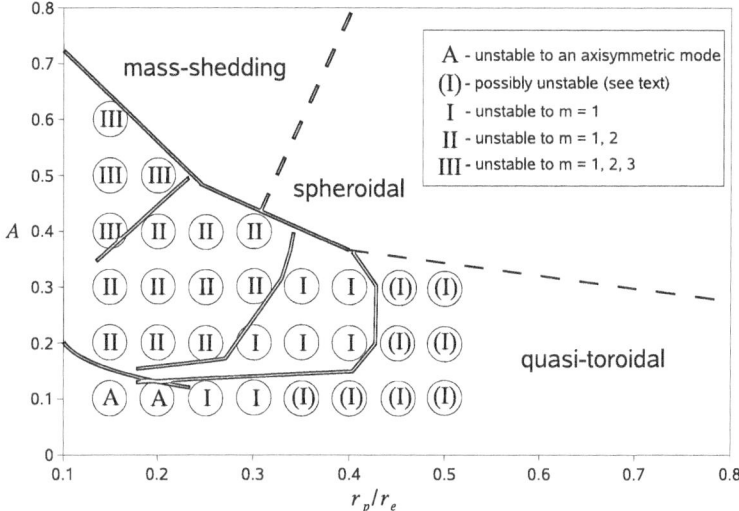

Fig. 5.8 Stability of quasi-toroidal models with $\rho_c = 10^{-7}$ [3]. A Latin number denotes the highest azimuthal order of the unstable modes, i.e., "I" for $m = 1$ unstable, "II" for $m = 1, 2$ unstable, and "III" for $m = 1, 2, 3$ unstable. Models denoted by "(I)" are either long-term unstable with growth times $\tau \gg t_{dyn}$ or stable, and models denoted by "A" exhibit an axisymmetric instability. The line in the lower left corner is the approximate location of the sequence $J/M^2 = 1$, and the three lines inside the quasi-toroidal region are the approximate locations of sequences with $T/|W| = 0.14$ (*right*), $T/|W| = 0.18$ (*middle*), and $T/|W| = 0.26$ (*left*)

Figure 5.8 gives an overview of the stability properties of the selected models. The Latin numbers "I" to "III" refer to the highest m with an unstable mode, i.e., in addition to the reference polytrope, which belongs to the class "II," there exist models which are unstable to an $m = 3$ perturbation, and models which appear to be stable against $m = 2$. The models denoted with an "A" were found to be unstable to an axisymmetric mode and collapse before any non-axisymmetric instability develops. Finally, the models marked with "(I)" could be either stable or long-term unstable with a growth time $\tau \gg t_{dyn}$. Each model was evolved for up to $10t_{dyn}$ to determine its stability. This limit is arbitrary, but imposed by the significant resource requirements of these simulations. If no mode amplitude exceeds the level of the $m = 4$ noise during this time, the model is marked with a "(I)" (this does not imply that the model is actually stable, it could be unstable to an $m = 1$ mode with slow growth rate). The additional lines in Fig. 5.8 are approximate isolines of the functions $T/|W|$ for the values $0.14, 0.18$, and 0.26 and of the function J/M^2 for the value 1. As long as the models do not rotate too differentially, $T/|W|$ still seems to be a reasonable indicator of the non-axisymmetric stability of the polytropes, even though they are quasi-toroidal and relativistic.

The nature of the non-linear behaviour of models exhibiting a non-axisymmetric instability is indicated in Fig. 5.9, where the evolution of the minimum of the lapse function is used to classify the models. Models denoted by "B" have a global mini-

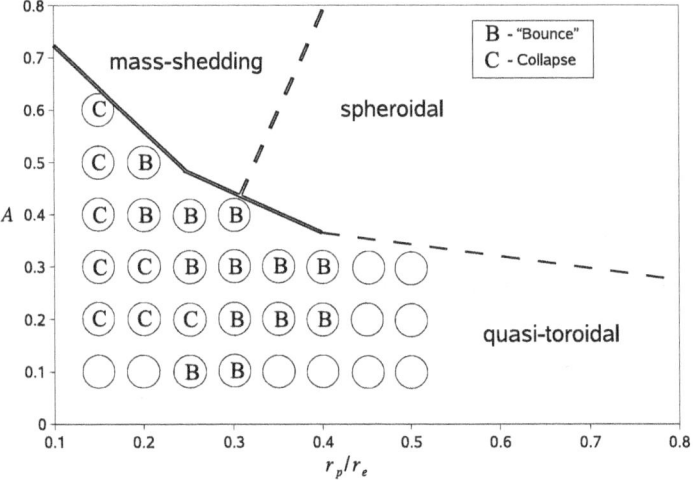

Fig. 5.9 Remnants of the models in Fig. 5.8, which are unstable with respect to non-axisymmetric modes [3]. The non-linear behaviour has been analyzed by observing the evolution of the function min α. Models which show a minimum in this function are marked by "B" for "bounce," while models exhibiting an exponential collapse of the lapse are marked by "C" for "collapse"

mum in the lapse, while models denoted by "C" do not. Given that the compactness of the models increases with smaller axes ratios in this figure, one can expect that a black hole forms for each member of the "C" class. To determine this uniquely, each of these models should be tested using, e.g., the adaptive mesh-refinement technique presented in [2]. Note that, models denoted by "B" might actually form a black hole by delayed collapse.

5.4.5 Nature of the Non-axisymmetric, Dynamical Instability

In order to determine whether the non-axisymmetric, dynamical instability is related to its pattern speed coinciding with the local angular velocity of matter somewhere inside the star, one can define a coordinate angular velocity of matter in the initial equilibrium model, by

$$\Omega(\varpi) \equiv \alpha v^\phi - \beta^\phi. \tag{5.26}$$

This can be compared to the mode pattern speed $1/m \, d\phi/dt$ (approximately valid for the whole star) to determine whether a certain mode has a corotation point. In Fig. 5.10, the angular velocity is plotted in addition to the approximate location of the $m = 1$ and $m = 2$ pattern speeds, for the reference model. Both modes have corotation points: the $m = 1$ mode near the radius of highest density at $0.25 \, R_e$ and the $m = 2$ mode near 0.5–$0.6 \, R_e$. This gives support to the arguments presented in

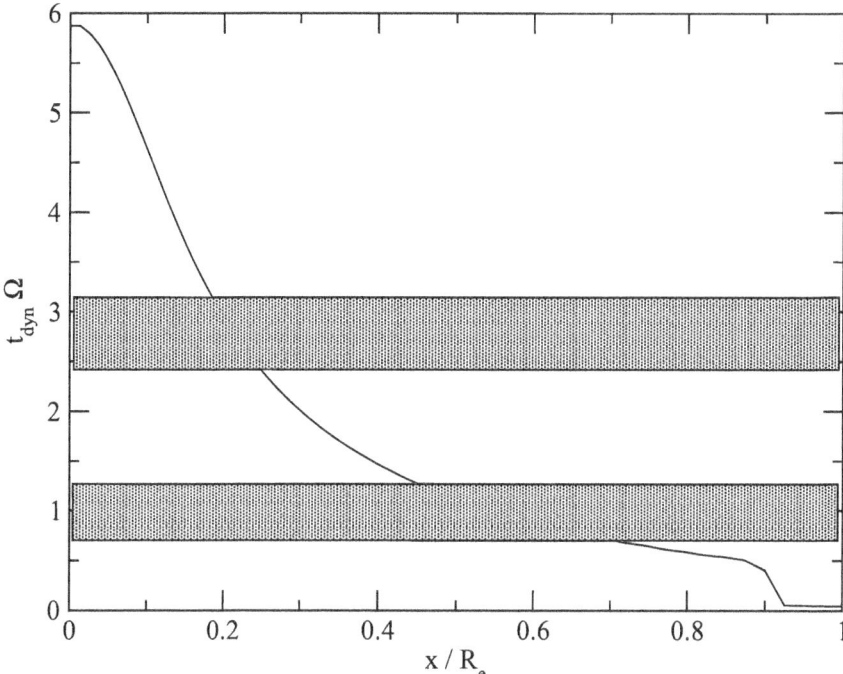

Fig. 5.10 Angular velocity of the reference polytrope in the equatorial plane, and approximate location (with error bar) of the pattern speed of the $m = 1$ mode (*upper rectangle*), and the $m = 2$ mode (*lower rectangle*) [3]. Both modes have corotation points inside the star, where non-axisymmetric instabilities could develop

[43], where the existence of low-$T/|W|$ and spiral-arm instabilities in differentially rotating polytropes are connected to the existence of corotation points.

Acknowledgments I would like to thank my collaborators, Luca Baiotti, Toni Font, Ian Hawke, Frank Löffler, Pedro Montero, Ewald Müller, Christian Ott, Luciano Rezzolla, Ed Seidel, Erik Schnetter, and Burkhard Zink, for their original contributions in the work presented here (which is a summary of joint work). The author acknowledges partial support by an IKYDA 2006/08 grant between IKY (Greece) and DAAD (Germany).

References

1. L. Baiotti, I. Hawke, P. J. Montero, F. Löffler, L. Rezzolla, N. Stergioulas, J. A. Font and E. Seidel, Phys. Rev. D **71**, 024035 (2005).
2. B. Zink, N. Stergioulas, I. Hawke, C. D. Ott, E. Schnetter and E. Müller, Phys. Rev. Lett. **96**, 161101 (2006).
3. B. Zink, N. Stergioulas, I. Hawke, C. D. Ott, E. Schnetter and E. Müller, Phys. Rev. D **76**, 024019 (2007).

4. T. Nakamura, K. I. Oohara and Y. Kojima, Prog. Theor. Phys. Suppl. **90**, 1 (1987).
5. M. Shibata and T. Nakamura, Phys. Rev. D **52**, 5428 (1995).
6. T. W. Baumgarte and S. L. Shapiro, Phys. Rev. D **59**, 024007 (1999).
7. M. Shibata, T. W. Baumgarte and S. L. Shapiro, Phys. Rev. D **61**, 044012 (2000).
8. M. Shibata, Phys. Rev. D **67**, 024033 (2003).
9. M. Alcubierre, B. Brügmann, T. Dramlitsch, J. Font, P. Papadopoulos, E. Seidel, N. Stergioulas and R. Takahashi, Phys. Rev. D **62**, 044034 (2000).
10. J. A. Font, T. Goodale, S. Iyer, M. Miller, L. Rezzolla, E. Seidel, N. Stergioulas, W. M. Suen and M. Tobias, Phys. Rev. D **65**, 084024 (2002).
11. G. Yoneda and H. Shinkai, Phys. Rev. D **66**, 124003 (2002).
12. M. Kawamura, K. Oohara and T. Nakamura, astro-ph/0306481 (2003).
13. E. Seidel and W. M. Suen, Phys. Rev. Lett. **69**, 1845 (1992).
14. S. Brandt, R. Correll, R. Gómez, M. F. Huq, P. Laguna, L. Lehner, P. Marronetti, R. A. Matzner, D. Neilsen, J. Pullin, E. Schnetter, D. Shoemaker and J. Winicour, Phys. Rev. Lett. **85**, 5496 (2000).
15. M. Alcubierre and B. Brügmann, Phys. Rev. D **63**, 104006 (2001).
16. L. E. Kidder, M. A. Scheel, S. A. Teukolsky, E. D. Carlson and G. B. Cook, Phys. Rev. D **62**, 084032 (2000).
17. L. E. Kidder, M. A. Scheel and S. A. Teukolsky, Phys. Rev. D **64**, 064017 (2001).
18. M. Alcubierre, B. Brügmann, D. Pollney, E. Seidel and R. Takahashi, Phys. Rev. D **64**, 61501 (R) (2001).
19. H. J. Yo, T. W. Baumgarte and S. Shapiro, Phys. Rev. D **66**, 084026 (2002).
20. P. Laguna and D. Shoemaker, Class. Quant. Grav. **19**, 3679 (2002).
21. M. W. Choptuik, E. W. Hirschmann, S. L. Liebling and F. Pretorius, Phys. Rev. D **68**, 044007 (2003).
22. U. Sperhake, K. L. Smith, B. Kelly, P. Laguna and D. Shoemaker, Phys. Rev. D **69**, 024012 (2004).
23. M. Nakamura and M. Sasaki, Phys. Lett. **106 B**, 69 (1981).
24. J. M. Bardeen and T. Piran, Phys. Reports **96**, 205 (1983).
25. R. F. Stark and T. Piran, Phys. Rev. Lett. **55**, 891 (1985).
26. R. F. Stark and T. Piran, in *Proceedings of the Fourth Marcell Grossman Meeting on General Relativity*, ed. by R. Ruffini (Elsevier Science Publisher, 1985), p. 327.
27. R. F. Stark, Comp. Phys. Rep. **5**, 221 (1987).
28. M. Shibata, Prog. Theor. Phys. **104**, 325 (2000).
29. M. Shibata, K. Taniguchi and K. Uryū, Phys. Rev. D **68**, 084020 (2003).
30. M. D. Duez, S. L. Shapiro and H. J. Yo, Phys. Rev. D **69**, 104016 (2004).
31. L. Baiotti, I. Hawke, L. Rezzolla and E. Schnetter, Phys. Rev. Lett. **94**, 131101 (2005).
32. L. Baiotti, I. Hawke and L. Rezzolla, Class. Quant. Grav. **24**, S187 (2007).
33. K. S. Thorne, Rev. Mod. Astrono. **10**, 1 (1997).
34. J. L. Houser, J. M. Centrella and S. C. Smith, Phys. Rev. Lett. **72**, 1314 (1994).
35. S. Chandrasekhar, Astrophys J. **140**, 417 (1964).
36. W. Fowler, Astrophys J. **144**, 180 (1966).
37. J. L. Friedman, J. R. Ipser and R. D. Sorkin, Astrophys. J. **325**, 722 (1988).
38. M. Saijo, T. Baumgarte, S. Shapiro and M. Shibata, Astrophys. J. **569**, 349 (2002).
39. P. Bodenheimer and J. Ostriker, Astrophys. J. **180**, 159 (1973).
40. K. C. B. New and S. L. Shapiro, Astrophys. J. **548**, 439 (2001).
41. K. C. B. New and S. L. Shapiro, Class. Quant. Grav. **18**, 3965 (2001).
42. M. Saijo, Astrophys. J. **615**, 866 (2004).
43. A. L. Watts, N. Andersson and D. I. Jones, Astrophys. J. **618**, L37 (2005).
44. N. Stergioulas, Living Rev. Relativity **6**, 3 (2003).
45. M. Rampp, E. Müller and M. Ruffert, Astron. Astrophys. **332**, 969 (1998).
46. J. M. Centrella, K. C. B. New, L. L. Lowe and J. D. Brown, Astrophys. J. **550**, L193 (2001).
47. A. L. Watts, N. Andersson, H. Beyer and B. F. Schutz, Mon. Not. R. Astron. Soc. **342**, 1156 (2003).
48. A. L. Watts, N. Andersson and R. L. Williams, Mon. Not. R. Astron. Soc. **350**, 927 (2004).

49. S. L. Shapiro and S. A. Teukolsky, *Black Holes, White Dwarfs, and Neutron Stars* (John Wiley & Sons, New York, 1983).
50. G. M. Fuller, S. E. Woosley and T. A. Weaver, Astrophys. J. **307**, 675 (1986).
51. K. C. B. New and J. M. Centrella, in *AIP Conf. Proc. 575: Astrophysical Sources for Ground-Based Gravitational Wave Detectors* (2001), p. 221.
52. M. Shibata, T. Baumgarte and S. L. Shapiro, Astrophys. J. **542**, 453 (2000).
53. M. Saijo, M. Shibata, T. Baumgarte and S. L. Shapiro, Astrophys. J. **548**, 919 (2001).
54. M. Shibata and Y. I. Sekiguchi, Phys. Rev. D **71**, 024014 (2005).
55. M. Saijo, Phys. Rev. D **71**, 104038 (2005).
56. L. Baiotti, R. De Pietri, G. M. Manca and L. Rezzolla, submitted to Phys. Rev. D (2006).
57. M. Shibata and S. L. Shapiro, Astrophys. J. **572**, L39 (2002).
58. M. D. Duez, P. Marronetti, S. L. Shapiro and T. Baumgarte, Phys. Rev. D **67**, 024004 (2003).
59. T. W. Baumgarte and S. L. Shapiro, Astrophys. J. **585**, 930 (2003).
60. M. D. Duez, Y. T. Liu, S. L. Shapiro, M. Shibata and B. C. Stephens, Phys. Rev. Lett. **96**, 031101 (2006).
61. M. Shibata, S. Karino and Y. Eriguchi, Mon. Not. R. Astron. Soc. **334**, L27 (2002).
62. M. Shibata, S. Karino and Y. Eriguchi, Mon. Not. R. Astron. Soc. **343**, 619 (2003).
63. M. Saijo and S. Yoshida, Mon. Not. R. Astron. Soc. **368**, 1429 (2006).
64. http://www.aei.mpg.de/~hawke/Whisky.html.
65. J. A. Font, Living Rev. Relativity **6**, 4 (2003).
66. European RTN on Sources of Gravitational Waves, www.eu-network.org.
67. Cactus. www.cactuscode.org.
68. L. Baiotti, I. Hawke, P. Montero and L. Rezzolla, in *Computational Astrophysics in Italy: Methods and Tools*, vol. 1, ed. by R. Capuzzo-Dolcetta (Mem. Soc. Astron. It. Suppl., Trieste, 2003), vol. 1, p. 327.
69. P. Colella and P. R. Woodward, J. Comput. Phys. **54**, 174 (1984).
70. A. Harten, B. Engquist, S. Osher and S. R. Chakrabarty, J. Comput. Phys. **71**, 231 (1987).
71. A. Harten, P. D. Lax and B. van Leer, SIAM Rev. **25**, 35 (1983).
72. M. A. Aloy, J. M. Ibáñez, J. M. Martíand E. Müller, Astroph. J. Supp. **122**, 151 (1999).
73. J. M. Ibáñez, M. A. Aloy, J. A. Font, J. M. Martí, J. A. Miralles and J. A. Pons, *Godunov Methods: Theory and Applications* (Kluwer Academic/Plenum Publishers, 2001).
74. M. A. Aloy, J. A. Pons and J. M. Ibáñez, Comput. Phys. Commun. **120**, 115 (1999).
75. R. Arnowitt, S. Deser and C. W. Misner, in *Gravitation: An Introduction to Current Research*, ed. by L. Witten (John Wiley, New York, 1962), pp. 227–265.
76. E. Gourgoulhon, arXiv:gr-qc/0703035 (2007).
77. R. D. Richtmyer and K. W. Morton, *Difference Methods for Initial Value Problems* (Interscience Publishers, New York, 1967).
78. J. W. York, in *Sources of Gravitational Radiation*, ed. by L. L. Smarr (Cambridge University Press, Cambridge, UK, 1979), p. 83.
79. M. Alcubierre, B. Brügmann, P. Diener, M. Koppitz, D. Pollney, E. Seidel and R. Takahashi, Phys. Rev. D **67**, 084023 (2003).
80. O. Sarbach, G. Calabrese, J. Pullin and M. Tiglio, Phys. Rev. D **66**, 064002 (2002).
81. C. Bona, T. Ledvinka, C. Palenzuela and M. Zacek, Phys. Rev. D **69**, 064036 (2003).
82. G. Nagy, O. E. Ortiz and O. A. Reula, Phys. Rev. D **70**, 044012 (2004).
83. M. Alcubierre, G. Allen, B. Brügmann, E. Seidel and W. Suen, Phys. Rev. D **62**, 124011 (2000).
84. J. A. Font, M. Miller, W. M. Suen and M. Tobias, Phys. Rev. D **61**, 044011 (2000).
85. J. Balakrishna, G. Daues, E. Seidel, W. M. Suen, M. Tobias and E. Wang, Class. Quant. Grav. **13**, L135 (1996).
86. C. Bona, J. Massó, E. Seidel and J. Stela, Phys. Rev. Lett. **75**, 600 (1995).
87. M. Alcubierre, Phys. Rev. D **55**, 5981 (1997).
88. M. Alcubierre and J. Massó, Phys. Rev. D **57**, 4511 (1998).
89. L. Smarr, J. York, Phys. Rev. D **17**, 2529 (1978).
90. J. M. Martí, J. M. Ibáñez and J. A. Miralles, Phys. Rev. D **43**, 3794 (1991).

91. F. Banyuls, J. A. Font, J. M. Ibáñez, J. M. Martí and J. A. Miralles, Astrophys. J. **476**, 221 (1997).
92. R. J. Leveque, *Computational Methods for Astrophysical Fluid Flow* (Springer-Verlag, 1998).
93. C. B. Laney, *Computational Gasdynamics* (Cambridge University Press, 1998).
94. E. F. Toro, *Riemann Solvers and Numerical Methods for Fluid Dynamics* (Springer-Verlag, 1999).
95. J. Thornburg, Class. Quant. Grav. **21**, 743 (2004).
96. I. Hawke, F. Löffler and A. Nerozzi, Phys. Rev. D **71**, 104006 (2005).
97. M. J. Berger and J. Oliger, J. Comput. Phys. **53**, 484 (1984).
98. E. Schnetter, S. H. Hawley and I. Hawke, Class. Quant. Grav. **21**(6), 1465 (2004).
99. http://www.carpetcode.org.
100. J. E. Tohline, R. H. Durisen and M. McCollough, Astrophys. J. **298**, 220 (1985).
101. N. Stergioulas and J. L. Friedman, Astrophys. J. **444**, 306 (1995).
102. N. Stergioulas and J. A. Font, Phys. Rev. Lett. **1148**, 2001 (86).
103. J. A. Font, N. Stergioulas and K. D. Kokkotas, Mon. Not. R. Astron. Soc. **313**, 678 (2000).
104. G. B. Cook, Living Rev. Relativity **5**, 5 (2000).
105. S. E. Woosley, in *Proceedings of the International Workshop held in Rome, CNR headquarters*, ed. by E. Costa, F. Frontera, J. Hjorth (Springer, Berlin Heidelberg, 2000), p. 257.
106. P. Diener, Class. Quant. Grav. **20**, 4901 (2003).
107. J. L. Tassoul, *Theory of Rotating Stars* (Princeton University Press, 1978).
108. M. Saijo, Astrophys. J. **595**, 352 (2003).
109. C. Bona, J. Massó, E. Seidel and J. Stella, Phys. Rev. Lett. **75**, 600 (1995).
110. M. Shibata, Astrophys. J. **595**, 992 (2003).
111. B. F. Schutz and E. Verdaguer, Mon. Not. R. Astron. Soc. **202**, 881 (1983).
112. K. C. B. New, J. M. Centrella and J. E. Tohline, Phys. Rev. D **62**, 064019 (2000).
113. J. De Villiers and J. Hawley, Astrophys. J. **577**, 866 (2002).
114. J. Thornburg, Class. Quant. Grav. **21**, 743 (2004).
115. M. Tiglio, L. Lehner and D. Neilsen, Phys. Rev. D. **70**, 104018 (2004).

Part II
Higher-Dimensional Black Holes

Chapter 6
Black Holes in Higher-Dimensional Gravity

N.A. Obers

Abstract This article reviews some of the recent progress in uncovering the phase structure of black hole solutions in higher-dimensional vacuum Einstein gravity. The two classes on which we focus are Kaluza–Klein black holes, i.e., static solutions with an event horizon in asymptotically flat spaces with compact directions and stationary solutions with an event horizon in asymptotically flat space. Highlights include the recently constructed multi-black hole configurations on the cylinder and thin rotating black rings in dimensions higher than five. The phase diagram that is emerging for each of the two classes will be discussed, including an intriguing connection that relates the phase structure of Kaluza–Klein black holes with that of asymptotically flat rotating black holes.

6.1 Introduction and Motivation

The study of the phase structure of black objects in higher-dimensional gravity (see, e.g., the reviews [1–4]) is interesting for a wide variety of reasons. First of all, it is of intrinsic interest in gravity where the spacetime dimension can be viewed as a tunable parameter. In this way one may discover which properties of black holes are universal and which ones show a dependence on the dimension. We know, for example, that the laws of black hole mechanics are of the former type, while, as will be illustrated in this lecture, properties such as uniqueness and horizon topology are of the latter type. In particular, recent research has revealed that as the dimension increases the phase structure becomes increasingly intricate and diverse. In this context, another interesting phenomenon that has been observed is the existence of critical dimensions above which certain properties of black holes can change drastically. Uncovering the phases of black holes is also relevant for the issue of classical stability of black hole solutions as well as gravitational phase transitions between different solutions, such as those that involve a change of topology of the event horizon. Furthermore, information about the full structure of the static or stationary

N.A. Obers (✉)
Niels Bohr Institute, Blegdamsvej 17, DK-2100 Copenhagen, Denmark
obers@nbi.dk

Obers, N.A.: *Black Holes in Higher-Dimensional Gravity.* Lect. Notes Phys. **769**, 211–258 (2009)
DOI 10.1007/978-3-540-88460-6_6　　　　　　　　　　　　　ⓒ Springer-Verlag Berlin Heidelberg 2009

phases of the theory can provide important clues about the time-dependent trajectories that interpolate between different phases.

Going beyond pure Einstein gravity, there are also important motivations originating from string theory. String/M-theory at low energies is described by higher-dimensional theories of gravity, namely various types of supergravities. As a consequence, black objects in pure gravity are often intimately related to black hole/brane solutions in string theory. These charged cousins and their near-extremal limits play an important role in the microscopic understanding of black hole entropy [5] and other physical properties (see also, e.g., the reviews [6, 7]) . A related application is in the gauge/gravity correspondence [8, 9], where the near-extremal limits of these black branes give rise to phases in the corresponding dual non-gravitational theories at finite temperature. In this way, finding new black objects can lead to the prediction of new phases in these thermal non-gravitational theories (see, e.g., [10, 11]). Finally, if large extra dimensions [12, 13] are realized in Nature, higher-dimensional black holes would be important as possible objects to be produced in accelerators or observed in the Universe (see, e.g., the review [14]).

In the past 7 years, the two classes that have been studied most intensely are

– stationary solutions with an event horizon in asymptotically flat space
– static solutions with an event horizon in asymptotically flat spaces with compact directions

For brevity, we will often refer in this lecture to the first type as *rotating black holes* and the second type as *Kaluza–Klein black holes*. In this nomenclature, the term 'black hole' stands for any object with an event horizon, regardless of its horizon topology (i.e., not necessarily spherical). We also allow for the possibility of multiple disconnected event horizons, to which we refer as *multi-black hole solutions*.

For rotating black holes most progress in recent years has been in five dimensions. Here, it has been found that in addition to the Myers–Perry (MP) black holes [15], there exist rotating black rings [2, 16] and multi-black hole solutions like black Saturns and multi-black rings [17–20] including those with two independent angular momenta [21–23]. All of these are exact solutions which have been obtained with the aid of special ansätze [24, 25] based on symmetries and inverse scattering techniques [26–29]. We refer in particular to the review [2] for further details on the black ring in five dimensions and [18] for a discussion of the phase diagram in five dimensions for the case of rotating black holes with a single angular momentum. Moreover, the very recent review [4] provides a pedagogical overview of black holes in higher dimensions, including the more general phase structure of five-dimensional stationary black holes and solution-generating techniques.

Only recently has there been significant progress in exploring the phase structure of stationary solutions in six and more dimensions [30]. This includes the explicit construction of thin black rings in six and higher dimensions [30] based on a perturbative technique known as matched asymptotic expansion [31–35]. Furthermore, in [30] the correspondence between ultra-spinning black holes [36] and black membranes on a two-torus was exploited, to take steps toward qualitatively completing the phase diagram of rotating blackfolds with a single angular momentum. That has led to the proposal that there is a connection between MP black holes and black rings, and between MP black holes and black Saturns, through merger transitions involving two kinds of 'pinched' black holes. More generally, this analogy suggests an infinite number of pinched black holes of spherical topology leading to a complicated pattern of connections and mergers between phases. The proposed phase diagram was obtained by importing the present knowledge of phase of Kaluza–Klein black holes on a two-torus.

For Kaluza–Klein (KK) black holes, most progress has been for the simplest KK space, namely Minkowski space times a circle. The simplest static solution of Einstein gravity (in five or more dimensions) in this case is the uniform black string, which has a factorized form consisting of a Schwarzschild–Tangherlini black hole and an extra flat (compactified) direction. But there are many more phases of KK black holes, which in recent years have been uncovered by a combination of perturbative techniques (matched asymptotic expansion), numerical methods and exact solutions. These phases include non-uniform black strings (see [37–42] for numerical results), localized black holes (see [31–35, 43–45] for analytical results and [46–48] for numerical solutions) and bubble-black hole sequences [49]. Here recent progress [35] includes the construction of small mass multi-black hole configurations localized on the circle which in some sense parallel the multi-black hole configurations obtained for rotating black holes.

All of these static, uncharged phases can be depicted in a two-dimensional phase diagram [50–52] parameterized by the mass and tension. Mapping out this phase structure has consequences for the endpoint of the Gregory–Laflamme instability [53, 54] of the neutral black string, which is a long wavelength instability that involves perturbations with an oscillating profile along the direction of the string. The non-uniform black string phase emerges from the uniform black string phase at the Gregory–Laflamme point, which is determined by the (time-independent) threshold mode where the instability sets in. An interesting property that has been found in this context is the existence of a critical dimension [39] where the transition of the uniform black string into the non-uniform black string changes from first order into second order. Moreover, it has been shown [41, 55–57] that the localized black hole phase meets the non-uniform black string phase in a horizon-topology changing merger point. Turning to more recent developments, we note that the new multi-black hole configurations of [35] raise the question of existence of new non-uniform black strings. Furthermore, analysis of the three black hole configuration [35] suggests the possibility of a new class of static lumpy black holes in Kaluza–Klein space.

Many of the insights obtained in this simplest case are expected to carry over as we go to Kaluza–Klein spaces with higher-dimensional compact spaces [3, 58, 59], although the degree of complexity in these cases will increase substantially.

In summary, recent research has shown that in going from four to higher dimensions in vacuum Einstein gravity a very rich phase structure of black holes is observed with fascinating new properties, such as symmetry breaking, new horizon topologies, merger points and in some cases infinite non-uniqueness. Obviously one of the reasons for this richer phase structure is that as the dimension increases there are many more degrees of freedom for the metric. Furthermore, for stationary solutions every time the dimension increases two units, there is one more orthogonal rotation plane available. Another reason is the existence of extended objects in higher dimensions, such as black p-branes (including the uniform black string for $p = 1$). Finally, allowing for compact directions introduces extra scales, and hence more dimensionless parameters in the problem.

The reasons that make the phase structure so rich, such as the increase of the degrees of freedom and the appearance of fewer symmetries, are those that also make it hard to uncover. As the overview above illustrates, there has been remarkable progress in recent years, but we have probably only seen a glimpse of the full phase structure of black holes in higher-dimensional gravity. However, the cases considered so far will undoubtedly provide essential clues toward a more complete picture and will form the basis for further developments into this fascinating subject.

The outline of these lectures is as follows. To set the stage, we first give in Sect. 6.2 a brief introduction to known uniqueness theorems for black holes in pure gravity and some prominent cases of non-uniqueness in higher dimensions. We also give a short overview of some of the most important techniques that have been used to obtain black hole solutions beyond four dimensions. Then we review the current status for Kaluza–Klein black holes in Sects. 6.3 and 6.4. In particular, Sect. 6.3 presents the main results for black objects on the cylinder with one event horizon as well as results for Kaluza–Klein black holes on a two-torus, which will be relevant in the sequel. Section 6.4 discusses the recently constructed multi-black hole configurations on the cylinder. Then the focus will be turned to rotating black holes in Sects. 6.5 and 6.6. The five-dimensional case will be very briefly reviewed, but most attention will be given to the recent progress for six and higher dimensions, including the construction of thin black rings in Sect. 6.5. We then discuss in Sect. 6.6 the proposed phase structure for rotating black holes in six and higher dimensions with a single angular momentum. The resulting picture builds on an interesting connection to the phase structure of Kaluza–Klein black holes discussed in the first part. We end with a future outlook for the subject in Sect. 6.7.

6.2 Uniqueness Theorems and Going Beyond Four Dimensions

In this section we first review known black hole uniqueness theorems in Einstein gravity as well as the most prominent cases of non-uniqueness of black holes in higher dimensions. We also give an overview of some of the most important techniques that have been used in finding black hole solutions beyond four dimensions.

6.2.1 Black Hole (Non-)uniqueness

The purpose of this lecture is to explore possible black hole solutions of the vacuum Einstein equations $R_{\mu\nu} = 0$ in dimensions $D \geq 4$. In four-dimensional vacuum gravity, a black hole in an asymptotically flat spacetime is uniquely specified by the ADM mass M and angular momentum J measured at infinity [60–63]. In particular, in the static case the unique solution is the four-dimensional Schwarzschild black hole solution and for the stationary case it is the Kerr black hole:

$$ds^2 = -dt^2 + \frac{\mu r}{\Sigma}(dt + a\sin^2\theta d\phi)^2 + \frac{\Sigma}{\Delta}dr^2 + \Sigma d\theta^2 + (r^2 + a^2)\sin^2\theta d\phi^2,$$

$$(6.1a)$$

$$\Sigma = r^2 + a^2\cos^2\theta \,, \quad \Delta = r^2 - \mu r + a^2 \,, \quad \mu = 2GM \,, \quad a = \frac{J}{M} \,. \qquad (6.1b)$$

For $J = 0$ this clearly reduces to the Schwarschild black hole, and the angular momentum is bounded by a critical value $J \leq GM^2$ (the Kerr bound) beyond which there appears a naked singularity. The bound is saturated for the extremal Kerr solution which is non-singular. The uniqueness in four dimensions fits nicely with the fact that black holes in four dimensions are known to be classically stable [64–66] (for further references see also the lecture [67] at this school).

The generalization of the Schwarschild black hole to arbitrary dimension D was found by Tangherlini [68] and is given by the metric

$$ds^2 = -fdt^2 + f^{-1}dr^2 + r^2 d\Omega_{D-2}^2 \,, \quad f = 1 - \frac{r_0^{D-3}}{r^{D-3}} \,. \qquad (6.2)$$

Here $d\Omega_{D-2}^2$ is the metric element of a $(D-2)$-dimensional unit sphere with volume $\Omega_{D-2} = 2\pi^{(D-1)/2}/\Gamma[(D-1)/2]$. Since the Newtonian potential Φ in the weak-field regime $r \to \infty$ can be obtained from $g_{tt} = -1 - 2\Phi$, this shows that $\Phi = -r_0^{D-3}/(2r^{D-3})$. The mass of the black hole is then easily obtained as

$$M = \frac{\Omega_{D-2}(D-2)}{16\pi G}r_0^{D-3} \,, \qquad (6.3)$$

by using $\nabla^2\Phi = 8\pi G\frac{D-3}{D-2}T_{tt}$ and $M = \int dx^{D-1}T_{tt}$ where T_{tt} is the energy density. Uniqueness theorems [69, 70] for D-dimensional ($D > 4$) asymptotically flat spacetimes state that the Schwarzschild–Tangherlini black hole solution is the only static black hole in pure gravity. The classical stability of these higher-dimensional black hole solutions was addressed in [71–73].

The generalization of the Kerr black hole (6.1) to arbitrary dimension D was found by Myers and Perry [15], who obtained the metric of a rotating black hole with angular momenta in an arbitrary number of orthogonal planes. The Myers–Perry (MP) black hole is thus specified by the mass and angular momenta J_k where $k = 1 \ldots r$ with $r = \mathrm{rank}(SO(D-2))$. For MP black holes with a single angular momentum, there is again a Kerr bound $J^2 < 32GM^3/(27\pi)$ in the five-dimensional case, but for six and more dimensions the angular momentum is unbounded, and the black hole can be ultra-spinning. This fact will be important in Sects. 6.5 and 6.6. When there are more than one angular momenta one needs at least one or two zero angular momenta to have an ultra-spinning regime depending on whether the dimensions are even or odd [36].

Despite the absence of a Kerr bound in six and higher dimensions, it was argued in [36] that in six or higher dimensions the Myers–Perry black hole becomes unstable above some critical angular momentum, thus recovering a dynamical Kerr bound. The instability was identified as a Gregory–Laflamme instability by showing that in a large angular momentum limit the black hole geometry becomes that of an unstable black membrane. This result is also an indication of the existence of new rotating black holes with spherical topology, where the horizon is distorted by ripples along the polar direction. This will be discussed in more detail in Sect. 6.6. Finally, we note that all of the black hole solutions discussed so far in this section have an event horizon of spherical topology S^{D-2}.

Contrary to the static case, there are no uniqueness theorems for non-static black holes in pure gravity with $D > 4$.[1] On the contrary, there are known cases of non-uniqueness. The first example of this was found by Emparan and Reall [16] and occurs in five dimensions for stationary solutions in asymptotically flat spacetime: for a certain range of mass and angular momentum there exist both a rotating MP black hole with S^3 horizon [15] and rotating black rings with $S^2 \times S^1$ horizons [16].

As mentioned in the introduction, following the discovery of the rotating black ring [16], further generalizations of these to black Saturns and multi-black rings have been found in five dimensions. It is possible that essentially all five-dimensional black holes (up to iterations of multi-black rings) with two axial Killing vectors have been found by now,[2] but the study of non-uniqueness for rotating black holes in six and higher dimensions has only recently begun (see Sects. 6.5 and 6.6).

Another case where non-uniqueness has been observed is for Kaluza–Klein black holes, in particular for black hole solutions that asymptote to Minkowski space $\mathcal{M}^{D-1}$ times a circle S^1. Here, the simplest solution one can construct is the uniform black string which is the $(D-1)$-dimensional Schwarzschild–Tangherlini black hole (6.2) plus a flat direction, which has horizon topology $S^{D-3} \times S^1$. However, at least for a certain range of masses, there are also non-uniform black strings and black holes that are localized on the circle, both of which are non-translationally invariant along the circle direction. All of these solutions, which have in common that they posses an $SO(D-2)$ symmetry, will be further discussed in Sect. 6.3. If one allows for disconnected horizons, then also multi-black hole configurations localized on the circle are possible, giving rise to an infinite non-uniqueness. These will be discussed in Sect. 6.4. In addition there are more exotic black hole solutions, called bubble-black hole sequences [49], but for simplicity these will not be further dealt with in this lecture.

More generally, for black hole solutions that asymptote to Minkowski space $\mathcal{M}^{D-p}$ times a torus $\mathbb{T}^p$, the simplest class of solutions with an event horizon is black p-branes. The metric is that of a $(D-p)$-dimensional Schwarzschild–Tangherlini black hole (6.2) plus p flat directions. Beyond that there will exist many more phases, which have only been partially explored. As an example, we discuss in Sect. 6.3.4 the phases of KK black holes on $\mathbb{T}^2$ that follow by adding a flat direction

---

[1] See [74] for recent progress in this direction.

[2] See [75, 76] for work on how to determine uniquely the black hole solutions with two symmetry axes.

to the phases of KK black holes on S^1. These turn out to be intimately related to the phase structure of rotating black holes for $D \geq 6$, as we will see in Sect. 6.6.

6.2.2 Overview of Solution Methods

We briefly describe here the available methods that have been employed in order to find the new solutions that are the topic of this lecture. The main techniques for finding new solutions are as follows.

6.2.2.1 Symmetries and Ansätze

It is often advantageous to use symmetries and other physical inputs to constrain the form of the metric for the putative solution. In this way one may be able to find an ansatz for the metric that enables to solve the vacuum Einstein equations exactly. This often involves also a clever choice of coordinate system, adapted to the symmetries of the problem. This ingredient is also important in cases where the Einstein equations can only be solved perturbatively around a known solution (see below).

As an example we note the generalized Weyl ansatz [24, 25] for static and stationary solutions with $D - 2$ commuting Killing vectors, in which the Einstein equations simplify considerably. For the static case, this ansatz is, for example, relevant for bubble-black hole sequences [49] in five- and six-dimensional KK space. For the stationary case, it is relevant for rotating black ring solutions in five-dimensional asymptotically flat space. Another example relevant for black holes and strings on cylinders is the $SO(D-2)$-symmetric ansatz of [43, 52, 56] based on coordinates that interpolate between spherical and cylindrical coordinates [43]. This has been used to obtain the metric of small black holes on the cylinder [31, 35].

6.2.2.2 Solution-Generating Techniques

Given an exact solution there are cases where one can use solution-generating techniques, such as the inverse scattering method, to generate other new solutions. See, for example, [26–29] where this method was first used for stationary black hole solutions in five dimensions, and [77] for a further solution-generating mechanism.

6.2.2.3 Matched Asymptotic Expansion

In some cases one knows the exact form of the solution in some corner of the moduli space. Then one may attempt to find the solution in a perturbative expansion around this (limiting) known solution. This method, called matched asymptotic expansion

[30–35], has been very successful. It applies to problems that contain two (or more) widely separated scales. In particular for black holes, this means that one solves Einstein equations perturbatively in two different zones, the asymptotic zone and the near-horizon zone and one thereafter matches the solution in the overlap region. One example is that of small black holes on a circle, where the horizon radius of the black holes is much smaller than the size of the circle (see in particular Sect. 6.4). Another example is that of thin black rings, where the thickness of the ring is much smaller than the radius of the ring (see Sect. 6.5).

6.2.2.4 Numerical Techniques

Since in many cases the Einstein equations become too complicated to be amenable to analytical methods, even after using symmetries and ansätze, the only way to proceed in the non-linear regime is to try to solve them numerically. For KK black holes, especially, these techniques have been successfully applied for non-uniform black strings [37–42] and localized black holes [46–48] (see Sect. 6.3).

6.2.2.5 Classical Effective Field Theory

There exists also a classical effective field theory approach for extended objects in gravity [78]. This can be used as a systematic low-energy (long-distance) effective expansion which gives results only in the region away from the black hole and so it does not provide the corrections to the metric near the horizon, but enables one to compute perturbatively corrected asymptotic quantities. This has been successfully applied in [44] to obtain the second-order correction to the thermodynamics of small black holes on a circle. Recently, it was shown [45] that this method is equivalent to matched asymptotic expansion where the near-horizon zone is replaced by an effective theory. Reference [45] also contains an interesting new application of the method to the corrected thermodynamics of small MP black holes on a circle.

6.3 Kaluza–Klein Black Holes

In this section we give a general description of the phases of Kaluza–Klein (KK) black holes (see also the reviews [3, 79]). A $(d + 1)$-dimensional Kaluza–Klein black hole will be defined here as a pure gravity solution with at least one event horizon that asymptotes to d-dimensional Minkowski space times a circle ($\mathcal{M}^d \times S^1$) at infinity. We will discuss only static and neutral solutions, i.e., solutions without charges and angular momenta. Obviously, the uniform black string is an example of a Kaluza–Klein black hole, but many more phases are known to exist. In particular, we discuss here the non-uniform black string and the localized black hole phase.

Finally, in anticipation of the connection with the phase structure of rotating black holes (discussed in Sect. 6.6) we also discuss part of the phases of KK black holes on Minkowski space times a torus ($\mathscr{M}^{D-2} \times \mathbb{T}^2$).

6.3.1 Setup and Physical Quantities

For any spacetime which asymptotes to $\mathscr{M}^d \times S^1$ we can define the mass M and the tension $\mathscr{T}$. These two asymptotic quantities can be used to parameterize the various phases of Kaluza–Klein black holes in a (μ, n) phase diagram, as we review below.

The Kaluza–Klein space $\mathscr{M}^d \times S^1$ consists of the time t and a spatial part which is the cylinder $\mathscr{R}^{d-1} \times S^1$. The coordinates of $\mathscr{R}^{d-1}$ are $x^1, ..., x^{d-1}$ and the radius $r = \sqrt{\Sigma_i (x^i)^2}$. The coordinate of the S^1 is denoted by z and its circumference is L. It is well known that for static and neutral mass distributions in flat space $\mathscr{R}^d$ the leading correction to the metric at infinity is given by the mass. For a cylinder $\mathscr{R}^{d-1} \times S^1$ we instead need two independent asymptotic quantities to characterize the leading correction to the metric at infinity.

6.3.1.1 Mass and Tension

Consider a static and neutral distribution of matter which is localized on a cylinder $\mathscr{R}^{d-1} \times S^1$. Assume a diagonal energy momentum tensor with components T_{tt}, T_{zz} and T_{ii}. Here T_{tt} depends on (x^i, z) while T_{zz} depends only on x^i because of momentum conservation. We can then write the mass and tension as

$$M = \int \mathrm{d}x^d T_{tt} , \quad \mathscr{T} = -\frac{1}{L} \int \mathrm{d}x^d T_{zz} . \tag{6.4}$$

From these definitions and the method of equivalent sources, one can obtain expressions for M and $\mathscr{T}$ in terms of the leading $1/r^{d-3}$ behavior of the metric components g_{tt} and g_{zz} around flat space [50, 51]. See also [11, 80–84] for more on the gravitational tension of black holes and branes.

For a neutral Kaluza–Klein black hole with a single connected horizon, we can find the temperature T and entropy S directly from the metric. Together with the mass M and tension $\mathscr{T}$, these quantities obey the Smarr formula [50, 51]

$$(d-1)TS = (d-2)M - L\mathscr{T} \tag{6.5}$$

and the first law of thermodynamics [51, 52, 83]

$$\delta M = T\delta S + \mathscr{T}\delta L . \tag{6.6}$$

This equation includes a 'work' term (analogous to $p\delta V$) for variations with respect to the size of the circle at infinity.

It is important to note that there are also examples of Kaluza–Klein black hole so-lutions with more than one connected event horizon [35, 49, 52]. The Smarr formula (6.5) and the first law of thermodynamics (6.6) generalize also to these cases.

6.3.1.2 Dimensionless Quantities

Since for KK black holes we have an intrinsic scale L it is natural to use it in order to define dimensionless quantities, which we take as

$$\mu = \frac{16\pi G}{L^{d-2}}M \ , \quad \mathfrak{s} = \frac{16\pi G}{L^{d-1}}S \ , \quad \mathfrak{t} = LT \ , \quad n = \frac{\mathscr{T}L}{M} \ . \tag{6.7}$$

Here μ, $\mathfrak{s}$ and $\mathfrak{t}$ are the rescaled mass, entropy and temperature, respectively, and n is the relative tension. The relative tension satisfies the bound $0 \leq n \leq d-2$ [50]. The upper bound is due to the strong energy condition whereas the lower bound was found in [85, 86]. The upper bound can also be understood physically in a more direct way from the fact that we expect gravity to be an attractive force. For a test particle at infinity it is easy to see that the gravitational force on the particle is attractive when $n < d-2$ but repulsive when $n > d-2$.

The program set forth in [50, 52] is to plot all phases of Kaluza–Klein black holes in a (μ, n) diagram. Note that it follows from the Smarr formula (6.5) and the first law of thermodynamics (6.6) that given a curve $n(\mu)$ in the phase diagram, the entire thermodynamics $\mathfrak{s}(\mu)$ of a phase can be obtained [50]. We also note that the (μ, n) phase diagram appears to be divided into two separate regions [49]. Here, the region $0 \leq n \leq 1/(d-2)$ contains solutions without Kaluza–Klein bubbles, and the solutions have a local $SO(d-1)$ symmetry and reside in the ansatz proposed in [43, 87] and proven in [52, 56]. Solutions of this type, also referred to as black holes and strings on cylinders, will be reviewed in Sect. 6.3.2. Because of the $SO(d-1)$ symmetry there are only two types of event horizon topologies: S^{d-1} for the black hole on a cylinder branch and $S^{d-2} \times S^1$ for the black string. The region $1/(d-2) < n \leq d-2$ contains solutions with Kaluza–Klein bubbles. This part of the phase diagram, which is much more densely populated with solutions compared to the lower part, is the subject of [49].

6.3.1.3 Alternative Dimensionless Quantities

The typical dimensionless quantities used for KK black holes in D dimensions are those defined in (6.7). Instead of these, [30] introduced the following new dimen-sionless quantities, more suitable for the analogy with rotating black holes (see Sect. 6.6), by defining

$$\ell^{D-3} \propto \frac{L^{D-3}}{GM} \ , \quad a_H^{D-3} \propto \frac{S^{D-3}}{(GM)^{D-2}} \ , \quad \mathfrak{t}_H \propto (GM)^{\frac{1}{D-3}}T \ . \tag{6.8}$$

In particular, the relation to the dimensionless quantities in (6.7) is given by

$$\ell = \mu^{-\frac{1}{D-3}} \ , \quad a_H = \mu^{-\frac{D-2}{D-3}} \mathfrak{s} \ , \quad \mathfrak{t}_H = \mu^{-\frac{1}{D-3}} \mathfrak{t} \ . \tag{6.9}$$

In the KK black hole literature, entropy plots are typically given as $\mathfrak{s}(\mu)$. Instead of these one can also use (6.9) to consider the area function $a_H(\ell)$, which is obtained as

$$a_H(\ell) = \ell^{D-2} \mathfrak{s}(\ell^{-D+3}) \ . \tag{6.10}$$

We will employ these alternative quantities when we discuss KK black holes on a torus in Sect. 6.3.4.

6.3.2 Black Holes and Strings on Cylinders

We now discuss the main three types of KK black holes that have $SO(d-1)$ symmetry, to which we commonly refer as black holes and strings on cylinders. These are the uniform black string, the non-uniform black string and the localized black hole. In Sect. 6.4 we will discuss in more detail the recently obtained multi-black hole configurations on the cylinder.

6.3.2.1 Uniform Black String and Gregory–Laflamme Instability

The metric for the uniform black string in $D = d + 1$ spacetime dimensions is

$$ds^2 = -f dt^2 + f^{-1} dr^2 + r^2 d\Omega_{d-2}^2 + dz^2 \ , \quad f = 1 - \frac{r^{d-3}}{r_0^{d-3}} \ , \tag{6.11}$$

where $d\Omega_{d-2}^2$ is the metric element of a $(d-2)$-dimensional unit sphere. The metric (6.11) is found by taking the d-dimensional Schwarzschild–Tangherlini static black hole (6.2) solution [68] and adding a flat z direction, which is the direction parallel to the string. The event horizon is located at $r = r_0$ and has topology $S^{d-2} \times \mathscr{R}$.

6.3.2.2 Gregory–Laflamme Instability

Gregory and Laflamme found in 1993 a long wavelength instability for black strings in five or more dimensions [53, 54]. The mode responsible for the instability propagates along the direction of the string and develops an exponentially growing time-dependent part when its wavelength becomes sufficiently long. The Gregory–Laflamme mode is a linear perturbation of the metric (6.11) that can be written as

$$g_{\mu\nu} + \epsilon h_{\mu\nu} \ . \tag{6.12}$$

Here $g_{\mu\nu}$ stands for the components of the unperturbed black string metric (6.11), ϵ is a small parameter and $h_{\mu\nu}$ is the metric perturbation:

$$h_{\mu\nu} = \Re \left\{ \exp \left(\frac{\Omega t}{r_0} + i \frac{kz}{r_0} \right) P_{\mu\nu}(r/r_0) \right\} , \tag{6.13}$$

where the symbol $\Re$ denotes the real part. The statement that the perturbation $h_{\mu\nu}$ of $g_{\mu\nu}$ satisfies the Einstein equations of motion can be stated as the differential operator equation

$$\Delta_L h_{\mu\nu} = 0 , \tag{6.14}$$

where $(\Delta_L)_{\mu\nu\rho\sigma} = -g_{\mu\rho}g_{\nu\sigma}D_\kappa D^\kappa + 2R_{\mu\nu\rho\sigma}$ is the Lichnerowitz operator for the background metric $g_{\mu\nu}$. The resulting Einstein equations for the GL mode can be found, e.g., in the appendix of the review [3].[3] Solution of these equations [53, 54] shows that there is an unstable mode for any wavelength larger than the critical wavelength

$$\lambda_{GL} = \frac{2\pi r_0}{k_c} \tag{6.15}$$

for which $\Omega = 0$ in (6.13). The values of k_c for $d = 4, ..., 14$, as obtained in [37, 39, 53], are listed, e.g., in Table 1 of [3]. The critical wave number k_c marks the lower bound of the possible wavelengths for which there is an unstable mode and is called the threshold mode. It is a time-independent mode of the form $h_{c,\mu\nu} \sim \exp(ik_c z/r_0)$. In particular, this suggests the existence of a static non-uniform black string.

6.3.2.3 GL Mode of the Compactified Uniform Black String

Since we wish to consider the uniform black string in KK space, we now discuss what happens to the GL instability when z is a periodic coordinate with period L. The Gregory–Laflamme mode (6.13) cannot obey the correct periodic boundary condition on z if $L < \lambda_{GL}$, with λ_{GL} given by (6.15). On the other hand, for $L > \lambda_{GL}$, we can fit the Gregory–Laflamme mode into the compact direction with the frequency and wave number Ω and k in (6.13) determined by the ratio r_0/L. Translating this in terms of the mass of the neutral black string, one finds the critical Gregory–Laflamme mass

$$\mu_{GL} = (d-2)\Omega_{d-2} \left(\frac{k_c}{2\pi} \right)^{d-3} . \tag{6.16}$$

For $\mu < \mu_{GL}$ the Gregory–Laflamme mode can be fitted into the circle, and the compactified neutral uniform black string is unstable. For $\mu > \mu_{GL}$, on the other hand, the Gregory–Laflamme mode is absent, and the neutral uniform black string is stable. For $\mu = \mu_{GL}$ there is a marginal mode which signals the emergence of a

---

[3] Various methods and different gauges have been employed to derive the differential equations for the GL mode. See [88] for a nice summary of these, including a new derivation (see also [89]).

new branch of black string solutions which are non-uniformly distributed along the circle. See, e.g., Table 2 in [3] for the values of μ_{GL} for $4 \leq d \leq 14$.

The large d behavior of μ_{GL} was examined numerically in [39] and analytically in [58]. We also note that there is an interesting correspondence between the Rayleigh–Plateau instability of long fluid cylinders and the Gregory–Laflamme instability of black strings [90, 91]. In particular, the critical wave numbers k_{RP} and k_c agree exactly at large dimension d (scaling both as $\sqrt{d}$ for $d \gg 1$).

6.3.2.4 Non-uniform Black String

It was realized in [37] (see also [92]) that the classical instability of the uniform black string for $\mu < \mu_{GL}$ implies the existence of a marginal (threshold) mode at $\mu = \mu_{GL}$, which again suggests the existence of a new branch of solutions.

The new branch, which is called the non-uniform string branch, has been found numerically in [37–39]. This branch of solutions has the same horizon topology $S^1 \times S^{d-2}$ as the uniform string, which is expected since the non-uniform string is continuously connected to the uniform black string. In particular, it emerges from the uniform black string in the point $(\mu, n) = (\mu_{GL}, 1/(d-2))$ and has $n < 1/(d-2)$. Moreover, the solution is non-uniformly distributed in the circle direction z since there is an explicit dependence in the marginal mode in this direction.

More concretely, considering the non-uniform black string branch for $|\mu - \mu_{GL}| \ll 1$ one obtains for the relative tension the behavior

$$n(\mu) = \frac{1}{d-2} - \gamma(\mu - \mu_{GL}) + \mathscr{O}((\mu - \mu_{GL})^2) . \tag{6.17}$$

Here γ is a number representing the slope of the curve that describes the non-uniform string branch near $\mu = \mu_{GL}$ (see Table 3 in [3] for the values of γ for $4 \leq d \leq 14$ obtained from the data of [37–39, 53, 54]).

The qualitative behavior of the non-uniform string branch depends on the sign of γ. If γ is positive, then the branch emerges at the mass $\mu = \mu_{GL}$ with increasing μ and decreasing n. If instead γ is negative the branch emerges at $\mu = \mu_{GL}$ with decreasing μ and decreasing n. To see what this means for the entropy we note that from (6.17) and the first law of thermodynamics one finds that

$$\frac{\mathfrak{s}_{nu}(\mu)}{\mathfrak{s}_u(\mu)} = 1 - \frac{(d-2)^2}{2(d-1)(d-3)^2} \frac{\gamma}{\mu_{GL}} (\mu - \mu_{GL})^2 + \mathscr{O}((\mu - \mu_{GL})^3) , \tag{6.18}$$

where $\mathfrak{s}_u(\mu)$ ($\mathfrak{s}_{nu}(\mu)$) refers to the rescaled entropy of the uniform (non-uniform) black string branch. It turns out that γ is positive for $d \leq 12$ and negative for $d \geq 13$ [39]. Therefore, as discovered in [39], the non-uniform black string branch has a qualitatively different behavior for small d and large d, i.e., the system exhibits a critical dimension $D = 14$. In particular, for $d \leq 12$ the non-uniform branch near the GL point has $\mu > \mu_{GL}$ and lower entropy than that of the uniform phase, while for

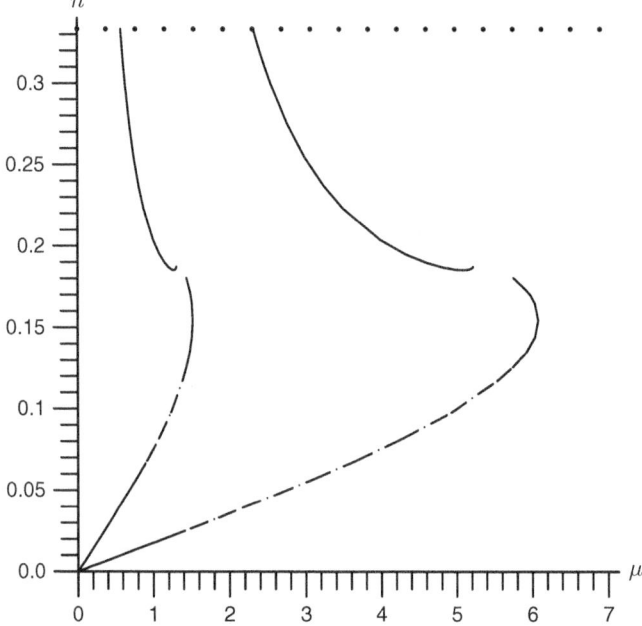

Fig. 6.1 Black hole and string phases for $d = 5$, drawn in the (μ, n) phase diagram. The *horizontal (dotted) line* is the uniform string branch. The rightmost solid branch emanating from this at the Gregory–Laflamme point is the non-uniform string branch and the rightmost dashed branch starting in the origin is the localized black hole branch. The solid and dashed branches to the left are the $k = 2$ copies of the non-uniform and localized branches. The results strongly suggest that the black hole and non-uniform black string branches meet

$d > 13$ it has $\mu < \mu_{\mathrm{GL}}$ and higher entropy. It also follows from (6.18) that for all d the curve $\mathfrak{s}_{\mathrm{nu}}(\mu)$ is tangent to the curve $\mathfrak{s}_{\mathrm{u}}(\mu)$ at the GL point.

A large set of numerical data for the non-uniform branch, extending into the strongly non-linear regime, have been obtained in [38, 48] for six dimensions (i.e., $d = 5$), in [40] for five dimensions (i.e., $d = 4$) and for the entire range $d \leq 5 \leq 10$ in [41]. For $d = 5$, these data are displayed in the (μ, n) phase diagram [50] of Fig. 6.1.

6.3.2.5 Localized Black Holes

On physical grounds, it is natural to expect a branch of neutral black holes in the spacetime $\mathcal{M}^d \times S^1$ with event horizon of topology S^{d-1}. This branch is called the localized black hole branch, because the S^{d-1} horizon is localized on the S^1 of the Kaluza–Klein space.

Neutral black hole solutions in the spacetime $\mathcal{M}^3 \times S^1$ were found and studied in [93–96]. However, the study of black holes in the spacetime $\mathcal{M}^d \times S^1$ for $d \geq 4$

is relatively new. The complexity of the problem stems from the fact that such black holes are not algebraically special [97] and moreover from the fact that the solution cannot be found using a Weyl ansatz since the number of Killing vectors is too small.

In [31, 32, 34] the metric of small black holes, i.e. black holes with mass $\mu \ll 1$, was found analytically using the method of matched asymptotic expansion. The starting point in this construction is the fact that as $\mu \to 0$, one has $n \to 0$ so that the localized black hole solution becomes more and more like a $(d+1)$-dimensional Schwarzschild black hole in this limit. For $d = 4$, the second-order correction to the metric and thermodynamics have been studied in [33]. More generally, the second-order correction to the thermodynamics was obtained in [44] (see also [45]) for all d using an effective field theory formalism in which the structure of the black hole is encoded in the coefficients of operators in an effective worldline Lagrangian.

The first-order result of [31] and second-order result of [44] can be summarized by giving the first- and second-order corrections to the relative tension n of the localized black hole branch as a function of μ

$$n = \frac{(d-2)\zeta(d-2)}{2(d-1)\Omega_{d-1}}\mu - \left(\frac{(d-2)\zeta(d-2)}{2(d-1)\Omega_{d-1}}\mu\right)^2 + \mathcal{O}(\mu^3)\,, \qquad (6.19)$$

where $\zeta(p) = \sum_{n=1}^{\infty} n^{-p}$ is the Riemann zeta function. The corresponding correction to the thermodynamics can be found e.g.in (3.18) of [3].

The black hole branch has been studied numerically for $d = 4$ in [46, 48] and for $d = 5$ in [47, 48]. For small μ, the impressively accurate data of [48] are consistent with the analytical results of [31–33]. The results of [48] for $d = 5$ are displayed in a (μ, n) phase diagram in Fig. 6.1.

6.3.3 Phase Diagram and Copied Phases

In Fig. 6.1 the (μ, n) diagram for $d = 5$ is displayed, which is one of the cases where most information is known. We have shown the complete non-uniform branch, as obtained numerically by Wiseman [38], which emanates at $\mu_{\text{GL}} = 2.31$ from the uniform branch that has $n = 1/3$. These data were first incorporated into the (μ, n) diagram in [50]. For the black hole branch we have plotted the numerical data of Kudoh and Wiseman [48]. It is evident from the figure that this branch has an approximate linear behavior for a fairly large range of μ close to the origin and the numerically obtained slope agrees very well with the analytic result (6.19).

6.3.3.1 Merger Point

The figure strongly suggests that the localized black hole branch meets with the non-uniform black string branch in a topology changing transition point, which is the scenario earlier suggested by Kol [55] (see [52] for a list of scenarios). For this reason, it seems reasonable to expect that the localized black hole branch is connected with the non-uniform string branch in any dimension. This means that we can go from the uniform black string branch to the localized black hole branch through a connected series of static classical geometries. The point in which the two branches are conjectured to meet is called the merger point.

6.3.3.2 Copied Phases

In [52] it was shown that one can generate new solutions by copying solutions on the circle several times, following an idea of Horowitz [98]. This works for solutions which vary along the circle direction (i.e., in the z direction), so it works for both the black hole branch and the non-uniform string branch. Let k be a positive integer. Then if we copy a solution k times along the circle we get a new solution with the following parameters:

$$\tilde{\mu} = \frac{\mu}{k^{d-3}} \; , \quad \tilde{\mathfrak{s}} = \frac{\mathfrak{s}}{k^{d-2}} \; , \quad \tilde{\mathfrak{t}} = k\mathfrak{t} \; , \quad \tilde{n} = n \; . \tag{6.20}$$

See [52] for the corresponding expression of the metric of the copies in the $SO(d-1)$-symmetric ansatz. Using the transformation (6.20), one easily sees that the non-uniform and localized black hole branches depicted in Fig. 6.1 are copied infinitely many times in the (μ, n) phase diagrams, and we have depicted the $k = 2$ copy in this figure.

6.3.3.3 General Dimension

The six-dimensional phase diagram displayed in Fig. 6.1 is believed to be representative for the black string/localized black hole phases on $\mathcal{M}^{D-1} \times S^1$ for all $5 \leq D \leq 13$. Here the upper bound follows from the fact that, as mentioned above, there is a critical dimension $D = 13$ above which the behavior of the non-uniform black string phase is qualitatively different [39]. The phase diagram for $D \geq 14$ is much poorly known in comparison, since there are no data like Fig. 6.1 available for the localized and non-uniform phases, only the asymptotic behaviors. However, we do know from (6.17) that the non-uniform branch will extend to the left (lower values of μ) as it emerges from the GL point and on general grounds is expected to merge again with the localized black hole branch.

6.3.4 KK Phases on $\mathbb{T}^2$ from Phases on S^1

We show here how one can translate the known results for KK black holes on the circle (i.e., on $\mathcal{M}^{D-2} \times S^1$) to results for KK black holes on the torus (i.e., on

$\mathscr{M}^{D-2} \times \mathbb{T}^2$). The resulting phases are relevant in connection with the phases of rotating black holes in asymptotically flat spacetime, as shown in Sect. 6.6.

We recall first the definitions of dimensionless quantities in (6.7). While these quantities were originally introduced in [11, 50] for black holes on a KK circle of circumference L, we may similarly use these definitions for KK black holes in D dimensions with a square torus of side lengths L, to which we restrict in the following. Likewise, we can use the alternative dimensionless quantities (6.9) for that case.

6.3.4.1 Map from Circle to Torus Compactification

We first want to establish a map for these dimensionless quantities from KK black holes on $\mathscr{M}^{D-2} \times S^1$ (denoted with hatted quantities) to those for KK black holes on $\mathscr{M}^{D-2} \times \mathbb{T}^2$ (denoted with unhatted quantities), obtained by adding an extra compact direction of size L. Suppose we are given an entropy function $\hat{s}(\hat{\mu})$ for a phase of KK black holes on $\mathscr{M}^{D-2} \times S^1$. Any such phase lifts trivially to a phase of KK black holes on $\mathscr{M}^{D-2} \times \mathbb{T}^2$ that is uniform in one of the torus directions. We show below how to obtain the function $a_H(\ell)$ for the latter in terms of $\hat{s}(\hat{\mu})$ of the former. In the following we will use the notation $D = n + 4$.

It is not difficult to see that in terms of the original dimensionless quantities (6.7) we have the simple mapping

$$\mu = \hat{\mu} , \quad s = \hat{s} , \quad t = \hat{t} . \tag{6.21}$$

It then follows from (6.9) and (6.10) that the area function $a_H(\ell)$ of KK black holes on $\mathscr{M}^{D-2} \times \mathbb{T}^2$ is obtained via the mapping relation

$$a_H(\ell) = \ell^{n+2} \hat{s}(\ell^{-n-1}) . \tag{6.22}$$

6.3.4.2 Application to Known Phases

Using now the entropy function $\hat{s}_{\mathrm{uni}}(\hat{\mu}) \sim \hat{\mu}^{\frac{n}{n-1}}$ of the uniform black string in $\mathscr{M}^{n+2} \times S^1$ we get from (6.22) the result

$$a_H^{\mathrm{ubm}}(\ell) \sim \ell^{-\frac{2}{n-1}} \tag{6.23}$$

for the uniform black membrane (ubm) $4 + n$ dimensions. Furthermore, using that for small μ (or equivalently large ℓ) the entropy of the localized black hole in $\mathscr{M}^{n+2} \times S^1$ is $\hat{s}_{\mathrm{loc}}(\hat{\mu}) \sim \hat{\mu}^{\frac{n+1}{n}}$ we find via the map (6.22) the result

$$a_H^{\mathrm{lbs}}(\ell) \sim \ell^{-\frac{1}{n}} \qquad (\ell \to \infty) \tag{6.24}$$

for the large ℓ limit of the localized black string (lbs) in $4 + n$ dimensions.

Finally, for the non-uniform string in $\mathscr{M}^{n+2} \times S^1$ dimensions we use (6.18) to obtain

$$a_H^{\text{nubm}}(\ell) = a_H^{\text{ubm}}(\ell)\left[1 - \frac{n^2(n+1)}{2(n-1)^2}\frac{\gamma_{n+2}}{\ell_{\text{GL}}^{n+4}}(\ell - \ell_{\text{GL}})^2 + \mathscr{O}\left((\ell - \ell_{\text{GL}})^3\right)\right] \quad (6.25)$$

for the non-uniform black membrane (nubm). Here, $\ell_{\text{GL}} = (\mu_{\text{GL},n+2})^{-\frac{1}{n+1}}$ is the critical GL wavelength in terms of the dimensionless GL mass $\mu_{\text{GL},d}$ given in (6.16) and γ_d the coefficient in (6.17).

6.3.4.3 Copies

As remarked in Sect. 6.3.3 the localized black hole and non-uniform black string phase on $\mathscr{M}^{n+2} \times S^1$ have copied phases with multiple non-uniformity or multiple localized black objects. From the map (6.20) we then find using (6.21) and the definitions (6.9) that corresponding copied phases of KK black holes on the torus obey the transformation rule

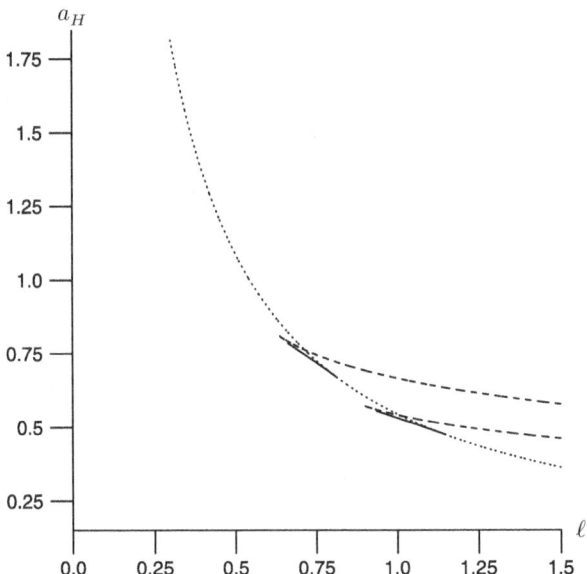

Fig. 6.2 $a_H(\ell)$ phase diagram in seven dimensions ($\mathscr{M}^5 \times \mathbb{T}^2$) for Kaluza–Klein black hole phases with one uniform direction. Shown are the uniform black membrane phase *(dotted)*, the non-uniform black membrane phase *(solid)* and the localized black string phase *(dashed)*. For the latter two phases, we have also shown their $k = 2$ copy. The non-uniform black membrane phase emanates from the uniform black membrane phase at the GL point $\ell_{\text{GL}} = 0.811$, while the $k = 2$ copy starts at the 2-copied GL point $\ell_{\text{GL}}^{(2)} = \sqrt{2}\ell_{\text{GL}} = 1.15$. This figure is representative of the phase diagram of phases on $\mathscr{M}^{D-2} \times \mathbb{T}^2$ for all $6 \leq D \leq 14$. Reprinted from [30]

$$\tilde{\ell} = k^{\frac{n-1}{n+1}} \ell \ , \quad \tilde{a}_H = k^{-\frac{2}{n+1}} a_H \ , \quad \tilde{t}_H = k t_H \ . \tag{6.26}$$

6.3.4.4 Seven-Dimensional Phase Diagram

As an explicit example, we give the mapping that can be used to convert the known results for KK black holes on $\mathcal{M}^5 \times S^1$ to KK black holes on $\mathcal{M}^5 \times \mathbb{T}^2$:

$$(\ell, a_H) = (\hat{\mu}^{-1/4}, \hat{\mu}^{-5/4} \hat{s}) \ . \tag{6.27}$$

This can be used to convert plots of points $(\hat{\mu}, \hat{s})$ (see, e.g., [3]) for six-dimensional KK black holes on a circle to the phase diagram in Fig. 6.2 for seven-dimensional KK black holes (with one uniform direction) on a torus. It includes the uniform black membrane, the black membrane with one uniform and one non-uniform direction, and the black string localized in one of the circles of $\mathbb{T}^2$. The figure also includes the $k = 2$ copies obtained from these data and the map (6.26). Both the uniform black membrane phase and the localized black string phase extend to $\ell \to \infty$ where they obey the behavior (6.23) and (6.24), respectively, with $n = 3$.

6.4 Multi-black Hole Configurations on the Cylinder

We now turn to the construction of multi-black hole configurations on the cylinder, recently obtained in [35]. In Sect. 6.3.3 we already encountered a special subset of these, namely the copied phases of the localized black hole branch, corresponding to multi-black hole configurations in which all black holes have the same mass. Here, we describe the main points of the construction of more general multi-black hole configurations [35] using matched asymptotic expansion. We also show how the thermodynamics of these configurations can be understood from a Newtonian point of view. Finally we comment on the consequences of these configurations for the phase diagram of KK black holes.

6.4.1 Construction of Multi-black Hole Solutions

The copies of the single-black hole localized on the circle correspond to a multi-black hole configurations of equal mass black holes that are spread with equal distance from each other on the circle. Beyond these, there exist more general multi-black hole configurations which have recently been considered in [35]. These solutions correspond to having several localized black holes of different sizes located at different points along the circle direction of the cylinder $\mathcal{R}^{d-1} \times S^1$. The location of each black hole is such that the total force on each of them is zero, ensuring that they are in equilibrium. It is moreover necessary for being in equilibrium that the black holes are all located in the same point in the $\mathscr{R}^{d-1}$ part of the cylinder.

The metric constructed in [35] are solutions to the Einstein equations to first order in the mass. More precisely, they are valid in a regime where the gravitational interaction between any one of the black holes and the others (and their images on the circle) is small. The solutions in [35] thus describe the small mass limit of these multi-black hole configurations on the cylinder, or equivalently they can be said to describe the situation where the black holes are far apart. The method used for solving the Einstein equations is the one of matched asymptotic expansion [31–35]. The particular construction follows the approach of [31] where it was used to find the metric of a small black hole on the cylinder based on an ansatz for the metric found in [43].

6.4.1.1 General Idea and Starting Point

We describe here the general idea behind constructing the new solutions for multi-black hole configurations on the d-dimensional cylinder $\mathscr{R}^{d-1} \times S^1$. The configuration under consideration is that of k black holes placed at different locations z_i^*, $i = 1,\ldots,k$, in the same point of the $\mathscr{R}^{d-1}$ part of the cylinder. We write M as the total mass of all of the black holes and define v_i as the fraction of mass of the ith black hole, i.e.,

$$M_i = v_i M, \qquad \sum_{i=1}^{k} v_i = 1, \tag{6.28}$$

where M_i is the mass of the ith black hole. Note that $0 < v_i \leq 1$.

The matched asymptotic expansion is suitable when there are two widely separated scales in the problem. Here they are the size (mass) of each of the black holes (all of which are taken of the same order) and the length of the circle direction. In particular, we assume that all black holes have a horizon radius (of the same order) which is small compared to the length of the circle.

The construction of the solution then proceeds in the following steps[4]:

- Step 1: Find a metric corresponding to the Newtonian gravitational potential sourced by a configuration of small black holes on the cylinder. This metric is valid in the region $R \gg R_0$.
- Step 2: Consider the Newtonian solution close to the sources, i.e., in the overlap region $R_0 \ll R \ll L$.
- Step 3: Find a general solution near a given event horizon and match this solution to the metric in the overlap region found in step 2. The resulting solution is valid in the region $R_0 \leq R \ll L$.

---

[4] Here we use the coordinate R which is part of the two-dimensional coordinate system (R,v) introduced in [43] that interpolates between cylindrical coordinates (r,z) and spherical coordinates (ρ,θ). In terms of $F(r,z)$ in (6.31) we have $R(r,z) \propto F(r,z)^{-1/(d-3)}$. Note that [31, 35, 43] set $L = 2\pi$, which we choose not to do here for pedagogical clarity.

With all these three steps implemented, we have a complete solution for all of the spacetime outside the event horizon. We refer to [35] for further details, including the explicit form of the first-order corrected metric and thermodynamics of the resulting multi-black hole configurations, but present some of the easy steps here.

6.4.1.2 Newtonian Potential

Following the discussion in Sect. 6.3.1 for static solutions on the cylinder the two relevant components of the stress tensor are T_{tt} and T_{zz}. These components source the two gravitational potentials [50]

$$\nabla^2 \Phi = 8\pi G \frac{d-2}{d-1} T_{tt} , \qquad \nabla^2 B = \frac{8\pi G}{d-1} T_{zz} , \qquad (6.29)$$

where G is the $(d+1)$-dimensional Newton constant. In the limit of small total mass, we have $B/(GM) \to 0$ for $M \to 0$ which means [31, 35] that we can neglect the binding energy potential B as compared to the mass density potential Φ. One thus only needs to consider the potential Φ, i.e., Newtonian gravity.

For the multi-black hole configuration described above, it is not difficult to find the solution for Φ using the method of images in terms of (r, z) coordinates of the cylinder. One finds

$$\Phi(r, z) = -\frac{8\pi GM}{(d-1)\Omega_{d-1}} F(r, z) , \qquad (6.30)$$

with

$$F(r, z) = \sum_{i=1}^{k} \sum_{m=-\infty}^{\infty} \frac{v_i}{\left[r^2 + (z - z_i^* - Lm)^2 \right]^{\frac{d-2}{2}}} , \qquad (6.31)$$

so that the potential (6.30) describes the Newtonian gravitational potential sourced by the multi-black hole configuration.

One can now study how the potential Φ looks when going near the sources. To achieve this it is useful to define for the ith black hole the spherical coordinates ρ and θ by

$$r = \rho \sin\theta , \quad z - z_i^* = \rho \cos\theta , \qquad (6.32)$$

where θ is defined in the interval $[0, \pi]$. In terms of these coordinates one finds that $F(r, z)$ in (6.31) can be expanded as

$$F(\rho, \theta) = v_i \rho^{-(d-2)} + \Lambda^{(i)} + \Lambda_1^{(i)} \cos\theta \, \rho + \mathcal{O}\left(\rho^2\right) , \qquad (6.33)$$

for $\rho \ll 1$, where

$$
\Lambda^{(i)} = \frac{1}{L^{d-2}} \left(v_i \, 2\zeta(d-2) \right.
$$
$$
\left. + \sum_{\substack{j=1 \\ j \neq i}}^{k} v_j \left[\tilde{z}_{ij}^{-(d-2)} + \zeta(d-2, 1-\tilde{z}_{ij}) + \zeta(d-2, 1+\tilde{z}_{ij}) \right] \right) . \quad (6.34)
$$

Here $\zeta(s, 1+a) = \sum_{m=1}^{\infty} (m+a)^{-s}$ is the generalized Riemann zeta function and $\tilde{z}_{ij} \equiv z_{ij}/L$ labels the distance in the z direction between the jth and ith black hole (see (2.24) of [35] for precise definitions).

Using now (6.33) with (6.30) one obtains the behavior of the Newtonian potential Φ near the ith black hole. This shows that the first term in (6.33) corresponds to the flat space gravitational potential due to the ith mass $M_i = v_i M$ and the second term is a constant potential due to its images and the presence of the other masses and their images. The quantity $\Lambda^{(i)}$ plays a crucial role in the explicit construction of the first-order corrected metric of multi-black hole configurations on the cylinder and also enters the first-order corrected thermodynamics (see Sect. 6.4.2).

6.4.1.3 Equilibrium Conditions

The third term in (6.33) is proportional to $\rho \cos \theta = z - z_i^*$ and therefore this term gives a non-zero constant term in $\partial_z \Phi$ if $\Lambda_1^{(i)}$ is non-zero. This therefore corresponds to the external force on the ith black hole, due to the other $k-1$ black holes. Indeed, $\Lambda_1^{(i)}$ can be written as a sum of the potential gradients corresponding to the gravitational force due to each of the $k-1$ other black holes on the ith black hole as

$$
\Lambda_1^{(i)} = \sum_{j=1, j \neq i}^{k} v_j V_{ij} , \quad (6.35)
$$

where V_{ij} corresponds to the gravitational field on the ith black hole from the jth black hole, given by

$$
V_{ij} = \frac{(d-2)}{L^{d-1}} \left\{ \tilde{z}_{ij}^{-(d-1)} - \zeta(d-1, 1-\tilde{z}_{ij}) + \zeta(d-1, 1+\tilde{z}_{ij}) \right\} , \quad (6.36)
$$

for $j \neq i$. Defining $F_{ij} \equiv v_i v_j V_{ij}$ as the Newtonian force on the ith mass due to the jth mass (and its images as seen in the covering space of the circle), the condition $\Lambda_1^{(i)} = 0$ can be written as the condition of zero external force on each of the k masses

$$
\sum_{j=1, j \neq i}^{k} F_{ij} = 0 , \quad (6.37)
$$

for $i = 1, ..., k$. As a check, note that it is not difficult to verify that Newton's law $F_{ij} = -F_{ji}$ is verified using an appropriate identity for the generalized zeta function (see (3.6) of [35]).

We thus conclude that for static solutions one needs to require the equilibrium condition $\Lambda_1^{(i)} = 0$ for all i, since otherwise the ith black hole would accelerate along the z axis. This gives conditions on the relation between the positions z_i^* and the mass ratios v_i, which are examined in detail in [35]. It is shown how to build such equilibrium configurations and a general copying mechanism is described that builds new equilibrium configurations by copying any given equilibrium configuration a number of times around the cylinder.

Note that this equilibrium is an unstable equilibrium, i.e., generic small disturbances in the position of one of the black holes will disturb the balance of the configuration and result in the merger of all of the black holes into a single black hole. As also argued in [35], it is expected that these equilibrium conditions are a consequence of regularity of the solution since with a non-zero Newtonian force present on the black hole the only way to keep it static is to introduce a counter-balancing force supported by a singularity. It turns out that the irregularity of the solution cannot be seen at the leading order since the binding energy, which accounts for the self-interaction of the solution, is neglected. It is therefore expected that singularities will appear at the second order in the total mass for solutions that do not obey the equilibrium condition mentioned above.

6.4.2 Newtonian Derivation of the Thermodynamics

It turns out that there is a quick route to determine the first-order corrected thermodynamics of the multi-black hole configurations, as explained in [35] following the method first found in [34]. Here one assumes the equilibrium condition (6.37) to be satisfied and all one needs is the quantity $\Lambda^{(i)}$ defined in (6.34), i.e., one does not need to compute the first-order corrected metric.

To start, we define for each black hole an 'areal' radius $\hat{\rho}_{0(i)}$, $i = 1, \ldots, k$, such that the individual mass, entropy and temperature of each black hole are given by

$$M_{0(i)} = \frac{(d-1)\Omega_{d-1}}{16\pi G}\hat{\rho}_{0(i)}^{d-2} , \quad S_{0(i)} = \frac{\Omega_{d-1}\hat{\rho}_{0(i)}^{d-1}}{4G} , \quad T_{0(i)} = \frac{d-2}{4\pi\hat{\rho}_{0(i)}} . \quad (6.38)$$

These are the intrinsic thermodynamic quantities associated to each black hole when they would be isolated in flat empty $(d+1)$-dimensional space.

If we now imagine placing the black holes on a circle at locations z_i^* each of them will experience a gravitational potential Φ_i. In particular, this is the Newtonian potential created by all images of the ith black hole as well as all other $k-1$ masses (and their images) as seen from the location of the ith black hole. It is not difficult to show that Φ_i is given by

$$\Phi_i = -\frac{\Lambda^{(i)}}{2v_i}\hat{\rho}_{0(i)}^{d-2} , \quad (6.39)$$

in terms of $\Lambda^{(i)}$ defined in (6.34). Taking into account this potential, we can now determine the thermodynamic quantities of the interacting system to leading order. By definition, the entropy $S_i = S_{0(i)}$ is unchanged. The temperature of each black hole, however, receives a redshift contribution coming from the gravitational potential Φ_i, so that

$$T_i = T_{0(i)}(1 + \Phi_i) . \tag{6.40}$$

The total mass of the configuration is equal to the sum of the individual masses when the black holes would be isolated plus the negative gravitational (Newtonian) potential energy that appears as a consequence of the black holes and their images. We thus have that the total mass is given by

$$M = M_0 + U_{\text{Newton}} , \tag{6.41}$$

where

$$M_0 \equiv \sum_{i=1}^{k} M_{0(i)} , \quad U_{\text{Newton}} \equiv \frac{1}{2} \sum_{i=1}^{k} M_{0(i)} \Phi_i . \tag{6.42}$$

From these Newtonian results one can then derive the formula for the relative tension simply by using the (generalized) first law of thermodynamics (see (6.6)):

$$\delta M = \sum_{i=1}^{k} T_i \delta S_i + \frac{nM}{L} \delta L , \tag{6.43}$$

from which one finds that

$$n = \frac{L}{M} \left(\frac{\partial M}{\partial L} \right)_{S_i} . \tag{6.44}$$

The condition of keeping S_i fixed means that we should keep fixed the mass $M_{0(i)}$ of each black hole, and hence also the total intrinsic mass M_0. It thus follows from (6.44) and (6.41) that

$$n = \frac{L}{M_0} \left(\frac{\partial U_{\text{Newton}}}{\partial L} \right)_{M_{0(i)}} = -\frac{1}{4M_0} \sum_{i=1}^{k} M_{0(i)} \frac{\hat{\rho}_{0(i)}^{d-2}}{v_i} L \frac{\partial \Lambda^{(i)}}{\partial L} = \frac{d-2}{4} \sum_{i=1}^{k} \Lambda^{(i)} \hat{\rho}_{0(i)}^{d-2} , \tag{6.45}$$

where we used $\Lambda^{(i)} \propto L^{-(d-2)}$ for fixed locations z_i^* (see (6.34)) and $M_{0(i)} = v_i M_0$. As shown in [35], the thermodynamics above agrees with the explicitly computed thermodynamic quantities from the first-order corrected metric.

We emphasize that these results are correct only to first order in the mass and note that in terms of the reduced mass (6.7) the expression (6.45) gives that n as a function of μ is given for the multi-black hole configurations by

$$n(\mu) = \frac{(d-2)(2\pi)^{d-2}}{4(d-1)\Omega_{d-1}} \sum_{i=1}^{k} v_i \Lambda^{(i)} \mu + \mathcal{O}(\mu^2) , \tag{6.46}$$

which generalizes the single-black hole result given in (6.17). In terms of the phase diagram in Fig. 6.1, it follows from this result that (at least for small masses) the

k black hole configurations correspond to points lying above the single-black hole phase and below the k-copied phase.

From the first-order corrected temperatures (6.40) one can show that the multi-black hole configurations are in general not in thermal equilibrium. The only configurations that are in thermal equilibrium to this order are the copies of the single-black hole solution studied previously [31, 52, 98]. As a further comment we note that Hawking radiation will seed the mechanical instabilities of the multi-black hole configurations. The reason for this is that in a generic configuration the black holes have different rates of energy loss and hence the mass ratios required for mechanical equilibrium are not maintained. This happens even in special configurations, e.g., when the temperatures are equal, because the thermal radiation is only statistically uniform. Hence asymmetries in the real-time emission process will introduce disturbances driving these special configurations away from their equilibrium positions.

6.4.3 Consequences for the Phase Diagram

The existence of the multi-black hole solutions has striking consequences for the phase structure of black hole solutions on $\mathcal{M}^d \times S^1$. It means that one can, for example, start from a solution with two equal size black holes, placed oppositely to each other on the cylinder, and then continuously deform the solution to be arbitrarily close to a solution with only one black hole (the other black hole being arbitrarily small in comparison). Thus, we get a continuous span of classical static solutions for a given total mass. In particular, a multi-black hole configuration with k black holes has k independent parameters. This implies a continuous non-uniqueness in the (μ, n) phase diagram (or for a given mass), much like the one observed for bubble-black hole sequences [49] and for other classes of black hole solutions [17–19, 99] (see also Sect. 6.6). In particular, this has the consequence that if we would live on $\mathcal{M}^4 \times S^1$ then from a four-dimensional point of view one would have an infinite non-uniqueness for static black holes of size similar to the size of the extra dimension, thus severely breaking the uniqueness of the Schwarzschild black hole.

6.4.3.1 New Non-uniform Strings?

Another consequence of the new multi-black hole configurations is for the connection to uniform and non-uniform strings on the cylinder. As discussed in Sect. 6.3.3, there is evidence that the black hole on the cylinder phase merges with the non-uniform black string phase in a topology changing transition point. It follows from this that the copies of black hole on the cylinder solution merge with the copies of non-uniform black strings. However, due to the multi-black hole configurations we now have a continuous span of solutions connected to the copies of the black hole on the cylinder. Therefore, it is natural to ask whether the new solutions also merge with non-uniform black string solutions in a topology changing transition point. If so, it probes the question whether there exist, in addition to having new black holes on the cylinder solutions, also new non-uniform black string solutions. Thus, these new solutions present a challenge for the current understanding of the phase diagram for black holes and strings on the cylinder. For a detailed discussion on this, see [35].

Another connection with strings and black holes on the cylinder is that a Gregory–Laflamme unstable uniform black string is believed to decay to a black hole on the cylinder (when the number of dimensions is less than the critical one [39]). However, the new multi-black hole solutions mean that one can imagine them as intermediate steps in the decay.

6.4.3.2 Lumpy Black Holes

Reference [35] also examines in detail configurations with two and three black holes. For two black holes this confirms the expectation that one maximizes the entropy by transferring all the mass to one of the black holes, and also that if the two black holes are not in mechanical equilibrium then the entropy is increasing as the black holes become closer to each other. These two facts are both in accordance with the general argument that the multi-black hole configurations are in an unstable equilibrium and generic perturbations of one of the positions will result in that all the black holes merge together in to a single black hole on the cylinder.

A detailed examination of the three black hole solutions suggests the possibility of further new types of black hole solutions in Kaluza–Klein spacetimes. In particular, this analysis suggests the possibility that new static configurations may exist that consist of a lumpy black hole, where the non-uniformities are supported by the gravitational stresses imposed by an external field. These new solutions were argued by considering a symmetric configuration of three black holes, with one of mass M_1 and two others of equal mass $M_2 = M_3$ at equal distance to the first one. Increasing the total mass of the system shows that it is possible that the two black holes (2 and 3) merge before merging with black hole 1. In this way one could end up with a static solution consisting of lumpy black hole (i.e., a 'peanut-like'-shaped black object) together with an ellipsoidal black hole.

6.4.3.3 Analogue Fluid Model

Finally we note that one may consider the multi-black hole configurations in relation to an analogue fluid model for the Gregory–Laflamme (GL) instability, recently proposed in [90]. There it was pointed out that the GL instability of a black string has a natural analogue description in terms of the Rayleigh–Plateau (RP) instability of a fluid cylinder. It turns out that many known properties of the gravitational instability have an analogous manifestation in the fluid model. These include the behavior of the threshold mode with d, dispersion relations, the existence of critical dimensions and the initial stages of the time evolution (see [90, 91, 100] for details). In the context of this analogue fluid model, [35] discusses a possible, but more speculative, relation of the multi-black hole configurations to configurations observed in the time evolution of fluid cylinders.

6.5 Thin Black Rings in Higher Dimensions

In this and the next section we turn our attention to rotating black holes. We start by reviewing the recent construction [30] of an approximate solution for an asymptotically flat neutral thin rotating black ring in any dimension $D \geq 5$ with horizon topology $S^{D-3} \times S^1$. As in Sect. 6.4, this construction uses the method of matched asymptotic expansion, and we only present the main points. We discuss in particular the equilibrium condition necessary for balancing the ring and how this enables to obtain the leading-order thermodynamics of thin rotating black rings. We also compare the thermodynamics of the thin black ring to that of the MP black hole. In this and the following section we denote the number of spacetime dimensions by $D = 4 + n$.

6.5.1 Thin Black Rings from Boosted Black Strings

Black rings in $(n + 4)$-dimensional asymptotically flat spacetime are solutions of Einstein gravity with an event horizon of topology $S^1 \times S^{n+1}$. As we briefly reviewed in Sects. 6.1 and 6.2 explicit solutions with this topology in five dimensions ($n = 1$) were first presented in [16] (see also [2] for a review).

In five dimensions, there is beyond the MP black hole and the black ring one more phase of rotating black holes if one restricts to phases with a single angular momentum that are in thermal equilibrium. This is the black Saturn phase consisting of a central MP black and one black ring around it, having equal temperature and angular velocity. If one abandons the condition of thermal equilibrium there are many more black Saturn phases with multiple rings as well as multi-black ring solutions. We refer to [18] and the recent review [4] for details on the more general phase structure for the five-dimensional case.

The construction of analogous solutions in more than five dimensions is considerably more involved, since for $D \geq 6$ these solutions are not contained in the generalized Weyl ansatz [24, 25, 101] because they do not have $D - 2$ commuting Killing symmetries. Furthermore the inverse scattering techniques of [26–29] do not extend to the asymptotically flat case in any $D \geq 6$.

Therefore, one way to make progress toward solving this problem can be achieved by first constructing thin black ring solutions in arbitrary dimensions as a perturbative expansion around circular boosted black strings. The idea that rotating thin black rings should be well approximated by boosted black strings is intuitively clear and already appears in earlier works [102–104]. This was used as a starting point in the explicit construction [30].

6.5.1.1 Boosted Black String

The zeroth-order solution is that of a *straight* boosted black string. The metric of this can easily be obtained from (6.11) by applying a boost in the (t, z) plane. The result is

$$
ds^2 = -\left(1 - \cosh^2 \alpha \frac{r_0^n}{r^n}\right) dt^2 - 2\frac{r_0^n}{r^n} \cosh \alpha \sinh \alpha \, dt dz + \left(1 + \sinh^2 \alpha \frac{r_0^n}{r^n}\right) dz^2
$$

$$
+ \left(1 - \frac{r_0^n}{r^n}\right)^{-1} dr^2 + r^2 d\Omega_{n+1}^2 , \tag{6.47}
$$

where r_0 is the horizon radius and α is the boost parameter. In general, we will take the z direction to be along an S^1 with circumference $2\pi R$, which means we can write z in terms of an angular coordinate ψ defined by $\psi = z/R$ ($0 \leq \psi < 2\pi$). At distances $r \ll R$, the solution (6.47) is the approximate metric of a thin black ring to zeroth order in $1/R$.

By definition, a thin black ring has an S^1 radius R that is much larger than its S^{n+1} radius r_0. In this limit, the mass of the black ring is small and the gravitational attraction between diametrically opposite points of the ring is very weak. So, in regions away from the black ring, the linearized approximation to gravity will be valid, and the metric will be well approximated if we substitute the ring by an appropriate delta-like distributional source of energy–momentum. The source has to be chosen so that the metric it produces is the same as that expected from the full exact solution in the region far away from the ring. Since the thin black ring is expected to approach locally the solution for a boosted black string, it is sensible to choose distributional sources that reproduce the metric (6.47) in the weak-field regime:

$$
T_{tt} = \frac{r_0^n}{16\pi G} \left(n \cosh^2 \alpha + 1\right) \delta^{(n+2)}(r) , \tag{6.48a}
$$

$$
T_{tz} = \frac{r_0^n}{16\pi G} n \cosh \alpha \sinh \alpha \, \delta^{(n+2)}(r) , \tag{6.48b}
$$

$$
T_{zz} = \frac{r_0^n}{16\pi G} \left(n \sinh^2 \alpha - 1\right) \delta^{(n+2)}(r) . \tag{6.48c}
$$

The location $r = 0$ corresponds to a circle of radius R in the $(n+3)$-dimensional Euclidean flat space, parameterized by the angular coordinate ψ. In this construction the mass and angular momentum of the black ring are obtained by integrating the energy and momentum densities:

$$M = 2\pi R \int_{S^{n+1}} T_{tt} , \quad J = 2\pi R^2 \int_{S^{n+1}} T_{tz} , \tag{6.49}$$

where S^{n+1} links the ring once.

6.5.1.2 Dynamical Equilibrium Condition

We now first show that the boost parameter α gets fixed by a dynamical equilibrium condition ensuring that the string tension is balanced against the centrifugal repulsion. To this end note that we are approximating the black ring by a distributional source of energy–momentum. The general form of the equation of motion for probe brane-like objects in the absence of external forces takes the form [105]

$$K_{\mu\nu}{}^\rho T^{\mu\nu} = 0, \tag{6.50}$$

where the indices μ, ν are tangent to the brane and ρ is transverse to it. The second fundamental tensor $K_{\mu\nu}{}^\rho$ extends the notion of extrinsic curvature to submanifolds of codimension possibly larger than one. The extrinsic curvature of the circle is $1/R$, so a circular linear distribution of energy–momentum of radius R will be in equilibrium only if

$$\frac{T_{zz}}{R} = 0, \tag{6.51}$$

i.e., for finite radius the pressure tangential to the circle must vanish. Hence, for the thin black ring with source (6.48), the condition that the ring be in equilibrium translates into a very specific value for the boost parameter

$$\sinh^2 \alpha = \frac{1}{n} , \tag{6.52}$$

which we will also refer to as the critical boost. For $D = 5$ ($n = 1$) this was already observed in [102] where the thin black string limit of five-dimensional black rings was first made explicit, but the connection with (6.50) was first noticed in [30].

6.5.1.3 Thermodynamics

Using (6.52) it is not difficult to obtain the physical quantities of the critically boosted black string, and hence the leading-order thermodynamics of thin black rings (see also [103, 106] for further details on boosted black strings and their thermodynamics). We find for the mass M, entropy S, temperature T, angular momentum J and angular velocity Ω the expressions [30]

$$M = \frac{\Omega_{n+1}}{8G} R r_0^n (n+2) \,, \quad S = \frac{\pi \Omega_{n+1}}{2G} R r_0^{n+1} \sqrt{\frac{n+1}{n}} \,, \quad T = \frac{n}{4\pi} \sqrt{\frac{n}{n+1}} \frac{1}{r_0} \,, \quad (6.53a)$$

$$J = \frac{\Omega_{n+1}}{8G} R^2 r_0^n \sqrt{n+1} \,, \qquad \Omega = \frac{1}{\sqrt{n+1}} \frac{1}{R} \,. \qquad (6.53b)$$

We also note that an equivalent but more physical form of the equilibrium equation (6.52) in terms of these quantities is

$$R = \frac{n+2}{\sqrt{n+1}} \frac{J}{M} \,. \qquad (6.54)$$

We thus see that the radius grows linearly with J for fixed mass.

It is remarkable that with the above reasoning one can already obtain the correct limiting thermodynamics of thin black rings to leading order, without having to solve for any metric. One finds from (6.53) that the entropy of thin black rings behaves as

$$S^{\text{ring}}(M,J) \propto J^{-\frac{1}{D-4}} M^{\frac{D-2}{D-4}} \,, \qquad (6.55)$$

whereas that of ultra-spinning MP black holes in $D \geq 6$ is given by [36]

$$S^{\text{hole}}(M,J) \propto J^{-\frac{2}{D-5}} M^{\frac{D-2}{D-5}} \,. \qquad (6.56)$$

This already shows the non-trivial fact that in the ultra-spinning regime of large J for fixed mass M the rotating black ring has higher entropy than the MP black hole (see also Sect. 6.5.3). Moreover, as explained in Sect. 6.5.2, it turns out that for $D \geq 6$ the results (6.53) are actually valid up to and including the next order in r_0/R, so receives only $O(r_0^2/R^2)$ corrections. This conclusion could already be drawn once one has convinced oneself that the first-order $1/R$ correction terms in the metric only involve dipole contributions which can easily be argued to give zero contribution to all thermodynamic quantities [30].

It is important to stress that the above reasoning relies crucially on the assumption that when the boosted black string is curved, the horizon remains regular. To verify this point, and also to obtain a metric for the thin black ring, [30] solves the Einstein equations explicitly by constructing an approximate solution for $r_0 \ll R$ using a matched asymptotic expansion. In this analysis one finds that the condition (6.51) appears as a consequence of demanding absence of singularities on the plane of the ring outside the horizon. Whenever $n \sinh^2 \alpha \neq 1$ with finite R, the geometry backreacts creating singularities on the plane of the ring. These singularities admit a natural interpretation. Since (6.50) is a consequence of the conservation of the energy–momentum tensor, when (6.51) is not satisfied there must be additional sources of energy–momentum. These additional sources are responsible for the singularities in the geometry. Alternatively, the derivation of (6.51) in [30] from the Einstein equations is an example of how general relativity encodes the equations of motion of black holes as regularity conditions on the geometry.

6.5.2 *Matched Asymptotic Expansion*

We now review the highlights of the perturbative construction of thin black rings using matched asymptotic expansion (see also Sect. 6.4.1). In the problem at hand, the two widely separated scales are the 'thickness' of the ring r_0 and the radius of the ring R, and the thin limit means that $r_0 \ll R$. There are therefore two zones, an asymptotic zone at large distances from the black ring, $r \gg r_0$, where the field can be expanded in powers of r_0. The other zone is the near-horizon zone which lies at scales much smaller than the ring radius, $r \ll R$. In this zone the field is expanded in powers of $1/R$. At each step, the solution in one of the zones is used to provide boundary conditions for the field in the other zone, by matching the fields in the 'overlap' zone $r_0 \ll r \ll R$ where both expansions are valid.

As already discussed in Sect. 6.5.1, the starting point is to consider the solution in the near-horizon zone to zeroth order in $1/R$, i.e., we take a boosted black string of infinite length, $R \to \infty$. The next steps in the construction are then as follows:

- Step 1: One solves the Einstein equations in the linearized approximation around flat space for a source corresponding to a circular distribution of a given mass and momentum density as given in (6.48). This metric is valid in the region $r \gg r_0$.
- Step 2: We consider the Newtonian solution close to the sources, i.e., in the overlap region $r_0 \ll r \ll R$.
- Step 3: We consider the near-horizon region of the ring and find the linear corrections to the metric of a boosted black string for a perturbation that is small in $1/R$; in other words, we analyze the geometry of a boosted black string that is now slightly curved into a circular shape. This solution is then matched to the metric in the overlap region found in step 2. The resulting solution is valid in the region $r_0 \leq r \ll L$.

To solve step 1 for a non-zero $T_{\psi\psi} = R^2 T_{zz}$ is not easy. It is therefore convenient to already assume that the equilibrium condition $T_{\psi\psi} = 0$ in (6.51) is satisfied. This then gives the solution of a black ring in linearized gravity [30]. Finding a more general solution with a source for the tension is much easier if one restricts to the overlap zone (step 2). In this regime we are studying the effects of locally curving a thin black string into an arc of constant curvature radius R. To this end it is convenient to introduce ring-adapted coordinates. These are derived in [30] and to first order in $1/R$ the flat space metric in these coordinates takes the form

$$ds^2(\mathbb{E}^{n+3}) = \left(1 + \frac{2r\cos\theta}{R}\right) dz^2 + \left(1 - \frac{2}{n}\frac{r\cos\theta}{R}\right) \left(dr^2 + r^2 d\theta^2 + r^2 \sin^2\theta d\Omega_n^2\right).$$

$$(6.57)$$

In terms of these coordinates the general form of the metric in the overlap region is then

$$g_{\mu\nu} \simeq \eta_{\mu\nu} + \frac{r_0^n}{r^n}\left(h_{\mu\nu}^{(0)}(r) + \frac{r\cos\theta}{R} h_{\mu\nu}^{(1)}(r)\right).$$

$$(6.58)$$

Solving Einstein equations to order $1/R$ then explicitly shows that regularity of the solution enforces vanishing of the tension T_{zz} (see (6.51)).

The technically most difficult part of the problem is to find the near-horizon solution in step 3. Physically, this corresponds to curving the black string into a circle of large but finite radius R. In effect, this means that we are placing the black string in an external potential whose form at large distances is that of (6.58) and which changes the metric $g_{\mu\nu}^{bbs}$ in (6.47) of the (critically) boosted black string by a small amount, i.e.,

$$g_{\mu\nu} \simeq g_{\mu\nu}^{bbs}(r;r_0) + \frac{\cos\theta}{R}h_{\mu\nu}(r;r_0) . \tag{6.59}$$

In [30] the Einstein equations to order $1/R$ are explicitly solved, showing that the perturbations $h_{\mu\nu}(r;r_0)$ can be expressed in terms of hypergeometric functions.

6.5.2.1 Corrected Thermodynamics

One can find the corrections to the thermodynamics as follows. First, one uses the near-horizon corrected metric (6.59) to find the corrections to the entropy S, temperature T and angular velocity Ω. Then one can use the first law,

$$dM = T\delta S + \Omega\delta J , \tag{6.60}$$

and the Smarr formula

$$(n+1)M = (n+2)(TS+\Omega J) , \tag{6.61}$$

to deduce the corrections to the mass and angular momentum.[5] Using now that the perturbations in (6.59) are only of dipole type, with no monopole terms, it follows that the area, surface gravity and angular velocity receive no modifications in $1/R$. The reason is that a dipole cannot change the total area of the horizon, only its shape. This is true of both the shape of the S^{n+1} and the length of the S^1, which can vary with θ but on average (i.e., when integrated over the horizon) remain constant. So S is not corrected. The surface gravity and angular velocity cannot be corrected either. They must remain uniform on a regular horizon, so, since the dipole terms vanish at $\theta = \pi/2$, no corrections to T and Ω are possible. It then follows from (6.60) and (6.61) that M and J are not corrected either.[6] So the function $S(M,J)$ obtained in (6.55) is indeed valid including the first order in $1/R$. It is interesting to observe that this conclusion could be drawn already when the asymptotic form of the metric (6.58) in the overlap zone, is seen to include only dipole terms at order $1/R$.

---

[5] This method was also used in [31, 35] for small black holes and multi-black holes on the cylinder.

[6] In five dimensions ($n = 1$) there *are* corrections to this order. Their origin is discussed in Appendix A of [30].

6.5.3 Black Rings Versus MP Black Holes

We now proceed by analyzing the thin black ring thermodynamics and compare it to that of ultra-spinning MP black holes. Recall that the thermodynamics of the thin black ring in the ultra-spinning regime is given by (6.53), which is valid up to $O(r_0^2/R^2)$ corrections.

6.5.3.1 Myers–Perry Black Hole

For the MP black hole, exact results can be obtained for all values of the rotation. The two independent parameters specifying the (single angular momentum) solution are the mass parameter μ and the rotation parameter a, from which the horizon radius r_0 is found as the largest (real) root of the equation

$$\mu = (r_0^2 + a^2) r_0^{n-1}. \tag{6.62}$$

In terms of these parameters the thermodynamics take the form [15]

$$M = \frac{(n+2)\Omega_{n+2}\,\mu}{16\pi G}, \qquad S = \frac{\Omega_{n+2}\,r_0\,\mu}{4G}, \qquad T = \frac{1}{4\pi}\left(\frac{2r_0^n}{\mu} + \frac{n-1}{r_0}\right), \tag{6.63a}$$

$$J = \frac{\Omega_{n+2}\,a\,\mu}{8\pi G}, \qquad \Omega = \frac{a\,r_0^{n-1}}{\mu}. \tag{6.63b}$$

Note the similarity between $a = \frac{n+2}{2}\frac{J}{M}$ and the black ring relation (6.54).

An important simplification occurs in the ultra-spinning regime of $J \to \infty$ with fixed M, which corresponds to $a \to \infty$. Then (6.62) becomes $\mu \to a^2 r_0^{n-1}$ leading to simple expressions for (6.63) in terms of r_0 and a, which in this regime play roles analogous to those of r_0 and R for the black ring. Specifically, a is a measure of the size of the horizon along the rotation plane and r_0 a measure of the size transverse to this plane [36]. In fact, in this limit

$$M \to \frac{(n+2)\Omega_{n+2}}{16\pi G}\,a^2 r_0^{n-1}, \qquad S \to \frac{\Omega_{n+2}}{4G}\,a^2 r_0^n, \qquad T \to \frac{n-1}{4\pi r_0} \tag{6.64}$$

take the same form as the expressions characterizing a black membrane extended along an area $\sim a^2$ with horizon radius r_0. This identification lies at the core of the ideas in [36], which were further developed in [30] and summarized in Sect. 6.6. We note that the quantities J and Ω disappear since the black membrane limit is approached in the region near the axis of rotation of the horizon and so the membrane is static in the limit. Note furthermore that (6.64) is valid up to $O(r_0^2/a^2)$ corrections.

Finally, we remark that the transition to the membrane-like regime is signaled by a qualitative change in the thermodynamics of the MP black holes. At

$a/r_0 = \sqrt{(\frac{n+1}{n-1})}$ the temperature reaches a minimum and $\left(\partial^2 S/\partial J^2\right)_M$ changes sign. For a/r_0 smaller than this value, the thermodynamic quantities of the MP black holes such as T and S behave similarly to those of the Kerr solution and one should not expect any membrane-like behavior. However, past this point they rapidly approach the membrane results. We do not expect that the onset of thermodynamic instability at this point is directly associated to any dynamical instability. Rather, one expects a GL-like instability to happen at a larger value of a/r_0 [30, 36].

6.5.3.2 Dimensionless Quantities

Contrary to the case of KK black holes where we could use the circle length to define dimensionless quantities (cf. (6.7) or (6.8)) in this case we need to use one of the physical parameters of the solutions to define dimensionless quantities. We choose the mass M and thus introduce dimensionless quantities for the spin j, the area a_H, the angular velocity ω_H and the temperature t_H via

$$j^{n+1} \propto \frac{J^{n+1}}{GM^{n+2}}, \qquad a_H^{n+1} \propto \frac{S^{n+1}}{(GM)^{n+2}}, \tag{6.65a}$$

$$\omega_H \propto \Omega (GM)^{\frac{1}{n+1}}, \qquad t_H \propto (GM)^{\frac{1}{n+1}} T, \tag{6.65b}$$

where convenient normalization factors can be found in (7.9) of [30]. We take j as our control parameter and now study and compare the functions $a_H(j)$, $\omega_H(j)$ and $t_H(j)$ for black rings and MP black holes in the ultra-spinning regime. These asymptotic phase curves can now be obtained using (6.65) together with (6.53) and (6.64), respectively. In the following we denote the results for the thin black ring with (r) and for the ultra-spinning MP black holes with (h), and generally omit numerical prefactors.

6.5.3.3 Comparison of the Thermodynamics

Starting with the reduced area function we see that

$$a_H^{(r)} \sim \frac{1}{j^{1/n}}, \qquad a_H^{(h)} \sim \frac{1}{j^{2/(n-1)}}, \tag{6.66}$$

and so, for any $D = 4 + n \geq 6$, the area decreases faster for MP black holes than for black rings, so we immediately see that black rings dominate entropically in the ultra-spinning regime [30]. For illustration, Fig. 6.3 shows these curves in $D = 7$ ($n = 3$).

Including prefactors one finds for the angular velocities that

$$\omega_H^{(r)} \to \frac{1}{2j}, \qquad \omega_H^{(h)} \to \frac{1}{j}. \tag{6.67}$$

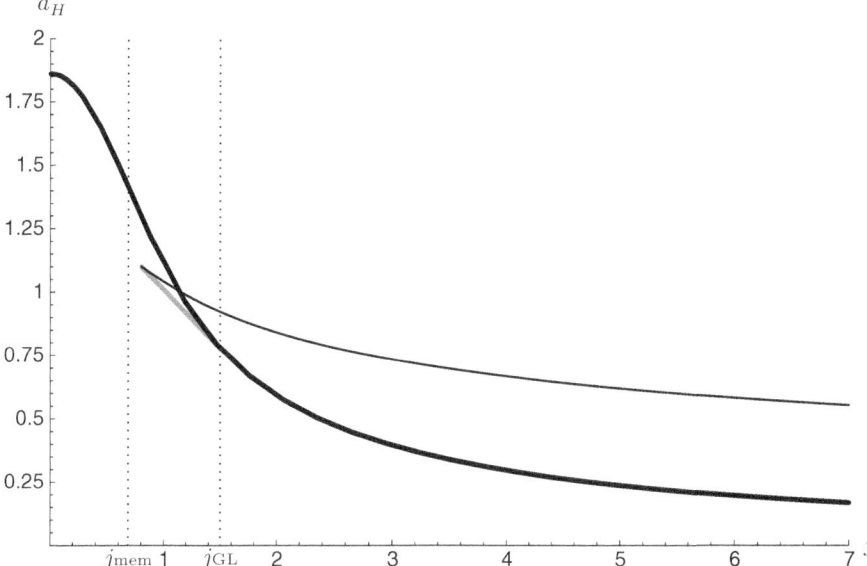

Fig. 6.3 Area versus spin for fixed mass, $a_H(j)$, in seven dimensions. For large j, the *thin curve* is the result for thin black rings and is extrapolated here down to $j \sim O(1)$. The *thick curve* is the exact result for the MP black hole. The *gray line* corresponds to the conjectured phase of pinched black holes (see Sect. 6.6), which branch off tangentially from the MP curve at a value $j_{GL} > j_{mem}$. At any given dimension, the phases should not necessarily display the swallowtail as shown in this diagram, but could also connect more smoothly via a pinched black hole phase that starts tangentially in j_{GL} and has increasing j. Reprinted from [30]

The ratio $\omega_H^{(h)}/\omega_H^{(r)} = 2$, which holds for all $D \geq 6$, is reminiscent of the factor of 2 in Newtonian mechanics between the moment of inertia of a wheel (i.e., a ring) and a disk (i.e., a pancake) of the same mass and radius, which implies that the disk must rotate twice as fast as the wheel in order to have the same angular momentum. Irrespective of whether this is an exact analogy or not, the fact that $\omega_H^{(r)} < \omega_H^{(h)}$ is clearly expected from this sort of picture. For the temperatures we find

$$t_H^{(r)} \sim j^{1/n} , \quad t_H^{(h)} \sim j^{2/(n-1)} , \tag{6.68}$$

so the thin black ring is colder than the MP black hole. In fact, since the temperature is inversely proportional to the thickness of the object the picture suggested above leads to the following argument: if we put a given mass in the shape of a wheel of given radius, then we get a thicker object than if we put it in the shape of a pancake of the same radius.

6.6 Completing the Phase Diagram

In this section we will discuss the phase structure of asympotically flat neutral rotating black holes in six and higher dimensions by exploiting a connection between, on one side, black holes and black branes in KK spacetimes and, on the other side, higher-dimensional rotating black holes. Building on the basic idea in [36], this phase structure was recently proposed in [30]. Part of this picture is conjectural, but is based on well-motivated analogies and appears to be natural from many points.

The curve $a_H(j)$ at values of j outside the domain of validity of the computations in Sect. 6.5 corresponds to the regime where the gravitational self-attraction of the ring is important. There are no analytical methods presently known to treat such values $j \sim O(1)$, and the precise form of the curve in this regime may require numerical solutions. However, as argued in [30] it is possible to complete the black ring curve and other features of the phase diagram, at least qualitatively. This is done by combining a number of observations and reasonable conjectures about the behavior of MP black holes at large rotation and using as input the presently known phase structure of Kaluza–Klein black holes (see Sect. 6.3).

6.6.1 GL Instability of Ultra-spinning MP Black Hole

In the ultra-spinning regime in $D \geq 6$, MP black holes approach the geometry of a black membrane $\approx \mathcal{R}^2 \times S^{D-4}$ spread out along the plane of rotation [36]. In Sect. 6.5.3 we have already observed that the extent of the black hole along the plane is approximately given by the rotation parameter a, while the 'thickness' of the membrane, i.e., the size of its S^{D-4}, is given by the parameter r_0. For a/r_0 larger than a critical value of order one we expect that the dynamics of these black holes is well approximated by a black membrane compactified on a square torus $\mathbb{T}^2$ with side length $L \sim a$ and with S^{D-4} size $\sim r_0$. The angular velocity of the black hole is always moderate, so it will not introduce large quantitative differences, but note that the rotational axial symmetry of the MP black holes translates into only one translational symmetry along the $\mathbb{T}^2$, the other one being broken.

Using this analogue mapping of membranes and fastly rotating MP black holes, [36] argued that the latter should exhibit a Gregory–Laflamme-type instability. Furthermore, as reviewed in Sect. 6.3 it is known that the threshold mode of the GL instability gives rise to a new branch of static non-uniform black strings and branes [37, 38, 92]. In correspondence with this, [36] argued that it is natural to conjecture the existence of new branches of axisymmetric 'lumpy' (or 'pinched') black holes, branching off from the MP solutions along the stationary axisymmetric zero-mode perturbation of the GL-like instability,

6.6.1.1 Map to Phases of KK Black Holes on the Torus

In [30] this analogy was pushed further by drawing a correspondence between the phases of KK black holes on the torus (see Sect. 6.3.4) and the phases of higher-dimensional black holes, as illustrated in Fig. 6.4. Here we have restricted to non-uniformities of the membrane along only of the two brane directions, since including non-uniformity in a second direction would not have a counterpart for rotating black holes. These would break axial symmetry and hence would be radiated away. Other limitations of the analogy are discussed in detail in [30].

Using the correspondence between the phases of the two systems, one can import, at least qualitatively, the known phase diagram of black membranes on $\mathscr{M}^{D-2} \times \mathbb{T}^2$ onto the phase diagram of rotating black objects in $\mathscr{M}^D$. To this end one needs to first establish the map between quantities on each side of this correspondence. For unit mass, the quantities ℓ (see (6.8)) and j (see (6.65)) measure

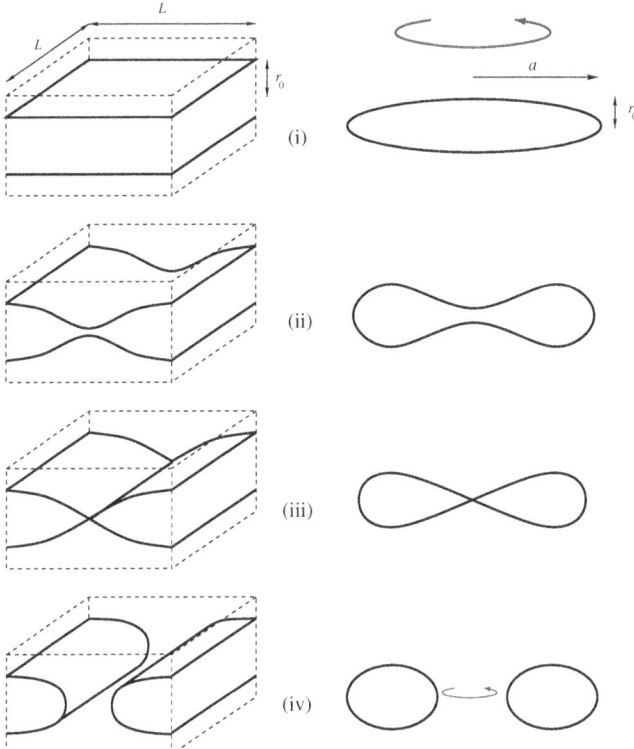

Fig. 6.4 Correspondence between phases of black membranes wrapped on a $\mathbb{T}^2$ of side L (*left*) and fastly rotating MP black holes with rotation parameter $a \sim L \geq r_0$ (*right*: must be rotated along a vertical axis): (**i**) uniform black membrane and MP black hole; (**ii**) non-uniform black membrane and pinched black hole; (**iii**) pinched-off membrane and black hole; (**iv**) localized black string and black ring. Reprinted from [30]

the (linear) size of the horizon along the torus or rotation plane, respectively. Then $a_H(\ell)$ for KK black holes on $\mathcal{M}^{n+2} \times \mathbb{T}^2$ is analogous (up to constants) to $a_H(j)$ for rotating black holes in $\mathcal{M}^{n+4}$.

More precisely, although the normalization of magnitudes in (6.65) and (6.8) is different, the functional dependence of a_H on ℓ or j must be parametrically the same in both functions, at least in the regime where the analogy is precise. As a check on this, note that the function $a_H(\ell)$ in (6.23) for the uniform black membrane exhibits exactly the same functional form (6.66) as $a_H(j)$ for the MP black hole in the ultra-spinning limit. Similarly, (6.24) for the localized black string shows the same functional form as (6.66) for the black ring in the large j limit. The most important application of the analogy, though, is to non-uniform membrane phases (see (6.25)), providing information about the phases of pinched rotating black holes and how they connect to MP black holes and black rings.

6.6.2 Phase Diagram of Neutral Rotating Black Holes on $\mathcal{M}^D$

We present here the main points of the proposed phase diagram [30] of neutral rotating black holes (with one angular momentum) in asymptotically flat space that follows from the analogy described above. To this end, we recall that the phases of KK black holes on a two-torus were discussed in Sect. 6.3.4 and depicted in the representative phase diagram in Fig. 6.2.

6.6.2.1 Main Sequence

The analogy developed above suggests that the phase diagram of rotating black holes in the range $j > j_{mem}$ where MP black holes behave like black membranes is qualitatively the same as that for KK black holes on the torus (see Fig. 6.2), with a pinched (lumpy) rotating black hole connecting the MP black hole with the black ring. This phase is depicted in Fig. 6.3 as a gray line emerging tangentially from the MP black hole curve at a critical value j_{GL} that is currently unknown. Arguments were given in [36] to the effect that $j_{GL} \geq j_{mem}$, consistent with the analogy. As one moves along the gray line in Fig. 6.3 in the direction away from the MP curve, the pinch at the rotation axis of these black holes grows deeper. Eventually, as depicted in Fig. 6.4, the horizon pinches down to zero thickness at the axis and then the solutions connect to the black ring phase. Note also that we may have the 'swallowtail' structure of first-order phase transitions (as depicted in Fig. 6.3), or instead that of second-order phase transitions (see Fig. 4 of [30]). It may not be unreasonable to expect that a swallowtail appears at least for the lowest dimensions $D = 6, 7, \ldots$, since this is in fact the same type of phase structure that appears for $D = 5$.

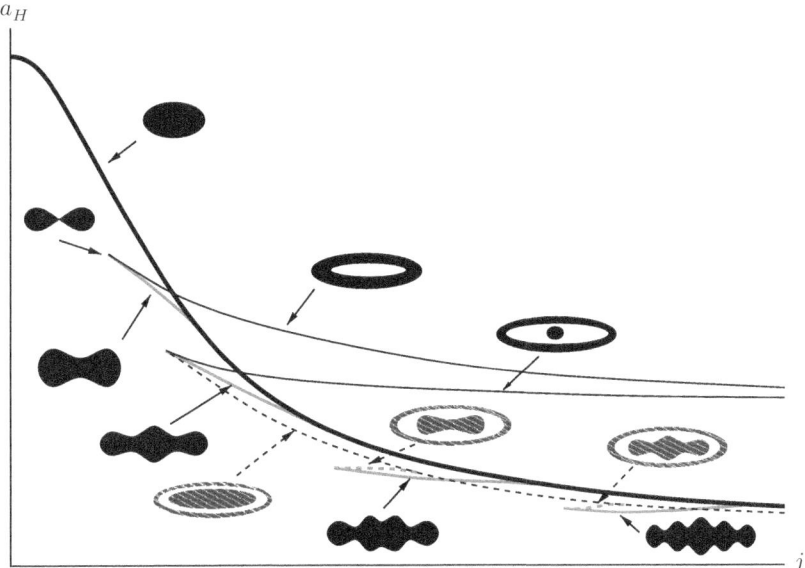

Fig. 6.5 Proposal for the phase diagram of thermal equilibrium phases of rotating black holes in $D \geq 6$ with one angular momentum. The *solid lines* and figures have significant arguments in their favor, while the *dashed lines* and figures might not exist and admit conceivable, but more complicated, alternatives. Some features have been drawn arbitrarily, e.g., at any given bifurcation and in any dimension one may either have smooth connections or swallowtails with cusps. If thermal equilibrium is not imposed, the whole semi-infinite strip $0 < a_H < a_H(j = 0), 0 \leq j < \infty$, is covered, and multi-rings are possible. Reprinted from [30]

Beyond this main sequence, [30] presents arguments for further completion of the phase diagram, which is summarized in Fig. 6.5. The most important features are as follows.

6.6.2.2 Infinite Sequence of Lumpy (Pinched) Black Holes

Another observation based on the membrane analogy is that the phase diagram of rotating black holes should also exhibit an infinite sequence [30, 36] of lumpy (pinched) black holes emerging from the curve of MP black holes at increasing values of j. These are the analogues of the k-copied phases in the phase diagram of KK black holes that appear at increasing ℓ according to (6.26). In this connection note that for the GL zero modes of MP black holes one must choose axially symmetric combinations, implying a change of basis from plane waves $\exp(ik_{GL}z)$ to Bessel functions. Axially symmetric modes have a profile $J_0(k_{GL}a\sin\theta)$ [36]. The main point here is that the wavelength λ_{GL} (see (6.15)) of the GL zero mode remains the same in the two analogue systems, to first approximation, even if the profiles are not the same. One is thus led to the existence of an infinite sequence of pinched black hole phases emanating from the MP curve at increasing values $j_{GL}^{(k)}$.

6.6.2.3 Black Saturn

If we focus on the first copy ($k = 2$), on the KK black hole side this corresponds to a non-uniform membrane on $\mathbb{T}^2$ with a GL zero-mode perturbation of the membrane with two minima, which grows to merge with a configuration of two identical black strings localized on the torus. For the MP black hole, the analogue is the development of a circular pinch, which then grows deeper until the merger with a black Saturn configuration in thermal equilibrium. Thermal equilibrium, i.e., equal temperature and angular velocity on all disconnected components of the event horizon, is in fact naturally expected for solutions that merge with pinched black holes, since the temperature and angular velocity of the latter should be uniform on the horizon all the way down to the merger, and we do not expect them to jump discontinuously there. These appear to be the natural higher-dimensional generalization of the five-dimensional black Saturn [17], and one may invoke the same arguments as those in [18]. When the size of the central black hole is small compared to the radius of the black ring, the interaction between the two objects is small and, to a first approximation, one can simply combine them linearly. It follows that, under the assumption of equal temperatures and angular velocities for the two black objects in the black Saturn, as j is increased a larger fraction of the total mass and the total angular momentum is carried by the black ring, and less by the central black hole. Then, this black Saturn curve must asymptote to the curve of a single black ring.

6.6.2.4 Pancaked and Pinched Black Saturns

The existence of these phases and their appearance in the phase diagram in Fig. 6.5 (in which they appear dashed) is based on comparatively less compelling arguments. Nevertheless, these conjectural phases provide a simple and natural way of completing the curves in the phase diagram that is consistent with the available information. We refer to [30] for further details on these phases.

It should also be noted that in the diagram of Fig. 6.5 only the thermal equilibrium phases among the possible multi-black hole phases are represented. The existence of multi-black rings, with or without a central black hole, in thermal equilibrium is not expected. In general one does expect the existence of multi-black ring configurations, possibly with a central black hole, in which the different black objects have different surface gravities and different angular velocities. These configurations can be seen as the analogue of the multi-localized string configurations on the torus that can be obtained from multi-black hole configurations on the circle [35] discussed in Sect. 6.4 by adding a uniform direction.

6.7 Outlook

We conclude by briefly presenting a number of important issues and questions for future research. See also the reviews [1–4] for further discussion and other open problems.

6.7.1 Stability

In both classes discussed in this lecture it would be interesting to further study the stability of the various solutions. For KK black holes, this includes the classical stability of the non-uniform black string and the localized black hole. For the rotating black hole case, we note that black rings at large j in any $D \geq 5$ are expected to suffer from a GL instability that creates ripples along the S^1 and presumably fragments the black ring into black holes flying apart [16, 103, 104]. This instability may switch off at $j \sim O(1)$. In analogy to the five-dimensional case [104, 107], one could also study turning points of j. If these are absent, pinched black holes would presumably be stable to radial perturbations.

6.7.2 Other Compactified Solutions

It would also be interesting to examine the existence of other classes of solutions with a compactified direction. For example in [108] a supersymmetric rotating black hole in a compactified spacetime was found and charged black holes in compactified spacetimes are considered in [109]. In another direction, new solutions with Kaluza–Klein boundary conditions for anti-de-Sitter spacetimes have recently been constructed in [110, 111]. Finally, rotating non-uniform solutions in KK space have been constructed numerically in [112] (see also [113]).

6.7.3 Numerical Solutions

For both classes of higher-dimensional black holes presented in this lecture, it would be interesting to attempt to further apply numerical techniques in order to construct the new solutions. For example, for multi-black hole configurations on the cylinder this could confirm whether there are multi-black hole solutions for which the temperatures converge when approaching the merger points (as discussed in [35]). Furthermore, one could try to confirm the existence of the conjectured lumpy black holes (see Sect. 6.4.3). Similarly, for rotating black holes, numerical construction of the entire black ring phase and of the pinched black hole phase would be very interesting.

6.7.4 Effective Field Theory Techniques

As mentioned in Sect. 6.2.2, an alternative to the matched expansion is the use of classical effective field theory [44, 45] to obtain the corrected thermodynamics of new solutions in a perturbative expansion. It would be interesting to use this method to go beyond the first order for the solutions discussed in this lecture and apply it to other extended brane-like black holes with or without rotation.

6.7.5 Other Black Rings

The method used to construct thin black rings in asymptotically flat space can also be used to study thin black rings in external gravitational potentials, yielding, e.g., black Saturn or black rings in AdS or dS spacetime.[7] Similarly, one could study black rings with charges [115, 116] and with dipoles [99]. In this connection we note that the existence of small supersymmetric black rings in $D \geq 5$ was argued in [117].

6.7.6 More Rotation Parameters

One may try to extend the analysis to black rings with horizon $S^1 \times S^{n+1}$ with rotation not only along S^1 but also in the S^{n+1}. Rotation in the S^{n+1} will introduce particularly rich dynamics for $n \geq 3$, since it is then possible to have ultra-spinning regimes for this rotation too, leading to pinches of the S^{n+1} and further connections to phases with horizon $S^1 \times S^1 \times S^n$, and so forth.

6.7.7 Blackfolds

Following the construction in Sect. 6.5, one can envision many generalizations. In this way one could study the possible existence of more general blackfolds, obtained by taking a black p-brane with horizon topology $\mathcal{R}^p \times S^q$ and bending $\mathcal{R}^p$ to form some compact manifold. One must then find out under which conditions a curved black p-brane can satisfy the equilibrium equation (6.50). This method is constructive and uses dynamical information to determine possible horizon geometries. In contrast, conventional approaches based on topological considerations are non-constructive and have only found very weak restrictions in six or more dimensions [118, 119].

6.7.8 Plasma Balls and Rings

There is also another more indirect approach to higher-dimensional black rings in AdS, using the AdS/CFT correspondence. In [120] stationary, axially symmetric spinning configurations of plasma in $\mathcal{N} = 4$ SYM theory compactified to $d = 3$ on a Scherk–Schwarz circle were studied. On the gravity side, these correspond to large rotating black holes and black rings in the dual Scherk–Schwarz compactified AdS_5 space. Interestingly, the phase diagram of these rotating fluid configurations, even if

---

[7] In [114] the existence of supersymmetric black rings in AdS is considered.

dual to black holes larger than the AdS radius, reproduces many of the qualitative features of the MP black holes and black rings in five-dimensional flat spacetime. Higher-dimensional generalizations of this setup give predictions for the phases of black holes in Scherk–Schwarz compactified AdS_D with $D > 5$. In this way, evidence was found [120] for rotating black rings and 'pinched' black holes in AdS_6, which can be considered as the AdS-analogues of the phases conjectured in [30, 36], discussed in Sects. 6.5 and 6.6.

6.7.9 Microscopic Entropy for Three-Charge Black Holes

One could extend the work of [121, 122] by applying the boost/U-duality map of [11] to the multi-black hole configurations of [35]. In particular, this would enable to compute the first correction to the finite entropy of the resulting three-charge multi-black hole configurations on a circle. It would be interesting to then try to derive these expressions from a microscopic calculation following the single three-charge black hole case considered in [121, 123].

6.7.10 Braneworld Black Holes

The higher-dimensional black holes and branes described in this lecture also appear naturally in the discussion of the braneworld model of large extra dimensions [13, 124]. In other braneworld models such as the one proposed by Randall and Sundrum [125, 126] the geometry is warped in the extra direction and the discovery of black hole solutions in this context has proven more difficult. It would be interesting to consider the higher-dimensional black hole solutions considered in this lecture in these contexts.

Acknowledgments I would like to thank the organizers, especially Elefteris Papantonopoulos, of the Fourth Aegean Summer School on Black Holes (Sept. 17–22, 2007, Mytiline, Island of Lesvos, Greece) for a stimulating and interesting school. I also thank Oscar Dias, Roberto Emparan, Troels Harmark, Rob Myers, Vasilis Niarchos and Maria Jose Rodriguez for collaboration on the work presented here.

References

1. B. Kol, The phase transition between caged black holes and black strings: A review. Phys. Rept. **422**, 119–165 (2006), `hep-th/0411240`.
2. R. Emparan and H. S. Reall, Black rings. Class. Quant. Grav. **23**, R169 (2006), `hep-th/0608012`.
3. T. Harmark, V. Niarchos, and N. A. Obers, Instabilities of black strings and branes. Class. Quant. Grav. **24**, R1–R90 (2007), `hep-th/0701022`.

4. R. Emparan and H. S. Reall, *Black Holes in Higher Dimensions*, arXiv:0801.3471 [hep-th].

5. A. Strominger and C. Vafa, Microscopic origin of the Bekenstein-Hawking entropy. Phys. Lett. **B379**, 99–104 (1996), hep-th/9601029.

6. S. D. Mathur, The fuzzball proposal for black holes: An elementary review. Fortsch. Phys. **53**, 793–827 (2005), hep-th/0502050.

7. S. D. Mathur, The quantum structure of black holes. Class. Quant. Grav. **23**, R115 (2006), hep-th/0510180.

8. J. M. Maldacena, The large N limit of superconformal field theories and supergravity. Adv. Theor. Math. Phys. **2**, 231–252 (1998), hep-th/9711200.

9. O. Aharony, S. S. Gubser, J. Maldacena, H. Ooguri, and Y. Oz, Large N field theories, string theory and gravity. Phys. Rept. **323**, 183 (2000), hep-th/9905111.

10. O. Aharony, J. Marsano, S. Minwalla, and T. Wiseman, Black hole - black string phase transitions in thermal 1+1 dimensional supersymmetric Yang-Mills theory on a circle. Class. Quant. Grav. **21**, 5169–5192 (2004), hep-th/0406210.

11. T. Harmark and N. A. Obers, New phases of near-extremal branes on a circle. JHEP **09**, 022 (2004), hep-th/0407094.

12. N. Arkani-Hamed, S. Dimopoulos, and G. R. Dvali, The hierarchy problem and new dimensions at a millimeter. Phys. Lett. **B429**, 263–272 (1998), hep-ph/9803315.

13. I. Antoniadis, N. Arkani-Hamed, S. Dimopoulos, and G. R. Dvali, New dimensions at a millimeter to a Fermi and superstrings at a TeV. Phys. Lett. **B436**, 257–263 (1998), hep-ph/9804398.

14. P. Kanti, Black holes in theories with large extra dimensions: A review. Int. J. Mod. Phys. **A19**, 4899–4951 (2004), hep-ph/0402168.

15. R. C. Myers and M. J. Perry, Black holes in higher dimensional space-times. Ann. Phys. **172**, 304 (1986).

16. R. Emparan and H. S. Reall, A rotating black ring in five dimensions. Phys. Rev. Lett. **88**, 101101 (2002), hep-th/0110260.

17. H. Elvang and P. Figueras, Black saturn. JHEP **05**, 050 (2007), hep-th/0701035.

18. H. Elvang, R. Emparan, and P. Figueras. Phases of five-dimensional black holes, JHEP **05**, 056 (2007), hep-th/0702111.

19. H. Iguchi and T. Mishima, Black di-ring and infinite nonuniqueness. Phys. Rev. **D75**, 064018 (2007), hep-th/0701043.

20. J. Evslin and C. Krishnan, The black di-ring: An inverse scattering construction, arXiv:0706.1231 [hep-th].

21. A. A. Pomeransky and R. A. Sen'kov, Black ring with two angular momenta, hep-th/0612005.

22. K. Izumi, Orthogonal black di-ring solution, arXiv:0712.0902 [hep-th].

23. H. Elvang and M. J. Rodriguez, Bicycling black rings, arXiv:0712.2425 [hep-th].

24. R. Emparan and H. S. Reall, Generalized Weyl solutions. Phys. Rev. **D65**, 084025 (2002), hep-th/0110258.

25. T. Harmark, Stationary and axisymmetric solutions of higher-dimensional general relativity. Phys. Rev. **D70**, 124002 (2004), hep-th/0408141.

26. V. A. Belinsky and V. E. Zakharov, Integration of the Einstein equations by the inverse scattering problem technique and the calculation of the exact soliton solutions. Sov. Phys. JETP **48**, 985–994 (1978).

27. V. A. Belinsky and V. E. Zakharov, Stationary gravitational solitons with axial symmetry. Sov. Phys. JETP **50**, 1 (1979).

28. V. Belinski and E. Verdaguer, *Gravitational Solitons* (University Press, Cambridge, UK, 2001), 258 p.

29. A. A. Pomeransky, Complete integrability of higher-dimensional Einstein equations with additional symmetry, and rotating black holes. Phys. Rev. **D73**, 044004 (2006), hep-th/0507250.

30. R. Emparan, T. Harmark, V. Niarchos, N. A. Obers, and M. J. Rodriguez, The phase structure of higher-dimensional black rings and black holes. JHEP **10**, 110 (2007), arXiv:0708.2181 [hep-th].

31. T. Harmark, Small black holes on cylinders. Phys. Rev. **D69**, 104015 (2004), hep-th/0310259.

32. D. Gorbonos and B. Kol, A dialogue of multipoles: Matched asymptotic expansion for caged black holes. JHEP **06**, 053 (2004), hep-th/0406002.

33. D. Karasik, C. Sahabandu, P. Suranyi, and L. C. R. Wijewardhana, Analytic approximation to 5 dimensional black holes with one compact dimension. Phys. Rev. **D71**, 024024 (2005), hep-th/0410078.

34. D. Gorbonos and B. Kol, Matched asymptotic expansion for caged black holes: Regularization of the post-Newtonian order. Class. Quant. Grav. **22**, 3935–3960 (2005), hep-th/0505009.

35. O. J. C. Dias, T. Harmark, R. C. Myers, and N. A. Obers, Multi-black hole configurations on the cylinder. Phys. Rev. **D76**, 104025 (2007), arXiv:0706.3645 [hep-th].

36. R. Emparan and R. C. Myers, Instability of ultra-spinning black holes. JHEP **09**, 025 (2003), hep-th/0308056.

37. S. S. Gubser, On non-uniform black branes. Class. Quant. Grav. **19**, 4825–4844 (2002), hep-th/0110193.

38. T. Wiseman, Static axisymmetric vacuum solutions and non-uniform black strings. Class. Quant. Grav. **20**, 1137–1176 (2003), hep-th/0209051.

39. E. Sorkin, A critical dimension in the black-string phase transition. Phys. Rev. Lett. **93**, 031601 (2004), hep-th/0402216.

40. B. Kleihaus, J. Kunz, and E. Radu, New nonuniform black string solutions. JHEP **06**, 016 (2006), hep-th/0603119.

41. E. Sorkin, Non-uniform black strings in various dimensions. Phys. Rev. **D74**, 104027 (2006), gr-qc/0608115.

42. B. Kleihaus and J. Kunz, Interior of nonuniform black strings, arXiv:0710.1726 [hep-th].

43. T. Harmark and N. A. Obers, Black holes on cylinders. JHEP **05**, 032 (2002), hep-th/0204047.

44. Y.-Z. Chu, W. D. Goldberger, and I. Z. Rothstein, Asymptotics of d-dimensional Kaluza-Klein black holes: Beyond the Newtonian approximation. JHEP **03**, 013 (2006), hep-th/0602016.

45. B. Kol and M. Smolkin, Classical effective field theory and caged black holes, arXiv:0712.2822 [hep-th].

46. E. Sorkin, B. Kol, and T. Piran, Caged black holes: Black holes in compactified spacetimes. II: 5d numerical implementation. Phys. Rev. **D69**, 064032 (2004), hep-th/0310096.

47. H. Kudoh and T. Wiseman, Properties of Kaluza-Klein black holes. Prog. Theor. Phys. **111**, 475–507 (2004), hep-th/0310104.

48. H. Kudoh and T. Wiseman, Connecting black holes and black strings. Phys. Rev. Lett. **94**, 161102 (2005), hep-th/0409111.

49. H. Elvang, T. Harmark, and N. A. Obers, Sequences of bubbles and holes: New phases of Kaluza-Klein black holes. JHEP **01**, 003 (2005), hep-th/0407050.

50. T. Harmark and N. A. Obers, New phase diagram for black holes and strings on cylinders. Class. Quant. Grav. **21**, 1709–1724 (2004), hep-th/0309116.

51. B. Kol, E. Sorkin, and T. Piran, Caged black holes: Black holes in compactified spacetimes. I: Theory. Phys. Rev. **D69**, 064031 (2004), hep-th/0309190.

52. T. Harmark and N. A. Obers, Phase structure of black holes and strings on cylinders. Nucl. Phys. **B684**, 183–208 (2004), hep-th/0309230.

53. R. Gregory and R. Laflamme, Black strings and p-branes are unstable. Phys. Rev. Lett. **70**, 2837–2840 (1993), hep-th/9301052.

54. R. Gregory and R. Laflamme, The instability of charged black strings and p-branes. Nucl. Phys. **B428**, 399–434 (1994), hep-th/9404071.

55. B. Kol, Topology change in general relativity and the black-hole black-string transition, hep-th/0206220.
56. T. Wiseman, From black strings to black holes. Class. Quant. Grav. **20**, 1177–1186 (2003), hep-th/0211028.
57. B. Kol and T. Wiseman, Evidence that highly non-uniform black strings have a conical waist. Class. Quant. Grav. **20**, 3493–3504 (2003), hep-th/0304070.
58. B. Kol and E. Sorkin, On black-brane instability in an arbitrary dimension. Class. Quant. Grav. **21**, 4793–4804 (2004), gr-qc/0407058.
59. B. Kol and E. Sorkin, LG (Landau-Ginzburg) in GL (Gregory-Laflamme). Class. Quant. Grav. **23**, 4563–4592 (2006), hep-th/0604015.
60. W. Israel, Event horizons in static vacuum space-times. Phys. Rev. **164**, 1776–1779 (1967).
61. B. Carter, Axisymmetric black hole has only two degrees of freedom. Phys. Rev. Lett. **26**, 331–333 (1971).
62. S. W. Hawking, Black holes in General Relativity. Commun. Math. Phys. **25**, 152–166 (1972).
63. D. C. Robinson, Uniqueness of the Kerr black hole. Phys. Rev. Lett. **34**, 905–906 (1975).
64. T. Regge, Stability of a Schwarzschild singularity. Phys. Rev. **108**, 1063–1069 (1957).
65. F. J. Zerilli, Gravitational field of a particle falling in a Schwarzschild geometry analyzed in tensor harmonics. Phys. Rev. **D2**, 2141–2160 (1970).
66. S. A. Teukolsky, Perturbations of a rotating black hole. 1. Fundamental equations for gravitational electromagnetic, and neutrino field perturbations. Astrophys. J. **185**, 635–647 (1973).
67. H. Kodama, Perturbations and stability of higher-dimensional black holes, arXiv:0712.2703 [hep-th].
68. F. R. Tangherlini, Schwarzschild field in n dimensions and the dimensionality of space problem. Nuovo Cimento **27**, 636 (1963).
69. G. W. Gibbons, D. Ida, and T. Shiromizu, Uniqueness and non-uniqueness of static vacuum black holes in higher dimensions. Prog. Theor. Phys. Suppl. **148**, 284–290 (2003), gr-qc/0203004.
70. G. W. Gibbons, D. Ida, and T. Shiromizu, Uniqueness and non-uniqueness of static black holes in higher dimensions. Phys. Rev. Lett. **89**, 041101 (2002), hep-th/0206049.
71. H. Kodama and A. Ishibashi, A master equation for gravitational perturbations of maximally symmetric black holes in higher dimensions. Prog. Theor. Phys. **110**, 701–722 (2003), hep-th/0305147.
72. A. Ishibashi and H. Kodama, Stability of higher-dimensional Schwarzschild black holes. Prog. Theor. Phys. **110**, 901–919 (2003), hep-th/0305185.
73. H. Kodama and A. Ishibashi, Master equations for perturbations of generalized static black holes with charge in higher dimensions. Prog. Theor. Phys. **111**, 29–73 (2004), hep-th/0308128.
74. S. Hollands, A. Ishibashi, and R. M. Wald, A higher dimensional stationary rotating black hole must be axisymmetric. Commun. Math. Phys. **271**, 699–722 (2007), gr-qc/0605106.
75. Y. Morisawa and D. Ida, A boundary value problem for the five-dimensional stationary rotating black holes. Phys. Rev. **D69**, 124005 (2004), gr-qc/0401100.
76. S. Hollands and S. Yazadjiev, Uniqueness theorem for 5-dimensional black holes with two axial Killing fields, arXiv:0707.2775 [gr-qc].
77. S. Giusto and A. Saxena, Stationary axisymmetric solutions of five dimensional gravity. Class. Quant. Grav. **24**, 4269–4294 (2007), arXiv:0705.4484 [hep-th].
78. W. D. Goldberger and I. Z. Rothstein, An effective field theory of gravity for extended objects. Phys. Rev. **D73**, 104029 (2006), hep-th/0409156.
79. T. Harmark and N. A. Obers, Phases of Kaluza-Klein black holes: A brief review, hep-th/0503020.
80. T. Harmark and N. A. Obers, General definition of gravitational tension. JHEP **05**, 043 (2004), hep-th/0403103.
81. R. C. Myers, Stress tensors and Casimir energies in the AdS/CFT correspondence. Phys. Rev. **D60**, 046002 (1999), hep-th/9903203.

82. J. H. Traschen and D. Fox, Tension perturbations of black brane spacetimes. Class. Quant. Grav. **21**, 289–306 (2004), `gr-qc/0103106`.
83. P. K. Townsend and M. Zamaklar, The first law of black brane mechanics. Class. Quant. Grav. **18**, 5269–5286 (2001), `hep-th/0107228`.
84. D. Kastor and J. Traschen, Stresses and strains in the first law for Kaluza-Klein black holes. JHEP **09**, 022 (2006), `hep-th/0607051`.
85. J. H. Traschen, A positivity theorem for gravitational tension in brane spacetimes. Class. Quant. Grav. **21**, 1343–1350 (2004), `hep-th/0308173`.
86. T. Shiromizu, D. Ida, and S. Tomizawa, Kinematical bound in asymptotically translationally invariant spacetimes. Phys. Rev. **D69**, 027503 (2004), `gr-qc/0309061`.
87. T. Harmark and N. A. Obers, Black holes and black strings on cylinders. Fortsch. Phys. **51**, 793–798 (2003), `hep-th/0301020`.
88. B. Kol, The power of action: 'The' derivation of the black hole negative mode, `hep-th/0608001`.
89. B. Kol, Perturbations around backgrounds with one non-homogeneous dimension, `hep-th/0609001`.
90. V. Cardoso and O. J. C. Dias, Rayleigh-Plateau and Gregory-Laflamme instabilities of black strings. Phys. Rev. Lett. **96**, 181601 (2006), `hep-th/0602017`.
91. V. Cardoso and L. Gualtieri, Equilibrium configurations of fluids and their stability in higher dimensions. Class. Quant. Grav. **23**, 7151–7198 (2006), `hep-th/0610004`.
92. R. Gregory and R. Laflamme, Hypercylindrical black holes. Phys. Rev. **D37**, 305 (1988).
93. R. C. Myers, Higher dimensional black holes in compactified space- times. Phys. Rev. **D35**, 455 (1987).
94. A. R. Bogojevic and L. Perivolaropoulos, Black holes in a periodic universe. Mod. Phys. Lett. **A6**, 369–376 (1991).
95. D. Korotkin and H. Nicolai, A periodic analog of the Schwarzschild solution, `gr-qc/9403029`.
96. A. V. Frolov and V. P. Frolov, Black holes in a compactified spacetime. Phys. Rev. **D67**, 124025 (2003), `hep-th/0302085`.
97. P.-J. De Smet, Black holes on cylinders are not algebraically special. Class. Quant. Grav. **19**, 4877–4896 (2002), `hep-th/0206106`.
98. G. T. Horowitz, Playing with black strings, `hep-th/0205069`.
99. R. Emparan, Rotating circular strings, and infinite non-uniqueness of black rings. JHEP **03**, 064 (2004), `hep-th/0402149`.
100. V. Cardoso, O. J. C. Dias, and L. Gualtieri, The return of the membrane paradigm? Black holes and strings in the water tap, `arXiv:0705.2777 [hep-th]`.
101. T. Harmark and P. Olesen, On the structure of stationary and axisymmetric metrics. Phys. Rev. **D72**, 124017 (2005), `hep-th/0508208`.
102. H. Elvang and R. Emparan, Black rings, supertubes, and a stringy resolution of black hole non-uniqueness. JHEP **11**, 035 (2003), `hep-th/0310008`.
103. J. L. Hovdebo and R. C. Myers, Black rings, boosted strings and Gregory-Laflamme. Phys. Rev. **D73**, 084013 (2006), `hep-th/0601079`.
104. H. Elvang, R. Emparan, and A. Virmani, Dynamics and stability of black rings. JHEP **12**, 074 (2006), `hep-th/0608076`.
105. B. Carter, Essentials of classical brane dynamics. Int. J. Theor. Phys. **40**, 2099–2130 (2001), `gr-qc/0012036`.
106. D. Kastor, S. Ray, and J. Traschen, The first law for boosted Kaluza–Klein black holes. JHEP **06**, 026 (2007), `arXiv:0704.0729 [hep-th]`.
107. G. Arcioni and E. Lozano-Tellechea, Stability and critical phenomena of black holes and black rings. Phys. Rev. **D72**, 104021 (2005), `hep-th/0412118`.
108. K.-i. Maeda, N. Ohta, and M. Tanabe, A supersymmetric rotating black hole in a compactified spacetime. Phys. Rev. **D74**, 104002 (2006), `hep-th/0607084`.
109. M. Karlovini and R. von Unge, Charged black holes in compactified spacetimes. Phys. Rev. **D72**, 104013 (2005), `gr-qc/0506073`.

110. K. Copsey and G. T. Horowitz, Gravity dual of gauge theory on $S^2 \times S^1 \times \mathbb{R}$. JHEP **06**, 021 (2006), hep-th/0602003.

111. R. B. Mann, E. Radu, and C. Stelea, Black string solutions with negative cosmological constant. JHEP **09**, 073 (2006), hep-th/0604205.

112. B. Kleihaus, J. Kunz, and E. Radu, Rotating nonuniform black string solutions. JHEP **05**, 058 (2007), hep-th/0702053.

113. B. Kleihaus, J. Kunz, and F. Navarro-Lerida, Rotating black holes in higher dimensions, arXiv:0710.2291 [hep-th].

114. H. K. Kunduri, J. Lucietti, and H. S. Reall, Do supersymmetric anti-de Sitter black rings exist?. JHEP **02**, 026 (2007), hep-th/0611351.

115. H. Elvang, A charged rotating black ring, hep-th/0305247.

116. H. Elvang, R. Emparan, D. Mateos, and H. S. Reall, A supersymmetric black ring. Phys. Rev. Lett. **93**, 211302 (2004), hep-th/0407065.

117. A. Dabholkar, N. Iizuka, A. Iqubal, A. Sen, and M. Shigemori, Spinning strings as small black rings. JHEP **04**, 017 (2007), hep-th/0611166.

118. C. Helfgott, Y. Oz, and Y. Yanay, On the topology of black hole event horizons in higher dimensions. JHEP **02**, 025 (2006), hep-th/0509013.

119. G. J. Galloway and R. Schoen, A generalization of Hawking's black hole topology theorem to higher dimensions. Commun. Math. Phys. **266**, 571–576 (2006), gr-qc/0509107.

120. S. Lahiri and S. Minwalla, Plasmarings as dual black rings, arXiv:0705.3404 [hep-th].

121. T. Harmark, K. R. Kristjansson, N. A. Obers, and P. B. Ronne, Three-charge black holes on a circle. JHEP **01**, 023 (2007), hep-th/0606246.

122. T. Harmark, K. R. Kristjansson, N. A. Obers, and P. B. Ronne, Entropy of three-charge black holes on a circle. Fortsch. Phys. **55**, 748–753 (2007), hep-th/0701070.

123. B. D. Chowdhury, S. Giusto, and S. D. Mathur, A microscopic model for the black hole - black string phase transition. Nucl. Phys. **B762**, 301–343 (2007), hep-th/0610069.

124. N. Arkani-Hamed, S. Dimopoulos, and G. R. Dvali, The hierarchy problem and new dimensions at a millimeter. Phys. Lett. **B429**, 263–272 (1998), hep-ph/9803315.

125. L. Randall and R. Sundrum, A large mass hierarchy from a small extra dimension. Phys. Rev. Lett. **83**, 3370–3373 (1999), hep-ph/9905221.

126. L. Randall and R. Sundrum, An alternative to compactification. Phys. Rev. Lett. **83**, 4690–4693 (1999), hep-th/9906064.

Chapter 7
Braneworld Black Holes

R. Gregory

Abstract In this article, I give an introduction to and overview of braneworlds and black holes in the context of warped compactifications. I first describe the general paradigm of braneworlds and introduce the Randall–Sundrum model. I discuss braneworld gravity, both using perturbation theory and also nonperturbative results. I then discuss black holes on the brane, the obstructions to finding exact solutions, and ways of tackling these difficulties. I describe some known solutions and conclude with some open questions and controversies.

7.1 Introduction

Nearly a century ago, Kaluza and Klein theorized that by adding an extra dimension to space, you could unify electromagnetism with gravity. Thus our first "unified theory" was born – at the price of an extra unseen dimension. Nowadays, extra dimensions are an integral part of fundamental theoretical physics, and the consequences of devising consistent means of hiding these extra dimension have led to an explosion of activity in recent years in string theory, cosmology, and phenomenology. Braneworlds are just part of this general story and represent a particular way of dealing with the extra dimensions that is empirical, but precise and calculable. They have proved indispensable for developing ideas and methods which have then been used in more esoteric but fundamentally grounded models in string theory. These lectures are about braneworlds and deal with the deeply interesting, but thorny issue of how to describe braneworld black holes.

Simply put, a braneworld is a slice through spacetime on which we live. We cannot (easily) see the extra dimensions perpendicular to our slice, as all of our standard physics is confined. We can, however, deduce those extra dimensions by carefully monitoring the behaviour of gravity. Confinement to a brane may at first sound counter-intuitive; however, it is in fact a common occurrence. The first braneworld scenarios [1–3] used topological defects to model the braneworld, with condensates

R. Gregory (✉)
Centre for Particle Theory, Durham University, South Road, Durham, DH1 3LE, UK
r.a.w.gregory@durham.ac.uk

Gregory, R.: *Braneworld Black Holes*. Lect. Notes Phys. **769**, 259–298 (2009)
DOI 10.1007/978-3-540-88460-6_7

and zero modes producing confinement. In string theory, D-branes have "confined" gauge theories on their worldvolumes [4, 5] and heterotic M-theory has a natural domain wall structure [6, 7].

The new phenomenology of braneworld scenarios is primarily located in the gravitational sector, with a particularly nice geometric resolution of the hierarchy problem [8–10]. The scenario has however far outgrown these initial particle phenomenology motivations and has proved a fertile testbed for new possibilities in cosmology, astrophysics, and quantum gravity. One of the most popular models with warped extra dimensions is that of Randall and Sundrum (RS) [11, 12], which consists of a domain wall universe living in 5D anti-de Sitter (adS) spacetime, and will be the setting for these lectures. Interestingly, although the RS model is an empirical braneworld set-up it can be related to, or motivated by, string theory in several ways. First of all, it is notionally similar to the heterotic M-theory set-up, in that the initial RS model had two walls at the end of an interval. However, this similarity is notional only, and calculationally, the gravitational spectrum of GR in five dimensions is very different from the spectrum of low-energy heterotic M-theory [6, 7]. A more fruitful and robust parallel occurs with type IIB string theory, where the RS model can (in some rough sense) be associated with the near horizon limit of a stack of D3-branes. Viewed in this context, the RS model provides an excellent opportunity to use and test ideas from the gauge/gravity or adS/CFT correspondence [13, 14].

The RS model is however particularly valuable as a concrete and explicit calculational testbed for any theory with extra dimensions in which gravity is able to probe and modify these hidden directions. One of the problems with having extra dimensions is that we have to hide evidence of their existence. We not only have to reproduce gravitational and standard model physics on the requisite scales, but also have to ensure that we do not create any additional unwanted physics. With RS, the gravitational physics is self-consistent and calculable. We can therefore compute the cosmological and astrophysical consequences of the extra dimension in a wide variety of physically interesting cases.

Black holes are perhaps the most interesting physical object to explore within the braneworld framework of extra dimensions. From the Kaluza–Klein point of view, extra dimensions show up as extra charges black holes can carry from the 4D point of view [15–18]; however, in these solutions the black hole is "smeared" along the extra dimension rather than localized. Braneworld scenarios are the antithesis of KK compactifications, consisting of highly localized and strongly warped extra dimensions, and therefore the implications of this strongly localized and gravitating brane for black hole physics are of particular physical and theoretical interest. We now have compelling evidence of the existence of black holes in nature, from stellar-sized black holes in binary systems, observed via X-ray emission from accretion discs [19], to supermassive black holes at the centre of galaxies [20, 21], which in the case of our own Milky Way can be seen quite clearly from stellar orbits [22]. As observational evidence accrues and becomes more robust, the bounds on the innermost stable orbit of the black hole (obtained from iron emission lines [23, 24]) may eventually start to confront the theoretical limit from the 4D Kerr metric, and possibly provide signatures of extra dimensions, for which the bounds can be quite different [25].

Turning to the small scale, and taking seriously the possibility that braneworlds can provide a resolution of the hierarchy problem via a geometric renormalization of the Newton constant [8–10], raises the possibility that mini black holes can be produced in particle collisions [26, 27]. Understanding the formation and decay of these highly energetic black holes will then allow us to predict signatures for black hole formation at the LHC [28–31] and is the topic of a companion set of lectures at this school [32].

Finally, there is also a compelling theoretical reason for studying braneworld black hole solutions, and that is the parallel between the RS model and the adS/CFT conjecture [33–37]. As we explore more concretely in Sect. 7.4, by taking the near horizon limit of a stack of D3-branes, the RS model can be thought of as cutting off the spacetime outside the D-branes; the adS curvature of the RS bulk is therefore given rather precisely in terms of the D3-brane charge and the string scale. Thus, we might expect a parallel between classical braneworld gravity and quantum corrections on the brane. The possibility of finding a calculational handle for computing the backreaction of Hawking radiation [38–40] is extremely attractive, and of course can potentially feed back into the issue of mini black holes at the LHC.

In these lectures, we will review the current status of black hole solutions in the Randall–Sundrum model, first reviewing the framework in some detail, concentrating on gravitational issues, and the link with adS/CFT and holography. We will see why it is so difficult to find an exact solution, before covering approximate methods and solutions for brane black holes. Finally, we describe objections to the holographic picture and some recent developments in the closely related Karch–Randall [41] set-up.

7.2 Some Randall–Sundrum Essentials

The Randall–Sundrum model has one (or two) domain walls situated as minimal submanifolds in adS spacetime. In its usual form, the spacetime is

$$ds^2 = e^{-2k|z|} \left[dt^2 - d\mathbf{x}^2 \right] - dz^2, \tag{7.1}$$

where $k = L^{-1}$ is the inverse curvature radius of the negatively curved 5D adS spacetime. Here, the spacetime is constructed so that there are 4D flat slices stacked along the fifth z-dimension, which have a z-dependent conformal prefactor known as the warp factor. Since this warp factor has a cusp at $z = 0$, this indicates the presence of a domain wall – the braneworld – which represents an exactly flat Minkowski universe. The reason for choosing this particular slicing of adS spacetime is to have a flat Minkowski metric on the brane, i.e. to choose the "standard vacuum".

The RS spacetime is an example of a codimension 1 braneworld, where we have one extra dimension. In this case, there is a well-defined prescription for finding gravitational solutions with an infinitesimally thin brane: the Israel equations [42], which are essentially a physicist's tool extracted from the Gauss–Codazzi formalism for the differential geometry of submanifolds. Since this formalism is so widely used, it is worth reviewing it briefly here (see also [43–45]).

In the Israel prescription, we rewrite our 5D spacetime as a 4D base space, with coordinates x^μ, plus a normal distance, z, from the "wall". The 4D coordinates remain constant along geodesics normal to the wall, thus giving a 5D coordinate system $\{x^\mu, z\}$. This coordinate system is valid within the radius of curvature of the wall, and splits the tangent space naturally into parallel and normal components, and the metric in general has the form

$$ds^2 = \gamma_{\mu\nu}(x,z)dx^\mu dx^\nu - dz^2. \tag{7.2}$$

Choosing the coordinates in this way results in the nontrivial content of the geometry being located in the 4D metric $\gamma_{\mu\nu}$, and the fifth metric component is always unity because z is the proper distance from the brane. $n^a = \delta^a_z$ is the normal to the brane, and $\gamma_{\mu\nu}|_{z=0}$ is the *intrinsic metric* on the brane. Note that $\gamma_{\mu\nu}$ lies in the tangent bundle of the brane as a manifold (i.e. is a 4D tensor) and has a 5D counterpart, the *first fundamental form* which we denote as

$$\hat\gamma_{ab} = g_{ab} + n_a n_b = \text{diag}(\gamma_{\mu\nu}, 0). \tag{7.3}$$

$\hat\gamma_{ab}$ is a 5D tensor, but acts as a projection, wiping out any components orthogonal to the brane. γ and $\hat\gamma$ contain the same physical information; the distinction is purely mathematical; however we will keep it for the purposes of this technical discussion. This particular coordinate or *gauge* choice is called the Gaussian normal (GN) gauge and is the space-like equivalent of the ADM synchronous gauge.

Surfaces can curve in the ambient manifold, whether or not that is itself curved (see Fig. 7.1). This is measured by the *extrinsic curvature* or *second fundamental form* and is defined via

$$K_{ab} = \hat\gamma^c_a \hat\gamma^d_c \nabla_b n_d. \tag{7.4}$$

We can use the Riemann identity $\nabla_c \nabla_d n^a - \nabla_d \nabla_c n^a = R^a{}_{bcd} n^d$ to get the Gauss–Codazzi relations:

$$^{(4)}R^a{}_{bcd} = \hat\gamma^a_e \hat\gamma^f_b \hat\gamma^g_c \hat\gamma^i_d R^e{}_{fgi} + K^a_d K_{bc} - K^a_c K_{bd} \tag{7.5}$$

$$\Rightarrow {}^{(4)}R_{bd} = \hat\gamma^a_b \hat\gamma^c_d \left(R_{ac} + R_{aecf} n^e n^f\right) + K_{ad} K^a_b - K K_{bd}. \tag{7.6}$$

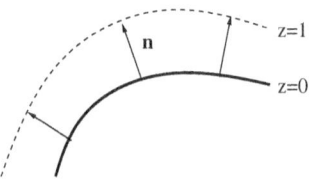

Fig. 7.1 An illustration of the curved brane and the Gaussian normal coordinate system. The brane is the solid line at $z=0$; moving out a uniform distance from the brane gives a new surface at $z=1$. The normal to the brane $\mathbf{n} = \partial/\partial z$ is indicated. As we move from $z=0$ to $z=1$, distances along the brane will change in general. This is reflected in the extrinsic curvature (7.4)

In this last relation, we have the 5D Ricci tensor, which we can replace with the energy momentum tensor via Einstein's equations; we also have a term which a second use of the Riemann identity allows us to write as a Lie derivative of the extrinsic curvature across the brane:

$$R_{aecf}n^e n^f = -n^b \nabla_b K_{ac} - K_c^d K_{ad} = -\mathcal{L}_n K_{ac} + K_a^d K_{cd} \,, \tag{7.7}$$

thus allowing us to rewrite the Gauss–Codazzi equations in terms of the extrinsic curvature and the energy momentum tensor:

$$\mathcal{L}_n K_{ab} = \hat{\gamma}_a^c \hat{\gamma}_b^d T_{cd} - \tfrac{1}{3} T \hat{\gamma}_{ab} + 2 K_a^c K_{bc} - K K_{ab} - {}^{(4)}R_{ab} \,. \tag{7.8}$$

Therefore, if we imagine our brane to be infinitesimally thin, having a distributional energy momentum, $T_{ab}\delta(z)$, then we see that the extrinsic curvature must have a jump across the brane. Integrating this out, we get the Israel equations:

$$\Delta K_{ab} = K_{ab}(z=0^+) - K_{ab}(z=0^-) = 8\pi G_5 \left(T_{ab} - \tfrac{1}{3} T \hat{\gamma}_{ab} \right). \tag{7.9}$$

Returning to the Randall–Sundrum metric, (7.1), we see that

$$K_{\mu\nu} = -\Gamma_{\mu\nu}^z n_z = \mp k e^{-2k|z|} \eta_{\mu\nu} \tag{7.10}$$

(where we are now dropping the distinction between the brane tangent space and the bulk tangent space, as the situation is physically clear). Using (7.9) we see that the brane has an energy momentum tensor proportional to the metric on the brane:

$$T_{\mu\nu} = \mathrm{E}_{RS} \eta_{\mu\nu} = \frac{6k}{8\pi G_5} \eta_{\mu\nu}. \tag{7.11}$$

Notice the very precise form of this energy momentum. First, because it is proportional to the intrinsic metric, this means that the brane has tension (rather than pressure) and this tension is exactly equal to its energy, $\mathrm{E} = \mathrm{T}$. Thus the brane energy momentum has exactly the same form as a cosmological constant term on the brane. Second, the actual value of this tension is precisely related to the bulk cosmological constant:

$$\Lambda = -6k^2 = -\frac{(8\pi G_5 \, \mathrm{E}_{RS})^2}{6}. \tag{7.12}$$

This is sometimes referred to as the fine-tuned or *critical* RS brane. As we will see later, this relation can be relaxed, leading to de Sitter or anti-de Sitter RS branes (the latter of which are also known as Karch–Randall (KR) branes [41]).

7.3 Gravity in the Randall–Sundrum Model

Obviously the RS model can only describe our real universe if it correctly reproduces gravitational physics at experimentally tested scales. This means we have to be able to reproduce Einstein gravity in our solar system, and the standard cosmological model for our universe. While the Israel equations give us the general formalism for getting our braneworld metric, finding actual solutions can be a far trickier matter, as indeed finding a general solution of Einstein's equations is a tricky matter! We therefore resort, as with standard gravity, to two main approaches:

- Local physics, or perturbation theory, and
- "Big Picture" or geometry, finding exact solutions assuming symmetries.

In either case, we have to accept that gravity on the brane is a projection of the full higher dimensional nature of gravity, and is therefore a *derived quantity*.

7.3.1 Perturbation Theory

In general relativity (GR), classical perturbation theory involves perturbing the metric

$$g_{ab} \rightarrow g_{ab} + h_{ab} \tag{7.13}$$

around a given background solution. There are three main issues to bear in mind:

1. h_{ab} is a perturbation and should therefore be "small". What does this mean? In practice we have to be careful about our coordinate system and always look at h in a regular system. For the Schwarzschild solution, for example, this means using Kruskal coordinates.
2. Gauge freedom: GR has a large gauge group – physics is invariant under general coordinate transformations (GCT's), and there are many gauge degrees of freedom in h_{ab}. For example, in 4D, h_{ab} has 10 independent components, but the graviton has only 2 physical degrees of freedom. Under a GCT

$$X^a \rightarrow X^a + \xi^a , \quad g_{ab} \rightarrow g_{ab} + \mathscr{L}_\xi g_{ab} , \tag{7.14}$$

hence

$$\delta g_{ab} = \xi_{a;b} + \xi_{b;a} \tag{7.15}$$

and we can use this to make a choice of gauge. A common choice for relativists is the harmonic gauge

$$\bar{h}^a_{b;a} = h^a_{b;a} - \tfrac{1}{2} h^a_{a;b} = 0 , \tag{7.16}$$

and in vacuo we can also choose $h^a_a = 0$: the transverse tracefree (TTF) gauge. Note that this does not uniquely specify the gauge, e.g. $\xi^a_{;a} = 0 = \nabla^2 \xi^a$ gives an allowed gauge transformation.

3. Finally, we need the perturbation of the Ricci tensor:

$$\delta R_{ab} = -\tfrac{1}{2} \nabla^2 h_{ab} - R_{acbd} h^{cd} + R^c_{(a} h_{b)c} + \nabla_{(a} \nabla^c h_{b)c} = -\tfrac{1}{2} \Delta_L h_{ab} \tag{7.17}$$

often called the Lichnerowicz operator.

The simplest way to perturb the brane system is to take a GN system, in which the brane stays at $z = 0$:

$$g_{zz} = -1 \quad g_{z\mu} = 0 .$$
(7.18)

The remaining gauge freedom allowed is

$$\xi^z = f(x^\mu) , \quad \xi^\mu = \int a^{-2} f_{,\nu} \eta^{\mu\nu} + \zeta^\mu(x^\mu) .$$
(7.19)

We can now input the purely 4D perturbation into the Lichnerowicz operator, and after some algebra, the perturbation equations reduce to

$$a^{-2} \left[a^2 \left(a^{-2} h^\lambda_\lambda \right)' \right]' = -\frac{16\pi G_5}{3} \delta(z) a^{-2} \mathscr{T}^\lambda_\lambda ,$$
(7.20)

$$\left[a^{-2} \left(h^\lambda_{\lambda,\mu} - h^\lambda_{\mu,\lambda} \right) \right]' = 0,$$
(7.21)

$$a^{-2} \partial^2 h_{\mu\nu} - a^{-2} \left[a^4 \left(a^{-2} h_{\mu\nu} \right)' \right]' - 2a^{-2} \bar{h}^\lambda_{(\mu,\nu)\lambda} ,$$

$$-aa' \eta_{\mu\nu} \left(a^{-2} h^\lambda_\lambda \right)' = -16\pi G_5 \delta(z) \left[\mathscr{T}_{\mu\nu} - \tfrac{1}{3} \mathscr{T}^\lambda_\lambda \eta_{\mu\nu} \right] , (7.22)$$

where brane indices are raised and lowered with $\eta_{\mu\nu}$, and we allow for a matter perturbation confined to the brane:

$$T_{\mu\nu} = \frac{6k}{8\pi G_5} + \mathscr{T}_{\mu\nu} .$$
(7.23)

It is easy to see that the RS gauge is only consistent for vacuum perturbations and that the zero modes have the behaviour $\sim a^2$ (the graviton [11, 46, 47]) and $\sim a^2 \int a^{-4}$ (the radion [48]).

A complete set of solutions to the free equations is readily found to be $h_{\mu\nu} \propto e^{ip\cdot x} u_m(z)$ with

$$u_m(z) = \sqrt{\frac{m}{2k}} \frac{J_1(\frac{m}{k}) N_2(m\zeta) - N_1(\frac{m}{k}) J_2(m\zeta)}{\sqrt{J_1(\frac{m}{k})^2 + N_1(\frac{m}{k})^2}} ,$$
(7.24)

where $\zeta = e^{kz}/k$, from which we can construct the Green's function:

$$G_R(x,x') = -\int \frac{d^4 p}{(2\pi)^4} e^{ip\cdot(x-x')} \left[\frac{ka^2(z)a^2(z')}{\mathbf{p}^2 - (\omega + i\epsilon)^2} + \int_0^\infty dm \frac{u_m(z) u_m(z')}{m^2 + \mathbf{p}^2 - (\omega + i\epsilon)^2} \right] .$$
(7.25)

This has the structure of a zero mode (the part proportional to a^2), and a continuum of KK states. This is seen more clearly by looking at the restriction to a perturbation in the brane induced by a particle on the brane, for which

$$G_R(x,0,x',0) = kD_0(x-x') + \int_0^\infty dm\, u_m(0)^2 D_m(x-x'). \qquad (7.26)$$

However, we have to remember that the RS gauge is only consistent in the absence of sources; in the presence of sources we have to fix the trace of the perturbation to satisfy the Lichnerowicz equation. Strictly speaking, we take the general metric perturbation, $h_{\mu\nu}$, and decompose it into its irreducible components with respect to the 4D Lorentz group (see [49]). This allows for a tensor (TTF) mode, a vector, and two scalars in general:

$$h_{\mu\nu} = h_{\mu\nu}^{\mathrm{TT}} + A_{\nu,\mu} + A_{\mu,\nu} + \phi_{,\mu\nu} - \frac{1}{4}\eta_{\mu\nu}\partial^2\phi + \frac{h}{4}\eta_{\mu\nu}. \qquad (7.27)$$

On shell, it can be shown that this reduces (up to purely 4D gauge transformations) to the following expression:

$$h_{\mu\nu} = h_{\mu\nu}^{\mathrm{TT}} - \frac{1}{k}f_{,\mu\nu} + 2ka^2 f\eta_{\mu\nu}, \qquad (7.28)$$

which physically corresponds to the TTF 4D tensor h^{TT}, and a scalar, $f(x^\mu)$, which can be interpreted as a bending of the brane with respect to an observer at infinity [50] (see Fig. 7.2a). This brane bending term couples to the trace of the energy momentum perturbation on the brane via (7.20), which implies a 4D equation for f:

$$\partial^2 f = \frac{8\pi G_5 \mathscr{T}_\lambda^\lambda}{6} \qquad \Rightarrow \qquad f = 8\pi G_5 \int D_0(x-x')\frac{\mathscr{T}_\lambda^\lambda(x')}{6}. \qquad (7.29)$$

Solving (7.22), and pulling all this information together, we can now write the solution on the brane:

$$h_{\mu\nu} = -16\pi G_5 \int G_R(x,0;x',0)[\mathscr{T}_{\mu\nu} - \tfrac{1}{3}\mathscr{T}\eta_{\mu\nu}] + 2k\eta_{\mu\nu}\int D_0(x-x')\frac{8\pi G_5 \mathscr{T}_\lambda^\lambda}{6}. \qquad (7.30)$$

At mid-to-long-range scales on the brane, the zero mode dominates the integral and so we get

$$h_{\mu\nu} = -16\pi G_5 k \int D_0(x-x')[\mathscr{T}_{\mu\nu} - \tfrac{1}{2}\mathscr{T}\eta_{\mu\nu}]. \qquad (7.31)$$

Thus, if we identify $G_N = G_5 k$ as the 4D Newton constant, we have precisely 4D perturbative Einstein gravity with the correct tensor structure.

The effect of the massive KK modes on the Newtonian potential is also easily extracted using asymptotics of Bessel functions:

$$u_m(0) = \sqrt{\frac{m}{2k}}\frac{J_1\left(\frac{m}{k}\right)N_2\left(\frac{m}{k}\right) - J_2\left(\frac{m}{k}\right)N_1\left(\frac{m}{k}\right)}{\left[J_1^2\left(\frac{m}{k}\right) + N_1^2\left(\frac{m}{k}\right)\right]^{1/2}} \sim -\sqrt{\frac{m}{2k}} \quad m/k \ll 1. \qquad (7.32)$$

To see how these feed into corrections to Einstein gravity, consider the effect of a point mass source $\mathscr{T}_{00} \sim M\delta(z)\delta^{(3)}(\mathbf{r})$:

Fig. 7.2 An illustration of
the effects of RS gravity.
On the *left*, the brane bends
in response to matter, and
on the *right*, a figurative
representation of the lines of
force for RS gravity

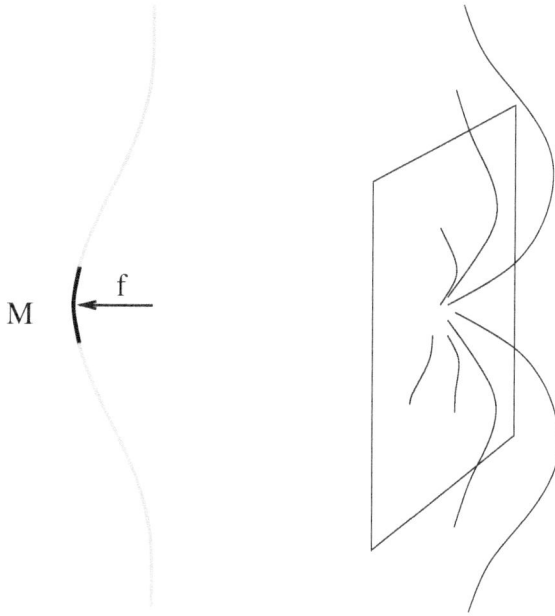

$$h_{\mu\nu} = -16\pi G_N \int \left(D_0(x-x')[\mathscr{T}_{\mu\nu} - \tfrac{1}{2}\mathscr{T}\eta_{\mu\nu}] \right.$$
$$\left. + \int_0^\infty \frac{m\,dm}{2k^2} D_m(x-x')[\mathscr{T}_{\mu\nu} - \tfrac{1}{3}\mathscr{T}\eta_{\mu\nu}] \right), \qquad (7.33)$$

giving

$$h_{tt} = -\frac{2G_N M}{r}\left(1 + \frac{2}{3k^2 r^2}\right), \qquad h_{ij} = -\frac{2G_N M}{r}\left(1 + \frac{1}{3k^2 r^2}\right)\delta_{ij}. \qquad (7.34)$$

Note this is in homogeneous gauge; transforming to the area gauge (where the area
of 2-spheres is $4\pi r^2$) we have to leading order in $G_N M$

$$ds^2 = \left(1 - \frac{2G_N M}{\hat{r}} - \frac{4G_N M}{3k^2 \hat{r}^3}\right)dt^2 - \frac{d\hat{r}^2}{\left(1 - \frac{2G_N M}{\hat{r}} - \frac{2G_N M}{k^2 \hat{r}^3}\right)} - \hat{r}^2 d\Omega^2. \qquad (7.35)$$

We can visualize RS gravity as lines of force spreading out from the brane, but being
"pushed back" by the negative bulk curvature. At small scales, the lines of force
leave the brane and gravity is 5D and weaker. At larger scales, the bulk curvature
bends the lines of force back onto the brane, and so gravity returns to being a 4D
force law (see Fig. 7.2b).

This is weak gravity, but what about strong gravity, such as black holes or
cosmology?

7.3.2 Cosmology

For a cosmological brane, we have to ask whether there are surfaces of lower dimensionality which have the interpretation of an expanding universe. Recall that in standard cosmology, homogeneity and isotropy give a simple model of the universe in which everything depends on a single scale factor $a(t)k$:

$$ds^2 = dt^2 - a^2(t)d\mathbf{x}_\kappa^2 \qquad (7.36)$$

where the spatial metric $d\mathbf{x}_\kappa^2$ is a surface of constant curvature $\kappa = 0, \pm1$, and in which $a(t)$ satisfies a simple first-order Friedman equation:

$$\left(\frac{\dot{a}}{a}\right)^2 + \frac{\kappa}{a^2} = \frac{8\pi G_N}{3}\rho \qquad (7.37)$$

where ρ is the energy density of the universe, typically modelled by a perfect fluid with some equation of state $p = w\rho$.

For the cosmological braneworld, homogeneity and isotropy will still imply a constant curvature spatial universe, but now our "scale factor" must depend not only on time but on the distance into the bulk. The remaining part of the metric in the t, z directions can be made conformally flat (any 2D metric can always be written in a conformally flat form) and so we may write the overall geometry as [51]

$$ds^2 = e^{2v(t,z)}(B(t,z))^{-2/3}(dt^2 - dz^2) - B^{2/3}\left[\frac{d\chi^2}{1-\kappa\chi^2} + \chi^2 d\Omega_{II}^2\right]. \qquad (7.38)$$

The rationale for this specific way of writing the scale factor becomes apparent once the Einstein equations are analysed. Here, z is representing the bulk coordinate away from the brane, though it no longer corresponds to proper distance. The brane sits at $z = 0$, which can always be chosen to be the location of the brane. (The conformal transformation $t' \pm z' = t \pm z \pm \zeta(t \pm z)$ maintains the form of the metric while taking an arbitrary wall trajectory $z' = \zeta(t')$ to $z = 0$.)

In addition, the presence of a cosmological fluid will alter the usual brane relation $E = T$ by adding additional energy, ρ, to E and subtracting pressure, p, from the tension T. Thus our brane energy momentum will now be

$$T_b^a = \delta(z)\,\mathrm{diag}\,(E + \rho, E - p, E - p, E - p, 0)\,. \qquad (7.39)$$

If we now compute the bulk Einstein equations, the reason for writing the metric in the slightly unusual form (7.38) becomes apparent. Using the lightcone coordinates

$$x_- = \frac{t-z}{2}, \qquad x_+ = \frac{t+z}{2}. \qquad (7.40)$$

the bulk equations are

$$B_{,+-} = \left(2\Lambda B^{1/3} - 6\kappa B^{-1/3}\right) e^{2v}, \tag{7.41}$$

$$v_{,+-} = \left(\frac{\Lambda}{3} B^{-2/3} + \kappa B^{-4/3}\right) e^{2v}, \tag{7.42}$$

$$B_{,\pm} [ln(B_{,\pm})]_{,\pm} = 2v_{,\pm} B_{,\pm} . \tag{7.43}$$

This system is completely integrable, giving, after a change of coordinates, the bulk solution[1]

$$ds^2 = \left(\kappa - \frac{\Lambda}{6}r^2 - \frac{\mu}{r^2}\right) dt^2 - \left(\kappa - \frac{\Lambda}{6}r^2 - \frac{\mu}{r^2}\right)^{-1} dr^2 - r^2 dx_\kappa^2 \tag{7.44}$$

which is clearly a black hole solution. The parameter μ is related to the mass of the bulk black hole via [54]

$$M = 3\pi\mu/8G_5 . \tag{7.45}$$

The change of coordinates results in a shift of the brane to

$$r = R(\tau), \tag{7.46}$$

where τ is the proper time of a brane observer.

Thus our cosmological brane is a slice of a black hole spacetime [33, 51, 55–61]. We can think of our brane as moving in the bulk, and as it moves through a warped background, the brane will experience contraction or expansion as the surrounding geometry contracts or expands. The Israel equations give the dynamical equations for the brane trajectory R, which can be massaged into the familiar Friedman form:

$$\left(\frac{\dot{R}}{R}\right)^2 + \frac{\kappa}{R^2} = \frac{[8\pi G_5(E+\rho)]^2 + \Lambda}{36} + \frac{\mu}{R^4} \tag{7.47}$$

(see [51] for details). For a critical RS brane, which has $E = 6k/8\pi G_5$, $\Lambda = -6k^2$, this gives

$$\left(\frac{\dot{R}}{R}\right)^2 + \frac{\kappa}{R^2} = \frac{8\pi G_N \rho}{3} + \frac{(8\pi G_5 \rho)^2}{36} + \frac{\mu}{R^4} . \tag{7.48}$$

As might have been expected from the calculation of linearized gravity, the dominant form of this equation for small ρ is indeed the standard Friedman equation. The effect of the brane shows up in the ρ^2 corrections, dubbed the *non-conventional* cosmology of the brane [58–61]. But most interesting from the point of view of these lectures is the presence of the last term, which is directly a result of the bulk black hole. This term, proportional to the mass parameter, takes the same functional form as a radiation source on the brane. Of course, the presence of the bulk black hole induces a periodicity in time in the Euclidean section, or, a finite temperature for any quantum field theory in the spacetime. Computing the background Hawking temperature of the black hole gives

---

[1] Although note that there is a special case $B = 1$, $2\Lambda = 6\kappa$, which is a near horizon limit of a black hole metric and a critical point of the Einstein equations [52, 53].

$$T_H = \frac{\sqrt{\kappa^2 + 4\mu k^2}}{2\pi r_h}, \tag{7.49}$$

where r_h is the location of the event horizon given by

$$k^2 r_h^2 = \frac{1}{2}\left[\sqrt{1 + 4\mu k^2} - 1\right]. \tag{7.50}$$

For the case of the RS model, for which $\kappa = 0$, this gives

$$\frac{\mu}{R^4} = \frac{(\pi T_H)^4}{k^6 R^4} = \frac{(\pi T)^4}{k^2}, \tag{7.51}$$

where T is now the comoving temperature on the brane. That this is suggestive of the Stefan law, $\rho \propto T^4$, is not a coincidence and is a theme we will pursue in the next section.

7.4 Black Holes and Holography

Both the linearized gravity result for an isolated mass and the brane cosmology metric suggest a somewhat deeper importance to braneworlds and black holes. The corrections to the Newtonian potential (7.33) in fact coincide precisely with the 1-loop corrections to the graviton propagator [36, 37, 62], and the cosmological dark radiation term in the brane Friedman equation corresponds (up to a factor) to the energy density of a conformal field theory at the Hawking temperature of the black hole [33]. These clues, and analogies with lower dimensional branes, have led to the black hole *holographic conjecture* of Emparan et al. [63] which states, loosely speaking, that if we have a classical solution to the RS model then we can interpret the braneworld as a quantum-corrected 4D spacetime. In the case of the black hole, this would mean that we have a quantum-corrected black hole.

The reason for putting forward such a conjecture is based on the adS/CFT conjecture [13, 14] of string theory. In string theory, D-branes arise as the physical manifestation of open-string Dirichlet boundary conditions. These D-branes are tangible objects carrying mass, Ramond–Ramond charge, and with worldvolume gauge theories to support the string endpoints [4, 5]. Furthermore, the supergravity solutions which correspond to the mass and charge of a particular type of D-brane must describe the same objects. The metric for a stack of N coincident D3-branes is given by

$$ds^2 = \left(1 + \frac{4\pi g N \alpha'^2}{r^4}\right)^{-1/2} dx_\parallel^2 - \left(1 + \frac{4\pi g N \alpha'^2}{r^4}\right)^{1/2} dx_\perp^2, \tag{7.52}$$

where g is the string coupling and $\alpha' = l_s^2$ the string length scale. $dx_\parallel^2$ and $dx_\perp^2$ are the cartesian metrics of the spaces, respectively, parallel and perpendicular to the brane and r is the radial coordinate in this latter space. We trust this supergravity solution in regions where the spacetime curvature is small, i.e. $L^2 \gg \alpha'$, where L

is the ambient spacetime curvature. Obviously, this will be true at large r in (7.52); however, at large r the effect of the branes is negligible. In order to trust the supergravity solution in regions where it is nontrivial, i.e. where $r \sim (4\pi gN\alpha'^2)^{1/4}$, we require $gN \gg 1$. In this case, we can effectively ignore the "1" in the prefactor, and (7.52) is approximately

$$ds^2 = \alpha' \left[\frac{(r/\alpha')^2}{\sqrt{4\pi gN}} dx_\parallel^2 - \frac{\sqrt{4\pi gN}}{r^2} dr^2 - \sqrt{4\pi gN} d\Omega_V^2 \right] . \tag{7.53}$$

This metric is adS$_5 \times S^5$. Thus, if we integrate over the S^5, and identify

$$L = k^{-1} = (4\pi gN)^{1/4} l_s \tag{7.54}$$

as the adS length scale, we can directly relate the near horizon régime of a stack of D3-branes with the RS model. Furthermore, the 5D Newton constant will be given in terms of the 10D Newton constant and the volume of the 5-sphere by

$$\hbar G_5 = \frac{\hbar G_{10}}{V_5} = \frac{g^2 \alpha'^4 (2\pi)^7}{16\pi^4 L^5} = \frac{\pi L^3}{2N^2} . \tag{7.55}$$

Thus we can relate finite and *classical* quantities in our 5D Einstein theory, such as the adS curvature scale, L, and the gravitational constant, G_5, to quantum mechanical quantities such as $\hbar$, and N, the number of D-branes. Indeed, we can potentially take a classical limit, $\hbar \to 0$, keeping our adS scale finite by simply simultaneously taking $N \to \infty$. On the other hand, this stack of N D3-branes has a low-energy $U(N)$ worldvolume conformal field theory, and taking $N \to \infty$ corresponds to the t'Hooft limit of the gauge theory. Since we have set $\hbar \to 0$, on the string side $\alpha' \to 0$ ensures that only this low-energy sector remains. This is the essence of the adS/CFT correspondence that certain strongly coupled conformal field theories are dual to string theory on certain anti-de Sitter spacetimes.

What does this mean for the RS model? As Gubser first noted [33], brane cosmology with a black hole in the bulk has the appearance of a radiation cosmology from the brane perspective. From the Hawking temperature of the bulk black hole, the dark radiation term has the form (7.51), $(\pi T)^4 L^2$. On the other hand, calculating the energy of a CFT at finite temperature (at weak coupling) gives

$$\rho = 2\pi^2 cT^4, \tag{7.56}$$

where $c = \hbar(N^2 - 1)/4$ is the coefficient for the trace anomaly in super Yang–Mills theory. Thus as $N \to \infty$,

$$\frac{8\pi G_N \rho_{CFT}}{3} = \frac{4}{3} \frac{G_N L}{2G_5} (\pi T)^4 L^2 . \tag{7.57}$$

Clearly, if we identify $G_N = 2G_5/L$, then we see that the classical bulk black hole has the effect on the brane of a thermal CFT at the comoving Hawking temperature of the black hole up to the conventional strong/weak coupling factor of $4/3$. Note that the factor of 2 in the definition of the 4D Newton's constant is due to the fact that in adS/CFT we have a bulk on only one side of the boundary, whereas in physical braneworlds, we have bulk on each side of the brane. This effectively halves the brane tension, which is the key factor in the relation between the brane and bulk gravitational constants.

This rather physical picture of the interplay between the RS model and the adS/CFT conjecture is further fleshed out by the work of Duff and Liu [36, 37], who note that the linearized corrections to the graviton propagator, calculated in (7.34), precisely agree with the 1-loop linearized corrections to flat space for a central mass [62]. These results are extremely suggestive that a fully nonlinear classical brane/bulk black hole solution would, from the brane point of view, correspond to a quantum-corrected black hole. Indeed, it was this perspective that first led Tanaka to conjecture that a braneworld black hole must therefore be time dependent, to agree with the thermal Hawking radiation from a Schwarzschild black hole [64]. Emparan, Fabbri, and Kaloper then pointed out that the issue of time dependence is linked to the choice of quantum vacuum and gave a comprehensive analysis of 3D brane black holes, together with options for the RS black hole.

Roughly, the picture is as follows. If we consider a closed universe with a bulk black hole, then the brane is precisely equidistant from the bulk black hole, and the radiation on the brane is precisely thermal. However, we could imagine displacing the brane slightly, which would introduce an inhomogeneity in the dark radiation on the brane. Moving one side of the brane even closer to the bulk black hole would then increase this distortion and would (hopefully!) correspond to a collapsing shell of warm radiation on the brane. This could then form its own black hole, which from the bulk perspective would correspond to the brane actually touching the black hole. The brane would remain glued to the black hole for a while, but eventually would separate, the process corresponding to black hole radiation (see Fig. 7.3).

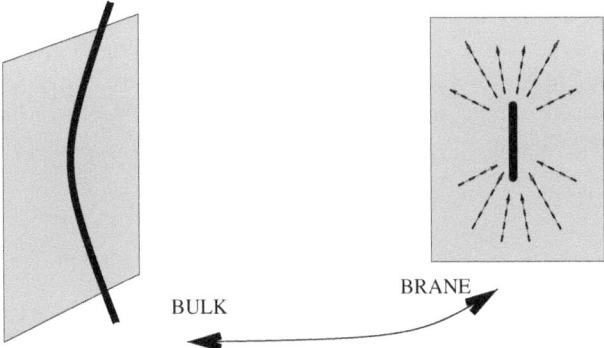

Fig. 7.3 A cartoon of the time-dependent radiating black hole from both the brane and the bulk perspectives. A bulk black hole moves towards the brane, touches, then eventually recoils back into the bulk. From the brane perspective this corresponds to anisotropic radiation steadily accreting around a central point which finally forms a black hole, persists for a while radiating, then finally evaporates

On the other hand, it is always possible that there does exist a static black hole solution, which asymptotically has the form of (7.35). Such a black hole would, according to EFK, necessarily have a singular horizon. This classical solution would be a 5D version of the C-metric [65], which is a solution representing two black holes accelerating away from each other. The black holes are being accelerated by two cosmic strings, one for each hole, which pull the black hole out to infinity. The exact purely gravitational solution has a conical deficit which can be smoothed out by a U(1) vortex [66–69], rendering the spacetime nonsingular apart from the central singularities of the black holes. It is then straightforward to slice this spacetime with a brane [70–72], thus producing a 3D braneworld with a black hole. A positive tension brane retains the bulk without the cosmic string; hence these braneworld black holes do not need any further regularization. It may seem strange that a static black hole on the brane is accelerating, but it is no more unusual than the fact that we are in an accelerating frame on the surface of the Earth. Geodesics in the RS bulk actually curve away from the brane:

$$2kz(t) \sim \ln(1 + k^2t^2).$$ (7.58)

Thus any observer glued to the brane is necessarily in an accelerating frame.

Moving up one dimension however changes the picture completely. The mathematics of the pure gravitational equations is now no longer amenable to analytic study, and no known C-metric exists. Even higher dimensional "cosmic strings" now have codimension three and are strongly gravitating [73, 74] with potentially singular geometries. We will now look at this problem in more detail.

7.5 Black Hole Metric

The first attempt to find a black hole on an RS brane was that of Chamblin, Hawking, and Reall (CHR) [75], in which they replaced the Minkowski metric in (7.1) by the Schwarzschild metric (indeed, we can replace η in (7.1) by *any* 4D Ricci-flat metric):

$$ds^2 = a^2(z)\left[\left(1 - \frac{2M}{r}\right)dt^2 - \left(1 - \frac{2M}{r}\right)^{-1}dr^2 - r^2 d\Omega_{II}^2\right] - dz^2 .$$ (7.59)

This is the only known exact solution looking like a black hole from the brane point of view. Unfortunately, it does not correspond to what we would expect for a brane black hole. If matter is confined to our brane, we would expect that any gravitational effect is localized near the brane. For a collapsed star, we would also intuitively expect that while the horizon might well extend out into the bulk, it too should be localized near the brane, and the singularity should not extend out into the bulk. The problem with the CHR black string, (7.59), is that it extends all the way out to the adS horizon; moreover, at this surface the black hole horizon actually becomes singular!

There is however another, more serious, problem with the CHR black string, and that is that it suffers from a classical instability [76]. Black string instabilities were first discovered in vacuum, [77, 78], for the Kaluza–Klein black string:

$$ds^2 = \left(1 - \frac{2M}{r}\right) dt^2 - \left(1 - \frac{2M}{r}\right)^{-1} dr^2 - r^2 d\Omega_{II}^2 - dz^2. \tag{7.60}$$

This has a cylindrical event horizon, with entropy $4\pi G_N M^2$. On the other hand, assuming a KK compactification scale of L_{KK}, a 5D black hole of the *same* mass as the string (7.60) has an entropy of $8\sqrt{2\pi L_{KK} G_N} M^{3/2}/3\sqrt{3}$. Thus, for small enough masses relative to the compactification scale ($G_N M \leq 2L_{KK}/27\pi$) a standard 5D black hole has higher entropy than the string, and thus the string should be either perturbatively or nonperturbatively unstable.

The existence of the instability can be confirmed by solving the Lichnerowicz equation:

$$\nabla^2 h^{ab} + 2R^a{}_c{}^b{}_d h^{cd} = 0. \tag{7.61}$$

There is a subtlety involving the initial data surface, which must be taken to touch the future event horizon (the black hole generically forms from gravitational collapse); however, there is an unstable s-mode with the form

$$h^{ab} = e^{i\mu z} e^{\Omega t} \begin{bmatrix} H^{tt}(h) & h(r) & 0 & 0 & 0 \\ h & H^{rr}(h) & 0 & 0 & 0 \\ 0 & 0 & K(h) & 0 & 0 \\ 0 & 0 & 0 & K/\sin^2\theta & 0 \\ 0 & 0 & 0 & 0 & 0 \end{bmatrix}. \tag{7.62}$$

(Note: this is written in Schwarzschild coordinates for convenience, but to check h is small, use Kruskals.) This mode is physical, since any gauge degree of freedom would have to be purely 4D, thus satisfying a massless 4D Lichnerowicz equation, whereas this mode satisfies a massive 4D Lichnerowicz equation. The effect of the instability is to cause the horizon to ripple.

For the CHR black string, the presence of the bulk cosmological constant might be supposed to change the technicalities of this analysis; however, the crucial feature of the black string instability is that it is a purely 4D (massive) tensor TTF mode – i.e. it satisfies the RS gauge! If we work out the perturbation equations for the CHR black string background they have the particularly simple form:

$$\left((\nabla^{(4)})^2 h_{\mu\nu} + 2R^{(4)}_{\mu\lambda\nu\rho} h^{\lambda\rho}\right) - \left[a^4 \left(a^{-2} h_{\mu\nu}\right)'\right]' = 0. \tag{7.63}$$

This means we can simply take the standard KK instability and substitute the appropriate massive z-dependent eigenfunction: $h_{\mu\nu} = \chi_{\mu\nu} u_m(z)$, so that $\chi_{\mu\nu}$ satisfies the equation of motion:

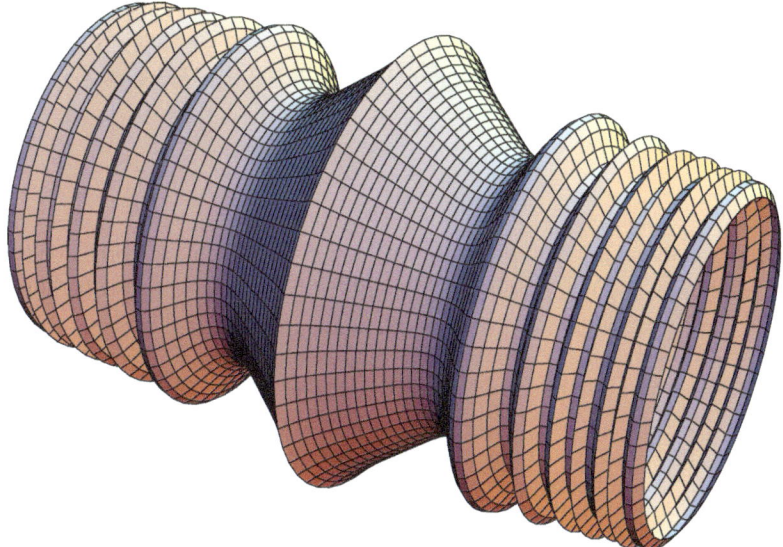

Fig. 7.4 The CHR black string horizon after the instability has set in. The brane is located at the central cusp

$$\left(\Delta_L^{(4)} + m^2\right) \chi_{\mu\nu} = 0, \qquad (7.64)$$

where $\Delta_L^{(4)}$ is the 4D Schwarzschild Lichnerowicz operator. In other words, we have the same 4D form for the instability, but a different z-dependence appropriate to the RS background. Figure 7.4 shows the effect of the instability on the black string horizon, which now ripples with ever-increasing frequency towards the adS horizon.

It is tempting to link the existence of this instability to the thermodynamic instability of black holes to Hawking evaporation; however, the timescales have rather different behaviour. Not only that, but the instability is a dynamical process, and the amplitude of the instability, $\mathscr{A}$, is essentially arbitrary. The thermal radiation from a black hole, however, is a quantum process with a well-defined amplitude. To see the difference, note that for a black hole emitting radiation into $O(N^2)$ states

$$\frac{dM}{dt} \propto \frac{\hbar N^2}{(G_N M)^2} = \frac{L^3}{2G_5 (G_N M)^2} = \frac{1}{G_N}\left(\frac{L}{G_N M}\right)^2. \qquad (7.65)$$

For the unstable black string, the mass function on the future horizon is given by an integral over the KK modes [79]:

$$M(v) = M_0 + \int_0^{m_{\max}} \frac{dm}{k} M_0 (2G_N M_0 \Omega - 1/2)\, e^{\Omega v} u_m(0), \qquad (7.66)$$

where v is the ingoing Eddington Finkelstein coordinate and Ω the half-life of the instability at m, which is well approximated by $\Omega(m) = m^2/M - m/2$ (see the plots of Ω vs. μ in [77, 78]). Given this approximation, we can compare the rate of mass loss of the black hole to that by evaporation, by simply taking $dM/dv|_{v=0}$:

$$\frac{dM}{dv} = \int_0^{\frac{1}{2G_N M}} \frac{dm}{2k} \sqrt{\frac{m}{2k}} (2m^2 - mM - \tfrac{1}{2})(2m^2 - mM) \propto \frac{1}{G_N}\left(\frac{L}{G_N M}\right)^{3/2} + \cdots \quad (7.67)$$

which clearly has a different dependence on L and M

It seems therefore that the holographic principle is not so straightforward to either confirm or implement and reinforces the need for an exact solution. A natural method to try would be to take a similar approach as in cosmology: use the symmetries of the spacetime and construct the most general metric. Clearly we have spherical symmetry around the black hole, but we also have a time translation symmetry (assuming a static solution). This introduces an additional degree of freedom into the system, which can be parametrized as follows [80]:

$$ds^2 = e^{2\phi/\sqrt{3}}dt^2 - e^{-\phi/\sqrt{3}}\left\{\alpha^{-1/2}e^{2\chi}(dr^2 + dz^2) + \alpha d\Omega_{II}^2\right\}. \quad (7.68)$$

The equations of motion then take the form

$$\Delta\alpha = -2\Lambda\alpha^{1/2}e^{2\chi - \phi/\sqrt{3}} + 2\alpha^{1/2}e^{2\chi}, \quad (7.69)$$

$$\Delta\phi + \nabla\phi \cdot \frac{\nabla\alpha}{\alpha} = -\frac{2\Lambda\alpha^{-1/2}e^{2\chi - \phi/\sqrt{3}}}{\sqrt{3}}, \quad (7.70)$$

$$\Delta\chi + \frac{1}{4}(\nabla\phi)^2 = -\frac{\Lambda\alpha^{-1/2}e^{2\chi - \phi/\sqrt{3}}}{2} - \frac{\alpha^{-1/2}e^{2\chi}}{2}, \quad (7.71)$$

$$\frac{\partial_\pm^2 \alpha}{\alpha} + \frac{1}{2}(\partial_\pm\phi)^2 - 2\partial_\pm\chi\frac{\partial_\pm\alpha}{\alpha} = 0, \quad (7.72)$$

where $2\partial_\pm = \partial/\partial(r \pm iz)$. This is clearly a fairly involved elliptic system, but unlike the cosmological equations, it is not integrable. What rendered the cosmological equations integrable was (7.43), of which (7.72) is the counterpart in this set of equations. The presence of the $(\partial_\pm\phi)^2$ in (7.72) means that we can no longer use this to integrate up the other equations. It is possible to classify the separable analytic solutions, [80]; however, none of these have the form expected of a brane black hole metric. The system can of course be integrated numerically; however, the typical method appropriate for elliptic systems (relaxation) is apparently extremely sensitive and has difficulty dealing with radically different scales for the black hole mass and the adS bulk length scales. The consensus seems to be that nonsingular solutions representing static braneworld black holes exist for horizon radii of up to a few adS lengths [81] (see also [82–84]). However, there is no convincing demonstration of the existence of nonsingular static astrophysical brane black holes.

7.6 Approximate Methods and Solutions for Brane Black Holes

Since we lack an exact solution, it is natural to attempt approximate methods to gain understanding of the system. There are two main approaches: One is to confine analysis to the brane and to try to find a self-consistent 4D solution. This has the advantage of only dealing with one variable (the radius), thereby reducing the problem to a set of ODEs. However, it has the clear disadvantage that it does not take into account the bulk spacetime, and therefore will not be closed as a system of equations – inevitably there will be some guesswork or approximation involved with terms that encode the bulk behaviour. The other main approach is to take a known bulk, such as the Schwarzschild–adS solution, and to explore what possible branes can exist. Within this method, the branes can be taken either as *probe* branes, i.e. branes which do not gravitationally backreact on the bulk black hole, or as fully gravitating solutions to the Israel equations, which will therefore have restricted trajectories.

Other approaches not reviewed here include allowing for more general bulk matter [85, 86], which moves beyond the RS model being considered here. Also, the extension of brane solutions into the bulk has been explored perturbatively [87], and numerically [88].

7.6.1 Brane Approach

The brane approach is based on the formalism of Shiromizu, Maeda, and Sasaki (SMS) [89], who showed how to project the 5D Einstein–Israel equations down to a 4D brane system. The SMS method uses the fact that the RS braneworld has $\mathbb{Z}_2$-symmetry and writes (7.6) at $z = 0^+$ using the bulk Einstein equations $R_{ab} = 4k^2 g_{ab}$ to replace the 5D Ricci tensor and the Israel equations (7.9) to replace the extrinsic curvature:

$$K_{ab}(0^+) = -k\gamma_{ab} + 4\pi G_5 (\mathscr{T}_{ab} - \tfrac{1}{3}\mathscr{T}\gamma_{ab}) \tag{7.73}$$

using (7.23) to define $\mathscr{T}_{ab}$. The only term that cannot be substituted by known quantities is the contraction of the Riemann tensor. Instead, SMS define an (unknown) Weyl term:

$$\mathscr{E}_{ab} = C_{acbd}n^c n^d = R_{acbd}n^c n^d - \frac{R}{12}\gamma_{ab} + \frac{1}{3}(R_{cd}\gamma_a^c\gamma_b^d - \gamma_{ab}R_{nn}) = R_{acbd}n^c n^d + k^2 \gamma_{ab}, \tag{7.74}$$

which is tracefree.

Using these substitutions, one arrives at a brane "Einstein" equation:

$$^{(4)}R_{ab} - \tfrac{1}{2}{}^{(4)}R\gamma_{ab} = 8\pi G_N \mathscr{T}_{ab} + \frac{(8\pi G_N)^2}{24k^2}\mathscr{Q}_{ab} + \mathscr{E}_{ab}, \tag{7.75}$$

where the tensor $\mathscr{Q}_{ab}$ is quadratic in $\mathscr{T}_{ab}$:

$$\mathscr{Q}_{ab} = 6\mathscr{T}_{ac}\mathscr{T}_b^c - 2\mathscr{T}\mathscr{T}_{ab} - 3\mathscr{T}_{cd}^2\gamma_{ab} + \mathscr{T}^2\gamma_{ab}. \tag{7.76}$$

Clearly these equations have an attractive simplicity, particularly if solving for an empty brane; however, it is important to note that the Weyl term (7.74) is a complete unknown and depends on the details of the bulk solution.

For the case of the black hole, however, one can use a method similar to that in brane cosmology [90] to decompose the Weyl term into two independent pieces:

$$\mathcal{E}_{\mu\nu} = \mathcal{U} \left(u_\mu u_\nu - \tfrac{1}{3} h_{\mu\nu} \right) + \Pi \left(r_\mu r_\nu + \tfrac{1}{3} h_{\mu\nu} \right), \tag{7.77}$$

where u^μ is a unit time vector, r^μ a unit radial vector, and $h_{\mu\nu} = \gamma_{\mu\nu} - u_\mu u_\nu$ is here the purely spatial part of the braneworld metric. This renders the vacuum brane equations (7.75) rather similar to the Einstein equations with a gravitating perfect fluid: the Tolman Oppenheimer Volkoff (TOV) equations. Of course, $\mathcal{U}$ and Π are complete unknowns and do not necessarily satisfy any conventional energy conditions; however, this notional similarity is very useful in understanding the physical system and in fact allows us to derive useful insight into braneworlds stars [91, 92], such as the fact that the exterior of a collapsing star is not, in fact, static and Schwarzschild.

The vacuum equations have been solved in many special cases, for example, Dadhich et al. [93] showed that there was an exact solution with $\Pi = -2\mathcal{U}$ having the form of a (zero mass) Reissner–Nordstrom metric on the brane. Other analytic solutions can be found by assuming a given form for the time or radial part of the metric [94–97]. However, a useful approach to solving these equations is to take an arbitrary spherically symmetric metric, in which we allow for a general area functional for the 2-spheres, then apply an equation of state between $\mathcal{U}$ and Π [98]:

$$\Pi = w\mathcal{U} . \tag{7.78}$$

The Einstein equations reduce to a 2D dynamical system from which it is relatively easy to extract general information about the system. Obviously, we do not expect that this unknown tensor will have such a simple equation of state as (7.78); however, just as in cosmology we approximate the energy momentum of the universe by various eras with fixed equations of state, it seems reasonable to approximate the near and far horizon behaviours by a fixed w.

At large r, we might expect the linearized solution (7.35) on the brane, which corresponds to $w = -5/4$. Closer to the horizon, however, it is possible that w could become very large and negative. There is in fact an exact solution for $w \to -\infty$ which displays features which are generic to solutions with $w < -2$ (the tidal Reissner–Nordstrom solution):

$$ds^2 = \left[(1+\epsilon)\sqrt{1 - \frac{2GM}{R}} - \epsilon \right]^2 dt^2 - \frac{dR^2}{1 - \frac{2GM}{R}} - R^2 d\Omega_{II}^2 . \tag{7.79}$$

This solution has a null singularity at $r = r_1$, the relic of an horizon, but also note that it actually has a "wormhole", i.e. the area of 2-spheres surrounding the origin actually has a minimum value outside the horizon (for $r_0 > r_1$) and is increasing

Fig. 7.5 A sketch of the
constant time surfaces of the
metric (7.79)

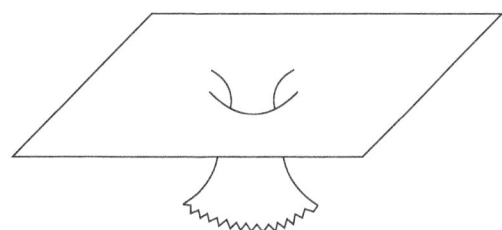

as the horizon is approached. A sketch of a constant time surface is shown in
Fig. 7.5. It is tempting to conjecture that in fact any such static solution is singular, but while suggestive, these general arguments are not a proof, and investigations
to determine the nature of the braneworld black hole horizon have been inconclusive
[99, 100].

7.6.2 Bulk Approach

The other main approach which can yield insight into brane solutions is to simplify
the problem by taking a known bulk and exploring the possibilities for a brane solution with internal spherical symmetry. With this approach the bulk is now known
(although rather rigidly fixed) and therefore the system has no "unknowns". The
first work in this area took the brane to be non-gravitating – a probe brane – and
determined the general trajectories and dynamics of the brane [101–104]. Although
this work is not gravitationally self-consistent, it is important in particular because
it gives insight into highly time-dependent and complicated processes and is to date
the only available study of the process of a black hole leaving the brane. This has
relevance for LHC black holes, as the main alternative to decay via Hawking evaporation is black hole recoil into the bulk, although the holographic point of view
would argue these are indistinguishable [105].

In these lectures, we are mostly concerned with the gravitational properties of
brane black holes and so want to keep the brane at finite tension and have a consistent backreaction. This is a far more complicated and restrictive problem; however it is possible to obtain a linearized metric for a black hole that has left the
brane [106]. This has the form of a shock wave of spherically symmetric outgoing
radiation on the brane. For a full nonlinear analysis, we have to look for spherically symmetric branes embedded in a 5D Schwarzschild–adS spacetime using the
Israel formalism. This leads to some interesting solutions, although the price to be
paid is that the brane is no longer empty: we require energy momentum on the
brane to source the gravitational field. This presentation is based on [107], but see
also [108, 109].

The basic idea is to use the Israel equations with a bulk metric of the general
Schwarzschild–adS form. The brane is spherically symmetric, with additional matter content corresponding to a homogeneous and isotropic fluid, in other words a

braneworld TOV system. Note however that here there is an actual energy momentum source on the brane, in addition to the Weyl term (7.74) which is now specified from the brane embedding in the bulk metric:

$$ds^2 = U(r)\,d\tau^2 - \frac{1}{U(r)}\,dr^2 - r^2(d\chi^2 + \sin^2\chi\,d\Omega_{II}^2), \qquad (7.80)$$

For the brane trajectory, consistent with the $SO(3)$ symmetry, we take a general axisymmetric slice $\chi(r)$. Finally, for the energy momentum tensor of the brane we take a general isotropic fluid source:

$$T_{\mu\nu} = [\mathrm{E}(r) - \mathrm{T}(r)]\,u_\mu u_\nu + \mathrm{T}(r)\,h_{\mu\nu}. \qquad (7.81)$$

It turns out to be convenient to write the Israel equations in terms of $\alpha = \cos\chi$,

$$U r \alpha' + \alpha = \frac{8\pi G_5\,\mathrm{E}\,r}{6}\sqrt{[r^2 U \alpha'^2 + 1 - \alpha^2]}, \qquad (7.82)$$

$$r^2 U \alpha'' + \frac{r^2 U'}{2}\alpha' + 2rU\alpha' + \frac{r^2 U \alpha'^2}{1-\alpha^2}\left(rU\alpha' + \alpha\right)$$
$$= \frac{8\pi G_5\,\mathrm{E}\,r}{6(1-\alpha^2)}\left[r^2 U \alpha'^2 + 1 - \alpha^2\right]^{3/2}, \qquad (7.83)$$

together with a conservation equation which determines T.

These equations can be completely integrated in terms of a modified radial variable

$$\tilde{r} = \int \frac{dr}{r\sqrt{U}} \qquad (7.84)$$

giving

$$\cos\chi = ae^{\tilde{r}} + be^{-\tilde{r}}, \qquad (7.85)$$

$$\mathrm{E}(r) = \frac{6}{8\pi G_5 r\sqrt{1-4ab}}\left[\sqrt{U}\left(ae^{\tilde{r}} - be^{-\tilde{r}}\right) + ae^{\tilde{r}} + be^{-\tilde{r}}\right], \qquad (7.86)$$

$$\mathrm{T}(r) = \frac{2}{3}\mathrm{E}(r) + \frac{U'}{8\pi G_5\sqrt{(1-4ab)U}}\left(ae^{\tilde{r}} - be^{-\tilde{r}}\right). \qquad (7.87)$$

Finally the induced metric on the brane is

$$ds^2 = U d\tau^2 - \frac{(1-4ab)dr^2}{U(1-\alpha^2)} - r^2(1-\alpha^2)d\Omega^2. \qquad (7.88)$$

So far, this is a completely general (implicit) exact solution which depends on an integral of the bulk Newtonian potential $U(r)$. Although this is an exact solution, the actual properties of the brane depend on the specifics of the relation between $\tilde{r}$ and r. Once this is determined, we have a solution describing a static, spherically symmetric distribution of an isotropic perfect fluid on the brane, i.e. a solution to the brane-TOV system, and therefore a candidate for a brane "star". The extent to which this is a physically realistic solution will depend on the energy and pressure profiles.

Note that the profiles $E(r)$ and $T(r)$ represent the full brane energy momentum and include the background brane tension. The relevant physical energy and pressure will be defined by

$$\rho = E - E_\infty, \qquad p = E_\infty - T, \tag{7.89}$$

where E_∞ is an appropriate background brane energy, which has to be identified on a case-by-case basis.

For a 5D adS bulk, $U(r) = 1 + k^2 r^2$, and the brane satisfies

$$r\cos\chi(r) = A\left(\sqrt{U} - 1\right) + B\left(\sqrt{U} + 1\right), \tag{7.90}$$

$$E = \frac{6k^2(A - B)}{8\pi G_5 \sqrt{1 - 4k^2 AB}}, \tag{7.91}$$

$$T(r) = E - \frac{2k^2(A + B)^2}{8\pi G_5 \sqrt{(1 - 4k^2 AB)U}}, \tag{7.92}$$

where $A = a/k$ and $B = b/k$ in terms of the parameters in (7.85). These brane trajectories are conic sections classified by $|A + B|$. For $|A + B| = k^{-1}$, the brane is a paraboloid with critical RS tension E_{RS} (7.12). For $|A + B| > (<)k^{-1}$, the brane is an ellipsoid (hyperboloid) with super- (sub-) critical tension. $A = -B$ is a special case, corresponding to a sub-critical Karch–Randall brane and is a straight line. Figure 7.6 shows sample brane configurations for various values of the integration parameters A and B.

Notice that the energy density is constant and requires $A > B$ to be positive. The tension on the other hand is clearly not constant unless $A = -B$. For the general brane we have a gravitating source composed purely of pressure! These branes clearly do not asymptote exact Randall–Sundrum or Karch–Randall branes. However, if $|kA|$ and $|kB|$ are large enough, the metric can be flat (or asymptotically (a)dS) over many orders of magnitude before the effect of the pressure kicks in.

It is also interesting also to look at a pure Schwarzschild bulk, $U(r) = 1 - \mu/r^2$, for which

$$\cos\chi = r\left[A\left(\sqrt{U} - 1\right) + B\left(\sqrt{U} + 1\right)\right], \tag{7.93}$$

$$E(r) = \frac{6}{8\pi G_5 \sqrt{1 + 4\mu AB}}\left[(B - A)(1 + U) + 2(B + A)\sqrt{U}\right], \tag{7.94}$$

$$T(r) = \frac{2}{8\pi G_5 \sqrt{1 + 4\mu AB}}\left[(B - A)(3 + U) + (B + A)(1 + 3U)/\sqrt{U}\right], \tag{7.95}$$

where now $A = -b/\sqrt{\mu}$ and $B = a/\sqrt{\mu}$ in terms of the general solution (7.85). Note that by construction, these trajectories are strictly only valid outside the event horizon of the black hole, since the definition of the $\tilde{r}$ coordinate involves a branch cut there. In contrast to the adS case, E is not constant for these branes. For our solution to correspond to a brane star or black hole, we require E to be positive, and to increase towards the centre of the brane.

Fig. 7.6 A selection of branes of varying coefficient A, for the case $B = 0, k = 1$ in a 5D anti-de Sitter bulk

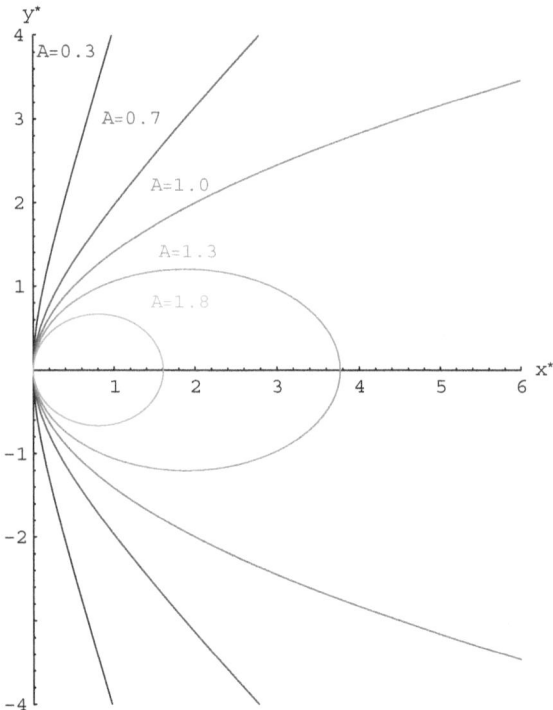

Looking at the large r behaviour of (7.93), we see that the brane can only reach large r if $B = 0$, otherwise the brane is either a bubble (enclosing the horizon or not, depending on A and B) or an arc touching the horizon. In general, the brane touches the horizon at a tangent, and the pressure diverges; however, for one special case $A = -B$, the brane slices through the horizon passing on to the singularity. Some sample closed trajectories are shown in Fig. 7.7.

The most physically interesting Schwarzschild trajectories are those which tend to infinity, for which $B = 0$, see Fig. 7.8. For these branes $E_\infty = 0$, and thus

$$\rho = E(r) = \frac{-6A}{8\pi G_5} \left(\sqrt{U} - 1\right)^2, \qquad p = -T = -\frac{\rho \left(\sqrt{U} - 1\right)}{3 \sqrt{U}}. \qquad (7.96)$$

For $A < 0$, these branes have positive energy and pressure, uniformly decreasing as $1/r^4$ and $1/r^6$, respectively. If $|A| > 1/\sqrt{\mu}$, the brane never touches the horizon and the pressure remains everywhere finite. Thus these correspond to asymptotically empty branes with positive mass sources. Plotting the energy and pressure for the brane shows that this does indeed correspond to a localized matter source, with the peak energy density dependent on the minimal distance from the horizon (see Fig. 7.9). The central energy and pressure can be readily calculated from this minimal radius, $r_m = \mu |A|/2 + 1/2|A|$:

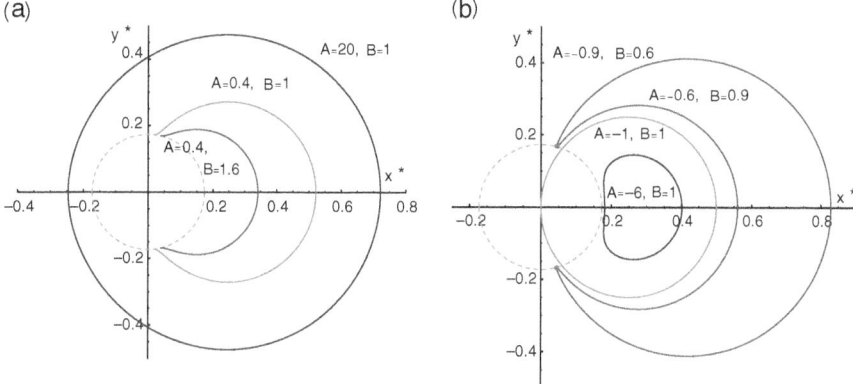

Fig. 7.7 A selection of branes (*solid lines*) for the case (**a**) $AB > 0$ and (**b**) $AB < 0$ in a 5D Schwarzschild bulk of fixed mass parameter $\mu = 0.03$. The *dashed line* denotes the corresponding horizon radius

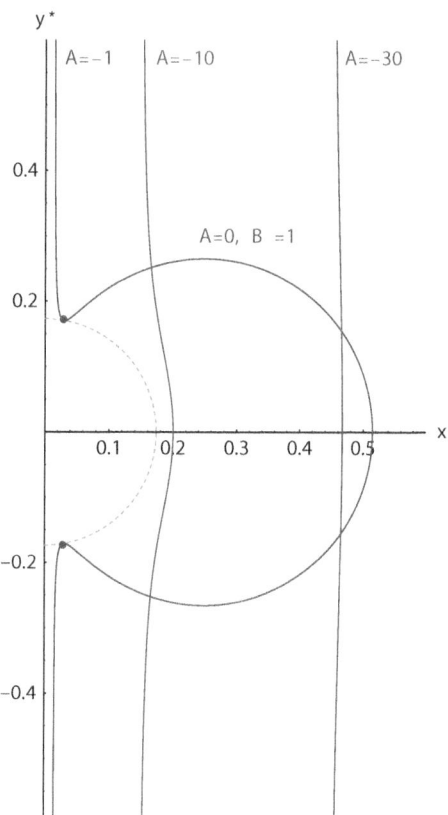

Fig. 7.8 A selection of branes for the case $AB = 0$ in a 5D Schwarzschild bulk of fixed mass parameter $\mu = 0.03$. The case $A = 0, B = 1$ is shown together with a set of branes with $B = 0$ and variable A. The *dashed line* denotes again the event horizon

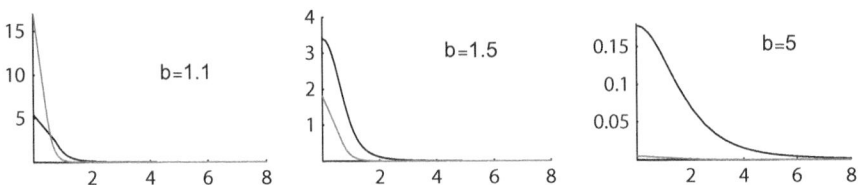

Fig. 7.9 The energy (*dark line*) and pressure (*grey line*) of brane stars with a pure Schwarzschild bulk as a function of the brane radial coordinate $\hat{r}$. The black hole mass is fixed at $\mu = 1$, and the distance of closest approach to the horizon increases across the plots

$$\rho_c = \frac{24|A|}{8\pi G_5(1+\mu A^2)^2}, \qquad p_c = \frac{16|A|}{8\pi G_5(\mu A^2-1)(1+\mu A^2)^2}, \qquad (7.97)$$

which shows that the central pressure diverges as $\mu A^2 \to 1$. This is analogous to the divergence of central pressure in the 4D TOV system, which is indicative of the existence of a Chandrasekhar limit for the mass of the star.

In these spacetimes, there is no actual black hole in the bulk, since it is the bulk to the right of the brane that is retained. Rather, it is the combination of the bulk Weyl curvature and the brane bending which produces the fully coupled gravitational solution. As the brane moves away from the horizon, the brane matter source spreads out, but the total mass changes very little, and is determined by the bulk black hole mass. The limit on mass is therefore not a true Chandrasekhar limit, but more a statement about an upper bound on the concentration of matter. The real reason there is no absolute upper bound is because, unlike the RS system with an adS bulk, gravity on the braneworld is not localized, nor is it four dimensional. Computing the induced metric on the brane in fact shows that it is the projection of the 5D Schwarzschild metric on the brane.

7.6.2.1 Braneworld Stars: A Schwarzschild–adS Bulk

For the true braneworld star, the appropriate bulk is expected to be Sch–adS bulk: $U(r) = 1 + k^2 r^2 - \frac{\mu}{r^2}$. Here $\tilde{r}$ has an exact analytic expression:

$$\tilde{r}(r) = \frac{1}{kr_h} \text{Elliptic F}\left[\text{Arcsin}\left(\frac{r}{r_-}\right), \frac{r_-^2}{r_h^2} \right], \qquad (7.98)$$

with r_h the black hole horizon, (7.50), and r_- is defined as

$$r_-^2 = \frac{-1 - \sqrt{1+4k^2\mu}}{2k^2}. \qquad (7.99)$$

Since the Randall–Sundrum model is a brane in adS spacetime, we expect that any consistent brane trajectories in Sch–adS will potentially correspond to brane stars or black holes. It is worth stressing that these solutions will not just be brane solutions, but full brane *and bulk* solutions, since the full Israel equations for the brane have been solved in a known bulk background.

From (7.86) the background brane tension is defined as

$$E_\infty = \frac{6k(a-b)}{8\pi G_5 \sqrt{1-4ab}}.$$ (7.100)

For large enough r, the geometry is dominated by the cosmological constant, therefore the pure adS solutions will be good approximations to any trajectories for large r. Also, if $\mu k^2 \ll 1$, i.e. if the black hole is much smaller than the adS scale, we expect that in the vicinity of the horizon the Schwarzschild solutions will be good approximations for the brane; therefore for small-mass black holes, we might expect brane trajectories to be well approximated by some combination of Schwarzschild and adS branes. Because the $\tilde{r}$-coordinate has been zeroed at infinity (for easy comparison with the pure adS limit) the range of $\tilde{r}$ in Sch–adS is finite and decreases sharply with increasing μ. This suggests that trajectories in large-mass Sch–adS black hole spacetimes are more finely tuned, and possibly more restricted than in small-mass black hole spacetimes.

Like adS spacetime, the Sch–adS trajectories can be classified according to whether they asymptote the adS boundary at nonzero χ, at $\chi = 0$, or do not reach the boundary at all, i.e. are closed bubbles. These correspond to sub-critical, critical, or super-critical branes ($a+b < 1$, $a+b = 1$, and $a+b > 1$), respectively. A sample of brane trajectories is shown in Fig. 7.10.

The super-critical branes are qualitatively similar to the pure Schwarzschild case; however, it is interesting to note that in each case there exists a purely empty spherical brane, equidistant from the horizon. This corresponds to the Einstein static universe [52], which from the brane perspective is a closed universe stabilized by a combination of the cosmological constant (the brane is super-critical) and the CFT dark radiation term. Using the holographic intuition, we might expect that by displacing this universe slightly we could mock up the start of gravitational collapse; however, a quick computation shows that displacing the brane relative to the black hole slightly sets up an energy *deficit* on the part of the brane closer to the black hole!

In the case of critical branes, $a+b = 1$, which means that the brane trajectories asymptote the adS boundary at exactly $\chi = 0$. The branes are thus open and may or may not touch the black hole horizon depending on the exact values of the parameters a and b. If

$$a-b < |\tanh \tilde{r}_+/2|,$$ (7.101)

the trajectory remains away from the horizon, otherwise it will touch the horizon and have a pressure singularity. A sample of critical trajectories in a Sch–adS background is shown in Fig. 7.11(left plot).

For branes that avoid the horizon the energy density is positive, peaking at the centre, and dropping rapidly to the background value, undershooting it very slightly to form an underdense region at very large r. The pressure also reaches its maximum value at the centre, but is uniformly decreasing with r, at a much slower rate,

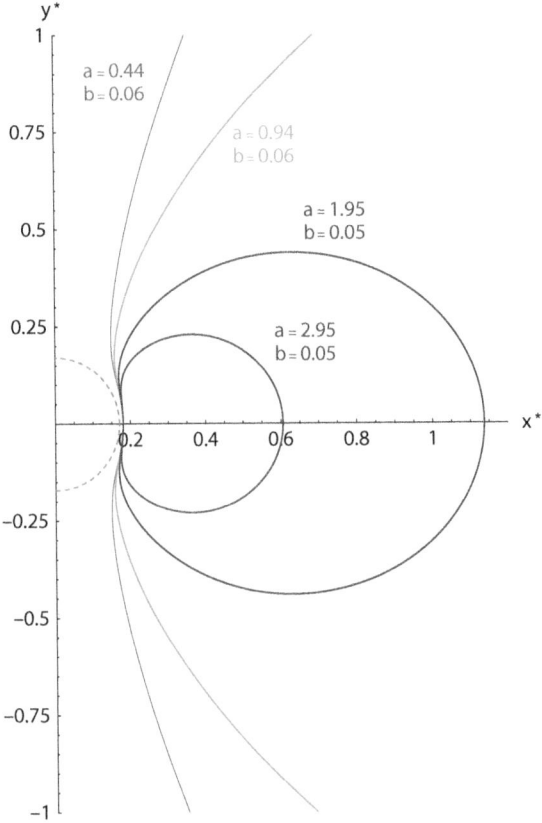

Fig. 7.10 A mixture of brane trajectories in a 5D Schwarzschild-anti-de Sitter background of fixed parameters $k = 1$ and $\mu = 0.03$. Note how these are deformed from those of Fig. 7.6 by the black hole horizon

consistent with the pressure excess observed for the pure adS branes. Apart from this pressure excess, the other main difference with pure Schwarzschild trajectories is that the brane matter can no longer universally satisfy the dominant energy condition (DEC) ($\rho \geq |p|$). In pure Schwarzschild, the DEC is satisfied except for branes which skirt extremely close to the horizon, where the local Weyl curvature causes the pressure to diverge. This phenomenon is also observed for the Sch–adS branes skimming close to the horizon; however, as we increase b the central energy dominates the pressure for only a finite range of b before once again dropping below the pressure. This is because the further we move away from the horizon the adS curvature becomes more important, and for pure adS branes, the effect of the adS curvature is to induce a pressure excess. In Fig. 7.11(right plots), the energy density and pressure of the matter on the brane are shown for a sequence of critical branes in a Sch–adS background displaced by an increasing distance from the horizon.

Sub-critical branes are largely similar to critical branes, and correspond to open trajectories that asymptote the adS boundary, although at nonzero χ in this case. The same bound as before, i.e. whether $|\cos \chi| \simeq |ae^{\tilde{r}_h} + be^{-\tilde{r}_h}| \leq 1$, will determine whether the brane terminates on the event horizon or remains on the RHS of it. The

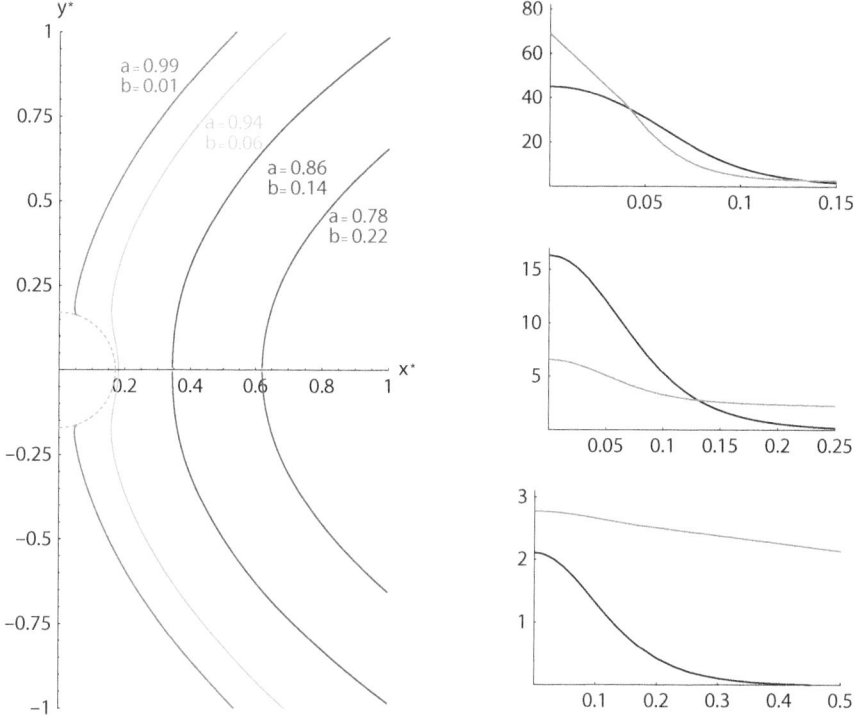

Fig. 7.11 (a) A sample of critical brane trajectories with $a + b = 1$ in a 5D Schwarzschild-anti-de Sitter background of fixed parameters $k = 1$ and $\mu = 0.03$. The *dashed line* denotes again the horizon. (**b**) A set of plots of the brane energy (*black line*) and pressure (*grey line*) for a sequence of critical branes moving away from the horizon

energy density and pressure profiles in this case are again similar to the ones found for critical branes. Once again, for a large family of parameters a and b, solutions with a positive energy excess at the centre of the brane may be easily found.

One special sub-critical trajectory found in the pure adS case was the Karch–Randall trajectory, $a + b = 0$. We can extend this to Sch–adS obtaining

$$\cos \chi = 2a \sinh \tilde{r}. \tag{7.102}$$

However, since $a > 0$ for a positive energy trajectory, this has $(\cos \chi)' > 0$, and hence the energy density is always increasing with r. Thus, whether or not these trajectories terminate on the horizon, they always correspond to energy deficits on the brane, and hence negative mass sources from the point of view of a brane observer.

To sum up: we can get static solutions to the brane-TOV equations, and hence static brane stars. Unfortunately, the restricted form of the bulk leads to unphysical asymptotic behaviour away from the star in the form of a pressure excess. One possible way of removing this would be to perturb the bulk slightly at large r to remove this excess. However, another interesting route to explore is to make the trajectory time dependent. In [91, 92] it was argued that the spacetime surrounding a collapsing brane star would be time dependent even though it was vacuum. In fact, the RS trajectory *is* time dependent when written in global adS coordinates, which of course are the coordinates used for the Schwarzschild–adS metric:

$$\sqrt{1 + k^2 r^2} \cos kt - kr \cos \chi = e^{-kz} = 1. \tag{7.103}$$

The RS wall is oscillatory because the spherical coordinates are the universal covering space of adS, and so the "wall" is actually an infinite family of walls, each in the local patch covered by the horospherical coordinates. Since $r = 0$ is a geodesic of the spherical adS spacetime, the image of $r = 0$ in the Randall–Sundrum spacetime, which is a hyperbola, will be a geodesic in the RS spacetime. Therefore, if we put a black hole at $r = 0$, it should look like a particle in the RS spacetime, at least to a first approximation.

We can generalize the brane trajectory to $\chi(r,t)$, compute the corresponding time-dependent versions of (7.82) and (7.83), and then find the energy momentum source required on the brane. The idea is that a time-dependent brane solution would describe a black hole forming from the collapse of radiation, and its subsequent evaporation; thus it is not clear whether we should expect a pure brane energy momentum solution; rather, a solution corresponding to the collapse of matter on the brane is perhaps more physically realistic. The energy momentum of a surface slicing the Sch–adS spacetime is given by the Israel junction conditions as

$$\mathcal{T}_{\mu\nu} = \frac{2}{8\pi G_5} \left(K_{\mu\nu} - K h_{\mu\nu} \right) + \frac{6k}{8\pi G_5} h_{\mu\nu}. \tag{7.104}$$

Clearly, since the trajectory is time dependent, the energy momentum will also be time dependent; however, since the largest effect of the bulk black hole will be represented by the $t = 0$ slice of the braneworld – the point of closest proximity – we evaluate the energy momentum at $t = 0$. For a pure RS trajectory, the black hole causes the energy of the brane to decrease from its critical value, whereas both the radial tension and azimuthal tension increase; thus the brane matter violates all the energy conditions! However, this was not unexpected as the RS trajectory was not modified, and the main feature of the static brane solutions was that they responded to the bulk black hole by bending. Indeed, in a definitive brane gravity paper [50], Garriga and Tanaka showed that a crucial part of obtaining 4D *Einstein* gravity (i.e. with the correct tensor structure) was what could be interpreted as a brane bending term. As shown in Sect. 7.3, the effect of matter on the brane is to "shift" the brane with respect to the acceleration horizon in the bulk (7.29). Clearly then, if a black hole forms on the brane, we would expect the brane to respond to this matter by bending.

A shift in the position of the brane corresponds to $kz \to 1 + k\delta z$, and trying a test function

$$\cos \chi(t,r) \simeq \frac{1}{r} \left(\sqrt{1 + r^2} \cos \tau - \frac{1}{1 - \frac{q}{r^p}} \right) \tag{7.105}$$

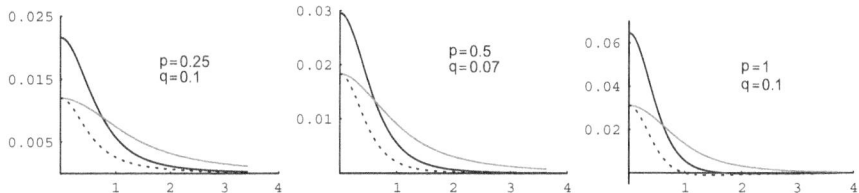

Fig. 7.12 A selection of plots of brane energy momentum with brane bending included for a range of amplitudes and powers of r. The brane energy is shown in units where $E_{RS} = 3$ and the radial brane distance in units of L. The *solid black line* is the energy, the *dashed line* the radial pressure, and the *gray line* the angular pressures

gives the behaviour shown in Fig. 7.12 for a range of p and q. (The brane bending of $1/|\mathbf{x}|$ corresponds approximately to $p = 1/2$.)

The brane energy momentum in Fig. 7.12 satisfies the WEC, however, not the DEC. If the brane is bent instead towards the black hole the brane WEC is violated. The excess of angular pressure is somewhat similar to the pressure excesses in the static brane trajectory; however, unlike the static trajectories, here the black hole actually is in the bulk, hence these are true candidates for black hole recoil into the bulk.

7.6.2.2 The Interaction of Black Holes and Branes

The main motivating factors for obtaining a time-dependent braneworld black hole are to gain insight into the backreaction of Hawking radiation on a quantum corrected 4D black hole and to understand the process of black hole recoil from a braneworld. Presumably the time-dependent process will be some perturbed version of a time-dependent brane trajectory in 5D Sch–adS spacetime. By allowing the brane to intersect the bulk black hole horizon, this would appear to describe black hole formation and evaporation via transport of a bulk black hole to the brane, and subsequent departure back into the bulk. When the brane hits the black hole, we might expect some part of it will be captured by the black hole, and will therefore remain behind the event horizon even when the black hole has left the brane, effectively having been chopped off from the rest of the brane. This feature is seen in the probe brane calculations of [103, 104], and we expect this to hold in the case of a fully gravitating brane. In support of this, we can appeal to the case of a cosmic string interacting with a black hole, where early work indicated that strings would be captured [110], and via self-intersection would leave some part behind in the black hole. Gravitational calculations of the fully coupled string/black hole system show explicitly how this ties in with the thermodynamic process of string capture and black hole entropy [111, 112].

The basic idea is that once part of a brane has fallen into the event horizon of a black hole, it can no longer leave. Thus, if the brane has enough kinetic energy to subsequently pull away from the black hole, the price it must pay is to leave

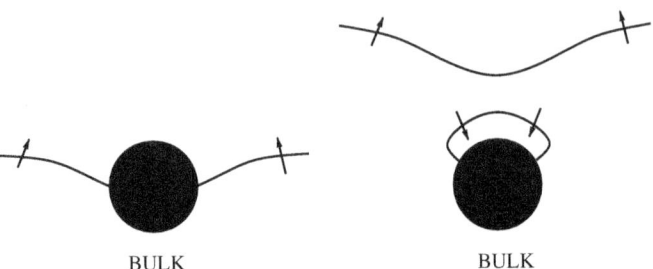

Fig. 7.13 An illustration of brane capture by a black hole. On the *left*, the black hole is on the brane, with the brane moving upwards. On the *right*, the brane has left the black hole by self-intersecting and cutting off a bubble which falls into the black hole

behind the part that has already been captured, see Fig. 7.13 [103, 104]. However, RS braneworlds are *not* probe branes, but are strongly gravitating objects, and therefore any dynamic process must also be gravitationally consistent. From the gravitational point of view, when the black hole captures part of the brane and excises it from the whole, the black hole must increase in mass. This interplay is seen particularly clearly in the related case of the cosmic string [111, 112], where a cosmic string piercing a black hole alters the thermodynamic relations between mass, entropy, and temperature. In that case, the (static) results are entirely consistent with the black hole having captured a length $4G_N M$ of cosmic string, thus increasing its mass. Just as in the cosmic string case, the capture of the codimension 1 RS brane by the black hole will turn out to be important in establishing the thermodynamic viability of the black hole recoil process.

At a first pass, it seems that in fact black hole recoil cannot occur in RS braneworlds due to a simple entropy argument [113]. In five dimensions, entropy is proportional to $M^{3/2}$; hence two black holes of mass $M/2$ have less entropy than a single black hole of mass M. However, this argument is both incorrect in the evaluation of the entropy and misses additional contributory factors such as brane bending and brane capture by the black hole.

To get a better estimate, first note that entropy is proportional to horizon area/volume, which for Sch–adS is not simply related to the mass, but also to the adS scale:

$$\mathscr{S} \propto 2\pi^2 r_h^3 = \frac{\pi^2}{\sqrt{2}k^3} \left(\sqrt{1 + \frac{32G_5 M k^2}{3\pi}} - 1 \right)^{3/2}. \tag{7.106}$$

Note that if $G_N M \geq 3.35L$, then the entropy of two black holes of mass $M/2$ will in fact be *greater* than that of a single black hole of mass M. Therefore, at least from this rather approximate entropic argument, black hole recoil would seem to be problematic only for small black holes. On the other hand, in any dynamic process, we must take into account the capture of part of the brane by the black hole. Consider the idealized situation where we have a black hole intersecting the brane along its equator; in this case, a volume of $4\pi r_h^3/3$ of brane has been captured by the black hole, with a corresponding mass of

$$\delta M = \frac{6k}{8\pi G_5} \frac{4}{3} \pi r_h^3 = \frac{1}{2\sqrt{2}G_N} \left[\sqrt{1 + 4\mu k^2} - 1 \right]^{3/2}. \tag{7.107}$$

Adding this mass to the recoiled black holes results in an order of magnitude improvement to the bound on M coming from the entropy: for $G_N M \geq 0.35L$, the entropy of the recoiled black holes becomes greater than that of the black hole on the brane.

Finally, however, the most crucial factor is the brane bending. For a mass on the brane, the brane bends away from the acceleration horizon, and (as we have seen) the brane tends to bend away from the black hole. This effect will be most marked for the smallest black holes. We therefore have to correct the entropy argument to allow for the fact that more than half of the black hole horizon is sticking out into the bulk (see Fig. 7.13). Ignoring the effect of the captured brane increasing the mass, a quick calculation shows that the effective mass of the intermediate black hole stuck on the brane is

$$M_{\text{int}} = \frac{2\pi M}{2\chi_0 - \sin 2\chi_0}, \tag{7.108}$$

where $\chi_0 > \pi/2$ is the minimal angle at which the brane touches the event horizon (assuming the black hole approaches from $\chi = \pi$). For $\chi_0 > 17\pi/30$, a rather modest amount of brane bending, the entropy of the recoiled black holes is always greater.

It is important to note that these arguments use the standard entropy of the isolated Sch–adS black hole. In other words, they assume a static solution with an event horizon at r_h. Clearly in the time-dependent spacetime there is some question about whether this approximation is valid, and entropy arguments should be used with caution, nonetheless, for small black holes, where we might expect them to be more reliable, taking into account brane bending and fragmentation shows that it is by no means entropically preferred for a black hole to stick to the brane.

7.7 Outlook

As we have seen, the problem of braneworld black hole solutions is rather complex, and extremely interesting. The holographic principle puts forward the tantalizing prospect that if we can find a classical brane black hole solution (be it time dependent or static) then this gives us invaluable information about the quantum-corrected black hole. The failure to find a classical solution so far can therefore be reinterpreted as the difficulty of consistently quantizing gravity. Yet the picture is not quite so clear. There have been several attempts to solve the brane black hole system numerically [82–84], but as yet no unequivocal result. As we have seen, finding classical solutions directly is extremely difficult, and the only progress that has been made is partial, either by ignoring the bulk or by relaxing the restrictions on the brane.

One interesting possibility, discussed in [114], is that the holographic principle is in fact not applicable to the RS model and that the lack of an exact solution is unrelated to any problem of quantum gravity. Fitzpatrick, Randall, and Wiseman (FRW) suggest that it is not appropriate to use the adS/CFT conjecture, as this refers to a quantum field theory at strong coupling, and the relation between the classical bulk solution and the quantum-corrected brane solution requires the relation (7.55) where the classical effect is related to the full N^2 degrees of freedom of the field theory. Since the field theory is strongly coupled, it is not obvious that we will indeed have access to all the N^2 states in all cases. For example, we do not see quarks or gluons outside the nucleus, so why should we expect to access the full range of states far away from a black hole?

Without an exact solution, there is no way of exploring which of these insights, the holographic picture of EFK discussed in Sect. 7.4, or the gluon analogy of FRW, is correct. FRW are of the opinion that there does exist a nonsingular, static braneworld black hole solution and proposed the CHR black string as a counterexample to the holographic conjecture. The main problem with this solution is that to render it stable a second brane is required in the bulk. This corresponds to an infrared cut-off in the CFT, and it is by no means clear how this additional complication affects the holographic argument.

There is however another option for exploring the physics of braneworld black holes, and that is to move to the Karch–Randall set-up [41]. The KR brane is slightly detuned from the critical RS value, and is sub-critical, with an effective negative cosmological constant residing on the brane. KR branes are thus adS slicings of adS. From the holographic point of view, this complicates the picture, as we are no longer in the near horizon limit of a stack of D3-branes; however, the KR brane can possibly be related to a defect CFT dual to the intersection of a probe D5-brane with a stack of D3-branes [115, 116]. The advantage of considering this slightly detuned situation is that black holes in adS can be thermodynamically stable [117], and therefore the backreaction due to Hawking radiation can, in principle, be computed. On the other hand, the adS black string in adS becomes stable once the mass is sufficiently high [118], which has been argued to be dual to the Hawking–Page transition [119]. Thus, for large–mass black holes on the KR brane, we can perform a direct comparison between the strong coupling holographic backreaction and the weak coupling Hawking radiation backreaction.

Such a comparison was made in [120] using Page's heat kernel method [40] for approximating the radiation back reaction. The physical set-up is that we have two KR branes stretching through the bulk, each with positive tension, and each cutting off the boundary of adS, hence each providing a UV cut-off CFT. The black string stretches between the two branes, and for large enough mass is stable (see Fig. 7.14):

$$ds^2 = \frac{L^2}{\cos^2 \theta} \left[\left(1 + k_4^2 r^2 - \frac{2G_N M}{r} \right) dt^2 - \frac{dr^2}{\left(1 + k_4^2 r^2 - \frac{2G_N M}{r} \right)} - r^2 d\Omega_{II}^2 - d\theta^2 \right],$$

$$(7.109)$$

where $k_4 = k \cos \theta_0$, with $\pm \theta_0$ being the location of the KR branes, is the 4D adS curvature scale.

Fig. 7.14 A sketch of the KR black string. The *black circle* is the adS boundary, which is excised from the braneworld spacetime. The string goes through the adS bulk between the two KR branes. Because the string has finite proper length relative to its mass, it can be stable for sufficiently large mass

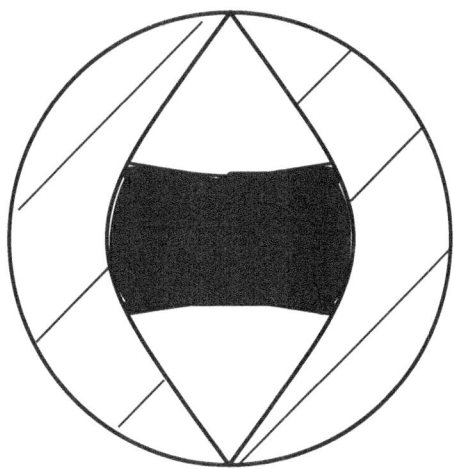

Restricting ourselves to a single brane, the geometry is that of 4D Sch–adS, and we can perform a standard weak coupling computation of the energy momentum tensor of the Hawking radiation. Figure 7.15 shows the energy and pressure of the thermal bath produced by the black hole (see [120] for details). Notice how at large r the energy and pressures asymptote the form of a cosmological constant.

On the other hand, we have a full brane+bulk classical solution, and we can directly compute the effective stress tensor on the brane. It is clear before starting, however, that this will not have the form of Fig. 7.15, as these varying energies and stresses will backreact on the spacetime to give a modification of the Sch–adS solution, whereas the classical solution is pure Sch–adS. On the other hand, although this is the classical brane solution, that does not mean that there is no backreaction on the brane energy momentum. In fact, the correction to the brane energy momentum is interpreted via the conventional 4D Einstein equation. From the brane point of view, we are unaware of the extra dimension, and therefore we interpret any deviation from the standard Einstein equation as additional energy momentum. Thus, while our KR brane energy momentum must have the form of

Fig. 7.15 The backreacted energy momentum tensor at weak coupling. The *solid black line* is the energy, the *dashed line* the radial tension, and the *dotted line* the angular tension

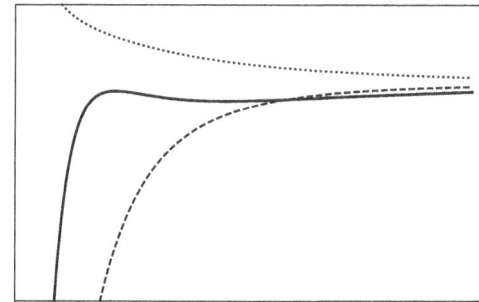

a cosmological constant, it is possible that this is renormalized from the expected bare value.

To see how this works, let the tension of the KR brane be

$$E = \frac{6k}{8\pi G_5} + \lambda = \frac{6k \sin \theta_0}{8\pi G_5}, \tag{7.110}$$

where $\lambda < 0$ is the bare tension on the brane. On the other hand, the *actual* 4D cosmological constant is given by

$$\Lambda_4 = -3k_4^2 = 8\pi G_4 \lambda_{\text{eff}}. \tag{7.111}$$

Note that in this case, the 4D gravitational constant is not labelled as G_N, since the relation between the brane and bulk gravitational constant is dependent on the brane tension, not the background adS curvature [121–123], and is altered from the critical RS relation:

$$G_4 = \frac{4\pi G_5}{3} E G_5. \tag{7.112}$$

From the definition of k_4 and (7.110) and (7.112), the value of the bare tension is

$$\lambda = \frac{3}{4\pi G_4} \left(k^2 - k_4^2 - k\sqrt{k^2 - k_4^2} \right). \tag{7.113}$$

Therefore, since the "expected" value of the cosmological constant is $8\pi G_4 \lambda$, we can compute the correction to the brane energy momentum as

$$\langle T_\nu^\mu \rangle = \frac{8\pi G_4 \lambda - 3k_4^2}{8\pi G_4} \delta_\nu^\mu = \frac{3(2k^2 - k_4^2 - 2k\sqrt{k^2 - k_4^2})}{8\pi G_5 \sqrt{k^2 - k_4^2}} \delta_\nu^\mu. \tag{7.114}$$

This is the precise (classical) braneworld result. We can obtain the holographic renormalization result [124] by taking the limit as the brane approaches the boundary or by approaching the critical RS limit $\lambda \to 0$. As $k_4 \to 0$, we get

$$\langle T_\nu^\mu \rangle = \frac{3k_4^4}{32\pi G_5} \tag{7.115}$$

which agrees with the strong coupling holographic result [120], up to the expected factor of two which arises from the braneworld set-up having two copies of the bulk, one on each side of the brane.

It is intriguing that the black hole apparently does not radiate in the strong coupling picture. This is a direct consequence of the fact that the bulk spacetime is foliated by conformal copies of the Schwarzschild–adS black hole. This "translation invariance" means that the classical KK graviton modes are not excited in the background solution, and geometrically the only possibility is renormalization of the cosmological constant. It is possible that the black string solution is not the correct black hole metric candidate; however, one might expect that for brane black holes with $r_h > L$, there is a unique stable regular black hole geometry, which this solution is.

Thus, the KR black string provides a counterexample to the expectation that a classical braneworld black hole corresponds to a quantum-corrected 4D black hole. There are of course many caveats to this claim. Clearly the KR brane is not the near horizon limit of a stack of pure D3-branes, and therefore we do not expect the CFT to be a simple SYM. However, the fact that the renormalization of the stress tensor is proportional to N^2, yet vanishes in the critical RS limit, is supportive of the arguments of [114]. Obviously this debate is far from over! (See [125–128] for some recent work.)

Hopefully these lectures have given an insight into the complex and fascinating topic of braneworld black holes. However as the field develops over the next few years, there are sufficient puzzles and unanswered questions to ensure that it will continue to be an active and exciting area.

Acknowledgments I would like to thank Elefteris Papantonopoulos for inviting me to such a lovely school, and also my collaborators throughout the years but in particular Simon Creek, Yiota Kanti, Bina Mistry, Simon Ross, Richard Whisker, and Robin Zegers. This work was partially supported by the EU FP6 Marie Curie Research and Training Network "UniverseNet" (MRTN-CT-2006-035863).

References

1. V. A. Rubakov and M. E. Shaposhnikov, Phys. Lett. B **125**, 139 (1983).
2. V. A. Rubakov and M. E. Shaposhnikov, Phys. Lett. B **125**, 136 (1983).
3. Akama, K.: Lect. Notes Phys. **176**, 267 (1982) [arXiv:hep-th/0001113].
4. J. Dai, R. G. Leigh and J. Polchinski, Mod. Phys. Lett. A **4**, 2073 (1989).
5. J. Polchinski, Phys. Rev. Lett. **75**, 4724 (1995) [arXiv:hep-th/9510017].
6. P. Horava and E. Witten, Nucl. Phys. B **475**, 94 (1996) [arXiv:hep-th/9603142].
7. A. Lukas, B. A. Ovrut, K. S. Stelle and D. Waldram, Phys. Rev. D **59**, 086001 (1999) [arXiv:hep-th/9803235].
8. N. Arkani-Hamed, S. Dimopoulos and G. Dvali, Phys. Lett. **B429**, 263 (1998) [hep-ph/9803315].
9. N. Arkani-Hamed, S. Dimopoulos and G. Dvali, Phys. Rev. D **59**, 086004 (1999) [hep-ph/9807344].
10. I. Antoniadis, N. Arkani-Hamed, S. Dimopoulos and G. Dvali, Phys. Lett. B **436**, 257 (1998) [hep-ph/9804398].
11. L. Randall and R. Sundrum, Phys. Rev. Lett. **83**, 3370 (1999) [arXiv:hep-ph/9905221].
12. L. Randall and R. Sundrum, Phys. Rev. Lett. **83**, 4690 (1999) [arXiv:hep-th/9906064].
13. J. M. Maldacena, Adv. Theor. Math. Phys. **2**, 231 (1998).
14. J. M. Maldacena, Int. J. Theor. Phys. **38**, 1113 (1999) [arXiv:hep-th/9711200].
15. G. W. Gibbons and D. L. Wiltshire, Ann. Phys. **167**, 201 (1986).
16. G. W. Gibbons and D. L. Wiltshire, Ann. Phys. **176**, 393 (1987) (Erratum).
17. D. Garfinkle, G. T. Horowitz and A. Strominger, Phys. Rev. D **43**, 3140 (1991).
18. D. Garfinkle, G. T. Horowitz and A. Strominger, Phys. Rev. D **45**, 3888 (1992) (Erratum).
19. N. I. Shakura and R. A. Sunyaev, Astron. Astrophys. **24**, 337 (1973).
20. D. Lynden-Bell, Nature **223**, 690 (1969).
21. J. Kormendy and D. Richstone, Ann. Rev. Astron. Astrophys. **33**, 581 (1995).

22. R. Schodel et al., Nature **419**, 694 (2002).
23. A. C. Fabian, K. Iwasawa, C. S. Reynolds and A. J. Young, Publ. Astron. Soc. Pac. **112**, 1145 (2000) [arXiv:astro-ph/0004366].
24. M. Gierlinski and C. Done, Mon. Not. Roy. Astron. Soc. **347**, 885 (2004) [arXiv:astro-ph/0307333].
25. E. G. Gimon and P. Horava, "Astrophysical violations of the Kerr bound as a possible signature of string theory," arXiv:0706.2873 [hep-th].
26. S. B. Giddings and S. D. Thomas, Phys. Rev. D **65**, 056010 (2002) [arXiv:hep-ph/0106219].
27. S. Dimopoulos and G. L. Landsberg, Phys. Rev. Lett. **87**, 161602 (2001) [arXiv:hep-ph/0106295].
28. C. M. Harris, P. Richardson and B. R. Webber, JHEP **0308**, 033 (2003) [arXiv:hep-ph/0307305].
29. C. M. Harris, M. J. Palmer, M. A. Parker, P. Richardson, A. Sabetfakhri and B. R. Webber, JHEP **0505**, 053 (2005) [arXiv:hep-ph/0411022].
30. G. L. Landsberg, J. Phys. G **32**, R337 (2006) [arXiv:hep-ph/0607297].
31. M. Cavaglia, R. Godang, L. Cremaldi and D. Summers, Comput. Phys. Commun. **177**, 506 (2007) [arXiv:hep-ph/0609001].
32. P. Kanti, Black holes at the LHC, arXiv:0802.2218 [hep-th].
33. S. S. Gubser, Phys. Rev. D **63**, 084017 (2001) [arXiv:hep-th/9912001].
34. H. L. Verlinde, Nucl. Phys. B **580**, 264 (2000) [arXiv:hep-th/9906182].
35. E. P. Verlinde and H. L. Verlinde, JHEP **0005**, 034 (2000) [arXiv:hep-th/9912018].
36. M. J. Duff and J. T. Liu, Phys. Rev. Lett. **85**, 2052 (2000).
37. M. J. Duff and J. T. Liu, Class. Quant. Grav. **18**, 3207 (2001) [arXiv:hep-th/0003237].
38. S. W. Hawking, Commun. Math. Phys. **43**, 199 (1975).
39. P. Candelas, Phys. Rev. D **21**, 2185 (1980).
40. D. N. Page, Phys. Rev. D **25**, 1499 (1982).
41. A. Karch and L. Randall, JHEP **05** (2001) 008, hep-th/0011156.
42. W. Israel, Nuovo Cimento Soc. Ital. Phys. B **44**, 4349 (1966).
43. D. Garfinkle and R. Gregory, Phys. Rev. D **41**, 1889 (1990).
44. B. Carter and R. Gregory, Phys. Rev. D **51**, 5839 (1995) [arXiv:hep-th/9410095].
45. B. Carter, Int. J. Theor. Phys. **40**, 2099 (2001) [arXiv:gr-qc/0012036].
46. A. Chamblin and G. W. Gibbons, Phys. Rev. Lett. **84**, 1090 (2000) [arXiv:hep-th/9909130].
47. S. B. Giddings, E. Katz and L. Randall, JHEP **0003**, 023 (2000) [arXiv:hep-th/0002091].
48. C. Charmousis, R. Gregory and V. A. Rubakov, Phys. Rev. D **62**, 067505 (2000) [arXiv:hep-th/9912160].
49. C. Charmousis, R. Gregory, N. Kaloper and A. Padilla, JHEP **0610**, 066 (2006) [arXiv:hep-th/0604086].
50. J. Garriga and T. Tanaka, Phys. Rev. Lett. **84**, 2778 (2000) [arXiv:hep-th/9911055].
51. P. Bowcock, C. Charmousis and R. Gregory, Class. Quant. Grav. **17**, 4745 (2000) [arXiv:hep-th/0007177].
52. L. A. Gergely and R. Maartens, Class. Quant. Grav. **19**, 213 (2002) [arXiv:gr-qc/0105058].
53. Z. Keresztes and L. A. Gergely, On the validity of the 5-dimensional Birkhoff theorem: The tale of a counterexample, arXiv:0712.3758 [gr-qc].
54. R. C. Myers and M. J. Perry, Annals Phys. **172**, 304 (1986).
55. H. A. Chamblin and H. S. Reall, Nucl. Phys. B **562**, 133 (1999) [arXiv:hep-th/9903225].
56. N. Kaloper, Phys. Rev. D **60**, 123506 (1999) [arXiv:hep-th/9905210].
57. P. Kraus, JHEP **9912**, 011 (1999) [arXiv:hep-th/9910149].
58. P. Binetruy, C. Deffayet and D. Langlois, Nucl. Phys. B **565**, 269 (2000) [arXiv:hep-th/9905012].
59. C. Csaki, M. Graesser, C. F. Kolda and J. Terning, Phys. Lett. B **462**, 34 (1999) [arXiv:hep-ph/9906513].
60. J. M. Cline, C. Grojean and G. Servant, Phys. Rev. Lett. **83**, 4245 (1999) [arXiv:hep-ph/9906523].
61. P. Kanti, I. I. Kogan, K. A. Olive and M. Pospelov, Phys. Lett. B **468**, 31 (1999) [arXiv:hep-ph/9909481];

62. M. J. Duff, Phys. Rev. D **9**, 1837 (1974).
63. R. Emparan, A. Fabbri and N. Kaloper, JHEP **0208**, 043 (2002) [arXiv:hep-th/0206155].
64. T. Tanaka, Prog. Theor. Phys. Suppl. **148**, 307 (2003) [arXiv:gr-qc/0203082].
65. W. Kinnersley and M. Walker, Phys. Rev. D **2**, 1359 (1970).
66. D. M. Eardley, G. T. Horowitz, D. A. Kastor and J. H. Traschen, Phys. Rev. Lett. **75**, 3390 (1995) [arXiv:gr-qc/9506041].
67. R. Emparan, Phys. Rev. Lett. **75**, 3386 (1995) [arXiv:gr-qc/9506025].
68. S. W. Hawking and S. F. Ross, Phys. Rev. Lett. **75**, 3382 (1995) [arXiv:gr-qc/9506020].
69. R. Gregory and M. Hindmarsh, Phys. Rev. D **52**, 5598 (1995) [arXiv:gr-qc/9506054].
70. R. Emparan, G. T. Horowitz and R. C. Myers, JHEP **0001**, 007 (2000) [arXiv:hep-th/9911043].
71. R. Emparan, G. T. Horowitz and R. C. Myers, JHEP **0001**, 021 (2000) [arXiv:hep-th/9912135].
72. R. Emparan, R. Gregory and C. Santos, Phys. Rev. D **63**, 104022 (2001) [arXiv:hep-th/0012100].
73. R. Gregory, Nucl. Phys. B **467**, 159 (1996) [arXiv:hep-th/9510202].
74. R. Gregory, JHEP **0306**, 041 (2003) [arXiv:hep-th/0304262].
75. A. Chamblin, S. W. Hawking and H. S. Reall, Phys. Rev. D **61**, 065007 (2000) [arXiv:hep-th/9909205].
76. R. Gregory, Class. Quant. Grav. **17**, L125 (2000) [arXiv:hep-th/0004101].
77. R. Gregory and R. Laflamme, Phys. Rev. Lett. **70**, 2837 (1993) [arXiv:hep-th/9301052].
78. R. Gregory and R. Laflamme, Nucl. Phys. B **428**, 399 (1994) [arXiv:hep-th/9404071].
79. A. Fabbri and G. P. Procopio, Class. Quant. Grav. **24**, 5371 (2007), 0704.3728 [hep-th].
80. C. Charmousis and R. Gregory, Class. Quant. Grav. **21**, 527 (2004) [arXiv:gr-qc/0306069].
81. H. Kudoh, T. Tanaka and T. Nakamura, Phys. Rev. D **68**, 024035 (2003) [arXiv:gr-qc/0301089].
82. T. Shiromizu and M. Shibata, Phys. Rev. D **62**, 127502 (2000) [arXiv:hep-th/0007203].
83. A. Chamblin, H. S. Reall, H. a. Shinkai and T. Shiromizu, Phys. Rev. D **63**, 064015 (2001) [arXiv:hep-th/0008177].
84. T. Wiseman, Phys. Rev. D **65**, 124007 (2002) [arXiv:hep-th/0111057].
85. P. Kanti and K. Tamvakis, Phys. Rev. D **65**, 084010 (2002) [arXiv:hep-th/0110298].
86. P. Kanti, I. Olasagasti and K. Tamvakis, Phys. Rev. D **68**, 124001 (2003) [arXiv:hep-th/0307201].
87. R. Casadio and L. Mazzacurati, Mod. Phys. Lett. A **18**, 651 (2003) [arXiv:gr-qc/0205129].
88. N. Tanahashi and T. Tanaka, JHEP **0803**, 041 (2008) [arXiv:0712.3799 [gr-qc]].
89. T. Shiromizu, K. i. Maeda and M. Sasaki, Phys. Rev. D **62**, 024012 (2000) [arXiv:gr-qc/9910076].
90. R. Maartens, Phys. Rev. D **62**, 084023 (2000) [arXiv:hep-th/0004166].
91. C. Germani and R. Maartens, Phys. Rev. D **64**, 124010 (2001) [arXiv:hep-th/0107011].
92. M. Bruni, C. Germani and R. Maartens, Phys. Rev. Lett. **87**, 231302 (2001) [arXiv:gr-qc/0108013].
93. N. Dadhich, R. Maartens, P. Papadopoulos and V. Rezania, Phys. Lett. B **487**, 1 (2000) [arXiv:hep-th/0003061].
94. R. Casadio, A. Fabbri and L. Mazzacurati, Phys. Rev. D **65**, 084040 (2002) [arXiv:gr-qc/0111072].
95. M. Visser and D. L. Wiltshire, Phys. Rev. D **67**, 104004 (2003) [arXiv:hep-th/0212333].
96. K. A. Bronnikov, V. N. Melnikov and H. Dehnen, Phys. Rev. D **68**, 024025 (2003) [arXiv:gr-qc/0304068].
97. T. Harko and M. K. Mak, Phys. Rev. D **69**, 064020 (2004) [arXiv:gr-qc/0401049].
98. R. Gregory, R. Whisker, K. Beckwith and C. Done, JCAP **0410**, 013 (2004) [arXiv:hep-th/0406252].
99. K. A. Bronnikov and S. W. Kim, Phys. Rev. D **67**, 064027 (2003) [arXiv:gr-qc/0212112].
100. D. Karasik, C. Sahabandu, P. Suranyi and L. C. R. Wijewardhana, Phys. Rev. D **70**, 064007 (2004) [arXiv:gr-qc/0404015].

101. V. P. Frolov, M. Snajdr and D. Stojkovic, Phys. Rev. D **68**, 044002 (2003) [arXiv:gr-qc/0304083].
102. D. Stojkovic, JHEP **0409**, 061 (2004) [arXiv:gr-qc/0409038].
103. A. Flachi and T. Tanaka, Phys. Rev. Lett. **95**, 161302 (2005) [arXiv:hep-th/0506145].
104. A. Flachi, O. Pujolas, M. Sasaki and T. Tanaka, arXiv:hep-th/0601174.
105. V. P. Frolov and D. Stojkovic, Phys. Rev. Lett. **89**, 151302 (2002) [arXiv:hep-th/0208102].
106. R. Gregory, V. A. Rubakov and S. M. Sibiryakov, Class. Quant. Grav. **17**, 4437 (2000) [arXiv:hep-th/0003109].
107. S. Creek, R. Gregory, P. Kanti and B. Mistry, Class. Quant. Grav. **23**, 6633 (2006) [arXiv:hep-th/0606006].
108. S. S. Seahra, Phys. Rev. D **71**, 084020 (2005) [arXiv:gr-qc/0501018].
109. C. Galfard, C. Germani and A. Ishibashi, arXiv:hep-th/0512001.
110. S. Lonsdale and I. Moss, Nucl. Phys. B **298**, 693 (1988).
111. A. Achucarro, R. Gregory and K. Kuijken, Phys. Rev. D **52**, 5729 (1995) [arXiv:gr-qc/9505039].
112. F. Bonjour, R. Emparan and R. Gregory, Phys. Rev. D **59**, 084022 (1999) [arXiv:gr-qc/9810061].
113. D. Stojkovic, Phys. Rev. Lett. **94**, 011603 (2005) [arXiv:hep-ph/0409124].
114. A. L. Fitzpatrick, L. Randall, and T. Wiseman, JHEP **11** (2006) 033, hep-th/0608208.
115. A. Karch and L. Randall, JHEP **0106**, 063 (2001) [arXiv:hep-th/0105132].
116. O. DeWolfe, D. Z. Freedman and H. Ooguri, Phys. Rev. D **66**, 025009 (2002) [arXiv:hep-th/0111135].
117. S. W. Hawking and D. N. Page, Commun. Math. Phys. **87**, 577 (1983).
118. T. Hirayama and G. Kang, Phys. Rev. D **64** (2001) 064010, hep-th/0104213.
119. A. Chamblin and A. Karch, Phys. Rev. D **72**, 066011 (2005) arXiv:hep-th/0412017.
120. R. Gregory, S. F. Ross and R. Zegers, arXiv:0802.2037 [hep-th].
121. U. Gen and M. Sasaki, Prog. Theor. Phys. **105**, 591 (2001) [arXiv:gr-qc/0011078].
122. R. Gregory and A. Padilla, Phys. Rev. D **65**, 084013 (2002) [arXiv:hep-th/0104262].
123. A. Padilla, Phys. Lett. B **528**, 274 (2002) [arXiv:hep-th/0111247].
124. S. de Haro, S. N. Solodukhin and K. Skenderis, Commun. Math. Phys. **217**, 595 (2001) [arXiv:hep-th/0002230].
125. A. Fabbri and G. P. Procopio, The holographic interpretation of hawking radiation, arXiv:0705.3363 [gr-qc].
126. T. Tanaka, Implication of classical black hole evaporation conjecture to floating black holes, arXiv:0709.3674 [gr-qc].
127. L. Grisa and O. Pujolas, Dressed domain Walls and holography, arXiv:0712.2786 [hep-th].
128. A. Flachi and T. Tanaka, Vacuum polarization in asymptotically anti-de Sitter black hole geometries, arXiv:0803.3125 [hep-th].

Chapter 8
Higher Order Gravity Theories and Their Black Hole Solutions

C. Charmousis

Abstract In this chapter, we will discuss a particular higher order gravity theory, Lovelock theory, that generalises in higher dimensions than 4, general relativity. After briefly motivating modifications of gravity, we will introduce the theory in question and we will argue that it is a unique, mathematically sensible, and physically interesting extension of general relativity. We will see, by using the formalism of differential forms, the relation of Lovelock gravity to differential geometry and topology of even-dimensional manifolds. We will then discuss a generic staticity theorem, quite similar to Birkhoff's theorem in general relativity, which will give us the charged static black hole solutions. We will examine their asymptotic behaviour, analyse their horizon structure and briefly their thermodynamics. For the thermodynamics we will give a geometric justification of why the usual entropy–area relation is broken. We will then examine the distributional matching conditions for Lovelock theory. We will see how induced four-dimensional Einstein–Hilbert terms result on the brane geometry from the higher order Lovelock terms. With the junction conditions at hand, we will go back to the black hole solutions and give applications for braneworlds: perturbations of codimension 1 braneworlds and the exact solution for braneworld cosmology as well as the determination of maximally symmetric codimension 2 braneworlds. In both cases, the staticity theorem evoked beforehand will give us the general solution for braneworld cosmology in codimension 1 and maximal symmetry warped branes of codimension 2. We will then end with a discussion of the simplest Kaluza–Klein reduction of Lovelock theory to a four-dimensional vector–scalar–tensor theory which has the unique property of retaining second-order field equations. We will comment briefly the non-linear generalisation of Maxwell's theory and scalar–tensor theory. We will conclude by listing some open problems and common difficulties.

C. Charmousis (✉)

LPT, Université de Paris-Sud, Bât. 210, 91405 Orsay CEDEX, France
Christos.Charmousis@th.u-psud.fr

Charmousis, C.: *Higher Order Gravity Theories and Their Black Hole Solutions.* Lect. Notes Phys. **769**, 299–346 (2009)
DOI 10.1007/978-3-540-88460-6_8

8.1 An Introduction to Lovelock Gravity

A convenient starting point for treating modifications of gravity is the fundamental building blocks of general relativity (GR) itself. According to Einstein's theory, gravitational interactions are described on a spacetime manifold by a symmetric metric tensor g endowed with a metric and torsion-free connection (by definition a Levi-Civita connection) that obeys Einstein's field equations. In component language these equations read as follows:

$$G_{ab} + \Lambda g_{ab} = 8\pi G T_{ab}, \qquad (8.1)$$

where the Einstein tensor $G_{ab} = R_{ab} - \frac{1}{2}g_{ab}R$ is given with respect to the Ricci curvature tensor R_{ab} and we have included Λ, the cosmological constant, and $T_{\mu\nu}$ the energy–momentum tensor. The field equations are acquired from the Einstein–Hilbert action,

$$S = \frac{1}{16\pi G} \int_{\mathcal{M}} d^4x\sqrt{-g}\, \mathcal{L}(\mathcal{M}, g, \nabla), \qquad (8.2)$$

where the Langrangian is the functional

$$\mathcal{L}(\mathcal{M}, g, \nabla) = -2\Lambda + R, \qquad (8.3)$$

by variation with respect to the metric g and adequate boundary conditions (see, for example, the appendix in [1]). The bare cosmological constant Λ is a free parameter of the theory.

We expect Einstein's theory to break down at very high energies close to the Planck scale, $m_{Pl}^2 = \frac{1}{16\pi G}$, where higher order curvature terms can no longer be neglected. Theories such as string theory or quantum loop gravity or again models, of extra dimensions, consider or model the effect of such modifications. GR on the other hand is very well tested at the solar system and by binary pulsar data in the regime of weak and strong gravity, respectively [2, 3]. However, recent cosmological experiments, or astrophysical data, such as galactic rotation curves, or even the Pioneer anomaly, appearing just beyond solar system scales, could question the validity of GR even at classical scales at large enough distances. In particular, recent cosmological evidence, coming essentially from type Ia supernovae explosions [4–6], points towards an *actually* accelerating universe. Looking at (8.1) there are three theoretical directions one could pursue in order to interpret this result. First, we can postulate the existence of an extremely small positive cosmological constant of value, $\Lambda_{now} \sim (10^{-3}eV)^4$, fixed by the actual Hubble horizon size, driving the acceleration in (8.1). To get an idea of how tiny this constant is note that this minute energy scale is most closely associated to the mass scale of neutrinos, 10^{-3} eV. Hence, although such a possibility[1] is the most economic of all, since we can fit actual multiple data with the use of a single parameter, it actually demands an enormous amount of fine-tuning. Indeed, from particle physics, the vacuum energy contributions to the total value of the cosmological constant are of the order of

---

[1] It is not an explanation until we find a precise mechanism of why it is there at all and why now.

the ultraviolet cut-off we impose on the QFT in question. It can therefore range as far up and close to the Planck scale (for discussions on the cosmological constant problem and ways to explain it see [7–9]). The "big" cosmological constant problem is precisely how all these vacuum energies associated to the GUT, SUSY, the standard model, etc. are fined-tuned each time to 0 by an exactly opposite in value bare cosmological constant Λ_{bare} appearing in (8.2). The unexplained small value of the cosmological constant Λ_{now} is then an additional two problems to add to the usual "big" cosmological constant problem, namely, why the cosmological constant is not cancelled exactly to zero and why do we observe it now.

A second alternative explanation one can consider is that the accelerated expansion is due to a cosmological fluid of as-yet-unknown matter, dubbed dark energy, such as a quintessence (scalar) field with some potential appearing in the right hand side of (8.2). One of the basic strengths of this approach is its simplicity and in some cases an interesting approach to the cosmological coincidence problem. Among its basic weaknesses, apart from the usual generic fine-tuning and stability problems to radiative corrections, is that if we sum up the as-yet-undiscovered matter sectors of the Universe, i.e. dark matter and dark energy, we conclude that only a mere 4% of the actual matter that constitutes our universe in its actual state has been discovered in ground-based accelerators! Although there exist theoretically motivated dark matter candidates, such as neutralinos or axions, stemming from well-motivated particle theories, our understanding of dark energy is rather poor. A third, far more ambitious alternative that is less well studied, far more constrained and admittedly less successful up to now is to modify the dynamics of geometry on the left hand side of (8.2) not only in the UV but also in the IR sector. This then would mean that Einstein's theory is also modified at large distances at the scale of the inverse Hubble scale of today as measured in a LFRW universe ($H_0/c = 7.566 \times 10^{-27} \mathrm{m}^{-1}$). This distance scale is enormous; to get an idea if we consider as our unit the distance of the earth to the sun (1 AU) we get[2] a present horizon distance of $10^{15} AU$!

Next question is, how do we modify gravity consistently? One can consider three basic types of modification which at the end of the day are not completely unrelated. Indeed, we can include additional fields or degrees of freedom, for example, scalar or vector, see, for example, [10–12], we can enlarge the parameter space where the theory evolves, for example the number of dimensions and the geometric connection in question (we include torsion, etc.) or again we can generalise the field equations.

In all cases, it is very important to fix basic consistency requirements for the modified gravity theory. To fix the discussion we can ask for three basic requirements: first we would like that the theory under consideration be consistent theoretically, for example we ask for sensible vacua of maximal symmetry, such as Minkowski, de Sitter or anti-de Sitter spacetime, and valid stable perturbation theory around these vacua. Second we need to satisfy all actual experimental constraints as for ordinary GR plus we need correct IR cosmological behaviour without the need for dark energy nor a cosmological constant. Third we want our theory to have the least number

---

[2] Astronomical units are interesting since most tests of general relativity are at distance scales of the solar system. Hence extrapolation to 10^{15} scales bigger of such experiments can be sometimes unjustified or at least questionable.

of degrees of freedom possible and to be naturally connected to GR theory.[3] For example, Brans–Dicke theory [10, 11] clearly passes the first and third tests whereas solar system constraints are rather restrictive [2, 3].

8.1.1 Lovelock's Theory

In these notes we will restrict our attention to a metric modification of gravity that generalises GR in higher dimensions. Remaining tangential to GR (principles) we consider a theory $\mathscr{L} = \mathscr{L}(M, g, \nabla)$, whose field variable is a single symmetric metric tensor g endowed with a Levi-Civita connection ∇. We ask for a divergence free geometric operator on the right hand side of (8.2), since we know that matter obeys the conservation equation $\nabla^\mu T_{\mu\nu} = 0$. Furthermore, in order to bypass perturbative stability constraints for the graviton, we ask for second-order field equations. These two properties are quite natural for our theory if we want to extend GR at the classical level but, we emphasise, not necessary at ultraviolet scales. Although higher derivatives generically introduce ghost degrees of freedom [13][4] around the vacuum [16], one may argue that these may disappear having correctly summed the infinite number of higher order corrections. This is precisely the case in string theory which although is a ghost free theory of two-dimensional surfaces embedded in 10 dimensions, at the effective action level, acquires (unphysical) ghost degrees of freedom because of the effective cut-off we impose. They are in general cured by arranging for the appearance of the relevant Lovelock term [17, 18] to the relevant order.

In $D = 4$ the only two-derivative metric modification to Einstein's theory is the addition of a cosmological constant term! In other words, any higher order curvature invariant either gives a pure divergence term, not contributing to the field equations, or adds higher order derivatives to the field equations. In higher dimensions this no-go extension theorem to GR is no longer true. It was the object of Lovelock's theorem [19] (see [20, 21] for the $D = 5$ case) to prove back in the 1970s that there exist theories containing precise higher order curvature invariants that actually modify Einstein's field equations (8.2) while satisfying $\nabla^\mu T_{\mu\nu} = 0$, in the face of modified Bianchi identities, and while keeping the order of the field equations down to second order in derivatives. The theory in question will be the subject of this brief study and gives in $D = 4$ precisely GR with a cosmological constant and in five and six dimensions reduces to Einstein–Gauss–Bonnet theory (EGB). Lovelock theory in a nutshell is the generalisation of general relativity in higher dimensions while keeping the full generality of GR in $D = 4$.

---

[3] The first and third requirements are not absolute, but one needs to be aware at least when a theory does not validate one of these.

[4] An exception to this rule is $f(R)$ theories [14] since they involve only functionals of the Ricci scalar. These theories have been known since a long time to be conformally equivalent to scalar–tensor theories; see, for example, [15].

Following Lovelock's proof of the uniqueness theorem [19] (see also the neat derivation of [22] using differential forms) significant interest developed in these higher dimensional relativity theories in the 1980s, with motivations originating from string theory [23, 24] and others originating from Kaluza–Klein cosmology [25, 26]. Initial interest in string theory was triggered by Zwiebach [16] who noted that second-order corrections to the Einstein–Hilbert action, other than the Gauss–Bonnet invariant, introduced a graviton ghost when considering perturbations around flat spacetime. Effective action calculations of certain string theories [17, 18] found that the leading (tree-level in g_S) α' string tension corrections could give rise, modulo field redefinitions to this order, to the Gauss–Bonnet invariant. Several nice papers appeared uncovering properties and analysing exact solutions of EGB [27–30] while slightly later tackling full Lovelock theory exact solutions [31]. More recently there have been a few exact solutions discussed [32–34] and some solution-generating techniques developed (see, for example, [35]). Discussions on issues of energy, stability and the Hamiltonian formalism have been carried out in [36–40].

Interest in Lovelock theory and in particular its five- and six-dimensional version, EGB theory, has attracted quite a lot of attention recently in the context of braneworlds (see Ruth Gregory's lecture notes [41]). Indeed from the braneworld point of view it would seem important to consider the general bulk theory rather than just GR in five or six dimensions and investigate if the four-dimensional braneworld picture remained GR like. In a nutshell (we will uncover the details later on) the Gauss–Bonnet term in the bulk action is similar in nature as is the induced gravity term [42, 43] to be added to the brane action. Loosely speaking, it thus enhances GR-type effects on the brane, adding also quite naturally a UV modification to the usual one identified by the fifth dimension. Perturbation theory in the bulk is exactly the same around a maximally symmetric spacetime and the main difference is the boundary conditions on the brane which become mixed [44, 45], similarly for those for induced gravity. In order to evaluate the correct boundary conditions, which give the braneworld gravitational spectrum, and hence determine the four-dimensional gravity and stability of the setup, the important difficulty one has to face is finding the extension of the Israel junction conditions [46–48] in the context of EGB theory. In fact the junction conditions can be calculated directly, for each solution in question, by a careful calculation of the distributional terms[5] [50–58] and [59] in the context of braneworld cosmology (see also [60]). The full covariant solution to the problem was first found by Davis, Gravannis and Willisson [61, 62] where it was realised that careful variation of the bulk metric with respect to the boundary term to this theory, discovered by Myers back in the 1980s [63], would give rise to the correct junction conditions. This is exactly similar to what happens when one considers careful variation of the Gibbons–Hawking boundary term, thus obtaining Israel's junction conditions. Using the junction conditions it was found that negative tension branes induced tachyonnic instabilities to braneworlds [44, 45] as well as important changes in the tensor perturbation amplitudes for braneworld

---

[5] As it was pointed out first in [49] distribution theory does not allow for ordinary multiplication and this can lead to erroneous junction conditions.

inflation [64]. Braneworld cosmology was further studied with particular focus on inflation [65] and IR modifications [39, 40, 66]. For codimension 2 braneworlds the relevant matching conditions were shown to give [67] precisely induced gravity terms on the brane plus extrinsic curvature corrections. However, up to now exact solutions or braneworld cosmology have not been found. We will discuss briefly here how one can obtain the maximally symmetric braneworld solutions in the context of EGB [68]. The full matching conditions of Lovelock theory, irrespective of codimension, were given in covariant formalism in [69, 107], and recently maximally symmetric braneworld examples to codimension 4 were found by Zegers [70].

In this review we will therefore study the basic properties and important characteristics of Lovelock theory. In the next section we will begin by introducing the theory in differential form language as it is the most adequate way to recognise its nice features and why it has unique properties. After this geometric parenthesis, we will study important exact solutions of this theory concentrating on the case of static black holes and solitons. We will see that a generalised version of Birkhoff's theorem holds as for GR, and we will then analyse the static black hole solutions, their thermodynamics and the solitonic solutions. After this we will discuss matching conditions of Lovelock theories and we will see that in this aspect Lovelock's theory is in principle a far richer extension to Einstein's theory in higher dimensions. Having done this we will discuss braneworld applications in codimension 1 and codimension 2. We will close by looking at the Kaluza–Klein reduction of these theories and the type of scalar–vector–tensor theories they predict [71–73].

8.1.2 Basic Definitions for Lovelock Theory: Differential Form Language

Our aim in this section is to construct the higher order curvature densities which will be the building blocks of Lovelock theory and explain what makes them special. Indeed, in component language these will turn out to be precise but seemingly ad hoc linear combinations or powers of the Riemann, Ricci tensor and Ricci scalar. We will thus use differential form language where we will see that they are indeed powers of the curvature 2 form with a precise and clear geometric interpretation (see also [74, 75]).

Let $(\mathcal{M}, g, \nabla)$ denote a D-dimensional spacetime manifold $\mathcal{M}$ endowed with a smooth[6] spacetime metric g. The connection ∇ is taken to be a Levi-Civita connection. To every point P of spacetime M, we associate a local orthonormal basis[7] of the tangent space $T_P M$, (e_A), with $A = 1, ..., D$ such that

$$g(e_A, e_B) = \eta_{AB}. \tag{8.4}$$

---

[6] By smooth we mean here metrics of at least C^2 regularity. This will be relaxed to piecewise C^1 when we look at braneworlds, allowing for distributional matter sources.

[7] For a more complete account on the geometrical notions used here and precise examples, see [76, 77].

Equivalently the metric can be expressed as $g = \eta_{AB}\theta^A \otimes \theta^B = g_{ab}dx^a dx^b$ where the 1-forms θ_A are precisely dual to the basis vectors e_A, since $\theta^A(e_B) = \delta^A_B$. The metric components in a coordinate frame dx^a (we use small-case Latin letters for a coordinate frame and upper-case latin letters for an orthonormal frame) are thus $g_{ab} = \theta^A_a \theta^B_b \eta_{AB}$ where $\theta^A_a dx^a = \theta^A$. The dual 1-forms θ^A form a natural basis of the vector space of 1-forms $\Omega^{(1)}(TM)$. In turn the antisymmetric product of 1-forms θ^A can be used in order to construct a basis of the higher order forms acting on TM. For any k-form w in $\Omega^{(k)}(TM)$, where $0 \leq k \leq D$, can be written as

$$w = w_{A_1 \cdots A_k}\theta^{A_1} \wedge \cdots \wedge \theta^{A_k}, \tag{8.5}$$

with $w_{A_1 \cdots A_k}$ some smooth function. Following this simple recipe we can define a $(D-k)$-form,

$$\theta^\star_{A_1 \cdots A_k} = \frac{1}{(D-k)!}\epsilon_{A_1 \cdots A_k A_{k+1} \cdots A_D}\theta^{A_{k+1}} \wedge \cdots \wedge \theta^{A_D}, \tag{8.6}$$

where $\epsilon_{A_1 \cdots \cdots A_D}$ is totally antisymmetric in its D indices and $\epsilon_{12 \cdots D} = 1$. This quantity is called the Hodge dual of the basis $\theta^{A_1} \wedge \cdots \wedge \theta^{A_k}$ of $\Omega^{(k)}(TM)$. It defines a dual basis of forms in $\Omega^{(D-k)}(TM)$. We can therefore write the Hodge dual of any k-form as

$$\star : \qquad\qquad \Omega^{(k)}(TM) \qquad\qquad \rightarrow \qquad\qquad \Omega^{(n-k)}(TM)$$
$$\omega = \omega_{A_1 \cdots A_k}\theta^{A_1} \wedge \cdots \wedge \theta^{A_k} \qquad \rightarrow \qquad \star\omega = \omega^{A_1 \cdots A_k}\theta^\star_{A_1 \cdots A_k}. \tag{8.7}$$

The wedge product of any form with its dual is a D-form, which is by construction proportional to the volume element of spacetime, which we note as $\theta^\star$. Obviously for $k > D$ all forms are identically zero. A useful identity is

$$\theta^B \wedge \theta^\star_{A_1 \cdots A_k} = \delta^B_{A_k}\theta^\star_{A_1 \cdots A_{k-1}} - \delta^B_{A_{k-1}}\theta^\star_{A_1 \cdots A_{k-2}A_k} + \cdots + (-1)^{k-1}\delta^B_{A_1}\theta^\star_{A_2 \cdots A_k}. \tag{8.8}$$

Having constructed this tower of k-forms, for $0 \leq k \leq D$ we now need define two quantities: the connection 1-form and the curvature 2-form. The connection 1-form of M, $\omega^A{}_B$ which replaces the usual Christophel symbols in coordinate language, is defined by

$$d\theta^A = -\omega^A{}_B \wedge \theta^B \tag{8.9}$$

since we have assumed a torsionless connection. On the other hand, the spacetime curvature 2-form is linked to the connection 1-form via the (second Cartan structure) equation,

$$\mathscr{R}^A{}_B = d\omega^A{}_B + \omega^A{}_C \wedge \omega^C{}_B. \tag{8.10}$$

The ambient curvature 2-form is related to the Riemann tensor components by

$$\mathscr{R}^A{}_B = \frac{1}{2} R^A{}_{BCD} \theta^C \wedge \theta^D, \tag{8.11}$$

with respect to the spacetime Riemann tensor.[8]

Langrangian densities appear under spacetime integrals; they are therefore by definition going to be D-forms. Hence in order to build such densities out of the curvature 2-form (8.11) and its powers, symbolically $\mathscr{R}^k$, we need to construct D-forms by correctly completing with the relevant Hodge duals (8.7). The higher the dimension of $\mathscr{M}$, the higher the possible powers of curvature we can consider. In differential form language it is straightforward to build this way the kth Lovelock Lagrangian density $\mathscr{L}_{(k)}$ which is a D-form defined by

$$\mathscr{L}_{(k)} = \mathscr{R}^{A_1 B_1} \wedge \cdots \wedge \mathscr{R}^{A_k B_k} \wedge \theta^\star_{A_1 B_1 \dots A_k B_k} = \bigwedge_{i=1}^{k} \mathscr{R}^{A_i B_i} \wedge \theta^\star_{A_1 B_1 \dots A_k B_k}, \tag{8.12}$$

and we stress that k stands for the power of curvature. Clearly, $\mathscr{L}_{(0)}$ is the volume element, giving rise to a cosmological constant whereas it is easy to check using (8.8) that

$$\mathscr{L}_{(1)} = \mathscr{R}^{A_1 B_1} \wedge \theta^\star_{A_1 B_1} = R\theta^\star \tag{8.13}$$

is the Ricci scalar density and

$$\mathscr{L}_{(2)} = \mathscr{R}^{A_1 B_1} \wedge \mathscr{R}^{A_2 B_2} \wedge \theta^\star_{A_1 B_1 A_2 B_2} = (R^{ABCD} R_{ABCD} - 4 R^{AB} R_{AB} + R^2)\theta^\star \tag{8.14}$$

is the Gauss–Bonnet density which we will denote by $\hat{G}$. Note that for $k > D/2$, (8.12) vanishes so that if $D = 4$ say, then $\mathscr{L}_{(0)}$, $\mathscr{L}_{(1)}$ and $\mathscr{L}_{(2)}$ are the only terms present in the action (although $\mathscr{L}_{(2)}$, as we will now see, turns out to be trivial). According to Lovelock's theorem [19], given a metric theory, $\mathscr{L}_{(k)}$ are the unique densities, made out of $(\mathscr{M}, \nabla, g)$ as defined in the beginning of this section, which allow for energy conservation and second-order field equations.

8.1.3 Lovelock Densities and Their Geometric Interpretation

So what is special about these densities? The answer lies in differential geometry. Indeed, if D is even, for $k = D/2$, the Hodge dual is trivial and the Lovelock density $\mathscr{L}_{D/2}$ reduces to

$$\mathscr{L}_{(D/2)} = \bigwedge_{i=1}^{(D/2)} \mathscr{R}^{A_i B_i} \epsilon_{A_1 B_1 \dots A_{D/2} B_{D/2}}. \tag{8.15}$$

This D-form can be recognised as the generalised Euler density for an even-dimensional compact manifold $\mathscr{M}$ [78, 79]. It is a geometric quantity whose integral over $\mathscr{M}$ is a topological invariant (see, for example, [80]):

---

[8] As a word of caution, the Riemann tensor components are given here with respect to the orthonormal basis and may differ, as is the case for stationary spacetimes for example, to coordinate basis components of the same tensor.

$$\chi[\mathcal{M}] = \frac{1}{(4\pi)^{D/2}(D/2)!} \int_{\mathcal{M}} \mathcal{L}_{(D/2)} . \tag{8.16}$$

This relation has important consequences since it yields a relation between a geometric quantity involving curvature with the topology of the manifold $\mathcal{M}$. This relation is familiar for the case of $D = 2$ where the $k = 1$ density, i.e. the Ricci scalar or Gauss curvature, gives the usual Euler characteristic for two-dimensional compact surfaces with no boundary. This familiar formula then relates surface geometry with a topological number defining topologically equivalent classes of surfaces,

$$\chi[\mathcal{M}] = \frac{1}{(4\pi)} \int_{\mathcal{M}} R = 2 - 2h , \tag{8.17}$$

where h denotes the number of handles of $\mathcal{M}$ (see Fig. 8.1). The above relation gives in essence the Gauss–Bonnet theorem. This classification is also familiar from string theory where χ is related to the string coupling $g_s = e^{\chi\phi}$ giving the string surface diagrams (rather than Feynmann diagrams as for point particles) for orientable strings. Variation of this quantity with respect to the local frame θ^A and use of the Bianchi identities on the curvature 2-form give us that $\mathcal{L}_{(D/2)}$ is locally an exact form. This in turn tells us that variation of (8.15) gives no contribution to the field equations (supposing $\mathcal{M}$ compact). In turn for $D = 4$, $\mathcal{L}_{(2)}$ stands for the Gauss–Bonnet density whose integral is now in turn the four-dimensional Euler characteristic.

Therefore we see that the special feature of Lovelock theory is that its Langrangian densities are *dimensional continuations* of the Chern–Euler forms[9] which at lower even dimension are topological invariants. For completeness we note that for an even dimensional manifold with a boundary, the Chern–Gauss–Bonnet relation is corrected by a Chern boundary form [78, 79]:

$$\chi[\mathcal{M}] = \frac{1}{(4\pi)^{D/2}(D/2)!} \left[\int_{\mathcal{M}} \mathcal{L}_{(D/2)} + \int_{\partial\mathcal{M}} \Phi_{(D/2-1)} \right] , \tag{8.18}$$

where the $(D-1)$-form $\Phi_{(D/2-1)}$ reads [76–80] as

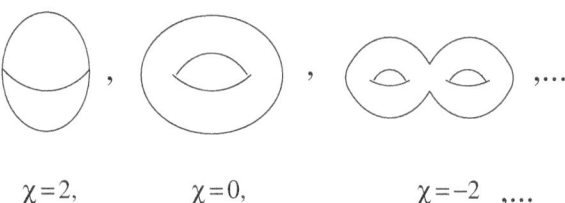

$\chi = 2,$ \qquad\qquad $\chi = 0,$ \qquad\qquad\qquad $\chi = -2$,...

Fig. 8.1 Any two-dimensional compact surface can be continuously deformed to one of its Euler classes parametrised by χ

---

[9] By dimensional continuation we mean that we take k powers of the curvature 2-form and conveniently "multiply" by the relevant Hodge dual as in (8.12).

$$\Phi_{(D/2-1)} = \sum_{m=0}^{D/2-1} \frac{D \cdot (D-2) \cdots 2(m+1)}{1 \cdot 3 \cdots (D-2m-1)} \epsilon_{\mu_1 \cdots \mu_{D-1}}$$

$$\left(\bigwedge_{l=0}^{m} \mathscr{R}^{\mu_{2l-1}\mu_{2l}} \right) \wedge \left(\bigwedge_{l=0}^{D-1} \mathscr{K}^{\mu_l}{}_N \right), \tag{8.19}$$

where $\mathscr{K}^{\mu_l}{}_N$ is a 1-form associated to the extrinsic curvature of the boundary $\partial \mathscr{M}$ [74], [69, 107]. It turns out that the relevant boundary terms for Lovelock theory are again the dimensional continuations of the Chern boundary forms which are multiples of the induced Riemann tensor on the brane and extrinsic curvature of the boundary. This is a rather useful statement when treating braneworlds since variation with respect to the local frame gives the codimension 1 junction conditions [69, 107] as we will see later on. The dimensionally continued boundary term associated to the Euler characteristic (8.13), $k = 1$, is the Gibbons–Hawking boundary term [81] whereas the boundary term associated to the Gauss–Bonnet invariant is the Myers boundary term [63]. For details see [69, 107].

The integral of the sum, for $k < D/2$, of all Lovelock densities $\mathscr{L}_{(k)}$ is the most general classical action for $\mathscr{M}$, yielding up to second-order field equations for the metric tensor. This is the Lovelock action

$$S_D = \int_M \sum_{k=0}^{[(D-1)/2]} \alpha_k \mathscr{L}_{(k)}, \tag{8.20}$$

where the brackets stand for the integer part. The first three terms of this sum

$$S_D = \int_M \left(\alpha_0 \theta^\star + \alpha_1 \mathscr{R}^{AB} \wedge \theta^\star_{AB} + \alpha_2 \mathscr{R}^{AB} \wedge \mathscr{R}^{CD} \wedge \theta^\star_{ABCD} + \cdots \right) \tag{8.21}$$

are, respectively, the cosmological constant, Einstein–Hilbert and Gauss–Bonnet terms, yielding the generalisation of the Einstein–Hilbert action in $D = 5$ and 6 dimensions.

A variation of the action (8.20) including matter, with respect to the frame, gives the Lovelock equations:

$$\sum_{k=0}^{[(D-1)/2]} \alpha_k \mathscr{E}_{(k)A} = -2T_{AB}\theta^{\star B}, \tag{8.22}$$

where $\mathscr{E}_{(k)A}$ is the kth Lovelock $(D-1)$-form:

$$\mathscr{E}_{(k)A} = \bigwedge_{i=1}^{k} \mathscr{R}^{A_i B_i} \wedge \theta^\star_{AA_1 B_1 \ldots A_k B_k}, \tag{8.23}$$

and we have chosen the normalisation according to

$$-\frac{1}{2}\mathscr{R}^{A_1 B_1} \wedge \theta^\star_{CA_1 B_1} = G^A_C \theta^\star_A,$$

so that the equations of motion read, in component formalism, $G_{AB} + \cdots = T_{AB}$. For $k = 2$ we get,

$$\mathcal{E}_{(2)A} = \mathcal{R}^{A_1 B_1} \wedge \mathcal{R}^{A_2 B_2} \theta^\star_{AA_1 B_1 A_2 B_2} = H^A_C \theta^\star_C, \tag{8.24}$$

where applying (8.8) and (8.11) we obtain in component language

$$H_{ab} = \frac{g_{ab}}{2}(R_{efcd}R^{efcd} - 4R_{cd}R^{cd} + R^2) - 2RR_{ab} + 4R_{ac}R^c_{\ b}$$
$$+ 4R_{cd}R^c_{\ a}{}^d_{\ b} - 2R_{acd\lambda}R_b^{\ cd\lambda}$$

the order 2 Lovelock tensor. In what follows we will explicitly set $\alpha_0 = -2\Lambda$, $\alpha_1 = 1$ (expect in the section where we will set $\alpha_1 = \zeta$) and $\alpha_2 = \hat{\alpha}$ for the Gauss–Bonnet coupling.

8.2 Exact Solutions

8.2.1 A Staticity Theorem

We argued in the previous section that Lovelock theory is the natural generalisation of GR in higher dimensions. Which of the classical properties of GR remain true when we switch on the extra Lovelock densities. One main difference is that for $k \geq 2$, i.e. once we allow for the Gauss–Bonnet term (8.14), the Weyl tensor appears in the field equations. Therefore Ricci flat solutions are no longer solutions of Lovelock gravity if they are not conformally flat. This means in particular that construction methods quite common in higher dimensional GR are not going to give generically solutions for Lovelock theory. For example, it is not clear how one can obtain a simple solution such as the black string [82], for Lovelock theory [35], and only a linear correction is known [83]. On the other hand we can question the status of classical GR theorems in Lovelock theory such as that of Birkhoff's theorem. Is a version of this theorem still true in Lovelock theory?

Consider the D-dimensional EGB action

$$S_{(2)} = \frac{M^{D-2}}{2} \int d^D x \sqrt{-g} \left[R - 2\Lambda + \hat{\alpha}\hat{G} \right], \tag{8.25}$$

where M is the fundamental mass scale of the D-dimensional theory and $\hat{G}$ is the Gauss–Bonnet density (8.14). We note by $2\Lambda = -k^2(D-1)(D-2)$ the bulk bare negative cosmological constant ($2\Lambda = a^2(D-1)(D-2)$ is the positive cosmological constant); $\hat{\alpha}$ is the Gauss–Bonnet coupling which has dimensions of length squared. We will set $\alpha = (D-3)(D-4)\hat{\alpha}$ to somewhat simplify notation. This action is that of Lovelock theory in component language for $D = 5$ or 6 dimensions. Let us for simplicity and without loss of generality [84] stick to $D = 5$ for the rest of this section in order to expose the staticity theorem. We will comment on the general result also involving an electric/magnetic field at the beginning of the next section.

Suppose spacetime has constant three-dimensional spatial curvature. This is the basic hypothesis we will make here. This is a slight generalisation from the common assumption of spherical symmetry in Birkhoff's theorem. A general metric anzatz for the given symmetry is

$$ds^2 = e^{2v(t,z)}B(t,z)^{-\frac{D-3}{D-2}}(-dt^2+dz^2)+B(t,z)^{\frac{2}{D-2}}\left(\frac{d\chi^2}{1-\kappa\chi^2}+\chi^2 d\Omega_{D-3}^2\right),$$
(8.26)

where $B(t,z)$ and $v(t,z)$ are the unknown component fields of the metric and $\kappa = 0,\pm 1$ is the normalised curvature of the three-dimensional homogeneous and isotropic surfaces. We choose to use the conformal gauge in order to take full advantage of the two-dimensional conformal transformations in the $t-z$ plane. The field equations we are seeking to solve are found by varying the above action (8.25) with respect to the background metric or by using (8.22) from the previous section and read as

$$\mathcal{E}_{ab} = G_{ab}+\Lambda g_{ab}-\alpha H_{ab} = 0,$$
(8.27)

where H_{ab} is given by (8.25).

It is rather useful to review and compare the equivalent system [85], [41] in Einstein gravity for $\alpha = 0$ where

$$R_{ab} = -\frac{(D-3)\Lambda}{(D-2)}g_{ab}.$$

Pass to light-cone coordinates,

$$u = \frac{t-z}{2}, \qquad v = \frac{t+z}{2},$$
(8.28)

and take the combination $R_{tt}+R_{zz}\pm 2R_{tz} = 0$; one obtains the equations which read as follows:

$$B_{,uu}-2v_{,u}B_{,u} = 0,$$
(8.29)
$$B_{,vv}-2v_{,v}B_{,v} = 0.$$
(8.30)

Note then that these are ordinary differential equations with respect to u and v. They are directly integrable giving

$$B = B(U+V) \qquad e^{2v} = B'U'V',$$
(8.31)

where $U = U(u)$ and $V = V(v)$ are arbitrary functions of u and v and a prime stands for the total derivative of the function with respect to its unique variable. We refer to (8.29) as the integrability conditions. Using a two-dimensional conformal transformation which is a symmetry of (8.26),

$$U = \frac{\bar{z} - \bar{t}}{2}, \qquad V = \frac{\bar{z} + \bar{t}}{2}$$

gives that the solution is locally static $B = B(\bar{z})$ and the equivalent of a generalised Birkhoff's theorem is therefore true. Starting from a general time- and space-dependant metric, spacetime has been shown to be locally static or equivalently that there exists a locally timelike Killing vector field (here $\frac{\partial}{\partial \bar{t}}$). By use of the remaining field equations we can then find the form of B, leading after coordinate transformation to a static black hole solution of horizon curvature κ.

Let us now take $\alpha \neq 0$. In analogy to the previous case let us take the combination, $\mathscr{E}_{tt} + \mathscr{E}_{zz} \pm 2\mathscr{E}_{tz} = 0$. On passing to light-cone coordinates (8.28) we get after some manipulations

$$\left(9B^{4/3} e^{2v} + 18\alpha\kappa B^{2/3} e^{2v} + 2\alpha B_{,u} B_{,v} \right) \left(B_{,uu} - 2v_{,u} B_{,u} \right) = 0,$$

$$\left(9B^{4/3} e^{2v} + 18\alpha\kappa B^{2/3} e^{2v} + 2\alpha B_{,u} B_{,v} \right) \left(B_{,vv} - 2v_{,v} B_{,v} \right) = 0 . \qquad (8.32)$$

Note how the Gauss–Bonnet terms factorise nicely leaving the integrability equations (8.29) we had in the absence of α.

The degenerate case where either $B_{,u} = 0$ or $B_{,v} = 0$ corresponds to flat solutions. For $B_{,u} \neq 0$ and $B_{,v} \neq 0$ the situation is clear: either we have static solutions and the staticity theorem holds as in the case above or we will have

$$e^{2v} = \frac{2\alpha(B_{,z}^2 - B_{,t}^2)}{9B^{2/3}(B^{2/3} + 4\alpha\kappa)} . \qquad (8.33)$$

Let us briefly examine the latter case, which we will call class I solution [28, 29] [59]. The two remaining field equations $\mathscr{E}_{\chi\chi} = 0$ and $\mathscr{E}_{tt} - \mathscr{E}_{zz} = 0$ are solvable iff we have the simple algebraic relation

$$4\alpha k^2 = 1. \qquad (8.34)$$

This is quite remarkable: if the coupling constants obey (8.34) then the B field is an *arbitrary* function of space and time; in other words the field equations do not determine the metric functions. Therefore, strictly speaking, Birkhoff's theorem does not hold for non-zero cosmological constant.[10] Setting $B(t,z) = R^3(t,z)$ the class I metric reads as follows:

$$ds^2 = \frac{R_{,z}^2 - R_{,t}^2}{\kappa + \frac{R^2}{2\alpha}}(-dt^2 + dz^2) + R^2 \left(\frac{d\chi^2}{1 - \kappa\chi^2} + \chi^2 d\Omega_{II}^2 \right) . \qquad (8.35)$$

This solution has generically a curvature singularity for $R_{,z} = \pm R_{,t}$. The class I static solutions are given by

---

[10] Note however that for a non-zero charge Q and spherical symmetry ($\kappa = 1$) Birkhoff's theorem is always true as was first shown by Wiltshire [30].

$$ds^2 = -\frac{A(R)^2}{\kappa + \frac{R^2}{2\alpha}}dt^2 + \frac{dR^2}{\kappa + \frac{R^2}{2\alpha}} + R^2\left(\frac{d\chi^2}{1-\kappa\chi^2} + \chi^2 d\Omega_{II}^2\right), \qquad (8.36)$$

with $A = A(R)$ now an arbitrary function of R.

In order to obtain t- and z-dependent solutions it suffices to take the functional R to be a non-harmonic function. Take for instance $R = exp(f(t)+g(z))$, with f and g arbitrary functions of a timelike and spacelike coordinate, respectively. Let us also assume $\kappa = 0$ for simplicity; the class I metric in proper time reads as

$$ds^2 = -d\tau^2 + \frac{2\alpha dg^2}{1+2\alpha f_{,\tau}^2} + e^{2(f+g)}\left(\frac{d\chi^2}{1-\kappa\chi^2} + \chi^2 d\Omega_{II}^2\right). \qquad (8.37)$$

Note here again that f is an arbitrary function of time and g an arbitrary function of space. The fine-tuning relation between α and k actually corresponds to a case of enhanced symmetry often refereed to as Chern–Simons gravity (for odd-dimensional spacetimes [86]).

On the other hand if (8.34) does not hold then Birkhoff's theorem remains true in the presence of the Gauss–Bonnet terms, i.e. *the general solution assuming the presence of a cosmological constant in the bulk and three-dimensional constant curvature surfaces is static if and only if (8.34) is not satisfied.* In this case the remaining two equations give the same ordinary differential equation for $B(U+V)$ which after one integration reads as

$$B' + 9B^{2/3}(k^2 B^{2/3} + \kappa) + 9\alpha\left(\frac{B'}{9B^{2/3}} + \kappa\right)^2 = 9\mu, \qquad (8.38)$$

where μ is an arbitrary integration constant. Then by making B the spatial coordinate and setting $B^{1/3} = r$ we get the solution discovered and discussed in detail by Boulware–Deser [27] ($\kappa = 1$) and Cai [87] ($\kappa = 0, -1$):[11]

$$ds^2 = -V(r)dt^2 + \frac{dr^2}{V(r)} + r^2\left(\frac{d\chi^2}{1-\kappa\chi^2} + \chi^2 d\Omega_{II}^2\right), \qquad (8.39)$$

where $V(r) = \kappa + \frac{r^2}{2\alpha}[1 \pm \sqrt{1 - 4\alpha k^2 + 4\frac{\alpha\mu}{r^4}}]$ and μ is an integration parameter related to the gravitational mass. The maximally symmetric solutions are obtained by setting $\mu = 0$. We will analyse in detail these solutions in the following section.

To close notice how (8.34) is a particular "end" point for (8.39) since the maximally symmetric solution is defined only for $1 \geq 4\alpha k^2$ (for $\alpha < 0$ there is no such restriction). We can deduce in all generality that for $1 \geq 4\alpha k^2$ there is a unique static solution (8.39). When (8.34) is satisfied and $\mu = 0$, the two branches coincide (we have an infinity of solutions) and $V = \kappa \pm \sqrt{\mu/\alpha} + \frac{r^2}{\alpha}$ is then a particular class I solution (8.34) very similar to the BTZ three-dimensional black holes. For $1 \leq 4\alpha k^2$ no solutions exist. We will come back to the six-dimensional version of these black

---

[11] We have kept the same label as in (8.26) for the rescaled time coordinate.

holes in a moment. Before doing so let us see briefly how they generalise for the full Lovelock theory.

8.2.2 Lovelock Black Holes

The staticity theorem we evoked in the previous section is generalised without major difficulty for the general Lovelock theory in arbitrary dimension and in the presence of an Abelian gauge field [84]. That is for the theory involving a Lovelock action (8.20) with a Maxwell gauge field:

$$S_D^{EM} = S_D - \frac{1}{4} \int d^D x \sqrt{-g} F_{ab} F^{ab}. \tag{8.40}$$

We obtain from the field equations that apart from the pathological cases of class 1 there is a unique solution given by [31]

$$ds^2 = -V(r)dt^2 + \frac{dr^2}{V(r)} + r^2 \left(\frac{d\chi^2}{1 - \kappa x^2} + \chi^2 d\Omega_{D-3}^2 \right), \tag{8.41}$$

where the electric field strength is $F = \frac{q^2}{4\pi r^{2(D-2)}} \, dt \wedge dr$. The metric potential reads $V = \kappa - r^2 f(r)$ where f is a solution of the kth order algebraic equation:

$$P(f) = \sum_{l=0}^{k} \hat{\alpha}_l f^l = \frac{\mu}{r^{D-1}} - \frac{Q^2}{r^{2D-4}}, \tag{8.42}$$

where $k = \left[\frac{D-1}{2} \right]$ is the order of the Lovelock theory. It is clear that the higher the dimension of spacetime, the more the terms in the Lovelock action, and hence the higher the order of the equation (8.42). The k possible roots of the polynomial $P(f)$ for $\kappa = 1$ actually give us the maximally symmetric vacua of the theory [31]. Note that even in the absence of a bare cosmological constant these vacua can be flat or of positive or negative curvature and their magnitude depends on the normalised Lovelock coupling constants:

$$\hat{\alpha}_0 = \frac{\alpha_0}{\alpha_1} \frac{1}{(D-1)(D-2)}, \qquad \hat{\alpha}_1 = 1,$$

$$\hat{\alpha}_l = \frac{\alpha_l}{\alpha_1} \prod_{n=3}^{2l} (D-n), \text{for} \quad l > 1 . \tag{8.43}$$

Positive roots correspond to de Sitter vacua whereas negative roots to anti-de Sitter vacua. It is interesting to note that in the presence of a bare cosmological constant $\alpha_0 \neq 0$ maximally symmetric vacua may not exist at all. For zero bare cosmological constant however the flat vacuum is always solution. Solutions of (8.42) have complex horizon structures and have been analysed by Myers and Simon [31] for $Q = 0$.

8.2.3 Einstein–Gauss–Bonnet Black Holes

For simplicity, let us now truncate Lovelock theory at $k = 2$, i.e. neglect higher order terms other than the Gauss–Bonnet invariant. We will follow for most of this section the analysis of [31]. We therefore consider the action (8.40) for up to $k = 2$:

$$S_2^{EM} = S_{(2)} - \frac{1}{4} \int d^D x \sqrt{-g} F_{ab} F^{ab}. \tag{8.44}$$

Take a $(D-2)$-dimensional space of maximal symmetry and therefore of constant curvature parametrised by $\kappa = 0, -1, 1$:

$$h^S = h_{\mu\nu}^S dx^\mu dx^\nu = \frac{d\chi^2}{1 - \kappa\chi^2} + \chi^2 d\Omega_{(D-3)}^2. \tag{8.45}$$

These are the maximal symmetry spaces we considered in the previous section and express the constant curvature geometry, normalised to κ, of the horizon surface. If on the other hand we perform a careful Wick rotation to h^S we can construct h^L, which is of Lorentzian signature, and give spacetimes which are $(D-2)$-dimensional sections of Minkowski, adS and dS. Therefore, the staticity theorem of the previous section tells us that any D-dimensional spacetime metric admitting $(D-2)$-dimensional maximal sub-spaces h^S (or sub-spacetimes h^L) which is a solution of the field equations emanating from (8.25) is locally isometric to

$$ds^2 = -V(r)dt^2 + \frac{dr^2}{V(r)} + r^2 h_{\mu\nu}^S dx^\mu dx^\nu, \tag{8.46}$$

with electric field strength

$$F = \frac{q^2}{4\pi r^{2(D-2)}} dt \wedge dr, \tag{8.47}$$

or, modulo a double Wick rotation, to

$$ds^2 = V(r)d\theta^2 + \frac{dr^2}{V(r)} + r^2 h_{\mu\nu}^L dx^\mu dx^\nu, \tag{8.48}$$

with magnetic field strength

$$F = \frac{p^2}{4\pi r^{2(D-2)}} d\theta \wedge dr. \tag{8.49}$$

In the latter case the theorem gives a local axial Killing vector ∂_θ and concerns locally axially symmetric solutions. This case has been explicitly treated for the case of general relativity $k = 1$ in [88]. We will come back to this simple yet powerful result to give the maximally symmetric cosmic string metrics in $D = 4$ and maximally symmetric braneworlds of codimension 2 in $D = 6$.

Since we want to search for black hole criteria we concentrate on the static case for the rest of this subsection. The potential reads (as can be easily verified from (8.42)) as

$$V(r) = \kappa + \frac{r^2}{2\alpha}\left[1 + \epsilon\sqrt{1 + 4\alpha\left(a^2 + \frac{\mu}{r^{D-1}} - \frac{Q^2}{r^{2(D-2)}}\right)}\right],\tag{8.50}$$

with integration constants

$$Q^2 = \frac{q^2}{2\pi(D-2)(D-3)}, \qquad \mu = \frac{16\pi GM}{(D-2)\Sigma_\kappa},\tag{8.51}$$

where q is the charge, M is the AD or ADM mass of the solution and Σ_κ is the horizon area.

First notable fact is the ambiguity of the vacuum which is parametrised by $\epsilon = \pm 1$ and gives rise to two branches of solutions. Indeed setting $\mu = Q^2 = 0$, $\kappa = 1$ gives us the possible vacua of the theory (8.25). In fact we note that if we set the bare cosmological constant to be zero $a = 0$, we note that we do not only obtain the flat vacuum. For $\epsilon = 1$, we actually asymptote anti-de Sitter space for $\alpha > 0$ and dS space for $\alpha < 0$. We will refer to this vacuum as the Gauss–Bonnet branch. Indeed the effective cosmological constant is $\Lambda_{eff} = \frac{(D-1)(D-2)}{\alpha}$ and hence α plays this role without a bare cosmological constant term in the action. Unlike what was first argued in [27] this branch is not, at least, classically unstable [89] in the sense that one needs to add positive energy to the system for it to roll off to a positive mass black hole solution. We will here caution the reader that this branch is still dangerous for stability and a further careful analysis is still needed to answer the question of the physical relevance of this branch [90]. It is an intriguing fact however that one can have an effective cosmological constant from higher order curvature terms as used in [91]. Any solution of this branch will not have an Einstein theory limit, although there maybe a relevant de Sitter or anti-de Sitter–Einstein-type solution mimicking (8.46) as we will see in a moment. The $\epsilon = -1$ branch gives the usual Minkowski vacuum in the absence of a bare cosmological constant Λ and for large r the solutions resemble the asymptotically Einstein black holes with the relevant mass and charge parameters. This branch therefore is perturbatively connected to Einstein theory and we can consider the action (8.44) as an effective action including a higher order correction as in actions for closed strings [17, 18]. For the Gauss–Bonnet branch $\epsilon = 1$ notice that this is no longer the case and small α yields an enormous effective cosmological constant. Hence it is in this case an erroneous statement to consider (8.44) as an effective action and one needs exact solutions rather than perturbative ones. This fact makes this branch either totally irrelevant or far more interesting than the Einstein branch since this is where we can expect novel effects. Given therefore the fact that the $\epsilon = -1$ case closely follows known solutions, we will check out mostly the Gauss–Bonnet branch for novel effects.

The "Chern–Simons" limit is obtained for $4\alpha k^2 = -4\alpha a^2 = 1$ (for α positive and negative, respectively). The two branches then become one and at this limit the combined Gauss–Bonnet and cosmological constant is of the same order as the Einstein term on the right hand side (whose coupling we have normalised to 1). For zero charge the potential simplifies to the simple function

$$V(r) = \kappa + \frac{r^2}{2\alpha} - \frac{m}{\sqrt{r^{(D-5)}}}, \tag{8.52}$$

where $m = -\sqrt{\frac{\mu}{\alpha}}$ is the mass integration constant. For $D = 5$ the potential is quite similar to the BTZ black holes of three dimensions since the mass plays the same role as curvature of the horizon κ. Furthermore note that we have a weaker gravitational force than in the Einstein case. Indeed set $D = 6$ and a planar black hole for $\alpha > 0$ exists with $r_h > (2m\alpha)^{2/5}$ whereas the Einstein horizon for the same mass m is its square root. Hence for the same mass the Chern–Simons black hole is of squared horizon radius compared to the Einstein one. Seemingly the closer we approach the Chern–Simons branch, the milder the curvature singularity in the bulk spacetime, an interesting fact to keep in mind.

So much for the branches of solutions. Let us now check out the curvature singularities and horizons. Generically, there are two possible singularities in the curvature tensor, the usual $r = 0$, but also a branch singularity at the maximal possible zero of the square root, say $r = r_1$. We have that $r = r_1$ is solution of

$$r^{2(D-2)}(1 + 4\alpha a^2) + 4\alpha(\mu r^{D-3} - Q^2) = 0, \tag{8.53}$$

and whenever $r_1 > 0$ this is the singular end of spacetime (8.46). We have a black hole solution if and only if there exists r spacelike with $r = r_h$ such that $V(r_h) = 0$ and $r_h > r_{min}$, where $r_{min} = max(0, r_1)$. Indeed the usual Kruskal extension

$$dv_\pm = dt \pm \frac{dr}{V(r)} \tag{8.54}$$

gives that (v_+, r) and (v_-, r) constitute a regular chart across the future and past horizons of (8.46) as in GR. It is straightforward to show the following criterion: $r = r_h$ is an event horizon for $r > r_h$ spacelike if

$-\ r_h > r_{min}$
$-\ \epsilon(2\alpha\kappa + r_h^2) \leq 0$
$-\ r_h$ is root[12] of $p_\alpha(x) = (-a^2 x^{2(D-2)} + \kappa x^{2(D-3)} + \alpha\kappa^2 x^{2(D-4)} - \mu x^{D-3} + Q^2)sign(\alpha)$

Notice therefore that for $\epsilon = 1$ we have $r_h^2 \leq -2\alpha\kappa$, i.e. the event horizon is bounded from above. This immediately yields that for $\kappa = 0$ there are no black hole

---

[12] For r spacelike we need $p(x) > 0$ away from the horizon for $\alpha \geq 0$ and the contrary for $\alpha < 0$.

solutions in this branch! Second, note that $p_{\alpha=0}$ is the usual polynomial for Λ-RN black hole of Einstein theory in D dimensions. In particular, α couples only to the horizon curvature. Hence for $\epsilon = -1$ and $\kappa = 0$ (the Einstein branch) the horizon positions are the same as for the GR planar black holes.

Let us now focus on some particular solutions for the Gauss–Bonnet branch (again we advise the interested reader to see the nice analysis of [31]). Thus take $\epsilon = 1$, $a = 0$, $\kappa = 1$. We have $r_h^2 \le -2\alpha$, hence take $|\alpha| = -\alpha > 0$. We also take $\mu < 0$ in order for (8.53) to be strictly positive and to have a correct definition of mass. Indeed the solution asymptotes a de Sitter–Schwarzschild solution with positive AD mass for negative μ. Now given that $p_\alpha(x) = -x^3(x^3 - x|\alpha| - \mu)$ we find $x_{min} = \sqrt{\frac{|\alpha|}{\sqrt{3}}}$ and we have two event horizons as long as $-\frac{2}{3\sqrt{3}}|\alpha|^{3/2} \le \mu < 0$. The bigger the mass of the black hole, the further we are allowed to stretch the α parameter of the action (see Fig. 8.2). The solution has the same structure as Schwarzschild–de Sitter, with an event horizon and a de Sitter horizon. If we switch on Q^2 we can get a three, horizon structures similar to the de Sitter RN black hole. When the two horizons come together at x_{min} we are in the extremal case of the Nariai solution.

Consider now a hyperbolic horizon, in other words $\kappa = -1$. Take $Q = a = 0$ and $\epsilon = 1$. We have $r_1^5 = -4\alpha\mu$ and $r_h^2 \ge -2\alpha$. Therefore for $\alpha > 0$ and $\mu > 0$ we have trivially the first two conditions verified. Furthermore, r_h is root of $-x^3 + \alpha x - \mu$ which is exactly the same polynomial as above for $\alpha \to -\alpha$ and $\mu \to -\mu$. Therefore in this case we will have a hyperbolic black hole with two event horizons.

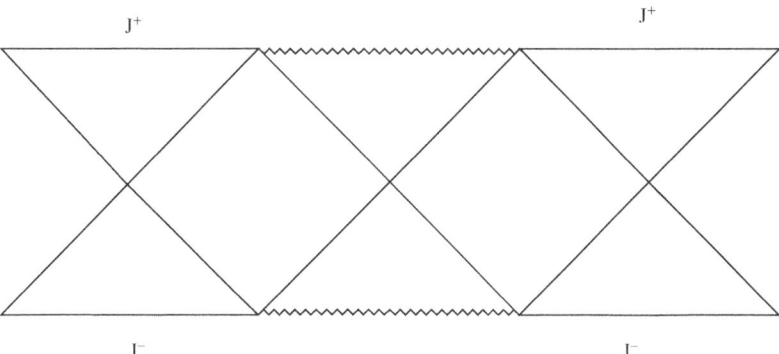

Fig. 8.2 Penrose diagram for de Sitter RN black hole

8.2.4 Thermodynamics and a Geometrical Explanation of the Horizon Area Formula

Our aim here is to evaluate the entropy of the black hole solutions discussed in the previous section. We therefore start by evaluating the mass and temperature. The mass of the black hole can be easily expressed in terms of the horizon radius, r_h, by

$$M = \frac{(D-2)\Sigma_\kappa}{16\pi G} \left(\kappa - r_h^2 a^2 + \frac{\alpha\kappa^2}{r_h^2} + \frac{Q^2}{\alpha r} \right) r_h^{D-3}. \tag{8.55}$$

In order to calculate the temperature of the black hole we follow the standard prescription. In summary, we start by Wick-rotating the time direction $t \to i\theta$. The resulting curved manifold of Riemannian signature has an axial Killing vector ∂_θ at its origin situated at $r = r_h$ as we will see in a moment. We then impose periodicity, say Π, in order for the angular coordinate to be well defined. As a result the Euclidean quantum field propagator, with the imposed periodic boundary conditions, describes a canonical ensemble of states in thermal equilibrium at a heat bath of temperature $T = \Pi^{-1}$ [92].

Indeed consider $t \to i\theta$ of (8.46). We have

$$ds^2 = V(r)d\theta^2 + \frac{dr^2}{V(r)} + r^2 h_{\mu\nu}^S dx^\mu dx^\nu, \tag{8.56}$$

$$V(r) = \kappa = \frac{r^2}{2\alpha} \left[1 + \epsilon \sqrt{1 + 4\alpha \left(a^2 + \frac{\mu}{r^{D-1}} + \frac{p^2}{r^{2(D-2)}} \right)} \right], \tag{8.57}$$

with (magnetic) field strength

$$F = \frac{p^2}{4\pi r^{2(D-2)}} d\theta \wedge dr \tag{8.58}$$

and the metric is then of Euclidean signature for $r > r_h$ and $p = iq$ ($P = iQ$) is the magnetic charge. Let us put the charge equal to zero for simplicity. The potential $V(r)$ admits at least one and at most two zeros for $0 < r_- < r_+$ where $r = r_+$ can be taken to be infinity for the single horizon case. Taking $x^\mu = constant$ and expanding around the zeros of the potential V we get,

$$ds^2 \sim \left(\tfrac{1}{4} V_{r_\pm}'^2 \right) \rho_\pm^2 d\theta^2 + d\rho_\pm^2, \tag{8.59}$$

with radial isotropic (or cylindrical) coordinate $\rho_\pm = \sqrt{\frac{2(r-r_\pm)}{V_{r_\pm}'}}$. Hence the periodicity condition reads as

$$\theta \sim \theta + \frac{4\pi}{V_{r_-}'} \sim \theta + \frac{4\pi}{V_{r_+}'} . \tag{8.60}$$

Generically we will have conical singularities. In fact the only "finite" regular solution exists when $V_{r_-}' = V_{r_+}'$, i.e. for a $T = 0$ temperature black hole (Nariai limit), and it corresponds to a regular gravitational instanton. In fact, one has to consider a simple coordinate transformation to show that space is of non-zero volume and corresponds to $S^4 \times S^2$ checkout [68, 93]. When we add charge it is a question of algebra to show that we can obtain warmer, $T \neq 0$ instantons.[13] Gravitational instantons describe the extremal spacelike path between two non-causally connected

---

[13] I thank Renaud Parentani for pointing this out.

regions of gravitational solutions [94], in other words their quantum tunnelling. In the case here the instanton describes decay of the vacuum into a pair of accelerated black holes [68] as in usual GR (for instanton solutions using boundaries, see [95]). The decay rate of course would have to be calculated here from scratch since it corresponds to a novel theory [68]. If it is enhanced it corresponds to a quantum instability of the vacuum. When $r_+ = \infty$ space rounds up smoothly and regularity is only imposed at $r = r_-$.

After this brief parenthesis let us now calculate the temperature of the black holes (8.46). The period is $\Pi = \frac{4\pi}{V'_{r_\pm}}$. It is easy to show the formula

$$V'_{r_\pm} = \frac{1}{r_\pm}\left[\frac{\mu(D-1)}{r_\pm^{D-5}(2\alpha\kappa + r_\pm^2)} - 2\kappa\right]. \tag{8.61}$$

The heat bath described by the thermal propagator is of temperature $T = \Pi^{-1}$. Therefore the temperature is found to be (using (8.55))

$$T = \frac{(D-1)k^2 r_\pm^{D-1} + (D-3)\kappa r_\pm^{D-3} + (D-5)\alpha\kappa^2 r_\pm^{D-5}}{4\pi^2 r_\pm(r_\pm^2 + 2\alpha\kappa)}. \tag{8.62}$$

Notice that the Gauss–Bonnet term α couples only to the horizon curvature and the temperature is the same for $\epsilon = \pm 1$. Also note that a Lovelock planar black hole will have same temperature as an Einstein one [87]. For the entropy we use here the standard recipe, $dM = TdS$ (see also [37, 38] for a direct calculation yielding the same result). Then in turn,

$$S = \int T^{-1}dM = \int_{r_{min}}^{r_h} T^{-1}\frac{\partial M}{\partial \bar{r}_h}d\bar{r}_h =$$
$$= \frac{r_h^{D-2}\Sigma_\kappa}{4G}\left(1 + \frac{2\alpha\kappa(D-2)}{(D-4)r_h^2}\right). \tag{8.63}$$

In order to evade erroneous conclusions it is important to note that $r_{min} = max(r_1, 0)$ [96], since in Lovelock black holes we can hit a singularity before reaching $r = 0$. Furthermore we see that the entropy is not equal to the area of the black hole horizon; we pick up a correction from the induced curvature of the horizon surface. One can understand this fact geometrically using the general formalism of Iyer and Wald [97, 98]. The entropy calculation in this case [96] gives the result

$$S = \frac{1}{4G}\int_{r_{min}}^{r_h} dx^{D-2}\sqrt{\tilde{h}_{\mu\nu}}(1 + 2\alpha\tilde{R}^{ind}), \tag{8.64}$$

where the tilded terms correspond to geometrical quantities of the horizon surface, $(r, \theta) = $ constant. The result is exactly the same as the effective action obtained for a codimension 2 matching condition (8.85) as we will see in the forthcoming section.[14] This is not too surprising with hindsight: in both cases we impose similar

---

[14] Here the extrinsic curvature quantities appearing in (8.85) cancel because the horizon surface is maximally symmetric.

regularity conditions which have to do with the temperature and hence the periodicity of the manifold. In one case conical singularities are accounted for by the
presence of branes of given tension whereas for black holes this precisely gives
the temperature of the horizon. The leading Lovelock correction appearing as an induced curvature term is therefore nothing but the extended Euler density of the horizon surface. On the other hand the Einstein term yields the tension of the horizon.
Therefore the failure to obey in this case a horizon area formula is most natural and a
simple geometric consequence. The entropy of the Lovelock black holes follows the
matching condition formula. For example if we were to take a $D = 8$ dimensional
Lovelock black hole we would get in addition the six-dimensional Gauss–Bonnet
correction of the horizon surface in the entropy formula, namely,

$$S = \frac{1}{4G} \int dx^{D-2} \sqrt{\tilde{h}_{\mu\nu}} (1 + 2\alpha \tilde{R}^{ind} + \alpha_2 \hat{G}^{ind}) . \qquad (8.65)$$

We will come back to these formulae in the next section.

8.2.5 The Lovelock Solitons and the Maximally Symmetric Cosmic Strings

So much for the Euclidean version of (8.46) and the thermodynamics (for details
check [96]). We now Wick-rotate the $D - 2$ horizon sections $h^S_{\mu\nu}$ to the Lorentzian
maximally symmetric sections $h^L_{\mu\nu}$ to construct soliton solutions (8.48) (the same
obviously holds for the Lovelock case (8.41)). Since the soliton is of axial symmetry we have $r_- \leq r \leq r_+$ where $r_\pm$ are the possible zeroes of V. One can always
locally go to the cylindrical anzatz (8.59), with $\rho_\pm$ the local cylindrical coordinate.
The soliton is of infinite proper distance ρ when $r_+ \to \infty$. A very nice and simple
interpretation of these solutions and a simple application of the staticity theorem reside in $D = 4$ general relativity. In this case the Lorentzian two-dimensional sections
in (8.48) have precisely the geometry of a maximally symmetric cosmic string, of
internal constant curvature given by κ, situated at $r = r_\pm$ once we allow for some
deficit angle in θ. These are the only solutions describing de Sitter, flat or anti-de
Sitter cosmic strings in a cosmological constant and (possibly) charged background.
We will not analyse these here, but as an example it is easy to note that in the presence of a bare negative cosmological constant the straight cosmic string bends the
ambient spacetime (this was first noted by Linet [99]). This is in complete contrast
to the flat string which is embedded in a locally flat spacetime. This is because the
resulting geometry is of non-trivial Weyl tensor,

$$ds^2 = \left(k^2 r^2 - \frac{\mu}{r} \right) \beta^2 d\theta^2 + \frac{dr^2}{\left(k^2 r^2 - \frac{\mu}{r} \right)} + r^2(-dt^2 + dz^2), \qquad (8.66)$$

precisely denoted here by the presence of a non-trivial μ parameter. In (8.66) (t,z)
are the string coordinates, the angular deficit is given by $\delta = 2\pi(1 - \beta)$ and the

string is of linear tension $T = \frac{\delta}{8\pi G_4}$. In other words a straight cosmic string mathematically at least corresponds to a double Wick-rotated black hole solution. Another way to understand the above result is that in pure AdS spacetime we cannot slice the geometry in a cylindrically symmetric anzatz. It is a simple exercise to classify the solutions, the case of de Sitter also presenting some unique characteristics due to the compactness of its spatial sections (cosmic strings will appear in pairs in this case). Furthermore, for $\kappa = 0$, D arbitrary and negative cosmological constant, we get the Lovelock AdS soliton analysed in [100] for the case of Einstein gravity. In a forthcoming section we will see that in $D = 6$ dimensions the metrics (8.48) will describe (with the help of suitable matching conditions) the generic bulk geometry of maximally symmetric four-dimensional braneworlds [68].

8.3 Matching Conditions for Distributional Sources

Most recent applications of Lovelock theory concern braneworld physics. In the braneworld picture our four-dimensional universe is part of a higher dimensional manifold and whereas matter is strictly confined on the braneworld, gravity propagates, according to the equivalence principle, in all dimensions. We therefore want to describe the motion or the evolution of a self-gravitating submanifold which for simplicity we take to be infinitesimally thin. In other words, we suppose that matter is confined on the braneworld via a Dirac distribution of dimension equal to the codimension $N = D - 4$ of the braneworld. Given that Lovelock theory is the general metric theory with second-order field equations, we can except some junction conditions quite similar to those of Einstein theory.[15] It turns out that in Lovelock theory we can even do better, at least in mathematical terms, and define matching conditions for higher codimension than that possible for Einstein theory [101–104].

Consider a p-brane Σ embedded in $D = p + 1 + N$ dimensions and suppose that Σ carries some localised energy–momentum tensor [16]:

$$T_{AB} = \begin{pmatrix} S_{\mu\nu} & 0 \\ 0 & 0 \end{pmatrix} \delta_\Sigma. \tag{8.67}$$

The N-dimensional Dirac distribution δ_Σ on Σ signifies that the brane is of zero thickness. The question we address is, What are the equations of motion for this self-gravitating p-brane sourced by the distributional energy–momentum tensor (8.67)? For codimension $N = 1$, the answer is given by the well-known Israel junction conditions [46–48] where, if the induced metric is continuous, a discontinuity in the first derivatives of the metric accounts for the Dirac charge in (8.67). Israel's junction conditions describe adequately a wide variety of GR problems and only when

---

[15] Fourth-order field equations would mean that distributional constraints would be imposed on the continuity of certain third-order directional derivatives which would mean a discontinuous limit to pure Einstein theory.

[16] Greek letters run through brane coordinates while capital Latin from the beginning of the alphabet through bulk coordinates.

we study junction problems of lesser symmetry do we run into shortcomings because we can no longer fulfil the continuity condition (for example we cannot match Kerr to flat spacetime). If the codimension is strictly greater than 2 then there are no distributional matching conditions in Einstein gravity. Thus, a finite thickness braneworld is needed in order to obtain non-trivial self-gravitating equations of motion for Σ, and the resulting equations of motion will generically depend on the regularisation scheme. This is hardly surprising. We know, for example, that far away from a gravitational source such as the sun, and by virtue of Birkhoff's theorem, we can approximate conveniently its gravitational field using a three dimensional Dirac distribution. What we mean by far away is precisely the Schwarzschild radius of the source in question which is about 3 km and which is far smaller than the actual size of the sun. In Einstein theory even when the codimension is equal to 2 (for $D = 4$ say), one only knows the self-gravitating field of a straight cosmic string, i.e. one induced by a pure tension matter tensor, which gives an overall conical deficit angle [101–104]. Hence even for codimension 2 distributional matching conditions give only a brute picture of the gravitational field and break down when the braneworld or defect is of lesser symmetry. The essential point for the discussion is the codimension of spacetime N, i.e. the number of spacelike or timelike, vectors defined normal to the brane (for lightlike junctions see [105]). If the codimension is 1 as it is for the usual junction conditions the p-brane is a hypersurface splitting the bulk in two and there is no geometry (other than that of real line) in the normal directions. Once $N > 1$ we have non-trivial geometry in the normal sections.

Lovelock's equations for the distributional energy–momentum tensor (8.67) are

$$\sum_{k=0}^{[(D-1)/2]} \alpha_k \mathscr{E}_{(k)\mu} = -2S_{\mu\nu}\theta^{*\nu}\delta_\Sigma. \tag{8.68}$$

Our task here involves finding the distributional part of the left hand side to be matched to the induced energy–momentum tensor $S_{\mu\nu}$. We will refer to such geometric distributional terms as those carrying Dirac charge. It turns out that not all Lovelock terms can carry a Dirac charge; we already know this from Einstein theory which is Lovelock theory in $D = 4$. Indeed we will find the simple inequality

$$N/2 \le k \le \left[\frac{D-1}{2}\right], \tag{8.69}$$

which selects in particular the Lovelock bulk terms $\mathscr{L}_{(k)}$ that can carry a Dirac charge. Indeed for $D = 4$ we have $N/2 \le k \le 1$ which tells precisely that in Einstein theory, $k = 1$ can carry up to codimension 2 Dirac charge. If $D > 4$ and we allow for the higher order Lovelock terms we see that we can go to higher codimension.

Without attempting to give a full proof of this result, given in [69, 107], we can however understand it geometrically in the following way: separate the geometry in normal and parallel sections as we commonly do in the Gauss–Codazzi formalism. Let e_μ be the $p + 1$ unit vectors that are everywhere tangent to the brane and n_I the N unit vectors that are everywhere normal to Σ, with the label $\mathscr{N}$ denoting the radial normal vector. Similarly, split locally the bulk θ^A into tangent 1-forms θ^μ and normal 1-forms θ^I. One can then deduce the first and second fundamental forms of the brane, respectively,

$$h = \eta_{\mu\nu}\theta^\mu \otimes \theta^\nu, \tag{8.70}$$

$$K_{I\mu\nu} = g(\nabla_{e_\mu} n_I, e_\nu), \tag{8.71}$$

describing the induced metric and extrinsic curvature of Σ. The Gauss–Codazzi equations can also be written in form formalism which is extremely useful in this context. Indeed the parallel projection along the brane coordinates gives the Gauss equation:

$$\mathscr{R}^\mu_{\parallel\nu} \equiv \frac{1}{2} R^\mu_{\nu\lambda\rho}\theta^\lambda \wedge \theta^\rho = \Omega^\mu_{\parallel\nu} - \mathscr{K}^\mu_{\parallel I} \wedge \mathscr{K}^I_{\parallel\nu}, \tag{8.72}$$

where

$$\mathscr{K}^I_{\parallel\mu} = K^I_{\nu\mu}\theta^\nu$$

is the extrinsic curvature 1-form and $\Omega^\mu_{\parallel\nu}$ is the *induced curvature 2-form* of the brane, associated with the induced metric $h_{\mu\nu}$. This geometrical identity relates the background curvature with respect to the induced and extrinsic curvature of Σ.

We look for geometric terms in the $\mu\nu$-components on the left hand side of (8.68) that can carry a Dirac charge. The situation is inherently different for odd and even codimension. Indeed for hypersurfaces (codimension 1), the distributional charge in Einstein theory is provided by a discontinuity of the extrinsic curvature of the metric $K^N_{\mu\nu}$. Thus we relate local geometry of the brane to matter. In even codimension defects, like cosmic strings (codimension 2), the normal sections to the brane have some non-trivial topology that can give rise to distributional terms. For the case of the straight cosmic string, we have

$$ds^2 = -dt^2 + dz^2 + d\rho^2 + L^2(\rho)d\theta^2 + d\rho^2, \tag{8.73}$$

and Einstein field equations are solved for $L(\rho) = \rho$ everywhere but at $\rho = 0$. We therefore set

$$L'(\rho) = 1, \; \rho \neq 0 \text{ and } L'(\rho) = \beta, \; \rho = 0.$$

The distributional part of (8.68) is then integrated over the normal section and reads as

$$\int_0^{2\pi} \int_0^r \mathrm{dist}(L'')r\,dr\,d\theta = -8\pi G_4 \int_\Sigma^\perp \frac{T}{2\pi}\delta^{(2)}\,r\,dr\,d\theta . \tag{8.74}$$

Given that $\mathrm{dist}(L'') = [L']\delta^{(2)} = (1-\beta)\frac{\delta(r)}{r}$, we get the well-known result $2\pi(1-\beta) = -8\pi G_4 T$ relating here topology (rather than geometry) to matter. It is important to note that the curvature of the normal section is precisely given by L''.

In each case we define locally a normal section, $\Sigma_\perp$, at each point of the brane, using the congruence of the normal vectors with specific regularity properties. We then integrate the $\mu\nu$-equations of motion over an arbitrary such section $\Sigma_\perp$. Since the Dirac distribution is by definition independent of regularisation, i.e. to the microscopic features of Σ, the only terms that can contribute are those *independent* of geometry deformations over $\Sigma_\perp$ as we take the limit of zero thickness. This boils down for codimension 2 to working out the distributional part as in (8.74). Mathematically, the only terms in (8.68) having this property are proven to be locally exact forms on $\Sigma_\perp$, basically made out of powers of extrinsic curvature or curvature of the normal section as for the example of the cosmic string above.[17]

What *multiply* these locally exact forms are now equated to the energy–momentum tensor of the brane and depend on the parallel geometry (since the normal geometry is integrated out) using essentially Gauss's equation (8.72). Since higher order Lovelock terms, $k > 1$, contain higher powers of the curvature tensor, we can expect the appearance of induced curvature and extrinsic curvature terms from (8.72). Indeed the relevant terms are sums in $0 \leq j \leq min([N/2], k - N + [N/2])$ of

$$\sigma^{//}_{(N,k,j)\mu} = \left(\bigwedge_{l=1}^{k-N+[N/2]-j} \mathscr{R}^{\lambda_l \nu_l}_{//} \right) \wedge \left(\bigwedge_{l=1}^{N-2[N/2]+2j} \mathscr{K}^{\rho_l}_{//N} \right) \wedge \theta^{\star //}_{\mu \lambda_l \nu_l \rho_l}, \qquad (8.75)$$

which are powers of the projected Riemann curvature two-form $\mathscr{R}_{//}$ and the extrinsic curvatures $\mathscr{K}_{//}$.

For the case of codimension 1 the junction conditions read as follows:

$$- \sum_{k=0}^{[(D-1)/2]} 8\alpha_k k! \left[\sigma^{//}_{(N,k)\mu} \right] = 2 S_\mu{}^\nu (P) \theta^{\star //}_\nu, \qquad (8.76)$$

and the Dirac charge will be provided by the jump of $\sigma^{//}_{(N,k)\mu}$. Here $1 \leq k \leq [(D-1)/2]$, and the higher the dimension, the more the Lovelock charges that contribute. For example in $p = 3$ we get in turn for the $k = 1$ Einstein and $k = 2$ Gauss–Bonnet terms,

$$\sigma^{//}_{(1,1)\mu} = 2 \mathscr{K}^\nu_{//N} \wedge \theta^{\star //}_{\mu \nu} = -2 (K^\nu_\mu - \delta^\nu{}_\mu K) \theta^{\star //}_\nu$$

$$\sigma^{//}_{(1,2)\mu} = 4 \mathscr{K}^\nu_{//N} \wedge \left(\Omega^{\rho \lambda}_{//} - \frac{1}{3} \mathscr{K}^\rho_{//N} \wedge \mathscr{K}^\lambda_{//N} \right) \wedge \theta^{\star //}_{\mu \nu \rho \lambda}$$

$$= -4 (3 J^\nu_\mu - J \delta^\nu_\mu - 2 P^\nu{}_{\lambda \rho \mu} K^{\lambda \rho}) \theta^{\star //}_\nu, \qquad (8.77)$$

where we have set

$$J_{\mu\nu} = \frac{1}{3} (2 K K_{\mu\lambda} K^\lambda{}_\nu + K_{\lambda\rho} K^{\lambda\rho} K_{\mu\nu} - 2 K_{\mu\lambda} K^{\lambda\rho} K_{\rho\nu} - K^2 K_{\mu\nu}),$$

$$P_{\mu\nu\lambda\rho} = R_{\mu\nu\lambda\rho} + 2 R_{\nu[\lambda} g_{\rho]\mu} - 2 R_{\mu[\lambda} g_{\rho]\nu} + R g_{\mu[\lambda} g_{\rho]\nu}.$$

---

[17] In codimension 2 the above statement boils down to the famous Gauss–Bonnet theorem. In higher codimension the relevant forms are precisely the Chern–Simons forms [78, 79].

and we have dropped the label N for the extrinsic curvature components. Replacing into the matching conditions (8.76) and taking the corresponding left and right limits one gets the well-known junction conditions for Einstein [46–48] and Einstein–Gauss–Bonnet gravity [61, 62]. The equations of motion can also be derived from the boundary action, after variation with respect to the induced frame θ^ν on the left and right sides of the brane. In differential form language the action is obtained trivially from (8.77) by removing the free index,

$$
S_\Sigma = 2\alpha_1 \int_\Sigma \mathcal{K}^\nu_{/\!/N} \wedge \theta^{\star/\!/}_\nu
$$
$$
+ 4\alpha_2 \int_\Sigma \mathcal{K}^\nu_{/\!/N} \wedge \left(\Omega^{\rho\lambda}_{/\!/} - \frac{1}{3}\mathcal{K}^\rho_{/\!/N} \wedge \mathcal{K}^\lambda_{/\!/N} \right) \wedge \theta^{\star/\!/}_{\nu\rho\lambda}, \tag{8.78}
$$

and agrees with the Gibbons–Hawking [81] and Myers [63] boundary terms. It is obvious now that for $D = 7$ a 5-brane will have an extra term contributing to the junction conditions which will be of fifth order in the extrinsic curvatures, etc. This can be understood intuitively since higher order Lovelock densities involve higher powers of curvature. For the case of higher odd codimension, see [106, 107].

Let us now concentrate on even codimension $N = 2n$. We will assume that the parallel sections are regular. Then when integrated out over $\Sigma_\perp$, the normal sections quite naturally yield the charge or topological defect of $(\chi - \beta)\mathrm{Area}(S_{2n-1})$. This is similar to the defect angle for the infinitesimal cosmic string in four dimensions (8.73) when we set $\chi = 1$, where χ is the Euler characteristic of the normal section.[18] In higher codimension one has to allow for non-trivial Euler character in $\Sigma_\perp$ in order to obtain removable-type singularities [70]. Due to their topological character, we call these topological matching conditions [106, 107]. The matching conditions read as follows:

$$
(\chi - \beta)\mathrm{Area}(S_{2n-1}) \sum_{\tilde{k}=0}^{\left[\frac{p}{2}\right]} \tilde{\alpha}_k \sigma^{/\!/}_{(n,\tilde{k})\mu}(P) = -2S^\nu_\mu(P)\theta^{\star/\!/}_\nu, \tag{8.79}
$$

where we have set $\tilde{\alpha}_k = 2^{2n-1}k!(n-1)!\alpha_k/(k-n)!$ and the smooth parallel forms $\sigma^{/\!/}_{(n,k)\mu}$ are given by

$$
\sigma^{/\!/}_{(n,k)\mu} = \sum_{j=0}^{\min(n-1,\tilde{k})} \binom{\tilde{k}}{j} \left(\bigwedge_{l=1}^{\tilde{k}-j} \mathcal{R}^{\lambda_l \nu_l}_{/\!/} \right) \wedge \left(\bigwedge_{l=1}^{2j} \mathcal{K}^{\rho_l}_{/\!/N} \right) \wedge \theta^{\star/\!/}_{\mu\lambda_1\nu_1\cdots\lambda_{\tilde{k}-j}\nu_{\tilde{k}-j}\rho_1\cdots\rho_{2j}}. \tag{8.80}
$$

Finally, $\tilde{k} = k - n$, which ranges between $0 \le \tilde{k} \le \left[\frac{p}{2}\right]$, will turn out to be the induced Lovelock rank of (8.80). The parallel forms (8.80) will dictate the dynamics of the brane. Their expressions involve powers of the projected Riemann tensor $\mathcal{R}_{/\!/}$ on the brane and even powers of the radial extrinsic curvature $\mathcal{K}_N$ of Σ. To see this, consider for each $\tilde{k}$ the first term in the sum, $j = 0$, which reads as follows:

---

[18] I thank Robin Zegers for pointing this out.

$$\sigma^{//}_{(\tilde{k},0)\mu} = \left(\bigwedge_{l=1}^{\tilde{k}} \mathscr{R}^{\lambda_l \nu_l}_{//} \right) \wedge \theta^{\star//}_{\mu\lambda_1\nu_1...\lambda_{\tilde{k}}\nu_{\tilde{k}}} . \tag{8.81}$$

Note the similarity to the bulk Lovelock densities (8.23). Clearly, using (8.72), we see the appearance of induced Lovelock densities involving $\Omega_{//}$ accompanied by even powers of the extrinsic curvature $\mathscr{K}_N$. The former describe the induced quantities of the brane and the latter how Σ is embedded in the bulk. In particular, for $\tilde{k} = 0$ we will have a pure tension term, for $\tilde{k} = 1$ an Einstein term and for $\tilde{k} = 2$ a Gauss–Bonnet term with extrinsic curvature terms. Note that the highest rank on the brane originates from the highest rank Lovelock term in the bulk. This agrees with the fact that in Einstein gravity, $k = 1$, there are no matching conditions beyond $n > 2$.

Let us compare our result with the work of [67]. Thus, let us consider the case of a 3-brane embedded in six-dimensional spacetime, for which there are only two terms in (8.80):

$$\sigma^{//}_{(1,0)\mu} = \theta^{\star//}_{\mu} , \qquad \sigma^{//}_{(1,1)\mu} = \mathscr{R}^{\nu\rho}_{//} \wedge \theta^{\star//}_{\mu\nu\rho} . \tag{8.82}$$

Thus, using (8.8), (8.11) and (8.72), we obtain from (8.79)

$$2\pi (1-\beta) \left\{ -\alpha_1 h_{\mu\nu} + 4\alpha_2 G^{(ind)}_{\mu\nu} - 4\alpha_2 W_{\mu\nu} \right\} = S_{\mu\nu} , \tag{8.83}$$

where [67]

$$W^{\lambda}_{\mu} = K_N K^{\lambda}_{\mu N} - K^{\nu}_{\mu N} K^{\lambda}_{\nu N} - \frac{1}{2} \delta^{\lambda}{}_{\mu} \left(K^2_N - K^{\nu}_{\rho N} K^{\rho}_{\nu N} \right) . \tag{8.84}$$

Note that the bulk Einstein term, $k = 1$, only allows for an effective cosmological constant on the brane, while the Gauss–Bonnet term, $k = 2$, induces the Einstein tensor for the brane's equation of motion. This equation is similar to the one found in [67], the difference being that here the extrinsic geometry is supposed perfectly regular. Indeed mathematically there is no reason to suppose that the extrinsic curvature has a jump, the topological defect carrying the necessary charge in complete analogy to the case of the cosmic string. Additionally, these matching conditions provide the maximal regularity for the bulk metric. Furthermore, since the induced and extrinsic geometries are smooth, the equations of motion can be seen to actually originate from a simple action taken over Σ as long as we suppose that $\beta = constant$. In other words, the degrees of freedom associated to the normal section are completely integrated out giving an exact action for the brane's motion. In differential form language, it is straightforward to read off the Langrangian density in question. Literally, taking out the free index for the charge in (8.81) and using Gauss equation (8.72), we get

$$S_{\Sigma}^{(p=3,n=1)} = 2\pi(1-\beta)\int_{\Sigma}\left(\tilde{\alpha}_1\,\theta^{\star//} + \tilde{\alpha}_2(\mathscr{R}_{//}^{\nu\rho}\wedge\theta_{\nu\rho}^{\star//})\right) + \int_{\Sigma}\mathscr{L}_{matter}$$

$$= 2\pi(1-\beta)\int_{\Sigma}\sqrt{-h}\left(\tilde{\alpha}_1 + \tilde{\alpha}_2(R^{ind} - K^2 + K_{\mu\nu}^2)\right)$$

$$+ \int_{\Sigma}\mathscr{L}_{matter}. \tag{8.85}$$

We see appearing the cosmological constant, the Einstein–Hilbert term with an extrinsic curvature term quite similar to the finite width corrections one obtains for cosmic strings in flat spacetime [108]! In other words, the only bulk quantity entering in the equations of motion is the topological defect fixing the overall mass scale and the extrinsic curvature of the surface giving a matter-like component in the action. Therefore the Gauss–Bonnet term quite naturally gives an induced gravity term on the brane for codimension 2. This does not mean that there is a localised 0-mode graviton on the brane location but much like in DGP one can expect a quasi-localised 0-mode or in other words ordinary four-dimensional gravity up to some crossover scale as in the DGP model [43]. It would seem that the topological quantity β would be giving us in this case a crossover scale for four-dimensional gravity to six-dimensional gravity. Clearly perturbation theory of well-defined warped backgrounds is needed in order to answer this question.

Equation (8.79) has another surprising property. Indeed, an important simplification takes place if we suppose that

$$[p/2] + 1 \leq n, \tag{8.86}$$

i.e. that the codimension of the brane is larger than its intrinsic dimension. In that case, using Gauss's equation, we have

$$\sigma_{(n,k)\mu}^{//} = \left(\bigwedge_{l=0}^{\tilde{k}}\Omega_{//}^{\lambda_l\nu_l}\right)\wedge\theta_{\mu\lambda_1\nu_1\cdots\lambda_{\tilde{k}}\nu_{\tilde{k}}}^{\star//}, \tag{8.87}$$

and the matching conditions (8.79) are simply the induced Lovelock equations on the brane with no extrinsic curvature terms! Therefore, the action for a distributional 3-brane embedded in $D = 8, 10, \ldots 2d$ dimensions is exactly the Einstein–Hilbert plus cosmological constant action:

$$S_{\Sigma}^{(p=3,n>1)} = (1-\beta)\text{Area}(S_{2n-1})\int_{\Sigma}\sqrt{-h}\left(\tilde{\alpha}_n + \tilde{\alpha}_{n+1}R^{ind}\right) + \int_{\Sigma}\mathscr{L}_{matter},$$

with Planck scale set by

$$M_{Pl}^2 = (1-\beta)\,\text{Area}(S_{2n-1})\,\tilde{\alpha}_{n+1}.$$

For a 4- or 5-brane, we will have in addition the Gauss–Bonnet term. etc. In other words, if the codimension verifies (8.86), all the extrinsic curvature corrections drop out and there is a complete Lovelock reduction from the bulk to the induced Love-lock terms on the brane. All the degrees of freedom originating from the bulk at zero thickness level are exactly integrated out giving the most general classical equations of motion for the brane. As a consequence we have in particular energy conservation on the brane (see also [109]).

8.4 Applications to Braneworlds

Braneworlds offer interesting and direct applications to Lovelock theory. This is due to the fact that Lovelock theory is the most general metric theory that can enjoy well-defined junction conditions. There are two reasons for this. First, the field equations being of second-order Dirac matter distributions still allow a continuous metric across the junction surface and therefore constraints are imposed on geometric quantities such as the extrinsic curvature and the induced curvature of the braneworld. They will lead inevitably to a self-gravitating equation of motion for the surface in codimension 1 and in interesting constraints for codimension 2. Furthermore, in higher codimension [110–112] Einstein theory does not possess the necessary richness in order to admit distributional sources, rather the finite thickness of the defect plays a necessary role or one has to admit the presence of non-removable curvature singularities at the brane location. Lovelock theory seemingly admits higher codimension distributions [69, 107] and simple examples of codimension 4 braneworlds are emerging [70]. Here we will concentrate in turn on codimension 1 and codimension 2 braneworlds.

8.4.1 Codimension 1 Braneworlds and Their Effective Four-Dimensional Gravity

In this section[19] we will consider the application of Lovelock theory to a four-dimensional braneworld embedded in a five-dimensional curved background space-time. We will take a flat single brane in adS, i.e. a Randall–Sundrum-type braneworld and its perturbations, and we will then give the relevant cosmological evolution equations. We will also include the induced gravity term so as to consider the most general five-dimensional configuration with a brane boundary. For simplicity we will also consider Z_2 symmetry although the asymmetric cases present a number of interesting features [39, 40]. Consider therefore the action,

---

[19] I thank Stephen Davis for unpublished collaboration throughout this section which in part can be found in his papers [113–115].

$$S = \frac{M^3}{2} \int_{\mathcal{M}} d^5 x \sqrt{-g} \left(\zeta R + \hat{\alpha} \hat{G} - 2\Lambda \right)$$
$$- M^3 \int_{\Sigma} d^4 x \sqrt{-h} \left[\zeta K + 2\hat{\alpha} (J - 2\hat{G}^{ab} K_{ab}) \right],$$
$$+ \frac{M^3}{2} \int_{\Sigma} d^4 x \sqrt{-h} \left(\beta \hat{R} - 2T \right)$$
$$- \int_{\mathcal{M}} d^5 x \sqrt{-g} \mathcal{L}_{mat} - \int d^4 x \sqrt{-h} \widehat{\mathcal{L}}_{mat} \qquad (8.88)$$

where the first line represents the bulk Lovelock theory, the second the Gibbons–Hawking and Myers boundary terms and the third the induced gravity and matter contributions. Notice that we have included a dimensionless number ζ in front of the Einstein–Hilbert term since we will take its zero limit in order to see the resulting four-dimensional brane gravity of a pure Gauss–Bonnet bulk term. Given the normal vector n_a, the five-dimensional projector

$$h_{ab} = g_{ab} - n_a n_b \qquad (8.89)$$

is the surface-induced metric and

$$K_{ab} = h^c{}_a \nabla_c n_b \qquad (8.90)$$

is the extrinsic curvature (strictly speaking the first and second fundamental forms of Σ). The caret denotes tensors constructed out of h_{ab}. $[X]$ as before denotes the jump in X across the brane.

The field equations in the bulk and on the brane are

$$\zeta G_{ab} + \hat{\alpha} H_{ab} + \Lambda g_{ab} = M^{-3} T_{ab}, \qquad (8.91)$$

where H_{ab} is the second-order Lovelock tensor (8.25). The bulk and brane energy–momentum tensors are, respectively, $T_{ab} = 2\delta \mathcal{L}_{mat} / \delta h^{ab} - h_{ab} \mathcal{L}_{mat}$ and $S_{ab} = 2\delta \widehat{\mathcal{L}}_{mat} / \delta h^{ab} - h_{ab} \widehat{\mathcal{L}}_{mat}$. According to (8.76) [61, 62] we have

$$\zeta [K_{ab} - K h_{ab}] + \alpha \left[3 J_{ab} - J h_{ab} + 2 \hat{P}_{acdb} K^{cd} \right] - \beta \hat{G}_{ab} = -M^{-3} S_{ab} + T h_{ab}. \quad (8.92)$$

For an adS bulk and a flat brane situated at $z = 0$ we use a Poincaré patch and the solution reads as follows:

$$ds^2 = a^2(z) \eta_{\mu\nu} dx^\mu dx^\nu + dz^2, \qquad (8.93)$$

where the warp factor is $a(z) = e^{-k|z|}$ and we have

$$\Lambda = -6k^2 (\zeta - 2\alpha k^2), \quad T = 2k(3\zeta - 2\alpha k^2) \qquad (8.94)$$

for the bare cosmological constant and the brane tension, respectively. The effective curvature scale k is given by $2\alpha k^2 = \zeta \mp \sqrt{\zeta^2 + 2\alpha \Lambda / 3}$ from solving the field equations in the bulk and on the boundary. The upper sign corresponds to the Einstein branch.

To obtain the effective four-dimensional gravity induced on the brane, we consider a general linear perturbation theory parametrised by $\gamma_{ab}(x,z)$, around the background solution (8.93):

$$ds^2 = a^2(z)(\eta_{\mu\nu} + \gamma_{\mu\nu})dx^\mu dx^\nu + 2\gamma_{\mu z}dx^\mu dz + (1 + \gamma_{zz})dz^2. \tag{8.95}$$

We will deal with gauge in a moment. We consider in particular the perturbed four-dimensional Ricci tensor corresponding to the linear perturbation $\gamma_{\mu\nu}$,

$$\mathscr{R}_{\mu\nu} = \frac{1}{2}\left(2\partial^\alpha \partial_{(\nu}\gamma_{\mu)\alpha} - \Box_4\gamma_{\mu\nu} - \partial_\mu\partial_\nu\gamma\right), \tag{8.96}$$

and we define $\mathscr{R} = \eta^{\mu\nu}\mathscr{R}_{\mu\nu}$ and $\mathscr{G}_{\mu\nu} = \mathscr{R}_{\mu\nu} - (1/2)\eta_{\mu\nu}\mathscr{R}$. These are the typical geometrical quantities encountered in perturbation theory of induced gravity terms such as DGP braneworlds [43].

Given our mirror symmetry bulk equations (for $z > 0$) are then

$$(\zeta - 2\alpha k^2)\left\{(6k\partial^\mu\gamma_{z\mu} - \mathscr{R})e^{2kz} - 3k(4k\gamma_{zz} + \partial_z\gamma)\right\} = \frac{2}{M^3}T_{zz}, \tag{8.97}$$

$$(\zeta - 2\alpha k^2)\left\{(\partial_\mu\partial^\nu\gamma_{zv} - \Box_4\gamma_{z\mu})e^{2kz} - 3k\partial_\mu\gamma_{zz} - \partial_z(\partial^\nu\gamma_{\mu\nu} - \partial_\mu\gamma)\right\} = \frac{2}{M^3}T_{z\mu}, \tag{8.98}$$

and

$$\begin{aligned}
(\zeta - 2\alpha k^2)&\Big[\left(2\mathscr{G}_{\mu\nu} + (\eta_{\mu\nu}\Box_4 - \partial_\mu\partial_\nu)\gamma_{zz}\right. \\
&- 2(\partial_z - 2k)(\eta_{\mu\nu}\partial^\alpha v_\alpha - \partial_{(\mu}v_{\nu)})\Big)e^{2kz} \\
&- (\partial_z^2 - 4k\partial_z)(\gamma_{\mu\nu} - \eta_{\mu\nu}\gamma) + 3k\eta_{\mu\nu}(\partial_z - 4k)\gamma_{zz}\Big] \\
&= \frac{2}{M^3}e^{2kz}T_{\mu\nu}.
\end{aligned} \tag{8.99}$$

The higher order Gauss–Bonnet contribution is recognised easily by the appearance of the coupling constant α while the usual Einstein terms are identified by ζ. Note first of all how the metric perturbation in the bulk is quasi-identical to those in Einstein theory apart from the overall coupling of α, with the warp factor k multiplying the differential operator. If there was no warp factor, for example in a Minkowski bulk, the higher order Gauss–Bonnet term would have had *no contribution in linear perturbation theory*. A general remark we can make is that since the coupling constant α is dimensionful $\alpha \sim [length]^2$ needs inevitably a bulk curvature scale to couple to. This can be related, for example, to a bare cosmological constant in the bulk or an effective cosmological constant on the brane [44, 45].[20] Also note that given our couplings to matter in action (8.88) we need to have $\zeta - 2\alpha k^2 \geq 0$. We otherwise quite clearly have a bulk ghost. If we switch off Einstein gravity in the bulk $\zeta = 0$ and choose $\alpha < 0$ we have Einstein-like perturbation theory. When $\zeta = 2\alpha k^2$ the linear perturbation operator switches off and seemingly we approach a strongly coupled limit. This is the case of Chern–Simons gravity which is very interesting by itself and we invite the interested reader to consult [86]. In turn the junction conditions at the brane $z = 0$ are

$$\left[(\zeta - 2\alpha k^2) \left\{ 3k\eta_{\mu\nu}\gamma_{zz} - 2\eta_{\mu\nu}\partial^\alpha v_\alpha + 2\partial_{(\mu}v_{\nu)} - \partial_z(\gamma_{\mu\nu} - \eta_{\mu\nu}\gamma) \right\} \right]$$
$$+ 2(\beta + 4\alpha k)\mathscr{G}_{\mu\nu} = \frac{2}{M^3}S_{\mu\nu}. \tag{8.100}$$

Here note that the higher order Lovelock term gives two contributions: first, a Neumann-type boundary term in the sense that it involves the first derivative of the perturbation metric and is exactly the same as the usual Einstein perturbation; second, an induced gravity-type term, which is accompanied by the induced gravity coupling β and which yields an induced Einstein term on the brane. Note that indices in the above equations are raised/lowered with $\eta^{\mu\nu}$ and that $T^a{}_a = T_{zz} + e^{-2kz}\eta^{\mu\nu}T_{\mu\nu}$.

Before solving the above equation let us first deal with gauge freedom. Any infinitesimal bulk transformation, $x^a \to x^a + \zeta^a$, gives the transformation law $\gamma_{ab} \to \gamma_{ab} + \mathscr{L}_\zeta g_{ab}$ which leaves the bulk equations invariant. Here take the infinitesimal shift,

$$z \to z + \epsilon(x, z)$$
$$x^\mu \to x^\mu + \xi^\mu(x, z) + \partial^\mu \xi(x, z) \tag{8.101}$$

(where $\partial_\mu \xi^\mu = 0$), and obtain the linear isometries:

$$h_{\mu y} \to h_{\mu y} - a(z)(\xi'_\mu + \partial_\mu \xi') - \partial_\mu \epsilon, \tag{8.102}$$
$$h_{yy} \to h_{yy} - 2\epsilon', \tag{8.103}$$
$$h_{\mu\nu} \to h_{\mu\nu} - \partial_\mu \xi_\nu - \partial_\nu \xi_\mu - 2\partial_\mu \partial_\nu \xi - 2a'(z)\epsilon \eta_{\mu\nu}, \tag{8.104}$$

which leave the bulk equations invariant. A bulk coordinate transformation however can also displace the brane by $F \to F + \epsilon(x, z = 0)$ where we call F the brane-bending mode. This is avoided as long as we choose $\epsilon(x, z = 0) = 0$ on the brane.

It is easy now to solve the equations (8.97) and (8.98) when $T_{ab} = 0$ and $k \neq 0$ by going to a fixed wall gauge where there is no brane-bending F and the wall is

---

[20] When spacetime is not conformally invariant as, for example, for a black hole, the perturbation operator picks up extra terms related to the background Weyl curvature that appears in the field equations for Lovelock theory for $k \geq 2$!

maintained at $z = 0$. In this gauge the brane boundary conditions are also invariant under the bulk transformations [42]. We get

$$\gamma_{zz} = -\frac{1}{4k}\partial_z\gamma, \tag{8.105}$$

$$\gamma_{z\mu} = -\frac{\text{sgn}(z)}{8k}\partial_\mu\gamma + B_\mu, \tag{8.106}$$

$$\partial^\mu\bar{\gamma}_{\mu\nu} = 0, \tag{8.107}$$

where $\bar{\gamma}_{\mu\nu} = \gamma_{\mu\nu} - (1/4)\gamma\eta_{\mu\nu}$ is the trace-free part of $\gamma_{\mu\nu}$ and the trace reads $\gamma = \eta^{\mu\nu}\gamma_{\mu\nu}$. We also have $\partial^\mu B_\mu = 0$ and $\Box_4 B_\mu = 0$ but we can choose gauge for which $B_\mu = 0$ as we will do from now.

The remaining bulk field equation is given by

$$\left(\zeta - 2\alpha k^2\right)\left(\partial_z^2 - 4k\partial_z + e^{2kz}\Box_4\right)\bar{\gamma}_{\mu\nu} = 0 \tag{8.108}$$

for $z > 0$. The boundary conditions at $z = 0$ give

$$2(\zeta - 2\alpha k^2)\partial_z\bar{\gamma}_{\mu\nu} + (\beta + 4\alpha k)\Box_4\bar{\gamma}_{\mu\nu} = -\frac{2}{M^3}\left\{S_{\mu\nu} - \frac{1}{3}\left(\eta_{\mu\nu} - \frac{\partial_\mu\partial_\nu}{\Box_4}\right)S\right\} \tag{8.109}$$

and

$$(\zeta + \beta k + 2\alpha k^2)\Box_4\gamma = \frac{4k}{3M^3}S. \tag{8.110}$$

It becomes clear here that, as we mentioned above, the equivalent of "brane-bending" effects are included in the tensionful part of the metric γ which is a genuine scalar mode since it couples to matter. Following Davis [61, 62] (see also [44, 45], [52–54]) we obtain the bulk solution (for $p^2 > 0$) which vanishes as $z \to \infty$:

$$\bar{\gamma}_{\mu\nu}(p,z) \sim e^{2k|z|}K_2\left(\frac{|p|e^{k|z|}}{k}\right) \tag{8.111}$$

for $k > 0$, where K_2 are special Bessel functions of the second kind. Therefore the boundary conditions imply

$$\bar{\gamma}_{\mu\nu} = \frac{e^{2kz}K_2(pe^{k|z|}/k)}{K_2(p/k)}\frac{2k}{p^2F(p/k)M^3}\left\{S_{\mu\nu} - \frac{1}{3}\left(\eta_{\mu\nu} - \frac{p_\mu p_\nu}{p^2}\right)S\right\}, \tag{8.112}$$

where

$$F(p/k) = 2(\zeta - 2\alpha k^2)\frac{K_1(p/k)}{p/kK_2(p/k)} + (k\beta + 4\alpha k^2) \tag{8.113}$$

and

$$\gamma = -\frac{k}{p^2}\frac{4}{3(\zeta + \beta k + 2\alpha k^2)M^3}S. \tag{8.114}$$

Any zeros of F with respect to momentum will correspond to tachyon modes on the brane (see [44, 45] for the 2-brane case) and this is essentially due to the fact that we have mixed boundary conditions. The condition for a perturbatively stable theory is to have no ghosts (brane or bulk) and no tachyons; therefore, $\zeta - 2\alpha k^2 > 0$ and $\beta k + 4\alpha k^2 \geq 0$.

We now keep k fixed. For large distances and very small momenta p/k,

$$F(p/k) = (\zeta + k\beta + 2\alpha k^2) + \frac{(\zeta - 2\alpha k^2)}{2k^2}p^2 \left(\ln \frac{p}{2k} + \gamma_E\right) + O(p^4/k^4), \quad (8.115)$$

where $\gamma_E \approx 0.577$ is Euler's constant. For small distances and very large momenta p/k we have

$$F(p/k) = (\beta k + 4\alpha k^2) + (\zeta - 2\alpha k^2)\left(\frac{2k}{p} - \frac{3k^2}{p^2} + O((p/k)^{-3})\right). \quad (8.116)$$

Hence for large distances (small p)

$$\mathscr{G}_{\mu\nu} = \frac{k}{(\zeta + k\beta + 2\alpha k^2)M^3}S_{\mu\nu} + O(p^2). \quad (8.117)$$

We obtain Einstein gravity with $M_{\mathrm{Pl}}^2 = M^3(\zeta + k\beta + 2\alpha k^2)/k$. This is true even if $\zeta = 0$.

For the short distance behaviour and large p, we define an effective scalar mode

$$\phi = -\frac{\zeta - 2\alpha k^2}{2k(\beta + 4\alpha k)}\gamma. \quad (8.118)$$

Then we obtain

$$\mathscr{G}_{\mu\nu} - 2(\eta_{\mu\nu}\Box_4 - \partial_\mu\partial_\nu)\phi = \frac{1}{(\beta + 4\alpha k)M^3}S_{\mu\nu} + O(p^{-1}) \quad (8.119)$$

and

$$\Box_4\phi = -\frac{2(\zeta - 2\alpha k^2)}{3(\zeta + \beta k + 2\alpha k^2)}\frac{1}{(\beta + 4\alpha k)M^3}S, \quad (8.120)$$

which is linearised Brans–Dicke gravity with $M_{\mathrm{Pl}}^2 = M^3(\beta + 8\alpha k)$ and

$$(2\omega + 3) = \frac{3(\zeta + \beta k + 2\alpha k^2)}{4(\zeta - 2\alpha k^2)}, \quad (8.121)$$

ω is a Brans–Dicke coupling. This is clearly an effect absent in conventional GR and is present due to the Gauss–Bonnet term and the induced gravity term on the brane parametrised by α and β, respectively. Solar system constraints dictate that $\omega > 4 \times 10^4$ in order to agree with time delay experiments as that of the Cassini spacecraft [116]. Therefore ω should be pretty big and hence we need to have $\zeta - 2\alpha k^2 \approx 0$, close to the Chern–Simons case.

8.4.2 Codimension 1, Brane Cosmology

Having looked at the perturbation theory of codimension 1 braneworld let us now study the exact solution for cosmology (see also Gregory's lecture notes [41]). Consider a four-dimensional cosmological 3-brane of induced geometry:

$$ds_{ind}^2 = -d\tau^2 + R^2(\tau)\left(\frac{d\chi^2}{1 - \kappa\chi^2} + \chi^2 d\Omega_{II}\right). \tag{8.122}$$

We suppose, following the symmetries of our metric, that matter on the brane is modelled by a perfect fluid of energy density $\mathscr{E}$ and pressure P. The brane is fixed at $z = 0$, and the energy–momentum tensor associated with the brane takes the form

$$S_\mu^{\nu(b)} = \text{diag}(-\mathscr{E}(\tau), P(\tau), P(\tau), P(\tau)).$$

We assume Z_2 symmetry across the location of the brane at $z = 0$ and set $M^3 = 1$ for the time being. The bulk symmetries are four-dimensional cosmological symmetries imposed from (8.122). In other words we have three-dimensional isotropy and homogeneity which lead us to the bulk metric (8.26). Using (8.26) and solving the field equations in the bulk we found two *bulk* gauge degrees of freedom, U and V. These leave the brane junction conditions invariant, or equivalently, the distributional part of the field equations (8.25) invariant, if and only if $U = V$. The one remaining bulk-brane physical degree of freedom can be traced after coordinate transformation to the expansion factor or brane trajectory $R = R(\tau)$ [85], [59] evolving in the black hole bulk (8.39). At the end of the day setting,

$$\mathscr{A} = H^2 + \frac{\kappa}{R^2},$$

where $H = \frac{\dot{R}}{R}$ the generalised Friedmann equation reads as follows:

$$\mathscr{E} - 3\beta\mathscr{A} = \sqrt{\frac{2\mathscr{A}\alpha + \zeta - U}{\alpha}} 2(4\mathscr{A}\alpha + 2\zeta + U), \tag{8.123}$$

where we have defined

$$U = \pm\sqrt{\zeta^2 + \frac{2\alpha\Lambda}{3} + \frac{4\alpha\mu}{R^4}},$$

and we set

$$U_0 = U(\mu = 0) = \zeta - 2\alpha k^2. \tag{8.124}$$

Thus to avoid bulk ghosts we must have $U_0 > 0$, and so the lower branch is ruled out. The standard conservation equation on the brane remains valid and reads as follows:

$$\frac{d\mathscr{E}}{d\tau} + 3H(P + \mathscr{E}) = 0. \tag{8.125}$$

Squaring (8.123) gives a third-order polynomial for $\mathscr{A}$:

$$\frac{\mathscr{E}^2}{2\alpha} = 32\alpha\mathscr{A}^3 + \frac{3}{2\alpha}\mathscr{A}^2(32\alpha\zeta - 3\beta^2) - \frac{3}{\alpha}(2U^2 - \mathscr{E}\beta - 8\zeta^2)\mathscr{A}$$
$$-\frac{1}{2\alpha^2}(U - \zeta)(U + 2\zeta)^2 . \tag{8.126}$$

Setting $\mathscr{E} = T + \rho$ and $\mu = 0$ we can read off the critical tension from (8.126). Critical means that the effective cosmological constant on the brane is zero:

$$T_c^2 = 2\frac{\zeta - U_0}{\alpha}(2\zeta + U_0)^2 = 4k^2(3\zeta - 2\alpha k^2)^2. \tag{8.127}$$

This condition has to be imposed if we want to analyse genuine geometric effect on the modified Friedmann equation. In turn this gives to linear order in ρ the reduced four-dimensional Planck mass

$$\frac{1}{M_{\mathrm{Pl}}^2} = \frac{1}{M^3}\frac{\sqrt{\zeta - U_0}}{\sqrt{2\alpha}(2\zeta - U_0) + \beta\sqrt{\zeta - U_0}}, \tag{8.128}$$

where we restore the fundamental mass scale M of the five-dimensional theory. This agrees with (8.117) obtained in the previous section.

Before proceeding to analysing specific cases it is useful to recast the Friedmann equation inputting T_c and the warp factor k. In order to avoid ghosts and tachyons we have $\beta > -8\alpha k/3$, so let us set $\beta = \beta_0 - 8\alpha k/3$. Then (8.123) for $\kappa = 0$ reads as follows:

$$\mathscr{E} = 3\beta_0 H^2 - 8\alpha k(1 - \mathscr{D})H^2 + (1 - \mathscr{M})\frac{8\alpha k^3}{3}\mathscr{D} + \frac{2\mathscr{D}}{3}T_c\left(1 + \frac{\mathscr{M}}{2}\right), \tag{8.129}$$

where

$$\mathscr{D} = \sqrt{\frac{H^2}{k^2} + \frac{1 + 2\mathscr{M}}{3} + \frac{T_c}{12\alpha k^3}(1 - \mathscr{M})}$$

and

$$\mathscr{M} = \sqrt{1 + \frac{4\alpha\mu}{R^4(\frac{T_c}{6k} - \frac{4\alpha k^2}{3})^2}}.$$

Now we see that in accord with perturbation theory the Gauss–Bonnet term yields an ordinary Friedmann term just as in induced gravity. This is parametrised by β_0. We emphasise that this term is due to the higher order Lovelock correction and not due to the higher order Einstein–Hilbert term. Therefore a naive expectation that an Einstein–Hilbert term of higher dimension gives four-dimensional gravity whereas a Gauss–Bonnet term differing phenomenology is not true. Quite the contrary, higher order Lovelock terms even in codimension 1 give naturally ordinary four-dimensional gravity at some scales. This can be also expected by the presence of the induced gravity term in the Myers boundary for Gauss–Bonnet theory.[21] Start-

---

[21] I thank Nemanja Kaloper for pointing this out.

ing from the above we can obtain the second FRW equation which will tell us about acceleration.

As an example let us now consider the case of zero tension. For a general analysis see also [82]. Although this means that we fine-tune the couplings, $2\alpha k^2 = 3\zeta$, in essence nothing special happens and the equations become easier to deal with. This means that in order not to have ghosts we take $\alpha < 0$, upon which $\beta_0 \sim M_{pl}^2$ (for M=1). The Friedmann equation (for $\kappa = 0$) simplifies to

$$\rho = 3\beta_0 H^2 + 8|\alpha|k(1 - \mathscr{Q})H^2 - (1 - \mathscr{M})\frac{8|\alpha|k^3}{3}\mathscr{Q}, \tag{8.130}$$

where now $\mathscr{Q} = \sqrt{\frac{H^2}{k^2} + \frac{1 + 2\mathscr{M}}{3}}$ and $\mathscr{M} = \sqrt{1 - \frac{9\mu}{4|\alpha|k^4 R^4}}$ with $\mathscr{M} = 1$ for a pure adS background. For late-time acceleration we need the second FRW equation. To obtain it, we differentiate (8.130) and successive use of (8.125) and (8.130) gives

$$
\begin{aligned}
& - (\rho + 3P) + \frac{8|\alpha|k^3}{\mathscr{Q}}\left(\frac{H^2}{k^2} + \frac{1 - \mathscr{M}}{3}\right)\left(-\frac{H^2}{k^2} + \frac{2(1 - \mathscr{M}^2)}{3\mathscr{M}}\right) \\
& - \frac{16(1 - \mathscr{M})\mathscr{Q}}{3\mathscr{M}}|\alpha|k^3 \\
& = 2\frac{\ddot{R}}{R}\left[\frac{\rho}{H^2} + \frac{8|\alpha|k^3}{3H^2}(1 - \mathscr{M})\mathscr{Q} - \frac{4|\alpha|k}{\mathscr{Q}}\left(\frac{H^2}{k^2} + \frac{1 - \mathscr{M}}{3}\right)\right].
\end{aligned} \tag{8.131}
$$

Setting $\alpha = 0$ gives ordinary Friedmann equations. This limit effectively kills all the five-dimensional part of the action for the special case we are treating. The second term in (8.130) can be interpreted as an α correction (of negative sign) and the third term a bulk black hole term (of positive sign). The terms on the left hand side of (8.131) tell us whether there is possible acceleration or not at late time for ordinary matter ($w = 0$) or radiation $w = 1/3$. This fact is true modulo the sign of the parenthesis of the right hand side which may also change sign as, we will see, inverting the accelerating equations of state but breaking the strong energy condition given (8.125).

It is now indicative to set $\mu = 0$ and study late-time effects. Then (8.130) simplifies to

$$\rho = 3\beta_0 H^2 + 16|\alpha|H^2(k - \sqrt{k^2 + H^2}) \tag{8.132}$$

and (8.131) to

$$-(\rho + 3P) - \frac{8|\alpha|H^4}{\sqrt{k^2 + H^2}} = 2\frac{\ddot{R}}{R}\left[\frac{\rho}{H^2} - \frac{4|\alpha|H^2}{\sqrt{H^2 + k^2}}\right]. \tag{8.133}$$

The first term on the right hand side of (8.132) is the usual FRW term (with the right Planck mass) but the second term is of negative sign which means that we cannot take H arbitrarily large. For late times cosmology approaches usual LFRW cosmology. Lastly let us set the bulk Einstein–Hilbert term to zero $\zeta = 0$ and take $\alpha < 0$. Then $T_c = -4\alpha k^3$ and

$$\mathcal{Q} = \sqrt{\frac{H^2}{k^2} + \mathcal{M}},$$

with

$$\mathcal{M} = \sqrt{1 + \frac{\mu}{\alpha k^4 R^4}}.$$

Then the effective LFRW equations reduce to

$$\mathcal{E} = 3\beta_0 H^2 - 8\alpha k(1 - \mathcal{Q})H^2 - 4\alpha k^3 \mathcal{M}\mathcal{Q}, \qquad (8.134)$$

which gives at late times ordinary four-dimensional cosmology despite the fact that there is no Einstein–Hilbert in the bulk action. This proves our claim in the introduction that the higher order terms of Lovelock theory enhance ordinary four-dimensional gravity.

8.4.3 Codimension 2 Braneworlds

Let us now look at a braneworld of codimension 2; in other words let us consider six-dimensional bulk spacetime with four-dimensional maximally symmetric subsections. The general bulk solutions, as we saw in Section 8.2, are solitons of manifest axial symmetry with $\partial/\partial\theta$ as the angular Killing vector,

$$ds^2 = V(r)d\theta^2 + \frac{dr^2}{V(r)} + r^2 d^2 K_4, \qquad (8.135)$$

and $d^2 K_4$ is the four-dimensional line element of adS, flat or dS spacetime, $\kappa = -1, 0, 1$, respectively. For suitable parameters in (8.57) the radial coordinate varies in between $r_- \leq r \leq r_+$, where $r = r_\pm$ are the former horizon positions for (8.46) and will be the possible brane locations, $r = r_\pm$ for (8.135). In particular $r = r_+$ can be infinity itself in which case the effective volume element in the (r, θ) direction is infinite (for the analysis in the Einstein case see [117–120]). This is the codimension 2 version of warped compactification. We see therefore that the six-dimensional soliton (8.135) possesses the correct spacetime symmetries to describe the general maximally symmetric four-dimensional braneworld of constant curvature. The Wick-rotated version of the staticity theorem tells us in particular that axial symmetry comes for free and need not be imposed when resolving the system of bulk equations. In order to introduce codimension 2 branes carrying some two-dimensional Dirac charge or tension, we need to reintroduce conical singularities at the relevant axis origins (8.59) which are at $r = r_h = r_-$ and $r = r_a = r_+$ by allowing for the presence of deficit angles $\beta_\pm$. Indeed from (8.60) we have the identification $\theta \sim \theta + \frac{2\pi - \beta_\pm}{\frac{1}{2}V'_\pm}$ which means that we have the relation

$$\frac{\beta_+}{\frac{1}{2}V_+'} = \frac{\beta_-}{\frac{1}{2}V_-'} \tag{8.136}$$

which relates the topological parameters $\beta_\pm$ with the geometrical quantities such as mass and charge of the soliton metric (8.135). In particular note that we can always get rid of one of the conical singularities (and the resulting tensionful brane) and thus construct warped spacetimes with finite volume element and a single brane. Since the extrinsic curvatures for the branes are zero, the brane junction conditions [67] are given by

$$2\pi(1-\beta_\pm)\left(\delta_\mu^\nu + 4\alpha G^\nu_{\mu\,\mathrm{ind}}\right) = S_\mu^\nu, \tag{8.137}$$

with $T_{\mu\nu}^{brane} = S_{\mu\nu}\delta^{(2)}(\rho) = S_{\mu\nu}\frac{\delta(\rho)}{2\pi\rho}$. We see the appearance of the induced Einstein tensor on the brane originating from the Gauss–Bonnet bulk term in the Lovelock action [69, 107]. Note that the warp factor is given by the value of $r_\pm^2$, and in particular for $\kappa = 1$, say, $G_{\mu\nu}^{ind} = -3H_\pm^2\gamma_{\mu\nu} = -\frac{3}{r_\pm^2}\gamma_{\mu\nu}$ where $\gamma_{\mu\nu}$ is the de Sitter-induced metric with curvature set to 1. The induced Newton's constant on the brane is given by

$$G_4^\pm = \frac{G_6}{8\pi\hat{\alpha}(1-\beta_\pm)}. \tag{8.138}$$

A complete analysis of these solutions is given in [68] where self-tuning and self-accelerating solutions are studied.

8.5 The Extended Kaluza–Klein Reduction

In the previous section we discussed warped compactifications of higher dimensional spacetimes. In this section we will discuss the case of Kaluza–Klein compactification. It is well known that the Kaluza–Klein (KK) reduction of Einstein theory gives us an Einstein–Maxwell dilaton theory (EMD) for specific KK couplings and some periodic boundary conditions. But what is the resulting theory for the KK reduction of a higher dimensional Lovelock theory? Is the resulting reduced theory going to give second-order field equations? Is the resulting theory unique as its higher dimensional counterpart? Let us take here for simplicity the Kaluza–Klein reduction of an Einstein–Gauss–Bonnet theory from $D = 5$ dimensions to $D = 4$ dimensions. The full analysis in arbitrary D and for Lovelock theory was carried out in [71–73] (see also [25, 26] for applications to cosmology). Starting with the relevant Lovelock theory in five-dimensions,

$$S = \frac{1}{16\pi G}\int_{\mathscr{M}} d^4x\,dy\,\sqrt{-g^{(5)}}\left(R^{(5)} - 2\Lambda + \alpha\hat{G}^{(5)}\right). \tag{8.139}$$

We consider the following ansatz for the five-dimensional metric,

$$ds^2 = g^{(5)}_{ab} dx^a dx^b$$
$$= (g_{\mu\nu} + e^{-4\phi} A_\mu A_\nu) dx^\mu dx^\nu + 2A_\mu e^{-2\phi} dx^\mu dy + e^{-4\phi} dy^2 \ . \quad (8.140)$$

As usual we are making the basic assumption that there is a Killing vector ∂_y in the fifth direction; in other words we do not consider here warped solutions; we are rather interested in integrating out directly and obtaining the resulting four-dimensional theory. We impose periodic boundary conditions on y. We expect that the resulting four-dimensional theory will be an extension of the Einstein–Maxwell dilaton (EMD) theory in four-dimensions with some exponential potential (in the presence of a five-dimensional cosmological constant). Indeed integrating out the y direction we obtain [71–73]

$$S_{\text{eff}} = \int_{\mathcal{M}_4} d^4x \sqrt{-g} e^{-2\phi} \left\{ R - (\nabla\phi)^2 - 2\Lambda e^{\gamma\phi} - \tfrac{1}{4}F^2 + \alpha \hat{G}^{(4)} \right.$$
$$\left. + \frac{3\alpha}{16} e^{-8\phi} \left[(F_{\mu\nu}F^{\mu\nu})^2 - 2F^\nu_\mu F^\lambda_\nu F^\kappa_\lambda F^\mu_\kappa \right] \right\} - S_{int}, \quad (8.141)$$

where note the presence of a non-trivial interaction term which reads as follows:

$$S_{\text{int}} = -\tfrac{1}{2} \int d^4x \sqrt{-g} e^{-6\phi} \left(F_{\mu\nu} F^{\kappa\lambda} R^{\mu\nu}_{\kappa\lambda} - 4F_{\mu\kappa} F^{\nu\kappa} R^\mu_\nu - F^2 R \right) \ . \quad (8.142)$$

The interesting result that one can prove is that the field equations obtained from variation of the fields are still of second order as for the higher dimensional Lovelock metric theory. In fact the higher order EMD theory in question is the most general second-order theory that has up to second-order partial derivatives. Any other numerical combination of the interaction terms, for example, would have given higher order derivatives. Hence we come to the interesting conclusion that higher dimensional Lovelock theories when dimensionally reduced via the KK formalism retain their nice properties dictated by Lovelock's theorem. In order to give the simple basic properties of the four-dimensional theories we freeze out in turn the degrees of freedom in (8.141).

Taking a constant dilaton the Gauss–Bonnet term drops out, since in four-dimensions it is a topological invariant, and we are left with a modified Einstein–Maxwell theory:

$$S_{\text{eff}} = \int_{\mathcal{M}_4} d^4x \sqrt{-g} \left\{ R - 2\Lambda - \tfrac{1}{4}F^2 + \tfrac{3\alpha}{16} \left[(F_{\mu\nu}F^{\mu\nu})^2 - 2F^\nu_\mu F^\lambda_\nu F^\kappa_\lambda F^\mu_\kappa \right] \right\}$$
$$- S_{int} \ . \quad (8.143)$$

Black hole solutions to this modified Einstein–Maxwell theory and for $\Lambda = 0$ have been studied (partially numerically) [121] and they are corrected Reissner–Nordstrom solutions. For flat spacetime in particular, i.e. setting $g_{\mu\nu} = \eta_{\mu\nu}$, we get a non-linear version of Maxwell's theory which reads as follows (see the nice paper by Kerner [122] whose notation we follow here):

$$S_{\text{eff}} = \int_{\mathcal{M}_4} d^4x \sqrt{-g} \left\{ -\tfrac{1}{4}F^2 + \tfrac{3\alpha}{16} \left[(F_{\mu\nu}F^{\mu\nu})^2 - 2F_\mu^\nu F_\nu^\lambda F_\lambda^\kappa F_\kappa^\mu \right] \right\}, \qquad (8.144)$$

with field equations ($\alpha = \tfrac{3\gamma}{16e^2}$),

$$\partial_\lambda \left(F^{\lambda\rho} - \tfrac{3\gamma}{2e^2}F^2 F^{\lambda\rho} \right) + \tfrac{3\gamma}{e^2}\partial_\lambda (F_{\mu\nu}F^{\lambda\mu}F^{\rho\nu}) = 0,$$

$$\partial_{[\mu}F_{\lambda\rho]} = 0. \qquad (8.145)$$

We see that the usual Maxwell equations are corrected by non-linear terms with coupling γ. It is interesting to recast everything with respect to the electric and magnetic fields, E, B. We then get

$$divE = -\frac{3\gamma}{e^2}B \cdot grad(E \cdot B),$$

$$rot(B) = \frac{\partial E}{\partial t} + \frac{3\gamma}{e^2}\left[B\frac{\partial(E \cdot B)}{\partial t} - E \wedge grad(E \cdot B) \right],$$

$$div(B) = 0,$$

$$rot(E) = -\frac{\partial B}{\partial t}. \qquad (8.146)$$

In this form we can make two obvious remarks. First of all whenever the electric and magnetic fields are perpendicular to each other, higher order terms drop out and hence usual EM solutions are unchanged. This holds in particular for electromagnetic wave solutions. However note that since we loose linearity one can no longer necessarily superimpose electromagnetic wave solutions if they are not perpendicular to each other. Furthermore, we can define an induced charge density and current

$$\rho_{ind} = -\frac{3\gamma}{e^2}B \cdot grad(E \cdot B)$$

$$j_{ind} = \frac{3\gamma}{e^2}\left[B\frac{\partial(E \cdot B)}{\partial t} - E \wedge grad(E \cdot B) \right] \qquad (8.147)$$

that simulate the higher order terms and verify the continuity equation, $\frac{\partial\rho_{ind}}{\partial t} + div(j_{ind}) = 0$.

Let us now in turn freeze the vector field strength $F_{\mu\nu} = 0$. Then we obtain a second-order scalar–tensor theory that has been studied quite a lot recently [61, 62]. Numerical black hole solutions to such theories were discussed early on in [123]. It is important to note that in this case the Gauss–Bonnet four-dimensional scalar is no longer redundant for the field equations and plays the role of the mediator in between Einstein gravity and the scalar sector. This is true in whatever frame we choose to go. For example in the Einstein frame, in other words even when the scalar does not couple to linear order with gravity we have that it does so with the Gauss–Bonnet term,

$$S_{\text{eff}} = \int \mathcal{M}_4 d^4x \sqrt{-g} \left\{ R - (\nabla \phi)^2 - 2\Lambda e^{\gamma \phi} + \alpha e^{\delta \phi} \hat{G} \right\}, \tag{8.148}$$

where δ, γ are some specific couplings depending on the dimensional reduction. Such models are higher order corrected dark energy-quintessence models that are used to evoke late-time acceleration of the universe. The essential point here is that although to leading order they do not affect solar system constraints, once one includes the Gauss–Bonnet term this is no longer true and their solar system constraints can be rather stringent [124–130]. Most stringent constraints arise from light time delay which is calculated from the Cassini spacecraft [116]. For cosmological constraints which are far weaker one can consult [127–130]. An interesting alternative has been put forward recently [131] in the case of a higher order generalisation of Brans–Dicke theory. It was there shown that the combined effect of the higher order corrections and the scalar sector can reduce or even eliminate the constraint on the Brans–Dicke parameter ω by imposing particular higher order coupling functions.

8.6 Concluding Remarks and Open Problems

In these short lectures we saw some of the basic properties of Lovelock theory. Some cosmological applications were given for codimension 1 and 2 braneworlds. There are a large number of open problems in these theories and certain results/issues which we have not treated here. One of the aims of this closing section is to list and comment on some of these open problems.

Let us start with exact solutions (check out the lecture notes of N. Obers for black holes in Einstein gravity). For definiteness we studied only the static ones here; a Taub-NUT version of these has been found in [32]. Stationary metrics for Lovelock theory have not been found despite efforts. Another important solution which is missing is that of the black string (see [83] for a perturbative treatment). Although this solution is trivially found in GR in Lovelock theory this is not the case. The reason is that Ricci flat solutions in four-dimensions (as is the four-dimensional black hole in the five-dimensional black string) are not vacuum solutions to Lovelock's equations. In fact the bulk Weyl tensor also contributes to the field equations (unlike in Einstein's equations!) for Lovelock order $k \geq 2$ [82, 132], i.e., once we switch on the Gauss–Bonnet invariant. Therefore a four-dimensional Ricci flat solution which is non-conformally flat will not solve the Lovelock equations (see the nice analysis of [35]). The absence of such simple solutions is maybe an accident but maybe it is also questioning the relevance of the black string type of solutions which are already questionable [133]. At the same time if such an exact solution was found for Lovelock theory it would without doubt be genuinely different from its Einstein version. What could we then say of its stability? Another generic technical problem is that even in the absence of a bare cosmological constant in the action we always expect one in the Lovelock solutions due to the presence of multiple branches which have no flat Einstein theory limit! Thus the candidate metric for resolution must be written in an anzatz suitable for a cosmological constant solution even in the absence of a bare cosmological constant! We know from studies in Einstein theory that beyond a certain symmetry a cosmological constant spoils integrability [134, 135]. One way around this is to start by looking for solutions at special cases as the Chern–Simons case. Furthermore additional technical problems, due to the absence of Ricci flatness, can arise if we simply translate Einstein results. For example, the diagonal Weyl anzatz in [134, 136] is a priori no longer true for $k > 2$ for in order to obtain it we use the fact that the background is Ricci flat![22]

Another subject that deserves further attention is that of higher codimension braneworlds. We saw that in the context of Lovelock theory one could in principle define higher codimension braneworlds with Dirac distributions. Only recently [70] did we obtain the first example of a codimension 4 defect having only a removable Dirac singularity. In particular the case of codimension 2 and its cosmology has yet to be elucidated. In this review we explained how one could obtain the exact solutions describing the maximally symmetric branes [68]. To what extend does Lovelock theory permit us to recover Einstein gravity on the brane [67]? Our understanding from [69, 107] is that Lovelock gravity gives us induced Einstein terms on the brane but not Einstein gravity on the brane at all scales. In other words similarly to DGP, only up to some crossover scale do we expect gravity on a codimension 2 brane to be four-dimensional. Localised four-dimensional gravity will occur only when there exists a localised four-dimensional zero mode graviton in the gravitational spectrum of perturbations. Therefore clearly, what is missing is a clear-cut way of developing perturbation theory in Lovelock gravity. We firmly believe that beyond the Gauss–Bonnet term one has to use differential form formalism which we highlighted here. Only then will we know the gravitational spectrum of higher codimension braneworlds and their four-dimensional phenomenology. We must point out that perturbations of Einstein–Gauss–Bonnet black holes have been carried out in [137–140] and causality issues have emerged in the context of the adS/CFT correspondence and EGB planar black hole perturbations [141].

We hope that with this manuscript we have communicated the fact that Lovelock theory is a technically challenging, interesting, well-motivated and exciting subject of research with numerous open problems that await resolution.

Acknowledgments It is a great pleasure to thank L. Amendola, S. C. Davis, J.-F. Dufaux, G. Kofinas, A. H. Padilla, A. Papazoglou and R. Z. Zegers for past and actual collaboration on some of the topics raised here. I also thank Nemanja Kaloper for many interesting and critical comments concerning numerous issues on Lovelock gravity. I thank Robin Zegers for taking time and reading through the manuscript and Renaud Parentani for discussions on gravitational instantons. Last but not least I thank the organisers of the Aegean black hole school for giving me the opportunity to participate and give this set of lectures.

---

[22] I thank Robin Zegers for pointing this out.

References

1. R. M. Wald, *General Relativity* (University Press, Chicago, USA 1984) 491p
2. C. M. Will, The confrontation between general relativity and experiment, gr-qc/0510072
3. C. M. Will, *Theory and Experiment in Gravitational Physics* (Cambridge University Press, Cambridge 1993).
4. A. G. Riess et al. and Supernova Search Team Collaboration, Observational evidence from supernovae for an accelerating universe and a cosmological constant, Astron. J. **116**, 1009 (1998) [astro-ph/9805201].
5. A. G. Riess et al. and Supernova Search Team Collaboration, Type Ia supernova discoveries at $z > 1$ from the Hubble space telescope: Evidence for past deceleration and constraints on dark energy evolution, Astrophys. J. **607**, 665 (2004) [astro-ph/0402512].
6. S. Perlmutter et al. and Supernova Cosmology Project Collaboration, Measurements of omega and lambda from 42 high-redshift supernovae, Astrophys. J. **517**, 565 (1999) [astro-ph/9812133].
7. S. M. Carroll, Living Rev. Rel. **4**, 1 (2001) [arXiv:astro-ph/0004075].
8. J. Polchinski, arXiv:hep-th/0603249.
9. R. Bousso, arXiv:0708.4231 [hep-th].
10. C. Brans and R. H. Dicke, Phys. Rev. **124**, 925 (1961).
11. T. Damour and G. Esposito-Farese, Class. Quant. Grav. **9**, 2093 (1992).
12. C. Eling, T. Jacobson and D. Mattingly, arXiv:gr-qc/0410001.
13. Woodard, R. P.: Lect. Notes Phys. **720**, 403 (2007) [arXiv:astro-ph/0601672].
14. S. M. Carroll, A. De Felice, V. Duvvuri, D. A. Easson, M. Trodden and M. S. Turner, Phys. Rev. D **71**, 063513 (2005) [arXiv:astro-ph/0410031].
15. B. Whitt, Phys. Lett. B **145**, 176 (1984).
16. B. Zwiebach, Phys. Lett. B **156**, 315 (1985).
17. D. J. Gross and J. H. Sloan, The quartic effective action for the heterotic string, Nucl. Phys. B **291**, 41 (1987).
18. R. R. Metsaev and A. A. Tseytlin, Order alpha-prime (two loop) equivalence of the string equations of motion and the sigma model Weyl invariance conditions: Dependence on the dilaton and the antisymmetric tensor, Nucl. Phys. B **293**, 385 (1987).
19. D. Lovelock, J. Math. Phys. **12**, 498 (1971).
20. C. Lanczos, Z. Phys. **73**, 147 (1932).
21. C. Lanczos, Ann. Math. **39**, 842 (1938).
22. B. Zumino, Gravity theories in more than four-dimensions, Phys. Rept. **137**, 109 (1986).
23. I. Antoniadis, J. Rizos and K. Tamvakis, Nucl. Phys. B **415**, 497 (1994) [arXiv:hep-th/9305025].
24. B. A. Campbell, M. J. Duncan, N. Kaloper and K. A. Olive, Nucl. Phys. B **351**, 778 (1991).
25. J. Madore, On the nature of the initial singularity in a Lanczos cosmological model, Phys. Lett. A **111**, 238 (1985).
26. N. Deruelle and J. Madore, The Friedmann universe as an attractor of a Kaluza-Klein cosmology, Mod. Phys. Lett. A **1**, 237 (1986).
27. D. G. Boulware and S. Deser, Phys. Rev. Lett. **55**, 2656 (1985).
28. J. T. Wheeler, Nucl. Phys. B **273**, 732 (1986).
29. J. T. Wheeler, Nucl. Phys. B **268**, 737 (1986).
30. D. L. Wiltshire, Phys. Lett. B **169**, 36 (1986).
31. R. C. Myers and J. Z. Simon, Phys. Rev. D **38**, 2434 (1988).
32. M. H. Dehghani and R. B. Mann, Phys. Rev. D **72**, 124006 (2005) [arXiv:hep-th/0510083].
33. M. H. Dehghani, Phys. Rev. D **67**, 064017 (2003) [arXiv:hep-th/0211191].
34. S. H. Hendi and M. H. Dehghani, arXiv:0802.1813 [hep-th].
35. D. Kastor and R. B. Mann, JHEP **0604**, 048 (2006) [arXiv:hep-th/0603168].
36. S. Deser and B. Tekin, Energy in generic higher curvature gravity theories, Phys. Rev. D **67**, 084009 (2003) [hep-th/0212292].
37. G. Kofinas and R. Olea, JHEP **0711**, 069 (2007) [arXiv:0708.0782 [hep-th]].

38. G. Kofinas and R. Olea, Phys. Rev. D **74**, 084035 (2006) [arXiv:hep-th/0606253].
39. J. P. Gregory and A. Padilla, Braneworld holography in Gauss-Bonnet gravity, Class. Quant. Grav. **20**, 4221 (2003) [hep-th/0304250].
40. A. Padilla, Class. Quant. Grav. **22**, 681 (2005) [arXiv:hep-th/0406157].
41. R. Gregory, arXiv:0804.2595 [hep-th].
42. H. Collins and B. Holdom, Phys. Rev. D **62**, 124008 (2000) [arXiv:hep-th/0006158].
43. G. R. Dvali, G. Gabadadze and M. Porrati, Phys. Lett. B **485**, 208 (2000) [arXiv:hep-th/0005016].
44. C. Charmousis and J. F. Dufaux, Phys. Rev. D **70**, 106002 (2004) [arXiv:hep-th/0311267].
45. M. Minamitsuji and M. Sasaki, Prog. Theor. Phys. **112**, 451 (2004) [arXiv:hep-th/0404166].
46. W. Israel, Nuovo Cim. B **44S10**, 1 (1966).
47. W. Israel, Nuovo Cim. B **48**, 463 (1967) (Erratum).
48. G. Darmois, *Mémorial des sciences mathématiques*, **XXV** (1927).
49. N. Deruelle and T. Dolezel, Phys. Rev. D **62**, 103502 (2000) [arXiv:gr-qc/0004021].
50. N. E. Mavromatos and J. Rizos, String inspired higher-curvature terms and the Randall-Sundrum scenario, Phys. Rev. D **62**, 124004 (2000) [hep-th/0008074].
51. N. E. Mavromatos and J. Rizos, Exact solutions and the cosmological constant problem in dilatonic domain wall higher-curvature string gravity, Int. J. Mod. Phys. A **18**, 57 (2003) [hep-th/0205299].
52. K. A. Meissner and M. Olechowski, Brane localization of gravity in higher derivative theory, Phys. Rev. D **65**, 064017 (2002) [hep-th/0106203].
53. Y. M. Cho, I. P. Neupane and P. S. Wesson, No ghost state of Gauss-Bonnet interaction in warped background, Nucl. Phys. B **621**, 388 (2002) [hep-th/0104227].
54. N. Deruelle and M. Sasaki, Newton's law on an Einstein 'Gauss-Bonnet' brane, gr-qc/0306032.
55. A. Jakobek, K. A. Meissner and M. Olechowski, New brane solutions in higher order gravity, Nucl. Phys. B **645**, 217 (2002) [hep-th/0206254].
56. K. A. Meissner and M. Olechowski, Domain walls without cosmological constant in higher order gravity, Phys. Rev. Lett. **86**, 3708 (2001) [hep-th/0009122].
57. P. Binetruy, C. Charmousis, S. C. Davis and J. F. Dufaux, Avoidance of naked singularities in dilatonic brane world scenarios with a Gauss-Bonnet term, Phys. Lett. B **544**, 183 (2002) [hep-th/0206089].
58. C. Charmousis, S. C. Davis and J. F. Dufaux, JHEP **0312**, 029 (2003) [arXiv:hep-th/0309083].
59. C. Charmousis and J. F. Dufaux, Class. Quant. Grav. **19**, 4671 (2002) [arXiv:hep-th/0202107].
60. N. Deruelle and J. Madore, arXiv:gr-qc/0305004.
61. S. C. Davis, Phys. Rev. D **67**, 024030 (2003) [arXiv:hep-th/0208205].
62. E. Gravanis and S. Willison, Phys. Lett. B **562**, 118 (2003) [arXiv:hep-th/0209076].
63. R. C. Myers, Phys. Rev. D **36**, 392 (1987).
64. J. F. Dufaux, J. E. Lidsey, R. Maartens and M. Sami, Phys. Rev. D **70**, 083525 (2004) [arXiv:hep-th/0404161].
65. J. E. Lidsey and N. J. Nunes, Phys. Rev. D **67**, 103510 (2003) [arXiv:astro-ph/0303168].
66. G. Kofinas, R. Maartens and E. Papantonopoulos, Brane cosmology with curvature corrections, hep-th/0307138.
67. P. Bostock, R. Gregory, I. Navarro and J. Santiago, Phys. Rev. Lett. **92**, 221601 (2004) [arXiv:hep-th/0311074].
68. C. Charmousis and A. Papazoglou, arXiv:0804.2121 [hep-th].
69. C. Charmousis and R. Zegers, Phys. Rev. D **72**, 064005 (2005) [arXiv:hep-th/0502171].
70. R. Zegers, arXiv:0801.2262 [gr-qc].
71. F. Mueller-Hoissen, Nucl. Phys. B **337**, 709 (1990).
72. F. Mueller-Hoissen, Phys. Lett. B **201**, 325 (1988).
73. F. Mueller-Hoissen, Class. Quant. Grav. **5**, L35 (1988).
74. R. C. Myers, Phys. Rev. D **36**, 392 (1987).

75. N. Deruelle and J. Madore, arXiv:gr-qc/0305004.
76. T. Eguchi, P.B. Gilkey and A.J. Hanson, Phys. Rep. **66**(6), 213–293 (1980).
77. P.B. Gilkey, *Invariance Theory, the Heat Equation and the Atiyah-Singer Index Theorem* (Publish or Perish Inc., Houston 1984) ISBN 0-914098-20-9.
78. S.-S. Chern, Ann. Math. **45**, 747–752 (1944).
79. S.-S. Chern, Ann. Math. **46**, 674–684 (1945).
80. M. Spivak, *A Comprehensive Introduction to Differential Geometry* (Publish or Perish, Houston 1999).
81. G. W. Gibbons and S. W. Hawking, Phys. Rev. D **15**, 2752 (1977).
82. C. Barcelo, R. Maartens, C. F. Sopuerta and F. Viniegra, Phys. Rev. D **67**, 064023 (2003) [arXiv:hep-th/0211013].
83. T. Kobayashi and T. Tanaka, arXiv:gr-qc/0412139.
84. R. Zegers, J. Math. Phys. **46**, 072502 (2005) [arXiv:gr-qc/0505016].
85. P. Bowcock, C. Charmousis and R. Gregory, Class. Quant. Grav. **17**, 4745 (2000) [arXiv:hep-th/0007177].
86. J. Crisostomo, R. Troncoso and J. Zanelli, Black hole scan, Phys. Rev. D **62**, 084013 (2001) [hep-th/0003271].
87. R. G. Cai, Phys. Rev. D **65**, 084014 (2002) [arXiv:hep-th/0109133].
88. R. Gregory and A. Padilla, Nested braneworlds and strong brane gravity, Phys. Rev. D **65**, 084013 (2002) [hep-th/0104262].
89. S. Deser and B. Tekin, Phys. Rev. D **67**, 084009 (2003) [arXiv:hep-th/0212292].
90. C. Charmousis and A. Padilla, arXiv:0807.2864
91. C. de Rham and A. J. Tolley, JCAP **0607**, 004 (2006) [arXiv:hep-th/0605122].
92. G. W. Gibbons and M. J. Perry, Proc. Roy. Soc. Lond. A **358**, 467 (1978).
93. P. H. Ginsparg and M. J. Perry, Nucl. Phys. B **222**, 245 (1983).
94. S. W. Hawking, G. T. Horowitz and S. F. Ross, Phys. Rev. D **51**, 4302 (1995) [arXiv:gr-qc/9409013].
95. C. Garraffo, G. Giribet, E. Gravanis and S. Willison, arXiv:0711.2992 [gr-qc].
96. T. Clunan, S. F. Ross and D. J. Smith, Class. Quant. Grav. **21**, 3447 (2004) [arXiv:gr-qc/0402044].
97. V. Iyer and R. M. Wald, Phys. Rev. D **50**, 846 (1994) [arXiv:gr-qc/9403028].
98. V. Iyer and R. M. Wald, Phys. Rev. D **52**, 4430 (1995) [arXiv:gr-qc/9503052].
99. B. Linet, J. Math. Phys. **27**, 1817 (1986).
100. G. T. Horowitz and R. C. Myers, Phys. Rev. D **59**, 026005 (1999) [arXiv:hep-th/9808079].
101. J. A. G. Vickers, Class. Quant. Grav. **4**, 1 (1987).
102. V. P. Frolov, W. Israel and W. G. Unruh, Phys. Rev. D **39**, 1084 (1989).
103. W. G. Unruh, G. Hayward, W. Israel and D. Mcmanus, Phys. Rev. Lett. **62**, 2897 (1989).
104. B. Boisseau, C. Charmousis and B. Linet, Phys. Rev. D **55**, 616 (1997) [arXiv:gr-qc/9607029].
105. C. Barrabes and W. Israel, Phys. Rev. D **43**, 1129 (1991).
106. C. Charmousis and R. Zegers, Phys. Rev. D **72**, 064005 (2005) [arXiv:hep-th/0502171].
107. C. Charmousis and R. Zegers, JHEP **0508**, 075 (2005) [arXiv:hep-th/0502170].
108. M. R. Anderson, F. Bonjour, R. Gregory and J. Stewart, Phys. Rev. D **56**, 8014 (1997) [arXiv:hep-ph/9707324].
109. G. Kofinas, arXiv:hep-th/0412299.
110. R. Gregory, Nucl. Phys. B **467**, 159 (1996) [arXiv:hep-th/9510202].
111. C. Charmousis, R. Emparan and R. Gregory, JHEP **0105**, 026 (2001) [arXiv:hep-th/0101198].
112. C. Charmousis and U. Ellwanger, JHEP **0402**, 058 (2004) [arXiv:hep-th/0402019].
113. S. C. Davis, AIP Conf. Proc. **736**, 147 (2005) [arXiv:hep-th/0410075].
114. S. C. Davis, Phys. Rev. D **72**, 024026 (2005) [arXiv:hep-th/0410065].
115. S. C. Davis, arXiv:hep-th/0408139.
116. B. Bertotti, L. Iess and P. Tortora, A test of general relativity using radio links with the Cassini spacecraft, Nature **425**, 374 (2003).

117. S. Mukohyama, Y. Sendouda, H. Yoshiguchi and S. Kinoshita, JCAP **0507**, 013 (2005) [arXiv:hep-th/0506050].
118. M. Peloso, L. Sorbo and G. Tasinato, Phys. Rev. D **73**, 104025 (2006) [arXiv:hep-th/0603026].
119. E. Papantonopoulos, A. Papazoglou and V. Zamarias, JHEP **0703**, 002 (2007) [arXiv:hep-th/0611311].
120. C. P. Burgess, D. Hoover and G. Tasinato, JHEP **0709**, 124 (2007) [arXiv:0705.3212 [hep-th]].
121. F. Mueller-Hoissen and R. Sippel, Class. Quant. Grav. **5**, 1473 (1988).
122. R. Kerner, C. R. Acad. Sc. Paris, t. 304, Série II, n° 12, 1987.
123. P. Kanti, N. E. Mavromatos, J. Rizos, K. Tamvakis and E. Winstanley, Dilatonic black holes in higher curvature string gravity, Phys. Rev. D **54**, 5049 (1996) [hep-th/9511071].
124. G. Esposito-Farese, Scalar-tensor theories and cosmology and tests of a quintessence-Gauss-Bonnet coupling, gr-qc/0306018.
125. G. Esposito-Farese, Tests of scalar-tensor gravity, AIP Conf. Proc. **736**, 35 (2004) [gr-qc/0409081].
126. L. Amendola, C. Charmousis and S. C. Davis, JCAP **0710**, 004 (2007) [0704.0175 [astro-ph]].
127. L. Amendola, C. Charmousis and S. C. Davis, Constraints on Gauss-Bonnet gravity in dark energy cosmologies, JCAP **0612**, 020 (2006) [hep-th/0506137].
128. T. Koivisto and D. F. Mota, Cosmology and astrophysical constraints of Gauss-Bonnet dark energy, Phys. Lett. B **644**, 104 (2007) [astro-ph/0606078].
129. T. Koivisto and D. F. Mota, Gauss-Bonnet quintessence: Background evolution, large scale structure and cosmological constraints, Phys. Rev. D **75**, 023518 (2007) [hep-th/0609155].
130. B. M. Leith and I. P. Neupane, Gauss-Bonnet cosmologies: Crossing the phantom divide and the transition from matter dominance to dark energy, hep-th/0702002.
131. L. Amendola, C. Charmousis and S. C. Davis, arXiv:0801.4339 [gr-qc].
132. B. Cuadros-Melgar, E. Papantonopoulos, M. Tsoukalas and V. Zamarias, arXiv:0804.4459 [hep-th].
133. R. Gregory and R. Laflamme, Phys. Rev. Lett. **70**, 2837 (1993) [arXiv:hep-th/9301052].
134. C. Charmousis and R. Gregory, Class. Quant. Grav. **21**, 527 (2004) [arXiv:gr-qc/0306069].
135. C. Charmousis, D. Langlois, D. Steer and R. Zegers, JHEP **0702**, 064 (2007) [arXiv:gr-qc/0610091].
136. R. Emparan and H. S. Reall, Phys. Rev. D **65**, 084025 (2002) [arXiv:hep-th/0110258].
137. M. Beroiz, G. Dotti and R. J. Gleiser, Phys. Rev. D **76**, 024012 (2007) [arXiv:hep-th/0703074].
138. R. J. Gleiser and G. Dotti, Phys. Rev. D **72**, 124002 (2005) [arXiv:gr-qc/0510069].
139. G. Dotti and R. J. Gleiser, Phys. Rev. D **72**, 044018 (2005) [arXiv:gr-qc/0503117].
140. G. Dotti and R. J. Gleiser, Class. Quant. Grav. **22**, L1 (2005) [arXiv:gr-qc/0409005].
141. M. Brigante, H. Liu, R. C. Myers, S. Shenker and S. Yaida, arXiv:0802.3318 [hep-th]. M. Brigante, H. Liu, R. C. Myers, S. Shenker and S. Yaida, arXiv:0712.0805 [hep-th].
142. S. Deser and B. Tekin, Phys. Rev. D **75**, 084032 (2007) [arXiv:gr-qc/0701140].

Chapter 9
Gravitational Waves from Braneworld Black Holes

S.S. Seahra

Abstract In this article, we present the black string model of a braneworld black hole and analyze its perturbations. We develop the perturbation formalism for Randall–Sundrum model from first principles and discuss the weak-field limit of the model in the solar system. We derive explicit equations of motion for the axial and spherical gravitational waves in the black string background. These are solved numerically in various scenarios, and the characteristic late-time signal from a black string is obtained. We find that if one waits long enough after some transient event, the signal from the string will be a superposition of nearly monochromatic waves with frequencies corresponding to the masses of the Kaluza–Klein modes of the model. We estimate the amplitude of the spherical component of these modes when they are excited by a point particle orbiting the string.

9.1 Introduction

Braneworld models hypothesize that our observable universe is a hypersurface, called the 'brane', embedded in some higher-dimensional spacetime. Standard model particles and fields are assumed to be confined to the brane, while gravitational degrees of freedom are free to propagate in the full higher-dimensional 'bulk'. The phenomenological implications of these models have been intensively studied by many different authors over the past decade, with great emphasis being placed on any observational consequences of the existence of large, possibly infinite, extra dimensions.

There are a number of different braneworld models, but perhaps one of the best studied is the Randall–Sundrum (RS) scenario [15, 16]. There are two variants of the model involving either one or two branes, but the common assumption in both setups is that there is a negative cosmological constant in the bulk characterized by a curvature scale ℓ. The great virtue of the model is that the gravity behaves like ordinary general relativity (GR) in 'weak-field' situations; i.e., when the density of

S.S. Seahra (✉)
Department of Mathematics & Statistics, University of New Brunswick, Canada
sseahra@unb.ca

Seahra, S.S.: *Gravitational Waves from Braneworld Black Holes*. Lect. Notes Phys. **769**, 347–386 (2009)
DOI 10.1007/978-3-540-88460-6_9   ⓒ Springer-Verlag Berlin Heidelberg 2009

matter is small or scale of interest is large. In particular, one recovers the Newtonian inverse-square law of gravitation in the RS model as long as the separation between the two bodies $\gg \ell$. This leads to a direct laboratory constraint on the bulk curvature scale, since Newton's law is known to be valid on scales larger than around $50\,\mu\text{m}$ [10].

The RS model is also consistent with various astrophysical tests of GR in the weak-field regime, including the solar system tests such as the perihelion shift of Mercury or time delay experiments using the Cassini spacecraft. On the cosmological side, one can also demonstrate that the RS predictions for the dynamics of the scale factor or the growth of fluctuations match the predictions of GR as long as the Hubble horizon H^{-1} is less than the AdS length scale ℓ. Hence, the RS model matches conventional theory in the low-energy universe.

The ability of the RS model to mimic GR in these cases is both fortuitous and somewhat surprising. The introduction of a large extra dimension is not a trivial modification of standard theory, and before the work of Randall & Sundrum the conventional wisdom was that such models could not be made to be consistent with the real measured behaviour of gravity. The fact that a fifth dimension can be made to conform to what we observe is part of the reason for the flurry of activity on the RS model since its inception. It also raises an interesting problem: The correspondence between the GR and the RS scenario must fail at some point, since at the end of the day they have very different geometric setups. In what situations does this breakdown occur, and are there any associated observational signatures that we can use to constrain the RS model?

We mentioned above that RS cosmology matches GR cosmology for $H\ell \lesssim 1$. Thus, we are led to look for deviations from standard theory in cosmological epochs with $H\ell \gtrsim 1$. This corresponds to the very high-energy radiation epoch, which is just after inflation and before nucleosynthesis. People have looked at modifications to the background expansion, dynamics of gravitational waves [9, 11, 17], and the growth of density perturbations in the high-energy epoch [1]. All of these phenomena show some departures from GR, but as of yet there has been no clean observational test proposed that could either rule out or rule in the RS model.

Hence, we need to look to other 'strong field' scenarios to test the model. One possibility is to look at black holes in the Randall–Sundrum model. We know that these objects are not describable in the Newtonian limit of GR, so one might expect that braneworld black holes to exhibit observable deviations from the ordinary Schwarzschild or Kerr solutions. However, there is a major problem with using black holes a probe of braneworld models: There is no known 'reasonable' brane-localized black-hole solution in the RS one brane scenario. The lack of a solution is not for lack of trying, many authors have attempted various techniques to find one. One of the first attempts was using the 5-dimensional black string solution as a bulk manifold [2]. However, it was demonstrated that such solutions were subject to the famous Gregory–Laflamme instability [8], which is a tachyonic mode with a long wavelength in the extra dimension. Others have tried to find brane black holes numerically [14], but success has been limited to small mass objects $GM \ll \ell$. Several have conjectured that the lack of a solution in the one brane case has to do with the AdS/CFT correspondence [5, 19].

However, the situation is somewhat better in the two brane case. It turns out that it is possible to find a stable braneworld model in this case and that the brane geometry is exactly 4-dimensional Schwarzschild [3, 4, 18]. Like the model considered in [2], this is based on the 5-dimensional black string. The Gregory–Laflamme instability is evaded by the infrared cutoff introduced by the second brane, i.e., the model is stable if the branes are close enough together. Because the geometry on the brane is identical to that of the Schwarzschild metric, the model is automatically in agreement with any test of GR sensitive to the background geometry only, such as light bending, perihelion shifts, time delays, etc.

Hence, we need to look at the perturbative aspects of the model to obtain differences with ordinary GR. In particular, we are interested in the gravitational waves (GWs) emitted from these black strings when they are displaced from their equilibrium configuration. Of primary importance is the issue of whether or not any deviations from the predictions of GR are observable by GW detectors such as LIGO or LISA. These issues are the subject of these lecture notes.

In Sect. 9.2 we introduce the RS model and the black string braneworld. In Sect. 9.3, we describe how to perturb the model and derive the relevant equations of motion. In Sect. 9.4, we show how to separate variables in the governing partial differential equations (PDEs) by introducing the Kaluza–Klein (KK) decomposition. In Sect. 9.5, we consider the limit under which we recover GR. In Sect. 9.6, we define the complete mode decomposition in terms of KK modes and spherical harmonics used in the rest of the notes. In Sect. 9.7, we consider homogeneous solutions to the axial equations of motion and determine (via simulations) the characteristic GW signal produced by the string. In Sect. 9.8, we consider the spherical sector of the GW spectrum excited by generic sources and discuss the Gregory–Laflamme instability in detail. In Sect. 9.9, we write down explicit equations of motion for the spherical GWs emitted by a point particle orbiting the black string and consider their numeric solution. In Sect. 9.10, we estimate the amplitude of Kaluza–Klein radiation emitted from the black string for a given point particle source. Finally, in Sect. 9.11 we give a brief summary and outline some open questions.

9.2 A Generalized Randall–Sundrum Two Brane Model

In this section, we present a generalized version of the Randall–Sundrum two brane model in a coordinate invariant formalism. We begin by outlining the geometry of the model, the action governing the dynamics, and the ensuing field equations. We then specialize to the black string braneworld model, which will be perturbed in the next section.

9.2.1 Geometrical Framework and Notation

Consider a (4+1)-dimensional manifold $(\mathcal{M}, g)$, which we refer to as the 'bulk'. One of the spatial dimensions of $\mathcal{M}$ is assumed to be compact; i.e., the 5-dimensional topology is $\mathbb{R}^4 \times S$. We place coordinates x^A on $\mathcal{M}$ so that the 5-dimensional line element reads:

$$ds_5^2 = g_{AB} dx^A dx^B. \tag{9.1}$$

We assume that there is a scalar function Φ that uniquely maps points in $\mathcal{M}$ into the interval $I = (-d, +d]$. Here, d is a constant parameter that is one of the fundamental length scales of the problem. The gradient of this mapping $\partial_A \Phi$ is spacelike,

$$\partial_A \Phi \, \partial^A \Phi > 0, \tag{9.2}$$

and is tangent to the compact dimension of $\mathcal{M}$. This scalar function defines a family of timelike hypersurfaces $\Phi(x^A) = Y$, which we denote by Σ_Y. The two submanifolds at the endpoints of I, Σ_d and Σ_{-d}, are periodically identified.

Let us now place 4-dimensional coordinates z^α on each of the Σ_Y hypersurfaces. These coordinates will be related to their 5-dimensional counterparts by parametric equations of the form: $x^A = x^A(z^\alpha)$. We then define the following basis vectors

$$e_\alpha^A = \frac{\partial x^A}{\partial z^\alpha}, \quad n^A = \frac{\partial^A \Phi}{\sqrt{\partial_B \Phi \, \partial^B \Phi}}, \quad n_A e_\alpha^A = 0, \quad n^A n_A = +1. \tag{9.3}$$

The tetrad e_α^A is everywhere tangent to Σ_Y, while n^A is everywhere normal to Σ_Y. The projection tensor onto the Σ_Y hypersurfaces is given by

$$q_{AB} = g_{AB} - n_A n_B, \quad n^A q_{AB} = 0. \tag{9.4}$$

From this, it follows that the intrinsic line element on each of the Σ_Y hypersurfaces is

$$ds_4^2 = q_{\alpha\beta} dz^\alpha dz^\beta, \quad q_{\alpha\beta} = e_\alpha^A e_\beta^B q_{AB} = e_\alpha^A e_\beta^B g_{AB}. \tag{9.5}$$

The object $q_{\alpha\beta}$ behaves as a tensor under 4-dimensional coordinate transformations $z^\alpha \to \bar{z}^\alpha(z^\beta)$ and is the induced metric on the Σ_Y hypersurfaces. It has an inverse $q^{\alpha\beta}$ that can be used to define e_A^α:

$$e_A^\alpha = g_{AB} q^{\alpha\beta} e_\beta^B, \quad \delta_\beta^\alpha = q^{\alpha\gamma} q_{\gamma\beta} = e_A^\alpha e_\beta^A. \tag{9.6}$$

Generally speaking, we define the projection of any 5-tensor T_{AB} onto the Σ_Y hypersurfaces as

$$T_{\alpha\beta} = e_\alpha^A e_\beta^B T_{AB}, \tag{9.7}$$

where the generalization to tensors of other ranks is obvious. The 4-dimensional intrinsic covariant derivative of $T_{\alpha\beta}$ is related to the 5-dimensional covariant derivative of T_{AB} by

$$[\nabla_\alpha T_{\mu\nu}]_q = e_\alpha^A e_\mu^M e_\nu^N \nabla_A q_M^B q_N^C T_{BC}, \tag{9.8}$$

where the notation $[\cdots]_q$ means that the quantity inside the square brackets is calculated with the $q_{\alpha\beta}$ metric.

Finally, the extrinsic curvature of each Σ_Y hypersurface is

$$K_{AB} = q_A^C \nabla_C n_B = \tfrac{1}{2}\pounds_n q_{AB} = K_{BA}, \quad n^A K_{AB} = 0,$$
$$K_{\alpha\beta} = e_\alpha^A e_\beta^B K_{AB} = e_\alpha^A e_\beta^B \nabla_A n_B. \tag{9.9}$$

9.2.2 The Action and Field Equations

We label the hypersurfaces at $Y = y_+ = 0$ and $Y = y_- = +d$ as the 'visible brane' Σ^+ and 'shadow brane' Σ^-, respectively. Our observable universe is supposed to reside on the visible brane. These hypersurfaces divide the bulk into two halves: the left-hand portion $\mathscr{M}_L$ which has $y \in (-d,0)$, and the right-hand portion which has $y \in (0,+d)$. The action for our model is

$$S = \frac{1}{2\kappa_5^2}\int_{\mathscr{M}_L}\left[{}^{(5)}R - 2\Lambda_5\right] + \frac{1}{2\kappa_5^2}\int_{\mathscr{M}_R}\left[{}^{(5)}R - 2\Lambda_5\right]$$
$$+ \sum_{\epsilon=\pm}\frac{1}{2}\int_{\Sigma^\epsilon}\left(\mathscr{L}^\epsilon - 2\lambda^\epsilon - \frac{1}{\kappa_5^2}[K]^\epsilon\right) + \frac{1}{2}\int_{\mathscr{M}_L}\mathscr{L}_L + \frac{1}{2}\int_{\mathscr{M}_R}\mathscr{L}_R. \tag{9.10}$$

In this expression, κ_5^2 is the 5-dimensional gravity matter coupling, $\Lambda_5 = -6k^2$ is the bulk cosmological constant, $\lambda^\pm = \pm 6k/\kappa_5^2$ are the brane tensions, and $\ell = 1/k$ is the curvature length scale of the bulk. Also, $\mathscr{L}^\pm$ is the Lagrangian density of matter residing on $\Sigma^\pm$, while $\mathscr{L}_L$ and $\mathscr{L}_R$ are the Lagrangian densities of matter living in the bulk. Note that the visible brane in our model has positive tension, while the shadow brane has negative tension.

The quantity $[K]^\pm$ is the jump in the trace of the extrinsic curvature of the Σ_Y hypersurfaces across each brane. To clarify, suppose that $\partial\mathscr{M}_L^\pm$ and $\partial\mathscr{M}_R^\pm$ are the boundaries of $\mathscr{M}_L$ and $\mathscr{M}_R$ coinciding with $\Sigma^\pm$, respectively. Then,

$$[K]^+ = \left. q^{\alpha\beta}K_{\alpha\beta}\right|_{\partial\mathscr{M}_R^+} - \left. q^{\alpha\beta}K_{\alpha\beta}\right|_{\partial\mathscr{M}_L^+}, \tag{9.11a}$$

$$[K]^- = \left. q^{\alpha\beta}K_{\alpha\beta}\right|_{\partial\mathscr{M}_L^-} - \left. q^{\alpha\beta}K_{\alpha\beta}\right|_{\partial\mathscr{M}_R^-}. \tag{9.11b}$$

We can now write down the field equations for our model. Setting the variation of S with respect to the bulk metric g^{AB} equal to zero yields that

$$G_{AB} - 6k^2 g_{AB} = \kappa_5^2\left[\theta(+y)T_{AB}^R + \theta(-y)T_{AB}^L\right],$$

$$T_{AB}^{L,R} = -\frac{2}{\sqrt{-g}}\frac{\delta\left(\sqrt{-g}\mathscr{L}_{L,R}\right)}{\delta g^{AB}}. \tag{9.12}$$

Meanwhile, variation of S with respect to the induced metric on each boundary yields

$$Q_{AB}^{\pm} = \left\{ [K_{AB}] \pm 2kq_{AB} + \kappa_5^2 (T_{AB} - \tfrac{1}{3} T q_{AB}) \right\}^{\pm} = 0, \tag{9.13a}$$

$$T_{AB}^{\pm} = e_A^{\alpha} e_B^{\beta} \left\{ -\frac{2}{\sqrt{-q}} \frac{\delta (\sqrt{-q} \mathscr{L})}{\delta q^{\alpha \beta}} \right\}^{\pm}. \tag{9.13b}$$

Here, the $\{ \cdots \}^{\pm}$ notation means that everything inside the curly brackets is evaluated at $\Sigma^{\pm}$. We see that (9.12) are the bulk field equations to be satisfied by the 5-dimensional metric g_{AB}, while (9.13) are the boundary conditions that must be enforced at the position of each brane. Of course, (9.13) are simply the Israel junction conditions for thin shells in general relativity.

In what sense is our model a generalization of the RS setup? The original Randall–Sundrum model exhibited a $\mathbb{Z}_2$ symmetry, which implied that $\mathscr{M}_L$ is the mirror image of $\mathscr{M}_R$. Also, in the RS model the bulk was explicitly empty. However, since we allow for an asymmetric distribution of matter in the bulk, we explicitly violate the $\mathbb{Z}_2$ symmetry and bulk vacuum assumption.

9.2.3 The Black String Braneworld

We now introduce the black string braneworld, which is a $\mathbb{Z}_2$ symmetric solution of (9.12) and (9.13) with no matter sources:

$$\mathscr{L}_L \doteq \mathscr{L}_R \doteq \mathscr{L}^{\pm} \doteq 0. \tag{9.14}$$

Here, we use $\doteq$ to indicate equalities that only hold in the black string background. The bulk geometry for this solution is given by

$$ds_5^2 \doteq a^2(y) \left[-f(r) dt^2 + \frac{1}{f(r)} dr^2 + r^2 d\Omega^2 \right] + dy^2, \tag{9.15a}$$

$$f(r) = 1 - 2GM/r, \quad a(y) = e^{-k|y|}. \tag{9.15b}$$

Here, M is the mass parameter of the black string and $G = \ell_{Pl}/M_{Pl}$ is the ordinary 4-dimensional Newton's constant. The function Φ used to locate the branes is trivial in this background:

$$\Phi(x^A) \doteq y, \tag{9.16}$$

which means that the $\Sigma^{\pm}$ branes are located at $y = 0$ and $y = d$, respectively. The $\Sigma_Y \doteq \Sigma_y$ hypersurfaces have the geometry of Schwarzschild black holes, and there is 5-dimensional line-like curvature singularity at $r = 0$:

$$R^{ABCD} R_{ABCD} \doteq \frac{48 G^2 M^2 e^{4k|y|}}{r^6} + 40 k^2. \tag{9.17}$$

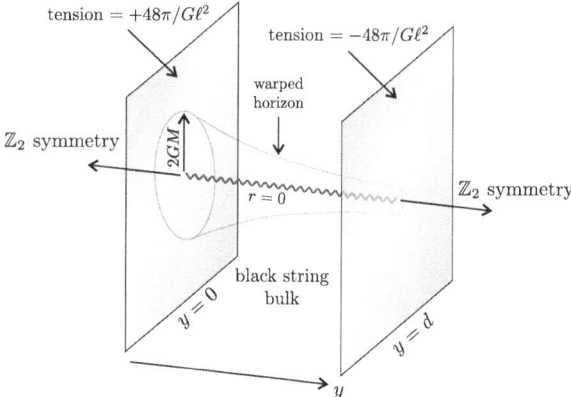

Fig. 9.1 A schematic illustration of the black string braneworld

Note that the other singularities at $y = \pm\infty$ are excised from our model by the restriction $y \doteq Y \in (-d, d]$, so we will not consider them further. An illustration of the black string braneworld background is given in Fig. 9.1.

We remark that it is actually possible to replace the 4-metric in square brackets in (9.15) by any 4-dimensional solution of $R_{\alpha\beta} = 0$ and still satisfy the 5-dimensional field equation. That is, we could have

$$ds_5^2 \doteq a(y)^2 ds_{\text{Kerr}}^2 + dy^2, \tag{9.18}$$

where ds_{Kerr}^2 is the line element corresponding to the Kerr solution for a rotating black hole. Such a solution is known as the rotating black string. The dynamics of perturbations of the rotating black string are still an open question due to the extreme complexity of the governing equations of motion.

Finally, note that the normal and extrinsic curvature associated with the Σ_Y hypersurfaces satisfies the following convenient properties:

$$n_A \doteq \partial_A y, \quad n^A \nabla_A n^B \doteq 0, \quad K_{AB} \doteq -k q_{AB}. \tag{9.19}$$

These expressions are used liberally below to simplify formulae evaluated in the black string background.

9.3 Linear Perturbations

We now turn our attention to perturbations of the black sting braneworld. We first describe the perturbative variable we use to describe the fluctuations of the system, then we linearize the bulk field equations and junction conditions. We finish this section by rewriting the perturbative equations of motion in a particularly useful form. Note that while we work from first principals in Sects. 9.3, 9.4, and 9.5, similar calculations and results have appeared many times in the literature, see the seminal works by Randall & Sundrum [15] and Garriga & Tanaka [6], for example.

9.3.1 Perturbative Variables

We are ultimately interested in the behavior of gravitational waves in this model, which are described by fluctuations of the bulk metric:

$$g_{AB} \rightarrow g_{AB} + h_{AB}, \tag{9.20}$$

where h_{AB} is understood to be a 'small' quantity. The projection of h_{AB} onto the visible brane is the observable that can potentially be measured in gravitational wave detectors. But it is not sufficient to consider fluctuations in the bulk metric alone—to get a complete picture, we must also allow for the perturbation of the matter content of the model as well as the positions of the branes.

Obviously, matter perturbations are simply described by the T^{L}_{AB}, T^{R}_{AB}, and $T^{\pm}_{AB}$ stress–energy tensors, which are considered to be small quantities of the same order as h_{AB}. On the other hand, we describe fluctuations in the brane positions via a perturbation of the scalar function Φ:

$$\Phi(x^A) \rightarrow y + \xi(x^A). \tag{9.21}$$

Here, ξ is a small spacetime scalar. Recall that the position of each brane is implicitly defined by $\Phi(x^A) = y_{\pm}$. Hence, the brane locations after perturbation are given by the solution of the following for y:

$$y + \xi \Big|_{y=y_{\pm}} + (y - y_{\pm}) \partial_y \xi \Big|_{y=y_{\pm}} + \cdots = y_{\pm}. \tag{9.22}$$

However, note that $y - y_{\pm}$ is of the same order as ξ, so at the linear level the new brane positions are simply given by

$$y = y_{\pm} - \xi \Big|_{y=y_{\pm}}. \tag{9.23}$$

Hence, the perturbed brane positions are given by the brane-bending scalars:

$$\xi^{\pm} = \xi \Big|_{y=y_{\pm}}, \quad n^A \partial_A \xi^{\pm} = 0. \tag{9.24}$$

Note that because ξ^+ and ξ^- are explicitly evaluated at the brane positions, they are essentially 4-dimensional scalars that exhibit no dependence on the extra dimension.

Having now delineated a set of variables that parameterize the fluctuations of the black string braneworld, we now need to determine their equations of motion.

9.3.2 Linearizing the Bulk Field Equations

First, we linearize the bulk field equations (9.12) about the black string solution. Notice that (9.12) only depends on the bulk metric and the bulk matter distribution. Hence, the linearized field equations will only involve h_{AB}, T_{AB}^{L}, and T_{AB}^{R}. The actual derivation of the equation proceeds in the same manner as in 4-dimensions, and we just quote the result

$$\nabla^C \nabla_C h_{AB} - \nabla^C \nabla_A h_{BC} - \nabla^C \nabla_B h_{AC} + \nabla_A \nabla_B h^C{}_C - 8k^2 h_{AB} = -2\kappa_5^2 \Sigma_{AB}^{\mathrm{bulk}}, \quad (9.25)$$

where

$$\Sigma_{AB}^{\mathrm{bulk}} = \Theta(+y)(T_{AB}^{\mathrm{R}} - \tfrac{1}{3}T^{\mathrm{R}} g_{AB}) + \Theta(-y)(T_{AB}^{\mathrm{L}} - \tfrac{1}{3}T^{\mathrm{L}} g_{AB}). \quad (9.26)$$

The wave equation (9.25) is valid for arbitrary choices of gauge and generic matter sources. If we specialize to the Randall–Sundrum gauge

$$\nabla^A h_{AB} = 0, \quad h^A{}_A = 0, \quad h_{AB} = e_A^\alpha e_B^\beta h_{\alpha\beta}, \quad (9.27)$$

equation (9.25) reduces to

$$\hat{\Delta}_{AB}{}^{CD} h_{CD} + (GMa)^2 (\pounds_n^2 - 4k^2) h_{AB} = -2(GMa)^2 \kappa_5^2 \Sigma_{AB}^{\mathrm{bulk}}, \quad (9.28)$$

where we have defined the operator

$$\begin{aligned}
\hat{\Delta}_{AB}{}^{CD} &= (GMa)^2 [q^{MN} \nabla_M q_N^P q_A^C q_B^D \nabla_P + 2^{(4)} R_A{}^C{}_B{}^D] \\
&= (GMa)^2 e_A^\alpha e_B^\beta \left[\delta_\alpha^\gamma \delta_\beta^\delta \nabla^\rho \nabla_\rho + 2 R_\alpha{}^\gamma{}_\beta{}^\delta \right]_q e_\gamma^C e_\delta^D \\
&= (GM)^2 e_A^\alpha e_B^\beta \left[\delta_\alpha^\gamma \delta_\beta^\delta \nabla^\rho \nabla_\rho + 2 R_\alpha{}^\gamma{}_\beta{}^\delta \right]_g e_\gamma^C e_\delta^D. \quad (9.29)
\end{aligned}$$

Here, $^{(4)}R_{ACBD}$ is the Riemann tensor on Σ_y, which can be related to the 5-dimensional curvature tensor via the Gauss equation

$$^{(4)}R_{MNPQ} = q_M^A q_N^B q_P^C q_Q^D R_{ABCD} + 2K_{M[P} K_{Q]N}. \quad (9.30)$$

On the second line of (9.29) the 4-tensor inside the square brackets is calculated using $q_{\alpha\beta}$. We can re-express this object in terms of the ordinary Schwarzschild metric $g_{\alpha\beta}$, which is conformally related to $q_{\alpha\beta}$ via the warp factor:

$$q_{\alpha\beta} = a^2 g_{\alpha\beta}, \quad (9.31\mathrm{a})$$

$$g_{\alpha\beta} dz^\alpha dz^\beta = -f \, dt^2 + f^{-1} dr^2 + r^2 d\Omega^2. \quad (9.31\mathrm{b})$$

The quantity in square brackets on the third line of (9.29) is calculated from $g_{\alpha\beta}$.[1] One can easily confirm that $\hat{\Delta}_{AB}{}^{CD}$ is 'y-independent' in the sense that it commutes with the Lie derivative in the n^A direction:

$$[^{(4)}\hat{\Delta}_{AB}{}^{CD}, \pounds_n] = 0. \tag{9.32}$$

In addition, the $(GM)^2$ prefactor makes $\hat{\Delta}_{AB}{}^{CD}$ dimensionless.

Notice that the left-hand side of (9.28) is both traceless and manifestly orthogonal to n^A, which implies the following constraints on the bulk matter:

$$\Sigma_{AB}^{\text{bulk}} = e_A^\alpha e_B^\beta \Sigma_{\alpha\beta}^{\text{bulk}}, \quad q^{\alpha\beta} \Sigma_{\alpha\beta}^{\text{bulk}} = 0. \tag{9.33}$$

In other words, our gauge choice is inconsistent with bulk matter that violates these conditions. If we wish to consider more general bulk matter, we cannot use the Randall–Sundrum gauge.

9.3.3 Linearizing the Junction Conditions

Next, we consider the perturbation of the junction conditions (9.13). These can be re-written as

$$Q_{AB}^\pm = \left\{ [\tfrac{1}{2}\nabla_{(A}n_{B)} - n_{(A|}n^C \nabla_C n_{|B)}] \pm kq_{AB} + \kappa_5^2 \left(T_{AB} - \tfrac{1}{3}Tq_{AB} \right) \right\}^\pm = 0. \tag{9.34}$$

We require that $Q_{AB}^\pm$ vanish before and after perturbation, so we need to enforce that the first-order variation $\delta Q_{AB}^\pm$ is equal to zero.

In order to calculate this variation, we can regard the tensors $Q_{AB}^\pm$ as functionals the brane positions (as defined by Φ), the brane normals n_A, the bulk metric, and the brane matter:

$$Q_{AB}^\pm = Q_{AB}^\pm(\Phi, n_M, g_{MN}, T_{MN}^\pm), \tag{9.35}$$

from which it follows that

$$\delta Q_{AB}^\pm = \left\{ \frac{\delta Q_{AB}}{\delta \Phi} \delta\Phi + \frac{\delta Q_{AB}}{\delta n_C} \delta n_C + \frac{\delta Q_{AB}}{\delta g_{CD}} \delta g_{CD} + \frac{\delta Q_{AB}}{\delta T_{CD}} \delta T_{CD} \right\}_0^\pm. \tag{9.36}$$

The $\{\cdots\}_0^\pm$ notation is meant to remind us that after we have calculated the variational derivatives, we must evaluate the expression in the background geometry at the *unperturbed* positions of the brane.

We now consider each term in (9.36). For simplicity, we temporarily focus on the positive tension visible brane and drop the $+$ superscript. The first term represents the variation of $Q_{AB}^\pm$ with brane position, which is covariantly given by the Lie derivative in the normal direction:

---

[1] Unless otherwise indicated, for the rest of the paper any tensorial expression with Greek indices should be evaluated using the Schwarzschild metric $g_{\alpha\beta}$.

$$\left\{\frac{\delta Q_{AB}}{\delta \Phi}\delta\Phi\right\}_0 = \{-\xi \pounds_n Q_{AB}\}_0. \tag{9.37}$$

But the Lie derivative of Q_{AB} vanishes identically in the background geometry, so this term is equal to zero.

The second term in (9.36) represents the variation of Q_{AB} with respect to the normal vector. Making note of the definition (9.3) of n^A in terms of Φ, as well as $\delta\Phi = \xi$ and $n^A\nabla_A\xi = 0$, we arrive at

$$\delta n_A = \nabla_A\xi, \quad n^A\delta n_A = 0. \tag{9.38}$$

Notice that since the normal itself must be continuous across the brane, we have $[\delta n_A] = 0$. After some algebra, we find that the variation of the junction conditions with respect to the brane normal is non-zero and given by

$$\left\{\frac{\delta Q_{AB}}{\delta n_C}\delta n_C\right\}_0 = 2q_A^C q_B^D\nabla_C\nabla_D\xi. \tag{9.39}$$

The third term in (9.36) is the variation with the bulk metric itself $\delta g_{AB} = h_{AB}$. Calculating this is straightforward, and the result is

$$\left\{\frac{\delta Q_{AB}}{\delta g_{CD}}\delta g_{CD}\right\}_0 = \frac{1}{2}[\pounds_n h_{AB}] + 2kh_{AB}. \tag{9.40}$$

The last variation we must consider is with respect to the brane matter fields, which is trivial:

$$\left\{\frac{\delta Q_{AB}}{\delta T_{CD}}\delta T_{CD}\right\}_0 = \kappa_5^2\left(T_{AB} - \frac{1}{3}Tq_{AB}\right). \tag{9.41}$$

So, we have the final result that

$$\delta Q_{AB}^{\pm} = \left\{2q_A^C q_B^D\nabla_C\nabla_D\xi + \frac{1}{2}[\pounds_n h_{AB}] \pm 2kh_{AB} + \kappa_5^2\left(T_{AB} - \frac{1}{3}Tq_{AB}\right)\right\}_0^{\pm} = 0. \tag{9.42}$$

If we take the trace of $\delta Q_{AB}^{\pm} = 0$, we obtain

$$q^{AB}\nabla_A\nabla_B\xi^{\pm} = \frac{1}{6}\kappa_5^2 T^{\pm}. \tag{9.43}$$

These are the equations of motion for the brane-bending degrees of freedom in our model, which are seen to be directly sourced by the matter fields on each brane.

9.3.4 Converting the Boundary Conditions into Distributional Sources

We can incorporate the boundary conditions $\delta Q_{AB}^{\pm} = 0$ directly into the h_{AB} equation of motion as delta-function sources. This is possible because the jump in the normal derivative of h_{AB} appears explicitly in the perturbed junction conditions. This procedure gives

$$\hat{\Delta}_{AB}{}^{CD} h_{CD} - \hat{\mu}^2 h_{AB} = -2(GMa)^2 \kappa_5^2 \left[\Sigma_{AB}^{\text{bulk}} + \sum_{\epsilon=\pm} \delta(y - y_\epsilon) \Sigma_{AB}^\epsilon \right]. \qquad (9.44)$$

Here, we have defined

$$\hat{\mu}^2 = -(GMa)^2 \left[\pounds_n^2 + \frac{2\kappa_5^2}{3} \sum_{\epsilon=\pm} \lambda^\epsilon \delta(y - y_\epsilon) - 4k^2 \right],$$

$$\Sigma_{AB}^\pm = \left(T_{AB}^\pm - \tfrac{1}{3} T^\pm q_{AB} \right) + \frac{2}{\kappa_5^2} q_A^C q_B^D \nabla_C \nabla_D \xi^\pm. \qquad (9.45)$$

If we integrate the wave equation (9.44) over a small region traversing either brane, we recover the boundary conditions (9.42).

Together with the gauge conditions,

$$n^A h_{AB} = q^{AC} \nabla_A h_{CB} = 0 = q^{AB} h_{AB}, \qquad (9.46)$$

equations (9.43) and (9.44) are the equations governing the perturbations of our model.

9.4 Kaluza–Klein Mode Functions

The metric fluctuation h_{AB} is governed by a system of partial differential equations (PDEs). As is common in all areas of physics, the best way to solve such equations is via a separation of variables. In this section, we separate the y variables from the conventional Schwarzschild variables on Σ_y. The part of the graviton wave function corresponding to the extra dimension satisfies an ODE boundary value problem, which implies that there is a discrete spectrum for h_{AB}.

9.4.1 Separation of Variables

As mentioned above, we have that

$$[\hat{\Delta}_{AB}{}^{CD}, \pounds_n] h_{CD} = 0; \qquad (9.47)$$

i.e., $\hat{\Delta}_{AB}{}^{CD}$ is independent of y when evaluated in the (t, r, θ, ϕ, y) coordinates. This suggests that we seek a solution for h_{AB} of the form

$$h_{AB} = Z \tilde{h}_{AB}, \quad \hat{\mu}^2 Z = \mu^2 Z, \qquad (9.48)$$

where,

$$0 = \pounds_n \tilde{h}_{AB} \text{ and } 0 = q_B^A \nabla_A Z; \tag{9.49}$$

that is, Z is an eigenfunction of $\hat{\mu}^2$ with eigenvalue μ^2. The existence of the delta functions in the $\hat{\mu}^2$ operator means that we need to treat the even and odd parity solutions of this eigenvalue problem separately.

9.4.2 Even Parity Eigenfunctions

If $Z(-y) = Z(y)$, we see that Z satisfies the following equations in the interval $y \in [0,d]$:

$$m^2 Z(y) = -a^2(y)(\partial_y^2 - 4k^2)Z(y),$$
$$0 = [(\partial_y + 2k)Z(y)]_\pm,$$
$$\mu = GMm. \tag{9.50}$$

There is a discrete spectrum of solutions to this eigenvalue problem that are labeled by the positive integers $n = 1,2,3\ldots$:

$$Z_n(y) = \alpha_n^{-1}[Y_1(m_n \ell)J_2(m_n \ell e^{k|y|}) - J_1(m_n \ell)Y_2(m_n \ell e^{k|y|})], \tag{9.51}$$

where α_n is a constant, and $m_n = \mu_n/GM$ is the nth solution of

$$Y_1(m_n \ell)J_1(m_n \ell e^{kd}) = J_1(m_n \ell)Y_1(m_n \ell e^{kd}). \tag{9.52}$$

There is also a solution corresponding to $m_0 = \mu_0 = 0$, which is known as the zero mode:

$$Z_0(y) = \alpha_0^{-1}e^{-2k|y|}, \quad \alpha_0 = \sqrt{\ell}(1 - e^{-2kd})^{1/2}. \tag{9.53}$$

Hence, there exists a discrete set of solutions for bulk metric perturbations of the form $h_{AB}^{(n)} = Z_n(y)\tilde{h}_{AB}^{(n)}(z^\alpha)$. When $n > 0$ these are called the Kaluza–Klein (KK) modes of the modes, and the mass of any given mode is given by the m_n eigenvalue. The α_n constants are determined from demanding that $\{Z_n\}$ forms an orthonormal set

$$\delta_{mn} = \int_{-d}^d dy\, a^{-2}(y)Z_m(y)Z_n(y). \tag{9.54}$$

These basis functions then satisfy:

$$\delta(y - y_\pm) = \sum_{n=0}^\infty a^{-2}Z_n(y)Z_n(y_\pm). \tag{9.55}$$

This identity is crucial to the model—inspection of (9.44) reveals that the brane stress-energy tensors appearing on the right-hand side are multiplied by one of $\delta(y - y_\pm)$. Hence, brane matter only couples to the even parity eigenmodes of $\hat{\mu}^2$.

Case 1: Light Modes

It is useful to have simple approximate forms of the Kaluza–Klein masses and normalization constants for the formulae that appear later on. There are straightforward to derive for modes that are 'light' compared to mass scale set by the AdS_5 length parameter:

$$m_n\ell \ll 1. \tag{9.56}$$

Let us define a set of dimensionless numbers x_n by:

$$x_n = m_n\ell e^{kd}. \tag{9.57}$$

Then for the light modes, we find that x_n is the nth zero of the first-order Bessel function:

$$J_1(x_n) = 0. \tag{9.58}$$

Also for light modes, the normalization constants reduce to

$$\alpha_n \approx 2\sqrt{\ell}\, e^{2kd}|J_0(x_n)|/\pi x_n, \quad n > 0. \tag{9.59}$$

Actually, it is more helpful to know the value of the KK mode functions at the position of each brane. We can parameterize these as

$$Z_n(y_\pm) = \sqrt{k}e^{-kd}z_n^\pm, \quad n > 0. \tag{9.60}$$

For the light Kaluza–Klein modes, the dimensionless $z_n^\pm$ are given by

$$z_n^\pm \approx \left\{ \begin{matrix} |J_0(x_n)|^{-1} \\ e^{in\pi} \end{matrix} \right\}. \tag{9.61}$$

Case 2: Heavy Modes

At the other end of the spectrum, we have the heavy Kaluza–Klein modes

$$m_n\ell \gg 1. \tag{9.62}$$

Under this assumption, we find[2]

$$x_n \approx \frac{n\pi}{1 - e^{-kd}}, \tag{9.63a}$$

$$Z_n(y) \approx \sqrt{\frac{ke^{-k|y|}}{e^{kd} - 1}} \cos\left[n\pi \frac{e^{k|y|} - 1}{e^{kd} - 1} \right], \tag{9.63b}$$

$$z_n^\pm \approx \frac{1}{\sqrt{1 - e^{-kd}}} \left\{ \begin{matrix} e^{kd/2} \\ e^{in\pi} \end{matrix} \right\}. \tag{9.63c}$$

---

[2] Strictly speaking, an asymptotic analysis leads to formulae with n replaced by another integer n' on the right-hand sides of (9.63). However, we note that for even parity modes, n counts the number of zeroes of $Z_n(y)$ in the interval $y \in (0, d)$, which allows us to deduce that $n' = n$.

Unlike the analogous quantities for the light modes, $z_n^\pm$ shows an explicit dependence on the dimensionless brane separation d/ℓ.

9.4.3 Odd Parity Eigenfunctions

As mentioned above, brane matter only couples to Kaluza–Klein modes with even parity. But a complete perturbative description must include the odd parity modes as well; for example, if we have matter in the bulk distributed asymmetrically with respect to $y = 0$ (i.e., $T_{AB}^L \neq T_{AB}^R$) modes of either parity will be excited. Hence, for the sake of completeness, we list a few properties of the odd parity Kaluza–Klein modes here.

Assuming $Z(-y) = -Z(y)$, we have:

$$m^2 Z(y) = -a^2(y)(\partial_y^2 - 4k^2)Z(y),$$
$$0 = Z(y_+) = Z(y_-). \tag{9.64}$$

Again, we have a discrete spectrum of solutions, this time labeled by half integers:

$$Z_{n+\frac{1}{2}}(y) = \alpha_{n+\frac{1}{2}}^{-1}\left[Y_2(m_{n+\frac{1}{2}}\ell)J_2(m_{n+\frac{1}{2}}\ell e^{k|y|}) - J_2(m_{n+\frac{1}{2}}\ell)Y_2(m_{n+\frac{1}{2}}\ell e^{k|y|})\right]. \tag{9.65}$$

The mass eigenvalues are now the solutions of

$$Y_2(m_{n+\frac{1}{2}}\ell)J_2(m_{n+\frac{1}{2}}\ell e^{kd}) = J_2(m_{n+\frac{1}{2}}\ell)Y_2(m_{n+\frac{1}{2}}\ell e^{kd}). \tag{9.66}$$

Proceeding as before, we define

$$x_{n+\frac{1}{2}} = m_{n+\frac{1}{2}}\ell e^{kd}. \tag{9.67}$$

For light modes with $m_{n+\frac{1}{2}}\ell \ll 1$, $x_{n+\frac{1}{2}}$ is the nth zero of the second-order Bessel function:

$$J_2(x_{n+\frac{1}{2}}) = 0. \tag{9.68}$$

Taken together, (9.58) and (9.68) imply the following for the light modes:

$$m_1 < m_{3/2} < m_2 < m_{5/2} < \cdots; \tag{9.69}$$

i.e., the first odd mode is heavier than the first even mode, etc.

Finally, we note that since the odd modes vanish at the background position of the visible brane, it is impossible for us to observe them directly within the context of linear theory. This can change at second order, since brane bending can allow us to directly sample regions of the bulk where $Z_{n+\frac{1}{2}} \neq 0$. However, this phenomenon is clearly beyond the scope of this paper.

9.5 Recovering 4-Dimensional Gravity

Let us now describe the limit in which we recover general relativity. We assume there are no matter perturbations in the bulk and on the hidden brane; hence, we may consistently neglect the odd parity Kaluza–Klein modes. By virtue of the brane-bending equation of motion (9.43), we can consistently set $\xi^- = 0$. Furthermore, (9.55) can be used to replace the delta function in front of Σ_{AB}^+ in (9.44). We obtain,

$$\hat{\Delta}_{AB}{}^{CD} h_{CD} - \hat{\mu}^2 h_{AB} = -2(GM)^2 \kappa_5^2 \Sigma_{AB}^+ \sum_{n=0}^{\infty} Z_n(y_+) Z_n(y). \tag{9.70}$$

We now note that for $e^{-kd} \ll 1$,

$$Z_0(y_+) = \sqrt{k}(1 - e^{-2kd})^{-1/2} \gg Z_n(y_+), \quad n > 0. \tag{9.71}$$

That is, the $n > 0$ terms in the sum are much smaller than the 0th order contribution. This motivates an approximation where the $n > 0$ terms on the right-hand side of (9.70) are neglected, which is the so-called 'zero-mode truncation'.

When this approximation is enforced, we find that h_{AB} must be proportional to $Z_0(y)$; i.e., there is no contribution to h_{AB} from any of the KK modes. Hence, we have $\hat{\mu}^2 h_{AB} = 0$. The resulting expression has trivial y dependence, so we can freely set $y = y_+$ to obtain the equation of motion for h_{AB} at the *unperturbed* position of the visible brane:

$$\hat{\Delta}_{AB}{}^{CD} h_{CD}^+ = -2(GM)^2 \kappa_5^2 \Sigma_{AB}^+ Z_0^2(y_+) \tag{9.72}$$

But we are not really interested in h_{AB}^+, the physically relevant quantity is the perturbation of the induced metric on the perturbed brane, which is defined as the variation of

$$q_{AB}^+ = [g_{AB} - n_A n_B]^+. \tag{9.73}$$

We calculate δq_{AB}^+ in the same way as we calculated δQ_{AB}^+ above (except for the fact that q_{AB} shows no explicit dependence on T_{AB}^+):

$$\delta q_{AB}^+ = \left\{ \frac{\delta q_{AB}}{\delta \Phi} \delta \Phi + \frac{\delta q_{AB}}{\delta n_C} \delta n_C + \frac{\delta q_{AB}}{\delta g_{CD}} \delta g_{CD} \right\}_0^+. \tag{9.74}$$

These variations are straightforward, and we obtain

$$\delta q_{AB}^+ \equiv \bar{h}_{AB}^+ = h_{AB}^+ + 2k\xi^+ q_{AB}^+ - (n_A \nabla_B + n_B \nabla_A)\xi^+, \tag{9.75}$$

where all quantities on the right are evaluated in the background and at the unperturbed position of the brane. Note that $\bar{h}_{AB} n^A \neq 0$, which reflects the fact that n_A is no longer the normal to the brane after perturbation.

We now define the 4-tensors

$$\bar{h}_{\alpha\beta}^+ = e_\alpha^A e_\beta^B \bar{h}_{AB}^+, \quad T_{\alpha\beta}^+ = e_\alpha^A e_\beta^B T_{AB}^+. \tag{9.76}$$

Here, $\bar{h}_{\alpha\beta}^{+}$ is the actual metric perturbation on the visible brane. Note that this perturbation is neither transverse or tracefree:

$$\nabla^{\gamma}\bar{h}_{\gamma\alpha}^{+} = 2k\nabla_{\alpha}\xi^{+}, \quad g^{\alpha\beta}\bar{h}_{\alpha\beta}^{+} = 8k\xi^{+}. \tag{9.77}$$

We can now re-express the equation of motion (9.72) in terms of $\bar{h}_{\alpha\beta}^{+}$ instead of h_{AB}^{+} using (9.75). Dropping the $+$ superscripts, we obtain

$$\nabla^{\gamma}\nabla_{\gamma}\bar{h}_{\alpha\beta} + \nabla_{\alpha}\nabla_{\beta}\bar{h}_{\gamma}^{\gamma} - \nabla^{\gamma}\nabla_{\alpha}\bar{h}_{\beta\gamma} - \nabla^{\gamma}\nabla_{\beta}\bar{h}_{\alpha\gamma}$$
$$= -2Z_{+}^{2}\kappa_{5}^{2}\left[T_{\alpha\beta} - \frac{1}{3}\left(1 + \frac{k}{2Z_{+}^{2}}\right)T^{\gamma}{}_{\gamma}g_{\alpha\beta}\right] + (6k - 4Z_{+}^{2})\nabla_{\alpha}\nabla_{\beta}\xi, \tag{9.78}$$

where we have defined

$$Z_{+}^{2} = Z_{0}^{2}(y_{+}) = k(1 - e^{-2kd})^{-1}. \tag{9.79}$$

In obtaining this expression, we have made use of the ξ equation of motion:

$$g^{\alpha\beta}\nabla_{\alpha}\nabla_{\beta}\xi = \frac{1}{6}\kappa_{5}^{2}g^{\alpha\beta}T_{\alpha\beta}. \tag{9.80}$$

Note that we still have the freedom to make a gauge transformation on the brane that involves an arbitrary 4-dimensional coordinate transformation generated by η_{α}:

$$\bar{h}_{\alpha\beta} \to \bar{h}_{\alpha\beta} + \nabla_{\alpha}\eta_{\beta} + \nabla_{\beta}\eta_{\alpha}. \tag{9.81}$$

We can use this gauge freedom to impose the condition

$$\nabla_{\beta}\bar{h}^{\beta}{}_{\alpha} - \frac{1}{2}\nabla_{\alpha}\bar{h}^{\beta}{}_{\beta} = (2Z_{+}^{2} - 3k)\nabla_{\alpha}\xi. \tag{9.82}$$

Then, the equation of motion for 4-metric fluctuations reads

$$\nabla^{\gamma}\nabla_{\gamma}\bar{h}_{\alpha\beta} + 2R_{\alpha}{}^{\gamma}{}_{\beta}{}^{\delta}\bar{h}_{\gamma\delta} = -16\pi G\left[T_{\alpha\beta} - \left(\frac{1 + \omega_{\mathrm{BD}}}{3 + 2\omega_{\mathrm{BD}}}\right)T^{\gamma}{}_{\gamma}g_{\alpha\beta}\right], \tag{9.83}$$

where we have identified

$$\omega_{\mathrm{BD}} = \frac{3}{2}(e^{2d/\ell} - 1), \quad G = \frac{\kappa_{5}^{2}}{8\pi\ell(1 - e^{-2d/\ell})}. \tag{9.84}$$

We see that (9.83) matches the equation governing gravitational waves in a Brans-Dicke theory with parameter ω_{BD}. Hence in the zero-mode truncation, the perturbations of the black string braneworld are indistinguishable from a 4-dimensional scalar–tensor theory.

Note that (9.83) must hold everywhere in our model, so we can consider the situation where our solar system is the perturbative brane matter located somewhere in the extreme far-field region of the black string. The forces between the various celestial bodies will be governed by (9.83) in the $R_{\alpha\beta\gamma\delta} \approx 0$ limit. In this scenario, solar system tests of general relativity place bounds on the Brans–Dicke parameter, and hence d/ℓ:

$$\omega_{\text{BD}} \gtrsim 4 \times 10^4 \quad \Rightarrow \quad d/\ell \gtrsim 5. \tag{9.85}$$

This lower bound on the dimensionless brane separation will be an important factor in the discussion below.

9.6 Beyond the Zero-Mode Truncation

In this section, we specialize to the situation where there is perturbative matter located on one of the branes and no other sources. Unlike Sect. 9.5, our interest here is to predict deviations from general relativity, so we will not use the zero-mode truncation. Just as in 4-dimensional black-hole perturbation theory, we introduce the tensor spherical harmonics to further decompose the equations of motion for a given KK mode into polar and axial parts.

9.6.1 KK Mode Decomposition

To begin, we make the assumptions

$$\Sigma_{AB}^{\text{bulk}} = 0, \text{ and } \Sigma_{AB}^+ = 0 \text{ or } \Sigma_{AB}^- = 0; \tag{9.86}$$

i.e., we set the matter perturbation in the bulk and one of the branes equal to zero. Note that due to the linearity of the problem we can always add up solutions corresponding to different types of sources; hence, if we had a physical situation with many different types of matter, it would be acceptable to solve for the radiation pattern induced by each source separately and then sum the results.

We decompose h_{AB} as

$$h_{AB} = \frac{\kappa_5^2 (GM)^2}{\mathscr{C}} e_A^\alpha e_B^\beta \sum_{n=0}^\infty Z_n(y) Z_n(y_\pm) h_{\alpha\beta}^{(n)}. \tag{9.87}$$

Here, $\mathscr{C}$ is a normalization constant (to be specified later) with dimensions of $(\text{mass})^{-4}$, and the expansion coefficients $h_{\alpha\beta}^{(n)}$ are dimensionless. We define a dimensionless brane stress–energy tensors and brane-bending scalars by

$$\Theta_{\alpha\beta}^\pm = \mathscr{C} e_\alpha^A e_\beta^B T_{AB}^\pm, \quad \tilde{\xi}^\pm = \frac{\mathscr{C}\xi^\pm}{(GM)^2 \kappa_5^2}. \tag{9.88}$$

Omitting the $\pm$ superscripts, we find that the equation of motion for $h^{(n)}_{\alpha\beta}$ is

$$(GM)^2 \left[\nabla^\gamma \nabla_\gamma h^{(n)}_{\alpha\beta} + 2R_{\alpha'\beta}{}^{\gamma}{}^{\delta} h^{(n)}_{\gamma\delta} \right] - \mu^2_n h^{(n)}_{\alpha\beta}$$
$$= -2 \left(\Theta_{\alpha\beta} - \tfrac{1}{3}\Theta g_{\alpha\beta} \right) - 4(GM)^2 \nabla_\alpha \nabla_\beta \tilde{\xi}, \quad (9.89)$$

while the equation of motion for $\tilde{\xi}$ is

$$\nabla^\alpha \nabla_\alpha \tilde{\xi} = \tfrac{1}{6}\Theta. \quad (9.90)$$

We also have the conditions

$$\nabla^\alpha h^{(n)}_{\alpha\beta} = \nabla^\alpha \Theta_{\alpha\beta} = 0 = g^{\alpha\beta} h^{(n)}_{\alpha\beta}. \quad (9.91)$$

Note that in all of these equations, all 4-dimensional quantities are to be calculated with the Schwarzschild metric $g_{\alpha\beta}$. In particular, $\Theta = g^{\alpha\beta}\Theta_{\alpha\beta}$.

9.6.2 The Multipole Decomposition

In addition to the decomposition of h_{AB} in terms of KK mode functions, the symmetry of the background geometry dictates that we decompose the problem in terms of spherical harmonics:

$$\tilde{\xi} = \sum_{l=0}^{\infty} \sum_{m=-l}^{l} Y_{lm} \tilde{\xi}_{lm}, \quad (9.92a)$$

$$h^{(n)}_{\alpha\beta} = \sum_{l=0}^{\infty} \sum_{m=-l}^{l} \sum_{i=1}^{10} [Y^{(i)}_{lm}]_{\alpha\beta}\, h^{(nlm)}_i, \quad (9.92b)$$

$$\Theta_{\alpha\beta} = \sum_{l=0}^{\infty} \sum_{m=-l}^{l} \sum_{i=1}^{10} [Y^{(i)}_{lm}]_{\alpha\beta}\, \Theta^{(lm)}_i. \quad (9.92c)$$

Here, $[Y^{(i)}_{lm}]_{\alpha\beta}$ are the tensorial spherical harmonics in four dimensions, which are the same quantities that appear in conventional black-hole perturbation theory. The tensor harmonics depend only on the angular coordinates $\Omega = (\theta, \phi)$, while the expansion coefficients depend on t and r:

$$\tilde{\xi}_{lm} = \tilde{\xi}_{lm}(t,r), \quad h^{(nlm)}_i = h^{(nlm)}_i(t,r), \quad \Theta^{(lm)}_i = \Theta^{(lm)}_i(t,r). \quad (9.93)$$

To define the tensor harmonics, first define the orthonormal 4-vectors

$$t^\alpha = f^{-1/2}\partial_t, \quad r^\alpha = f^{1/2}\partial_r, \quad \theta^\alpha = r^{-1}\partial_\theta, \quad \phi^\alpha = (r\sin\theta)^{-1}\partial_\phi. \quad (9.94)$$

The we define

$$\gamma_{\alpha\beta} = g_{\alpha\beta} + t_\alpha t_\beta - r_\alpha r_\beta = \theta_\alpha \theta_\beta + \phi_\alpha \phi_\beta, \quad t^\alpha \gamma_{\alpha\beta} = r^\alpha \gamma_{\alpha\beta} = 0, \quad (9.95)$$

which is the projection tensor onto the 2-spheres of constant r and t, and the anti-symmetric tensor $\epsilon_{\alpha\beta} = -\epsilon_{\beta\alpha}$

$$\epsilon_{\alpha\beta} = \theta_\alpha \phi_\beta - \phi_\alpha \theta_\beta. \tag{9.96}$$

Using these objects, the $[Y_{lm}^{(i)}]_{\alpha\beta}$ are defined in Table 9.1.[3]

Notice that we have divided the ten tensor harmonics into two groups labeled 'polar' and 'axial'. This division is based on how they transform under the parity, or space-inversion, operation $\mathbf{r} \to -\mathbf{r}$. In particular, under this type of operation, polar objects acquire a $(-1)^l$ factor, while axial quantities transform as $(-1)^{l+1}$.[4] It is useful to re-write the spherical harmonic decomposition of $h_{\alpha\beta}^{(n)}$ in terms of explicitly polar and axial parts:[5]

$$h_{\alpha\beta}^{(n)} = \underbrace{\sum_{l=0}^{\infty} \sum_{m=-l}^{l} \sum_{i=1}^{7} \mathbb{P}_{\alpha\beta}^{ilm}(\Omega)\, \mathscr{P}_{ilm}^{(n)}(t,r)}_{\text{polar contribution } h_{\alpha\beta}^{(n,\text{polar})}} + \underbrace{\sum_{l=0}^{\infty} \sum_{m=-l}^{l} \sum_{i=1}^{3} \mathbb{A}_{\alpha\beta}^{ilm}(\Omega)\, \mathscr{A}_{ilm}^{(n)}(t,r)}_{\text{axial contribution } h_{\alpha\beta}^{(n,\text{axial})}}. \tag{9.97}$$

In this expression and similar ones below, there is no summation over the spherical harmonic or i index unless indicated explicitly.

It is easy to confirm that the parity operation commutes with the $\hat{\Delta}_{AB}{}^{CD}$ and $\hat{\mu}^2$ operators in (9.44), or conversely commutes with the operator $\delta_\alpha^\gamma \delta_\beta^\delta \nabla^\lambda \nabla_\lambda + 2R_\alpha{}^\gamma{}_\beta{}^\delta$ in (9.89). Therefore, solutions of (9.89) that are eigenfunctions of the parity operator with different eigenvalues are decoupled from one another; i.e., we can solve for the dynamics of $h_{\alpha\beta}^{(n,\text{polar})}$ and $h_{\alpha\beta}^{(n,\text{axial})}$ individually. As is common for spherically symmetric systems, modes with different values of l and m are also decoupled.

Table 9.1 The spherical tensor harmonics $[Y_{lm}^{(i)}]_{\alpha\beta}$

Index i	Polar harmonics $\mathbb{P}_{\alpha\beta}^{ilm}$	Axial harmonics $\mathbb{A}_{\alpha\beta}^{ilm}$
1	$f^{-1} t_\alpha t_\beta Y_{lm}$	$2f^{-1/2} t_{(\alpha} \epsilon_{\beta)\gamma} \nabla^\gamma Y_{lm}$
2	$2t_{(\alpha} r_{\beta)} Y_{lm}$	$2f^{+1/2} r_{(\alpha} \epsilon_{\beta)\gamma} \nabla^\gamma Y_{lm}$
3	$f r_\alpha r_\beta Y_{lm}$	$\gamma_{\gamma(\alpha} \epsilon_{\beta)\delta} \nabla^\delta \nabla^\gamma Y_{lm}$
4	$-2t_{(\alpha} \gamma_{\beta)\gamma} \nabla^\gamma Y_{lm}$	$\cdots$
5	$+2r_{(\alpha} \gamma_{\beta)\gamma} \nabla^\gamma Y_{lm}$	$\cdots$
6	$r^{-2} \gamma_{\alpha\beta} Y_{lm}$	$\cdots$
7	$\gamma_{\alpha\gamma} \gamma_{\beta\delta} \nabla^\gamma \nabla^\delta Y_{lm}$	$\cdots$

---

[3] The definition of tensor harmonics is not unique; there are numerous other conventions in the literature.

[4] Alternatively, we can note that any tensor harmonic whose definition involves the pseudo-tensor ϵ_{ab} is automatically an axial object.

[5] A similar decomposition for $\Theta_{\alpha\beta}$ also exists.

Before moving on, we should mention that the decomposition of the brane-bending scalar $\tilde{\xi}$ is given entirely in terms of Y_{lm}; i.e., it is an explicitly polar quantity. It follows that $\nabla_\alpha \nabla_\beta \tilde{\xi}$ is also a polar quantity, which means that the brane-bending contribution in (9.89) only sources polar GW radiation.

9.7 Homogeneous Axial Perturbations

In this section, we present the equations of motion for the axial moments of $h_{\alpha\beta}^{(n)}$ in the absence of all matter sources. As mentioned above, the brane-bending contribution to (9.89) is a polar quantity. Therefore, the axial GW modes are completely decoupled from the brane-bending scalar. Hence, the equation we try to solve in this section is simply:

$$(GM)^2 \left[\nabla^\gamma \nabla_\gamma h_{\alpha\beta}^{(nlm,\text{axial})} + 2 R_\alpha{}^\gamma{}_\beta{}^\delta h_{\gamma\delta}^{(nlm,\text{axial})} \right] - \mu_n^2 h_{\alpha\beta}^{(nlm,\text{axial})} = 0, \qquad (9.98)$$

where the total axial contribution to $h_{\alpha\beta}^{(n)}$ is

$$h_{\alpha\beta}^{(n,\text{axial})} = \sum_{lm} h_{\alpha\beta}^{(nlm,\text{axial})}. \qquad (9.99)$$

In addition to this equation, remember that we also need to satisfy the gauge conditions (9.91).

Notice that (9.98) reduces to the graviton equation of motion in ordinary GR for $m_n = 0$, which corresponds to $n = 0$. It turns out that the $n = 0$ case must be handled separately from the $n \geq 0$ case due to an enhanced gauge symmetry present in the zero-mode sector. Therefore, for the purposes of this section we always assume $n \geq 0$.

9.7.1 High Angular Momentum $l \geq 2$ Radiation

In Table 9.1, notice that the axial harmonics are identically equation to zero for $l = 0$. Also note that for $l = 1$, the third harmonic vanishes $\mathbb{A}_{lm}^{3lm} = 0$. This means that there are no axial harmonics for $l = 0$ and that $l = 1$ is a special case. In this subsection, we concentrate on the $l \geq 2$ situation, where all of the axial tensor harmonics are non-trivial.

The decomposition of $h_{\alpha\beta}^{(nlm,\text{axial})}$ explicitly reads

$$h_{\alpha\beta}^{(nlm,\text{axial})} = \mathbb{A}_{\alpha\beta}^{1lm}(\Omega) \mathscr{A}_{1lm}^{(n)}(t,r) + \mathbb{A}_{\alpha\beta}^{2lm}(\Omega) \mathscr{A}_{2lm}^{(n)}(t,r) + \mathbb{A}_{\alpha\beta}^{3lm}(\Omega) \mathscr{A}_{3lm}^{(n)}(t,r). \qquad (9.100)$$

When this is substituted into the equation of motion (9.98) and gauge conditions (9.91), we get four PDEs that must be satisfied by the three expansion coefficients. These four equations are not independent, however, as the time derivative of one of them is a linear combination of the other three. Removing this equation, it is possible to use one of the other PDEs to algebraically eliminate $\mathscr{A}_{1lm}^{(n)}$ from the other two equations. Defining the 'master variables'

$$u_{nlm}(t,r) = f(r)\mathscr{A}_{2lm}^{(n)}(t,r), \quad v_{nlm}(t,r) = r^{-1}\mathscr{A}_{3lm}^{(n)}(t,r), \tag{9.101}$$

we eventually find that

$$0 = \left(\frac{\partial^2}{\partial t^2} - \frac{\partial^2}{\partial r_*^2}\right)\begin{pmatrix} u_{nlm} \\ v_{nlm} \end{pmatrix} + \mathbf{V}_{nl}\begin{pmatrix} u_{nlm} \\ v_{nlm} \end{pmatrix}. \tag{9.102}$$

Here, $\mathbf{V}_{nl}$ is a potential matrix, given by

$$\mathbf{V}_{nl} = f\begin{pmatrix} \dfrac{5f}{r^2} + \dfrac{f''}{2} - \dfrac{2f'}{r} + \dfrac{l(l+1)-1}{r^2} + m_n^2 & \dfrac{f'[2-l(l+1)]}{2r} \\ \dfrac{4}{r^2} & \dfrac{f'}{r} + \dfrac{l(l+1)-2}{r^2} + m_n^2 \end{pmatrix}, \tag{9.103}$$

and the well-known tortoise coordinate is defined by

$$r_* = r + 2GM\ln\left(\frac{r}{2GM} - 1\right). \tag{9.104}$$

Hence, to be able to describe homogeneous axial perturbations of the black string braneworld, one needs to specify initial data for u_{nlm} and v_{nlm}, solve the coupled wave equations (9.102), and then use the definitions (9.101) to obtain the original expansion coefficients $\mathscr{A}_{2lm}^{(n)}$ and $\mathscr{A}_{3lm}^{(n)}$. The last step is to integrate one of the original equations of motion,

$$\frac{\partial\mathscr{A}_{1lm}^{(n)}}{\partial t} = f^2\frac{\partial\mathscr{A}_{2lm}^{(n)}}{\partial r} + \frac{f(2f+f'r)}{r}\mathscr{A}_{2lm}^{(n)} + \frac{f[l(l+1)-2]}{2r^2}\mathscr{A}_{3lm}^{(n)}, \tag{9.105}$$

to obtain the other expansion coefficient $\mathscr{A}_{1lm}^{(n)}$. This procedure can be repeated for each individual value of n, l, and m. However, it should be noted that since the potential matrix does not explicitly depend on m, solutions that share the same values of n and l only really differ from one another by the choice of initial data.

Why are we interested in solving homogeneous problems like the one presented in this section? Recall that in the case of 4-dimensional black-hole perturbation theory, the numeric solution of the homogeneous axial wave equation lead to the discovery of quasinormal modes. In other words, by examining the solutions of equations such as (9.102), one can learn a lot about the characteristic behavior of a system when perturbed away from equilibrium, which is what we shall do in Sect. 9.7.3. The solution of the homogeneous problem can also have some direct observational significance, since it can describe how the system settles down into its equilibrium state after some event. That is, we expect the late-time axial gravitational wave signal from a black string to be described by the solutions of (9.102) after a black string is formed or undergoes some traumatic event.

Before moving on, it is worthwhile to note the asymptotic behavior of the potential matrix:

$$\lim_{r_* \to -\infty} \mathbf{V}_{nl} = 0, \quad \lim_{r \to +\infty} \mathbf{V}_{nl} = \begin{pmatrix} m_n^2 + \mathcal{O}(r^{-1}) & \mathcal{O}(r^{-3}) \\ \mathcal{O}(r^{-2}) & m_n^2 + \mathcal{O}(r^{-1}) \end{pmatrix}. \tag{9.106}$$

For $r_* \to -\infty$, which corresponds to the black-hole horizon, we see that u_{nlm} and v_{nlm} behave as free massless scalars. Conversely, far away from the black hole they behave as decoupled scalars of mass m_n. It turns out that the asymptotic form of $\mathbf{V}_{nl}$ as $r \to \infty$ is crucial in determining the characteristic GW signal from a black string, as we will see below.

9.7.2 Axial p-Waves

For the sake of completeness, we can write down the equations of motion governing the $l = 1$, or p-wave, sector. In this case, general fluctuations are described by

$$h_{\alpha\beta}^{(n1m,\text{axial})} = \mathbb{A}_{\alpha\beta}^{1,1,m}(\Omega)\,\mathscr{A}_{1,1,m}^{(n)}(t,r) + \mathbb{A}_{\alpha\beta}^{2,1,m}(\Omega)\,\mathscr{A}_{2,1,m}^{(n)}(t,r). \tag{9.107}$$

In this case, when we substitute this into the equation of motion (9.98), we find a single master equation

$$0 = (\partial_t^2 - \partial_{r_*}^2)u_{n1m} + V_{n1}u_{n1m}, \tag{9.108}$$

where

$$u_{n1m}(t,r) = f(r)\,\mathscr{A}_{2,1,m}^{(n)}(t,r), \tag{9.109}$$

and the potential is

$$V_{n1} = f\left(\frac{5f+1}{r^2} - \frac{2f'}{r} + \frac{f''}{2} + m_n^2\right). \tag{9.110}$$

Once this equation is solved and $\mathscr{A}_{2,1,m}^{(n)}$ is found, the remaining expansion coefficient is determined by a quadrature:

$$\frac{\partial \mathscr{A}_{1,1,m}^{(n)}}{\partial t} = f^2 \frac{\partial \mathscr{A}_{2,1,m}^{(n)}}{\partial r} + \frac{f(2f + f'r)}{r}\,\mathscr{A}_{2,1,m}^{(n)}. \tag{9.111}$$

Notice that this is identical to (9.105) with $l = 1$.

One comment on the $l = 1$ perturbations is in order before we proceed. In ordinary black-hole perturbation theory, there are no truly time-dependent p-wave perturbations of the Schwarzschild spacetime. This is because the $l = 1$ perturbations correspond to giving the black hole a small amount of angular momentum about some axis in 3-space; i.e., they represent the linearization of the Kerr solution about the Schwarzschild background and are hence time independent. In the black string case, however, the $l = 1$ perturbation can be viewed as endowing a small spin to the Schwarzschild 4-metrics on each Σ_y hypersurface. However, the amount of spin delivered to each hypersurface by each massive mode is not uniform, in fact it is easily shown that it is proportional to $Z_n(y)$ evaluated at that hypersurface. In other words, dipole perturbations give rise to a differentially rotating black string, where the amount of rotation varies with y. It turns out that there is no time-independent black string solution of this type, so we have dynamic perturbations. The exception is the zero mode $n = 0$, which gives rise to a uniform rotation of the black string; i.e., these perturbations give rise to the linearization of (9.18) about (9.15).

9.7.3 Numeric Integration of Quadrupole Equations

In Fig. 9.2, we present the results of some numerical solutions of (9.102) for the case of quadrupole radiation $l = 2$. In this plot, we assume that we have Gaussian

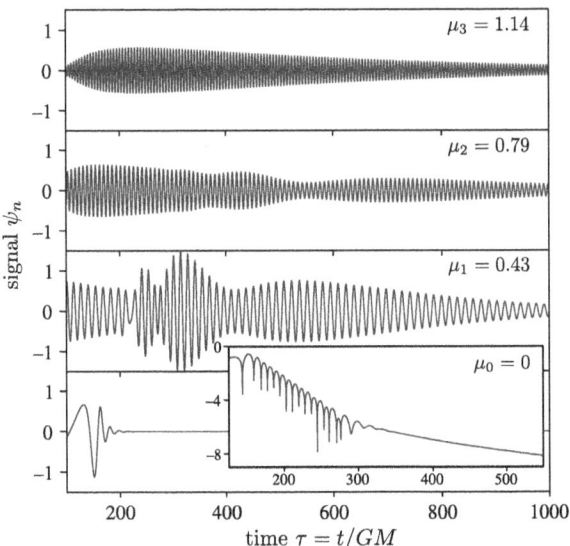

Fig. 9.2 Results of the integration of the quadrupole axial equations of motion. The waveforms are observed at $r_* = 100GM$ while the initial data was originally located at $r_* = 50GM$. We show results for the $n = 0, 1, 2, 3$ modes. The massive mode signals are characterized by a long-lasting oscillating tail; i.e., u_{n2m} and v_{n2m} are proportional to $(t/GM)^{-5/6} \sin(m_n t + \phi)$ at late times for $n > 0$ (here, ϕ is a phase angle). This is in contrast to the zero-mode result, which shows no oscillations and a power-law decay at late times (the inset shows the zero-mode result on a log-log scale)

initial data for u_{n2m} on some initial time slice and that $v_{n2m} = 0$ initially. It turns out that the particular choice of initial data does not much affect the outcome of the simulations; that is, changing the shape or location of the initial Gaussian, or taking $v_{n2m} \neq 0$, results in very similar waveforms.

The key feature of the displayed waveforms is the nature of the late-time signal. We see that each of the $n > 0$ waveforms exhibits very long-lived late-time oscillations.[6] This behavior is totally unlike the standard picture of black hole oscillations in GR, where one expects the late-time ringdown waveform to be a featureless power-law tail. This kind of signal is exhibited by the $n = 0$ zero-mode signal, which we already know corresponds exactly to the GR result. One of the most remarkable things about the massive mode signal is that it is present for all types of initial data, suggesting that it is a fundamental property of the black string as opposed to just some simulation fluke. In this sense the massive mode tail observed here is analogous to the quasinormal modes of standard 4-dimensional theory.

An exercise in curve fitting reveals that the late-time massive signal is well modeled by

$$\left\{ \begin{matrix} u_{n2m} \\ v_{n2m} \end{matrix} \right\} \sim \text{const} \times \left(\frac{t}{GM} \right)^{-5/6} \sin(m_n t + \phi). \tag{9.112}$$

That is, the frequency of oscillation matches the mass of the mode. The decay rate $\sim t^{-5/6}$ is much slower than the decay of the zero-mode signal, which decays at least as fast as t^{-4}. We can confirm via simulations that these result holds for other values of l. Hence, we are led to the following important conclusion: *Irrespective of the initial amplitudes of the various KK modes, if one waits long enough the GW signal from a perturbed black string will be dominated by a superposition of slowly-decaying massive modes.* A challenge for gravitational wave astronomy is to observe these massive mode signals directly. The actual prospects of doing this are discussed in Sect. 9.10.4.

9.8 Spherical Perturbations with Source Terms

We can re-write the decomposition (9.113) by explicitly pulling out the spherical contributions:

$$\tilde{\xi} = \frac{\xi^{(s)}}{\sqrt{4\pi}} + \sum_{l=1}^{\infty} \sum_{m=-l}^{l} Y_{lm} \tilde{\xi}_{lm}, \tag{9.113a}$$

$$h_{\alpha\beta}^{(n)} = \frac{h_{\alpha\beta}^{(n,s)}}{\sqrt{4\pi}} + \sum_{l=1}^{\infty} \sum_{m=-l}^{l} \sum_{i=1}^{10} [Y_{lm}^{(i)}]_{\alpha\beta} \, h_i^{(nlm)}, \tag{9.113b}$$

$$\Theta_{\alpha\beta} = \frac{\Theta_{\alpha\beta}^{(s)}}{\sqrt{4\pi}} + \sum_{l=1}^{\infty} \sum_{m=-l}^{l} \sum_{i=1}^{10} [Y_{lm}^{(i)}]_{\alpha\beta} \, \Theta_i^{(lm)}. \tag{9.113c}$$

---

[6] A mathematical rationalization of this is given is Sect. 9.10.2.

Here, $\xi^{(s)}$, $h_{\alpha\beta}^{(n,s)}$, and $\Theta_{\alpha\beta}^{(s)}$ represent the spherically symmetric parts of the brane-bending scalar, metric perturbation, and brane stress–energy tensor, respectively. In this section, we are going to concentrate on the dynamics of this sector when there are non-trivial matter sources on one of the branes sourcing gravitational radiation. The reason that we focus on the $l = 0$, or s-wave, sector is computational convenience; the equations of motion become rather involved for higher multipoles.

Before starting to calculate things, we note that some readers may be a little confused as to why we are even looking at spherically symmetric gravitational radiation. In general relativity, it is a well-known consequence of Birkhoff's theorem that there is no spherically symmetric radiation about a Schwarzschild black hole. This is because the theorem states that the only solutions to the Einstein equations with cosmological constant with structure $\mathbb{R}^2 \times S_d$ are $(d+2)$-dimensional Schwarzschild-de Sitter or Schwarzschild anti-de Sitter black holes. Since these are static solutions, any perturbation that respects the S_d symmetry of the background must also be static.[7] But the black string background has structure $(\mathbb{R}^2 \times S_2) \times S/\mathbb{Z}_2$. Birkhoff's theorem does not apply in this case and we can indeed have time-dependant solutions of $G_{AB} = 6k^2 g_{AB}$ with the same structure. Therefore, it is possible to have dynamical spherically symmetric radiation around a black string, which is what we study in this section.

9.8.1 Spherical Master Variables

We write the $l = 0$ contribution to the metric perturbation as

$$h_{\alpha\beta}^{(n,s)} = \mathsf{H}_1 \, t_\alpha t_\beta - 2\mathsf{H}_2 \, t_{(\alpha} r_{\beta)} + \mathsf{H}_3 \, r_\alpha r_\alpha + \mathsf{K} \gamma_{\alpha\beta}, \qquad (9.114)$$

where the 4-vectors and $\gamma_{\alpha\beta}$ are defined in (9.94) and (9.95), respectively. Each of the expansion coefficients is a function of t and r; i.e., $\mathsf{H}_i = \mathsf{H}_i(t,r)$ and $\mathsf{K} = \mathsf{K}(t,r)$. Notice that the condition that $h_{\alpha\beta}^{(n,s)}$ is tracefree implies

$$\mathsf{K} = \tfrac{1}{2}(\mathsf{H}_1 - \mathsf{H}_3). \qquad (9.115)$$

Before going further, it is useful to define dimensionless coordinates:

$$\rho = \frac{r}{GM}, \quad \tau = \frac{t}{GM}, \quad x = \rho + 2\ln\left(\frac{\rho}{2} - 1\right). \qquad (9.116)$$

Then, when our decompositions (9.113) are substituted into the equations of motion, we find that all components of the metric perturbation are governed by master variables

$$\psi = \frac{2\rho^3}{2 + \mu_n^2 \rho^3}\left(\rho \frac{\partial \mathsf{K}}{\partial \tau} - f\mathsf{H}_2\right), \quad \varphi = \rho \frac{\partial \xi^{(s)}}{\partial \tau}. \qquad (9.117)$$

---

[7] Static here means that one can find a gauge in which the perturbation does not depend on time.

Both $\psi = \psi(\tau, x)$ and $\varphi = \varphi(\tau, x)$ satisfy simple wave equations:

$$(\partial_\tau^2 - \partial_x^2 + V_\psi)\psi = \mathscr{S}_\psi + \hat{\mathscr{I}}\varphi, \tag{9.118a}$$

$$(\partial_\tau^2 - \partial_x^2 + V_\varphi)\varphi = \mathscr{S}_\varphi. \tag{9.118b}$$

The potential and matter source terms in the ψ equation are:

$$
\begin{aligned}
V_\psi = {} & \frac{f}{\rho^3(2+\rho^3\mu_n^2)^2}\Big[\mu_n{}^6\rho^9 + 6\mu_n{}^4\rho^7 - 18\mu_n{}^4\rho^6 \\
& - 24\mu_n{}^2\rho^4 + 36\mu_n{}^2\rho^3 + 8\Big],
\end{aligned}
\tag{9.119a}
$$

$$
\begin{aligned}
\mathscr{S}_\psi = {} & \frac{2f\rho^3}{3(2+\mu_n^2\rho^3)^2}\Big[\rho(2+\mu_n^2\rho^3)\partial_\tau(2\Lambda_1 + 3\Lambda_3) \\
& + 6(\mu_n^2\rho^3 - 4)f\Lambda_2\Big].
\end{aligned}
\tag{9.119b}
$$

Here, we have defined the following three scalars derived from the dimensionless stress–energy tensor $\Theta_{\alpha\beta}^{(s)}$:

$$\Lambda_1 = -\Theta_{\alpha\beta}^{(s)}g^{\alpha\beta}, \quad \Lambda_2 = -\Theta_{\alpha\beta}^{(s)}t^\alpha r^\beta, \quad \Lambda_3 = +\Theta_{\alpha\beta}^{(s)}\gamma^{\alpha\beta}. \tag{9.120}$$

The potential and source terms in the brane-bending equation are somewhat less involved:

$$V_\varphi = \frac{2f}{\rho^3}, \quad \mathscr{S}_\varphi = \frac{\rho f}{6}\partial_\tau\Lambda_1. \tag{9.121}$$

Finally, the interaction operator is

$$\hat{\mathscr{I}} = \frac{8f}{(2+\mu_n^2\rho^3)^2}\Big[6f\rho^2\partial_\rho + (\mu_n^2\rho^3 - 6\rho + 8)\Big]. \tag{9.122}$$

9.8.2 Inversion Formulae

Assuming that we can solve the wave equations (9.118) for a given source, we need formulae that allow us to express H_i, K in terms of ψ and φ in order to make gravitational wave prediction. This can be derived by inverting the master variable definitions (9.117) with the aid (9.118). The general formulae are actually very complicated and not particularly enlightening, so we do not reproduce them here. Ultimately, to make observational predictions it is sufficient to know the form of the metric perturbation far away from the black string and the matter sources, so we evaluate the general inversion formulae in the limit of $\rho \to \infty$ and with $\Lambda_i = 0$:

$$\partial_\tau H_1 = \frac{1}{\rho}\left[\left(\partial_\tau^2 + \frac{3}{\rho}\partial_\rho + \mu_n^2\right)\psi + \frac{4}{\mu_n^2}\left(\partial_\tau^2 - \frac{1}{\rho}\partial_\rho\right)\varphi\right],$$

$$H_2 = \frac{1}{\rho}\left[\left(\partial_\rho + \frac{2}{\rho}\right)\psi + \frac{4}{\mu_n^2}\left(\partial_\rho - \frac{1}{\rho}\right)\varphi\right],$$

$$\partial_\tau H_3 = \frac{1}{\rho}\left[\left(\partial_\tau^2 + \frac{1}{\rho}\partial_\rho\right)\psi + \frac{4}{\mu_n^2}\left(\partial_\tau^2 - \frac{2}{\rho}\partial_\rho\right)\varphi\right],$$

$$\partial_\tau K = \frac{1}{\rho}\left[\left(\frac{1}{\rho}\partial_\rho + \frac{\mu_n^2}{2}\right)\psi + \frac{4}{\mu_n^2\rho}\left(\partial_\rho - \frac{1}{\rho}\right)\varphi\right]. \qquad (9.123)$$

Note that these do not actually complete the inversion; in most cases, a quadrature is also required to arrive at the final form of the metric perturbation.

9.8.3 The Gregory–Laflamme Instability

We now discuss one extremely important consequence of the equation of motion (9.118). Note that we can always add-on a solution of the homogeneous wave equation:

$$0 = (\partial_\tau^2 - \partial_x^2 + V_\psi)\psi, \qquad (9.124)$$

to any particular solution ψ_p of (9.118a) generated by a given source. If we analyze this homogenous equation in Fourier space by setting $\psi(\tau,x) = e^{i\omega\tau}\Psi(x)$, we find that

$$\omega^2\Psi = -\frac{d^2\Psi}{dx^2} + V_\psi\Psi. \qquad (9.125)$$

This is identical to the time-independent Schrödinger equation from elementary quantum mechanics with ω^2 playing the role of the energy parameter. Now, suppose that the potential supports a bound state solution with negative energy $\omega^2 < 0$. That is, suppose we can find a solution of this ODE with $\Psi \to 0$ as $x \to \pm\infty$ with $\omega = -i\Gamma$, where $\Gamma > 0$. In such cases, $\psi \propto e^{\Gamma t}$ and we have an exponentially growing solution to the equations of motion, which represents a linear instability of the system. Since such a tachyonic mode ψ is spatially bounded and arbitrary small in the past, it is possible for any initial data with compact support to excite it.

Clearly, the black string braneworld cannot be a viable black-hole model if we can find such a tachyonic mode. It turns out that the potential V_ψ (9.119a) is not actually capable of supporting a negative energy bound state for all values of μ. There are numerous ways of demonstrating this; including the WKB method and direct numeric solution of (9.125). One finds that no bound state exists if

$$\mu_n > \mu_c \approx 0.4301 \text{ or } \mu_n = 0. \qquad (9.126)$$

That is, the zero-mode of the s-wave sector is stable,[8] and the high-mass modes are also stable. This implies that the black string braneworld is perturbatively stable if the smallest KK mass satisfies

$$\mu_1 = GMm_1 > \mu_c \approx 0.4301. \tag{9.127}$$

Under the approximation that the first mode is light ($x_1 e^{-kd} \ll 1$) and using $G = \ell_{\text{Pl}}/M_{\text{Pl}}$, this gives a restriction on the black string mass

$$\frac{M}{M_{\text{Pl}}} \gtrsim \frac{\ell}{\ell_{\text{Pl}}} \frac{\mu_c}{x_1} e^{kd}, \tag{9.128}$$

or equivalently,

$$\frac{M}{M_\odot} \gtrsim 8 \times 10^{-9} \left(\frac{\ell}{0.1 \text{ mm}} \right) e^{d/\ell}. \tag{9.129}$$

If we take $\ell = 0.1$ mm, then we see that all solar mass black holes will in actuality be stable black strings provided that $d/\ell \lesssim 19$. The stability of the black string braneworld is summarized in Fig. 9.3.

Before moving on, we have two final comments: First, we should note that all black strings are unstable if the distance between the branes becomes large $d \to \infty$. This essentially means that there is no stable black string solution when the extra dimension is infinite. This is the well-known Gregory–Laflamme instability of black strings [7, 8]. Second, if we denote the minimum mass stable black string to be M_{GL} for a given d/ℓ, note that we do note claim that black holes with $M < M_{\text{GL}}$ do not exist in this braneworld setup. Rather, such small mass black holes are not described by the black string bulk. They would instead be described by some localized black

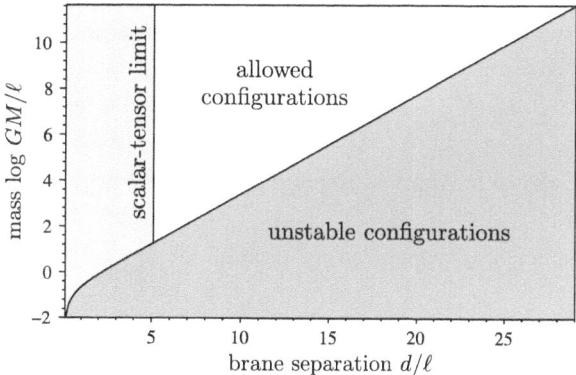

Fig. 9.3 The stability of the black string braneworld model. If the black string mass M, or the brane separation d is selected such that GM/ℓ and d/ℓ lies outside of the 'unstable configurations' configurations portion of parameter space, the model is stable. We have also indicated the $d/\ell \gtrsim 5$ limit imposed by the low-energy scalar-tensor limit of the model in the solar system (cf. Sect. 9.5)

---

[8] One can show that this is actually a gauge mode

hole solution that has yet to be obtained. It has been suggested in the literature that the transition between the localized black hole and the black string may be a violent first-order phase transition, an hence be a significant source of gravitational radiation [12].

9.9 Point Particle Sources on the Brane

Up until this point, we have either been discussing homogenous equations or generic sources. As an illustration of a more specific application of the formulae we have derived, we specialize to the situation where the perturbing brane matter is a 'point particle' located on one of the branes. Our goal is to explicitly write down the equations of motion for the GWs emitted by the particle. This is a situation of a significant astrophysical interest in 4-dimensions, because it is thought to be a good model of 'extreme-mass-ratio-inspirals' (EMRIs). This is a scenario when an object of mass M_p merges with a black hole of mass M. When $M_p \ll M$, it is a good approximation to replace the small body with a point particle, or delta-function, source. Our interest here is to generalize this standard 4-dimensional calculation to the black string background.

One caution is in order before we proceed: It is not entirely clear that the delta-function approximation is a good one to make in the braneworld scenario. In 4 dimensions, there are only two length scales in the problem: the two Schwarzschild radii $2GM$ and $2GM_p$.[9] Hence, an extreme scenario is well defined when one scale is much larger than the other. However, in the braneworld scenario there is an additional length scale ℓ. In typical situations, $\ell \ll 2GM_p \ll 2GM$. It is unclear whether or not it is valid to model the perturbing body as a point particle in this case, since a point particle always has a physical size less than ℓ. However, it the absence of a better source model, we will pursue the point particle description here, while always keeping this caveat in mind.

9.9.1 Point Particle Stress–Energy Tensor

We take the particle Lagrangian density to be

$$\mathscr{L}_p^{\pm} = \frac{M_p}{2} \left\{ \int \frac{\delta^4(z^\mu - z_p^\mu)}{\sqrt{-q}} q_{\alpha\beta} \frac{dz_p^\alpha}{d\eta} \frac{dz_p^\beta}{d\eta} d\eta \right\}^{\pm} . \tag{9.130}$$

In this expression, η is a parameter along the particle's trajectory as defined by the $q_{\alpha\beta}$ metric, z_p^μ are the four functions describing the particle's position on the brane,

---

[9] We generally consider cases where the physical size of the perturbing particle is close to its horizon radius, as for neutron stars, etc.

and M_p is the particle's mass parameter. Using (9.13), we find the stress–energy tensor

$$T_{\alpha\beta}^{\pm} = M_p \left\{ \int \frac{\delta^4(z^{\mu} - z_p^{\mu})}{\sqrt{-q}} q_{\alpha\rho} q_{\beta\lambda} \frac{dz_p^{\rho}}{d\eta} \frac{dz_p^{\lambda}}{d\eta} d\eta \right\}^{\pm}. \tag{9.131}$$

The contribution from the particle to the total action is

$$S_p^{\pm} = \frac{1}{2} \int_{\Sigma^{\pm}} \mathcal{L}_p^{\pm} = \frac{M_p}{4} \int q_{\alpha\beta}^{\pm} \frac{dz_p^{\alpha}}{d\eta} \frac{dz_p^{\beta}}{d\eta} d\eta. \tag{9.132}$$

Varying this with respect to the trajectory z_p^{α} and demanding that η is an affine parameter yields that the particle follows a geodesic along the brane:

$$\frac{d^2 z_p^{\alpha}}{d\eta^2} + \Gamma_{\beta\gamma}^{\alpha}[q^{\pm}] \frac{dz_p^{\beta}}{d\eta} \frac{dz_p^{\gamma}}{d\eta} = 0, \quad -1 = q_{\alpha\beta}^{\pm} \frac{dz_p^{\alpha}}{d\eta} \frac{dz_p^{\beta}}{d\eta}, \tag{9.133}$$

where $\Gamma_{\beta\gamma}^{\alpha}[q^{\pm}]$ are the Christoffel symbols defined with respect to the $q_{\alpha\beta}^{\pm}$ metric.

We note that the above formulae make explicit use of the induced brane metrics $q_{\alpha\beta}^{\pm}$. However, all of our perturbative formalism is in terms of the Schwarzschild metric $g_{\alpha\beta}$, especially the definition of the Λ_i scalars (9.120). Hence, it is useful to translate the above expressions using the following definitions:

$$\eta = a_{\pm} \lambda, \quad u^{\alpha} = \frac{dz_p^{\alpha}}{d\lambda}, \quad -1 = g_{\alpha\beta} u^{\alpha} u^{\beta}. \tag{9.134}$$

Then, the stress–energy tensor and particle equation of motion become

$$T_{\alpha\beta}^{\pm} = \frac{M_p}{a_{\pm}} \int \frac{\delta^4(z^{\mu} - z_p^{\mu})}{\sqrt{-g}} u_{\alpha} u_{\beta} \, d\lambda, \quad u^{\alpha} \nabla_{\alpha} u^{\beta} = 0. \tag{9.135}$$

Note that the only difference between the stress–energy tensors on the positive and negative tension branes is an overall division by the warp factor.

By switching over to dimensionless coordinates, transforming the integration variable to τ from λ, and making use of the spherical harmonic completeness relationship, we obtain

$$T_{\alpha\beta}^{\pm} = \frac{f}{\mathscr{C}_{\pm} E \rho^2} u_{\alpha} u_{\beta} \delta(\rho - \rho_p) \left[\frac{1}{4\pi} + \sum_{l=1}^{\infty} \sum_{m=-l}^{l} Y_{lm}(\Omega) Y_{lm}^*(\Omega_p) \right]. \tag{9.136}$$

Here, we have defined

$$\mathscr{C}_{\pm} = \frac{(GM)^3}{M_p e^{ky_{\pm}}}, \quad E = -g_{\alpha\beta} u^{\alpha} \xi_{(t)}^{\beta}, \quad \xi_{(t)}^{\alpha} = \partial_t. \tag{9.137}$$

As usual, E is the particle's energy per unit rest mass defined with respect to the time-like Killing vector $\xi_{(t)}^{\alpha}$.

9.9.2 The s-Wave Sector

Comparing (9.88) and (9.113c) with (9.136), we see that

$$\Theta^{(s)}_{\alpha\beta} = \frac{f}{\sqrt{4\pi E \rho^2}} u_\alpha u_\beta \, \delta[\rho - \rho_p(\tau)], \tag{9.138a}$$

$$\Lambda_1 = \frac{f}{\sqrt{4\pi E \rho^2}} \, \delta[\rho - \rho_p(\tau)], \tag{9.138b}$$

$$\Lambda_2 = \frac{E\dot{\rho}_p}{\sqrt{4\pi f \rho^2}} \, \delta[\rho - \rho_p(\tau)], \tag{9.138c}$$

$$\Lambda_3 = \frac{f\tilde{L}^2}{\sqrt{4\pi E \rho^4}} \, \delta[\rho - \rho_p(\tau)], \tag{9.138d}$$

where $\dot{\rho}_p = d\rho_p/d\tau$. Here, we have identified L as the total angular momentum of the particle (per unit rest mass), defined by

$$\frac{L^2}{r^2} = \gamma_{\alpha\beta} u^\alpha u^\beta, \quad \tilde{L} = \frac{L}{GM}. \tag{9.139}$$

Note that for particles traveling on geodesics, E and L are constants of the motion. These are commonly re-parameterized in terms of the eccentricity e and the semi-latus rectum p, both of which are non-negative dimensionless numbers:

$$E^2 = \frac{(p-2-2e)(p-2+2e)}{p(p-3-e^2)},$$
$$\tilde{L}^2 = \frac{p^2}{p-3-e^2}. \tag{9.140}$$

The orbit can then be conveniently described by the alternative radial coordinate χ, which is defined by

$$\rho = \frac{p}{1+e\cos\chi}. \tag{9.141}$$

Taking the plane of motion to be $\theta = \pi/2$, we obtain two first-order differential equations governing the trajectory

$$\frac{d\chi}{d\tau} = \left[\frac{(p-2-2e\cos\chi)^2(p-6-2e\cos\chi)}{\rho_p^4(p-2-2e)(p-2+2e)} \right]^{1/2},$$

$$\frac{d\phi}{d\tau} = \left[\frac{p(p-2-2e\cos\chi)^2}{\rho_p^4(p-2-2e)(p-2+2e)} \right]^{1/2}. \tag{9.142}$$

These are well behaved thorough turning points of the trajectory $d\rho_p/dt = 0$. When $e < 1$ we have bound orbits such that $p/(1+e) < \rho_p < p/(1-e)$, while for $e > 1$ we have unbound 'fly-by' orbits whose closest approach is $\rho_p = p/(1+e)$. To obtain orbits that cross the future event horizon of the black string, one needs to apply a Wick rotation to the eccentricity $e \mapsto ie$ and make the replacement $\chi \mapsto i\chi + \pi/2$. Then a radially infalling particle corresponds to $e = \infty$.

It is worthwhile to write out the associated source terms in the wave equation explicitly as a function of orbital parameters

$$\mathscr{S}_\psi = \frac{2f^2\dot{\rho}_p}{3\sqrt{4\pi}E(2+\mu^2\rho^3)}\left[-(2\rho^2+3\tilde{L}^2)\delta'[\rho-\rho_p(\tau)]\right.$$

$$\left. +\frac{6\rho E^2}{f}\left(\frac{\mu^2\rho^3-4}{\mu^2\rho^3+2}\right)\delta[\rho-\rho_p(\tau)]\right],$$

$$\mathscr{S}_\varphi = -\frac{f^2\dot{\rho}_p}{6\sqrt{4\pi}E\rho}\delta'[\rho-\rho_p(\tau)].\tag{9.143}$$

Note that

$$|\dot{\rho}_p| < f,$$

$$\dot{\rho}_p = 0 \Rightarrow \mathscr{S}_\psi = \mathscr{S}_\varphi = 0,\tag{9.144}$$

$$E \gg 1 \Rightarrow \mathscr{S}_\psi \gg \mathscr{S}_\varphi.$$

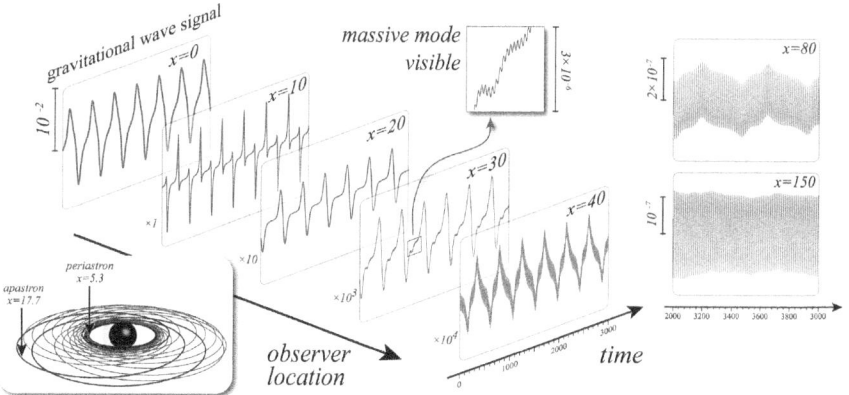

Fig. 9.4 The steady-state KK gravitational wave signal induced by a particle undergoing a periodic orbit around the black string with $\mu = 0.5$. The orbit (*bottom left*) has eccentricity $e = 0.5$ and angular momentum $p = 3.62$. The waveform of radiation falling into the black string is quite different than that of radiation escaping to infinity: The infalling signal precisely mimics the orbital profile of the source, while the outgoing signal is dominated by monochromatic radiation whose frequency is proportional to the KK mass μ

That is, the particle's speed is always less than unity, the sources wave equation vanish if the particle is stationary or in a circular orbit, and high-energy trajectories imply that the system's dynamics are not too sensitive to brane-bending modes $\psi \gg \varphi$.

Numeric solutions of the spherical equations of motion with a point particle source have been obtained elsewhere [4]. A major consideration in performing such simulations is that the sources in the save equations are distributional, and hence must be regulated in some way. In [4], the authors regulated the delta-functions by replacing them with thin Gaussians. In Fig. 9.4, we show the results of such a simulation when the perturbing particle is undergoing a periodic orbit. One observes that the GW signal for from the brane is essentially that of a pure massive mode signal.

9.10 Estimating the Amplitude of the Massive Mode Signal

We have seen in previous sections that if we consider a black string relaxing to its equilibrium configuration or if we look at the GWs emitted by a small particle orbiting the black string, the signal is dominated by massive mode oscillations. The question is: are these oscillations observable? The ability of a GW detector to see a given signal depends on that signal's frequency and its amplitude. The frequency of massive mode signals is well defined, it is simply given by the solution of the eigenvalue problem presented in Sect. 9.4. However, the amplitude is difficult to pin down unless we consider a specific situation. So in this section, we concentrate on the s-wave massive modes emitted by a particle in orbit about a black string. We will be interested in the entire massive mode spectrum; i.e., all values of n. To estimate the GW amplitude associated with heavy modes we will need to analyze the asymptotics of the Green's function solution of the coupled wave equations (9.118).

9.10.1 Green's Function Analysis

The formal solution to the coupled wave equations (9.118) can be written in terms of the Green's functions

$$(\partial_\tau^2 - \partial_x^2 + V_\psi)G(\tau;x,x') = \delta(\tau)\delta(x - x'), \tag{9.145a}$$

$$(\partial_\tau^2 - \partial_x^2 + V_\varphi)D(\tau;x,x') = \delta(\tau)\delta(x - x'). \tag{9.145b}$$

To preserve casuality in the model, we demand that G and D satisfy retarded boundary conditions. That is, they are identically zero if the field point (τ,x) is not contained within thefuture light cone the source point $(0,x')$.

In terms of these Green's functions, we have

$$\psi(\tau, x) = \psi_1(\tau, x) + \psi_2(\tau, x),$$

$$\psi_1(\tau, x) = \int d\tau' dx'\, G(\tau - \tau'; x, x') \mathscr{S}_\psi(\tau', x'),$$

$$\psi_2(\tau, x) = \int d\tau' dx'\, G(\tau - \tau'; x, x') \hat{\mathscr{I}}(\tau', x') \varphi(\tau', x'),$$

$$\varphi(\tau, x) = \int d\tau' dx'\, D(\tau - \tau'; x, x') \mathscr{S}_\varphi(\tau', x'). \tag{9.146}$$

Note the decomposition of ψ into a contribution ψ_1 from the matter source $\mathscr{S}_\psi$, and a contribution ψ_2 from from brane-bending φ. These expressions suggest that if we knew the two Green's functions explicitly, the gravitational wave master variable and brane-bending scalar would be given by quadrature.

Unfortunately, G and D are not known in closed form, so we have to resort to numeric computations to accurately calculate the values of ψ and φ induced by a particular source, and for a particular choice of μ. However, any given source will excite all the KK modes to some degree, so to rigorously model the spherical gravitational radiation we would need to do an infinite number of numeric simulations, one for each discrete value of μ. This is not practical, so our goal here is to use the asymptotic behavior of the propagators to determine the transcendental properties of the emitted radiation and how these scale with the dimensionless Kaluza–Klein mass.

9.10.2 Asymptotic Behaviour

In this subsection, we outline the behavior of the two retarded Green's functions G and D under the assumption that the the field point is deep within the future light cone of the source point and is also far away from the string. This is the relevant limit to take if we are interested in the 'late-time' gravitational wave signal seen by distant observers.

9.10.2.1 Brane-Bending Propagator

First, consider the brane-bending Green's function. Note that the brane-bending potential V_φ is identical to that for the $l = 0$ component of a spin-0 field propagating in the Schwarzschild spacetime. This is because the brane-bending equation of motion (9.43) is essentially that of a massless Klein–Gordon field. Fortunately, this propagator has been well studied in the literature, and one can show that

$$D(\tau; x, x') \sim \tau^{-3}, \quad \tau \gg x' - x > 0. \tag{9.147}$$

This result is most easily interpreted if one considers the initial value problem for φ. That is, we switch off the source in (9.43) and prepare the field in some initial state on a given hypersurface. Then, a distant observer measuring φ at late times would see the field amplitude decay in time as a power law with exponent -3.

9.10.2.2 Gravitational Wave Propagator

The retarded Green's function for potentials similar to V_ψ have also been considered in the literature. It turns out that the asymptotic character of the potential is the crucial issue. Koyama & Tomimatsu [13] have demonstrated that for potentials of the form

$$V_\psi \xrightarrow[\infty]{r} \mu_n^2 + \mathscr{O}\left(\frac{1}{r}\right),$$ (9.148)

the Green's function has the asymptotic form

$$G(\tau;x,x') \sim \mu_n^{-1/2}\tau^{-5/6}\sin[\mu_n\tau + \phi(\tau)], \quad \tau \gg x' - x > 0.$$ (9.149)

The form of this Green's function rationalizes the waveforms seen in Fig. 9.2, especially the $t^{-5/6}$ envelope of the late-time signal, despite the fact that the governing equations (9.102) were matrix valued. The key point is the asymptotic form of the potential matrix (9.106), which says that far from the string the two degrees of freedom are decoupled and governed by a potential of the form (9.148).

Comparing this expression to the asymptotic form of D above, we see that G decays much slower. This suggests that $\psi_1 \gg \psi_2$ at late times in (9.146); i.e., the portion of the GW signal sourced directly by the stress–energy tensor dominates the brane-bending contribution. Also note the overall $\mu_n^{-1/2}$ scaling of the Green's function with the KK mass of the mode. We will use this below.

9.10.3 Application to the Point Particle Case for $n \gg 1$: Kaluza–Klein Scaling Formulae

Let us now use the asymptotic Green's functions in the case where the perturbing matter is a point particle. Our goal is to estimate how the KK signal scales with n for the high-mass KK modes.

When the matter stress–energy tensor has delta-function support, the $\int d\tau' dx'$ integrals in (9.146) reduce to line integrals over the portion of the particle's worldline inside of the past light cone. Now, working in the late time far-field limit, we know that the brane-bending contribution to the signal is minimal. We also focus on the high n modes; i.e.,

$$\mu_n \gg 1.$$ (9.150)

Concentrating on the direct signal produced by the particle, we see that the source term $\mathcal{S}_\psi$ for a point particle (see section 9.9.2) seems to scale as μ_n^{-2}. However, note that the source also involves the derivative of a delta function, which means we must perform an integration by parts. This brings a derivative of G with respect to time into the mix. Again assuming that $\mu_n \gg 1$, we see $\partial_\tau G \sim \mu_n^{1/2} e^{i\mu_n \tau}$. The net result is that we expect

$$\psi \propto \mu_n^{-3/2}, \quad \mu_n \gg 1. \tag{9.151}$$

That is, all other things being equal, the spherical master variable for a given KK mode scales as $\mu_n^{-3/2}$.

But this is not the entire solution to the problem, since we do not actually observe ψ, we observe h_{AB}. So we need to use the inversion formulae (9.123) to obtain $h_{\alpha\beta}^{(n,s)}$ and then (9.87) to get the spherical part of h_{AB}. The detailed analysis leads to the following late-time/distant-observer approximation for the KK metric perturbations:

$$h_{AB}^{(n,s)} \approx h_n \mathcal{F}(t) \sin(\omega_n t + \phi_n) \, \mathrm{diag}\left(0, +1, -\tfrac{1}{2}r^2, -\tfrac{1}{2}r^2 \sin^2\theta, 0\right), \tag{9.152}$$

where $\mathcal{F}(t)$ is a slowly varying function of time that depends on the details of the initial data. The characteristic amplitudes h_n are given by

$$h_n = \sqrt{8\pi}\mathcal{A}\left(\frac{2GM_p}{r}\right)\left(\frac{2GM}{\ell}\right)^{-1/2} F_n(d/\ell). \tag{9.153}$$

Here, r is the distance between the observer and the string, and $\mathcal{A}$ is a dimensionless quantity that depends on the orbit of the perturbing particle but not on n or any other parameters; its value must be determined from simulations. $F_n(d/\ell)$ is a complicated expression involving Bessel functions with the following limiting behavior: When the perturbing matter is on our brane

$$F_n(d/\ell) \approx \begin{cases} \frac{1}{2} e^{-3d/2\ell}(n\pi^3)^{1/2}, & n \ll 2e^{d/\ell}/\pi^2, \\ e^{-d/2\ell}(n\pi)^{-1/2}, & n \gg 2e^{d/\ell}/\pi^2. \end{cases} \tag{9.154a}$$

On the other hand, for particles on the shadow brane:

$$F_n(d/\ell) \approx \begin{cases} e^{-d/2\ell}(\pi/2)^{1/2}, & n \ll 2e^{d/\ell}/\pi^2, \\ (n\pi)^{-1/2}, & n \gg 2e^{d/\ell}/\pi^2. \end{cases} \tag{9.154b}$$

Finally, to a good approximation, the KK frequencies are given by

$$\omega_n = 2\pi f_n \approx \frac{c}{\ell}\left(n + \tfrac{1}{4}\right)\pi e^{-d/\ell}. \tag{9.155}$$

We note that even though these formulae were derived in the context of the large n approximation, they are actually reasonable approximations to the small n case as well.

9.10.4 Observability of the Massive Mode Signal

We now have an expression (9.152) for the amplitude of the spherical massive modes in terms of a parameter $\mathscr{A}$ that can be determined from simulations with μ_n small. This amplitude varies with the type of orbit generating the GWs: it can be $\mathscr{O}(10^{-6})$ or smaller for periodic orbits, or as high as $\mathscr{O}(1)$ for 'zoom-whirl' orbits.[10]

In Fig. 9.5, we plot the characteristic amplitudes h_n as a function of their frequency for a scenario where a $1.4 M_\odot$ object is orbiting a $10 M_\odot$ black string at a distance of 1 kpc away. Several general trends are obvious:

- The amplitude of the GW signal decreases with increasing brane separation d/ℓ.
- The lowest frequency in the spectrum also decreases with increasing brane separation d/ℓ.
- For a source on the visible brane, the spectrum is peaked about a critical frequency given by

$$f_{\text{crit}} = \frac{1}{\pi^2 \ell} \sim 304\,\text{GHz} \left(\frac{\ell}{0.1\,\text{mm}}\right)^{-1} \qquad (9.156)$$

- When the perturbing particle is on the shadow brane, the spectrum is flat underneath the critical frequency f_{crit}
- In all cases, the signal from shadow particles is stronger than that of visible particles.

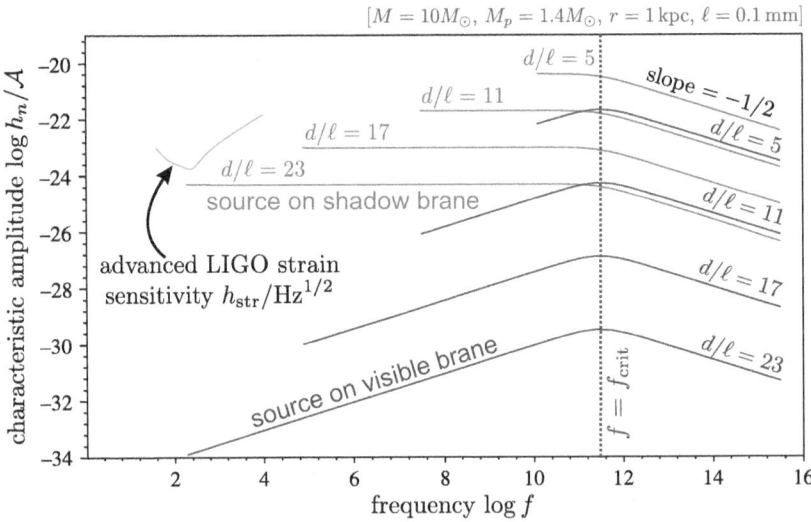

Fig. 9.5 Characteristic amplitudes of KK radiation emitted by point particles on the visible brane or the shadow brane as follows from (9.153). The particular parameters for this example are indicated just above the plot. Also shown is a dimensionally reduced version of the characteristic strain sensitivity of advanced LIGO for comparison

---

[10] These are orbits where the particle comes in from infinity, is briefly captured by the black string, and then escapes to infinity again.

In general, the peak amplitude $h_{\max}$ is the one corresponding to the critical frequency and is given by

$$h_n \le h_{\max} \sim \mathscr{A} \left(\frac{M_p}{M_\odot}\right) \left(\frac{r}{\text{kpc}}\right)^{-1} \left(\frac{M}{M_\odot}\right)^{-1/2} \left(\frac{\ell}{0.1\,\text{mm}}\right)^{1/2}$$
$$\times \begin{cases} 5.0 \times 10^{-22} e^{-(d-5\ell)/\ell}, & \text{visible source}, \\ 9.1 \times 10^{-21} e^{-(d-5\ell)/2\ell}, & \text{shadow source}. \end{cases} \tag{9.157}$$

Figure 9.5 illustrates the main problem with observing the KK signal from a black string. The frequencies in the KK spectrum are bounded below by

$$f_n \ge f_{\min} \sim 12\,\text{GHz} \left(\frac{\ell}{0.1\,\text{mm}}\right)^{-1} e^{-(d-5\ell)/\ell}. \tag{9.158}$$

This implies that the KK spectrum is usually in a higher waveband that the operation frequencies of LIGO and LISA, assuming that $\ell \lesssim 50\,\mu\text{m}$ in line with current experimental tests. The way to mitigate this is to push the branes farther apart, which reduces $f_{\min}$. But if one does this, the amplitude of the signal goes down exponentially. Clearly, the situation is much better for shadow particles, which have an intrinsically stronger GW signal. The detailed prospects of observing massive mode signal with realistic GW detectors is discussed in [4].

9.11 Summary and Outlook

In these lecture notes, we have introduced the black string braneworld, which is a candidate model for a brane black hole in the Randall–Sundrum scenario. At the background level, this model is indistinguishable from the Schwarzschild solution to brane observers, so we need to examine the perturbations of the model to find deviations from general relativity. We have developed the formalism necessary to calculate the gravitational wave signals emitted from black strings perturbed away from their equilibrium configurations. We have found that the late-time nature of these signals is somewhat independent of the nature of the mechanism which generated them, and is a long-lived superposition of discrete monochromatic massive modes. We have discussed how these massive modes could be produced by a point particle orbiting a black string and estimated what their amplitude might be.

There are a number of open issues that need to be addressed in this model. So far, we have only been able to estimate amplitudes by analyzing the scaling behavior of Green's functions and using point particle sources. We need to confirm our scaling results with direct simulations and we need to move beyond the point particle approximation to model realistic sources with size larger than ℓ. The phenomenon of localized black–hole–black string transitions must be looked at in quantitative detail. The possibility that such a phase transition can produce significant amounts of massive mode radiation and contribute to the gravitational wave background provides one of the best prospects for the actual detection of a black string.

Acknowledgments I would like to thank Chris Clarkson and Roy Maartens for the collaboration throughout this work. I also wish to thank NSERC of Canada and STFC of the UK for financial support. Finally, I must thank Elefteris Papantanopoulos for the kind invitation to a stimulating meeting in a gorgeous setting.

References

1. A. Cardoso, T. Hiramatsu, K. Koyama and S. S. Seahra, Scalar perturbations in braneworld cosmology. JCAP 0707:008 (2007).
2. A. Chamblin, S. W. Hawking and H. S. Reall, Brane-world black holes. Phys. Rev. **D61**, 065007 (2000).
3. C. Clarkson and R. Maartens, Gravity-wave detectors as probes of extra dimensions. Gen. Rel. Grav. **37**, 1681–1687 (2005).
4. C. Clarkson and S. S. Seahra, A gravitational wave window on extra dimensions. Class. Quant. Grav. **24**, F33–F40 (2007).
5. R. Emparan, A. Fabbri and N. Kaloper, Quantum black holes as holograms in ads braneworlds. JHEP **08**, 043 (2002).
6. J. Garriga and T. Tanaka, Gravity in the brane-world. Phys. Rev. Lett. **84**, 2778–2781 (2000).
7. R. Gregory and R. Laflamme, Black strings and p-branes are unstable. Phys. Rev. Lett. **70**, 2837–2840 (1993).
8. R. Gregory, Black string instabilities in anti-de sitter space. Class. Quant. Grav. **17**, L125–L132 (2000).
9. T. Hiramatsu, K. Koyama and A. Taruya, Evolution of gravitational waves in the high-energy regime of brane-world cosmology. Phys. Lett. **B609**, 133–142 (2005).
10. D. J. Kapner et al., Tests of the gravitational inverse-square law below the dark-energy length scale. Phys. Rev. Lett. **98**, 021101 (2007).
11. T. Kobayashi and T. Tanaka, The spectrum of gravitational waves in randall-sundrum braneworld cosmology. Phys. Rev. **D73**, 044005 (2006).
12. B. Kol, The phase transition between caged black holes and black strings: A review. Phys. Rept. **422**, 119–165 (2006).
13. H. Koyama and A. Tomimatsu, Asymptotic tails of massive scalar fields in schwarzschild background. Phys. Rev. **D64**, 044014 (2001).
14. H. Kudoh, T. Tanaka and T. Nakamura, Small localized black holes in braneworld: Formulation and numerical method. Phys. Rev. **D68**, 024035 (2003).
15. L. Randall and R. Sundrum, An alternative to compactification. Phys. Rev. Lett. **83**, 4690–4693 (1999).
16. L. Randall and R. Sundrum, A large mass hierarchy from a small extra dimension. Phys. Rev. Lett. **83**, 3370–3373 (1999).
17. S. S. Seahra, Gravitational waves and cosmological braneworlds: A characteristic evolution scheme. Phys. Rev. **D74**, 044010 (2006).
18. S. S. Seahra, C. Clarkson and R. Maartens, Detecting extra dimensions with gravity wave spectroscopy. Phys. Rev. Lett. **94**, 121302 (2005).
19. T. Tanaka, Classical black hole evaporation in randall-sundrum infinite braneworld. Prog. Theor. Phys. Suppl. **148**, 307–316 (2003).

Chapter 10
Black Holes at the Large Hadron Collider

P. Kanti

Abstract In these two lectures, we will address the topic of the creation of small black holes during particle collisions in a ground-based accelerator, such as LHC, in the context of a higher-dimensional theory. We will cover the main assumptions, criteria and estimates for their creation, and we will discuss their properties after their formation. The most important observable effect associated with their creation is likely to be the emission of Hawking radiation during their evaporation process. After presenting the mathematical formalism for its study, we will review the current results for the emission of particles both on the brane and in the bulk. We will finish with a discussion of the methodology that will be used to study these spectra and the observable signatures that will help us identify the black-hole events.

10.1 Introduction

These two lectures aim at offering an introduction to the idea that miniature black holes may be created during high-energy particle collisions at ground-based colliders. This scenario can only be realised in the context of higher-dimensional theories, i.e. theories that postulate the existence of additional spacelike dimensions in nature. An introduction to the two most important versions of these theories, namely the scenario with large extra dimensions and the one with warped extra dimensions, will be our starting point.

We will then proceed to introduce the idea of the possible creation of black holes at the laboratory. We will present some simple but illuminating geometrical criteria for this to happen. We will then discuss the boundary value problem whose solution determines whether a black hole has been formed out of two colliding particles. Certain aspects of the creation process will be studied in more detail, namely the amount of energy that is absorbed by the created black hole and the value of the production cross-section. We will finally discuss the properties of the produced black holes,

P. Kanti (✉)

Division of Theoretical Physics, Physics Department, University of Ioannina,
Ioannina GR-45110, Greece
pkanti@cc.uoi.gr

Kanti, P.: *Black Holes at the Large Hadron Collider.* Lect. Notes Phys. **769**, 387–423 (2009)
DOI 10.1007/978-3-540-88460-6_10   © Springer-Verlag Berlin Heidelberg 2009

such as the horizon value, temperature and lifetime, and compare with the ones of their four-dimensional analogues. The non-vanishing, in general, temperature of the black hole is associated with the emission of a thermal type of radiation from the black hole, i.e. the Hawking radiation. This has its source at the creation of a virtual pair of particles outside the horizon of the black hole (or, equivalently, the quantum tunnelling of a particle from within the black-hole horizon). We will finish our first lecture with a brief outline of the mathematical formalism that was developed for the study of the Hawking radiation.

The emission of Hawking radiation, i.e. of elementary particles with a thermal spectrum, takes place during the two intermediate phases in the life of a black hole. These are the spin-down phase and the Schwarzschild phase, in chronological order. Starting from the second, which has the simplest gravitational background, we will present a review of the results that have been derived in the literature related to the form of the radiation spectra and their most characteristic features, including their dependence on the dimensionality of spacetime and the relative emissivities of different species of fields. A similar task will then be taken for the spin-down phase during which the black hole carries a non-vanishing angular momentum. In this case, the radiation spectra will have an extra dependence on the angular momentum parameter of the black hole, as well as an angular distribution in space due to the existence of a preferred direction in space, that of the rotation axis. The most important part, from the phenomenological point of view, will be the emission of the black hole directly on the brane on which the standard model particles and the observers themselves are located. However, the bulk emission will also be considered as this will determine the amount of energy remaining for emission on the brane.

Having completed the theoretical study of the radiation spectra from a higher-dimensional black hole, we now need to address the question of what information we may deduce from these spectra, if one day we manage to detect them, and in what way. Certain properties of the produced black hole such as the mass and temperature need to be determined first. From these, one may then turn to the derivation of more fundamental parameters such as the dimensionality of the gravitational background, or even the value of the fundamental Planck scale and the cosmological constant. As we will see, this task is highly non-trivial and demands the close cooperation of theoretical studies and experimental skill. But if it works, it might provide answers to the most fundamental questions in theoretical physics.

10.2 Creation of Black Holes and Their Properties

During the first lecture, we will set the stage for the production and subsequent detection of higher-dimensional black holes. After a brief introduction to models with extra dimensions, we will discuss the possibility of the creation of a black hole during a particle collision and address certain questions related to this phenomenon. We will then turn to the evaporation process of the black hole, and we will briefly present the mathematical formalism for the study of the Hawking radiation.

10.2.1 Extra Dimensions

It is an amazing feature of the theory of general relativity that it can be straight forwardly extended to an arbitrary number of dimensions. Its main mathematical construction, Einstein's field equations

$$G_{\mu\nu} = R_{\mu\nu} - \frac{1}{2} g_{\mu\nu} R = \kappa^2 T_{\mu\nu} , \qquad (10.1)$$

is expressed in terms of second-rank tensors whose indices can take any values, depending on our assumptions for the dimensionality of spacetime, without its mathematical consistency to be in any danger. It comes therefore as no surprise that, only a few years after Einstein formulated his theory of gravity, Kaluza produced a gravitational model in five dimensions. The model was soon supplemented with further suggestions about the topology of the extra dimension by Klein, and it was the first attempt ever to derive a unification theory in which gravity played the fundamental role.

Klein pictured the extra spacelike dimension introduced by Kaluza as a regular, compact one with finite size $\mathcal{R}$. To avoid any conflicts with observational data, the size of the extra dimension was assumed to be much smaller than any observable length scale. The idea was extensively used decades later in the formulation of string theory: there, the size of the additional six spacelike dimensions, necessary for the mathematical and physical consistencies of the theory, was assumed to be $\mathcal{R} = l_P = 10^{-33}$ cm. However, all traditional ideas about the structure, size and use of the extra space radically changed in the 1990s. The start was made in the context of string theory, where the idea [1–4] that the string scale does not necessarily need to be tied to the Planck scale, $M_P \simeq 10^{19}$ GeV, was put forward. This soon led[1] to the construction of two, much simpler but extremely rich from the phenomenological point of view, gravitational models: the scenario with Large Extra Dimensions [10–12] and the one with Warped Extra Dimensions [13, 14].

The topological structure of the higher-dimensional spacetime in each case is shown in Figs. 10.1(a,b). In the scenario with Large Extra Dimensions, depicted in Fig. 10.1(a), a four-dimensional brane is embedded in a $(4 + n)$-dimensional flat space with (3+1) non-compact and n spacelike compact dimensions. All ordinary matter, made up of and interacting through standard model (SM) fields, is localised on the brane and experiences gravitational forces that become strong at Planck scale. On the other hand, gravitons, and possibly scalars or other fields not carrying any charges under the SM gauge group, can propagate in the full spacetime. The higher-dimensional, fundamental theory has a new scale for gravity, M_*, that is related to the effective four-dimensional one through the equation [10–12]

$$M_P^2 \simeq \mathcal{R}^n M_*^{2+n} . \qquad (10.2)$$

---

[1] For some early attempts to construct higher-dimensional gravitational models, see [5–9].

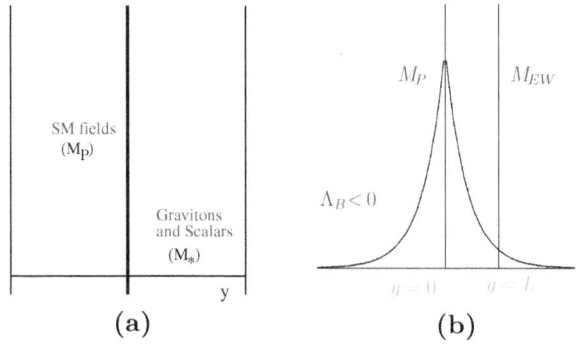

Fig. 10.1 (a) A 3-brane embedded in a $(4+n)$-dimensional flat spacetime. (b) Two 3-branes embedded in a five-dimensional anti-de Sitter spacetime

According to the above equation, if $\mathscr{R} \gg l_P$, the fundamental scale for gravity M_* can be significantly lower than the four-dimensional one. By inverting the above relation and using the definition $G_D = 1/M_*^{n+2}$ for the fundamental gravitational constant, we find that

$$G_4\,\mathscr{R}^n \simeq G_D. \tag{10.3}$$

This means that while, for $r \gg \mathscr{R}$, the Newtonian potential between two masses m_1 and m_2 is given by the well-known four-dimensional formula

$$V(r) = G_4\,\frac{m_1 m_2}{r}, \tag{10.4}$$

for $r \ll \mathscr{R}$, the corresponding potential is now not only a higher-dimensional one but a much stronger one for the same masses m_1 and m_2 and is written as

$$V(r) = G_D\,\frac{m_1 m_2}{r^{n+1}}. \tag{10.5}$$

In the case of the scenario with Warped Extra Dimensions, shown in Fig. 10.1(b), a four-dimensional brane is embedded in the higher-dimensional spacetime which now is five dimensional. The extra spacelike dimension is generically non-compact but it may be compactified at will if a second brane is introduced in the model. The *visible* brane, where all SM fields live, is placed at a finite distance $y = L$ from the *hidden* brane located at $y = 0$. If all fundamental scales at the hidden brane are of the order of M_*, then it may be shown that the electroweak symmetry breaking in the visible brane takes place at a scale [13, 14]

$$M_{EW} = e^{-kL}M_*, \tag{10.6}$$

where k is the curvature scale associated with the negative cosmological constant that fills the five-dimensional spacetime of the model. The effective Planck scale M_P is now related to the fundamental one M_* through the equation

$$M_P^2 = \frac{M_*^3}{k}\left(1 - e^{-2kL}\right). \tag{10.7}$$

In both scenarios, a low-scale gravitational theory can be realised in the context of the higher-dimensional model. As we will see, this will have important consequences for the creation and evaporation processes of black holes in these theories. In these lectures, we will concentrate on the scenario with Large Extra Dimensions; however, many of the arguments and results that will be presented hold for the scenario with Warped Extra Dimensions, too, under the assumption that the AdS radius $1/k$ is much larger than the horizon radius r_H of the corresponding black holes.

10.2.2 Creation of Black Holes

A summary of the most important – experimental, astrophysical and cosmological – limits on the fundamental energy scale M_* is presented in Table 10.1. From its entries one may see that, in general, the constraints become more relaxed as the number of additional spacelike dimensions increases. The most optimistic case is the one where the higher-dimensional Planck scale M_* is very close to the TeV scale – this case is still viable; however, one needs to introduce at least three additional spacelike dimensions. In this version of the model the hierarchy between the gravitational and the electroweak scales almost disappears. What is more important, the scale of quantum gravity, where gravitational and SM interactions become of the same magnitude, approaches the energy scale where present-day and future experiments operate. As a result, if M_* is of the order of a few TeV, then collider experiments with $E > M_*$ can probe the strong gravity regime and may witness the creation of heavy, extended objects!

Table 10.1 Current limits on the fundamental energy scale

Type of experiment/analysis	$M_* \geq$	$M_* \geq$
Collider limits on the production of real or virtual KK gravitons [15–17]	1.45 TeV ($n = 2$)	0.6 TeV ($n = 6$)
Torsion-balance experiments [18, 19]	3.2 TeV ($n = 2$)	($\mathcal{R} \leq 50\,\mu$m)
Overclosure of the Universe [20]	8 TeV ($n = 2$)	
Supernovae cooling rate [21, 22, 24]	30 TeV ($n = 2$)	2.5 TeV ($n = 3$)
Non-thermal production of KK modes [25]	35 TeV ($n = 2$)	3 TeV ($n = 6$)
Diffuse gamma-ray background [20, 26, 27]	110 TeV ($n = 2$)	5 TeV ($n = 3$)
Thermal production of KK modes [27]	167 TeV ($n = 2$)	1.5 TeV ($n = 5$)
Neutron star core halo [28–30]	500 TeV ($n = 2$)	30 TeV ($n = 3$)
Time delay in photons from GRBs [31]	620 TeV ($n = 1$)	
Neutron star surface temperature [28–30]	700 TeV ($n = 2$)	0.2 TeV ($n = 6$)
BH absence in neutrino cosmic rays [32]		1–1.4 TeV ($n \geq 5$)

The following question therefore arises naturally: can we then produce a black hole in a collider experiment on our brane? The idea was put forward in [33] very soon after the formulation of the two aforementioned models with extra dimensions. In there, it was argued that during a high-energy scattering process with $E > M_*$ and impact parameter b between the colliding particles, the following two cases should be expected: (i) if $b > r_H(E)$, elastic and inelastic processes will take place, dominated by the exchange of gravitons, while (ii) if $b < r_H(E)$, a black hole will be formed according to the Thorne's Hoop Conjecture[2] [34] and the colliding particles will disappear for ever behind the event horizon. In the above conjecture, $r_H(E)$ is the Schwarzschild radius that corresponds to the centre-of-mass energy E of the colliding particles.

Since gravity is higher dimensional, every gravitational object, including the produced black hole, will be generically higher dimensional. We thus expect the black hole not only to form on but also to extend off our brane. Under the assumption that the produced black hole has a horizon radius r_H much smaller than the size of the extra dimensions $\mathscr{R}$ – a case that can be indeed realised as we will see in the next section, it may be assumed that it lives in a spacetime with $(4+n)$ non-compact dimensions. The simplest such black hole is the spherically symmetric, neutral, higher-dimensional one described by the Schwarzschild–Tangherlini line element [36]

$$ds^2 = -\left[1 - \left(\frac{r_H}{r}\right)^{n+1}\right] dt^2 + \left[1 - \left(\frac{r_H}{r}\right)^{n+1}\right]^{-1} dr^2 + r^2 d\Omega_{2+n}^2, \quad (10.8)$$

where $d\Omega_{2+n}^2$ is the line element of a $(2+n)$-dimensional unit sphere

$$d\Omega_{2+n}^2 = d\theta_{n+1}^2 + \sin^2\theta_{n+1}\left(d\theta_n^2 + \sin^2\theta_n\left(\cdots + \sin^2\theta_2\left(d\theta_1^2 + \sin^2\theta_1 d\varphi^2\right)\ldots\right)\right). \quad (10.9)$$

By applying the Gauss law in $D = 4 + n$ dimensions, we find for the horizon radius the result [37]

$$r_H = \frac{1}{M_*}\left(\frac{M_{BH}}{M_*}\right)^{\frac{1}{n+1}}\left(\frac{8\Gamma\left(\frac{n+3}{2}\right)}{(n+2)\sqrt{\pi}^{(n+1)}}\right)^{1/(n+1)}. \quad (10.10)$$

The above expression reveals the, by now, well-known result that, in an arbitrary number of dimensions, the horizon radius of the black hole has a power-law dependence on its mass M_{BH} – the more familiar linear dependence is restored if one sets $n = 0$. More importantly, it is the fundamental Planck scale M_* that appears in the denominator instead of the four-dimensional one M_P, a feature that will play

---

[2] ...which says that "A black hole is formed when a mass M gets compacted into a region whose circumference in every direction is $\mathscr{C} \leq 2\pi r_H(E)$". A higher-dimensional version of this conjecture was developed in [35] where the "circumference" was substituted by the "area" $\mathscr{V}_{D-3}$ of the $(D-3)$-dimensional "surface" that now needs to be $\mathscr{V}_{D-3} \leq G_D M$.

Table 10.2 The values of the ratio $x_{min} = E/M_*$, necessary for the creation of a black hole, as a function of n

$n = 2$	$n = 3$	$n = 4$	$n = 5$	$n = 6$	$n = 7$
$x_{min} = 8.0$	$x_{min} = 9.5$	$x_{min} = 10.4$	$x_{min} = 10.9$	$x_{min} = 11.1$	$x_{min} = 11.2$

an important role in deciding whether black holes may be created at high-energy particle collisions.

Turning therefore to this question, the basic criterion for the creation of such a black hole is [38] that the Compton wavelength $\lambda_C = 4\pi/E$ of the colliding particle of energy $E/2$ must lie within the corresponding Schwarzschild radius $r_H(E)$. By using the expression for the horizon radius (10.10), the above relation becomes

$$\frac{4\pi}{E} < \frac{1}{M_*}\left(\frac{E}{M_*}\right)^{\frac{1}{n+1}}\left(\frac{8\Gamma\left(\frac{n+3}{2}\right)}{(n+2)\sqrt{\pi}^{(n+1)}}\right)^{1/(n+1)}. \tag{10.11}$$

This inequality can be solved to give the ratio $x_{min} = E/M_*$, necessary for the creation of the black hole. The results for x_{min} for various values of the number of extra dimensions n are given in Table 10.2. From these, we conclude that the centre-of-mass energy of the collision must be approximately one order of magnitude larger than the fundamental Planck scale M_*. Note that if the factor 4π is left out, as it was often done in earlier back-on-the-envelope calculations, the constraint on E comes out to be much more relaxed, i.e. $E \geq M_*$. As the maximum centre-of-mass energy that can be achieved at the Large Hadron Collider at CERN is 14 TeV, it seems that a window of approximately 5 TeV remains at our disposal to witness a strong gravity effect such as the creation of a black hole.

Moving beyond the classical criterion (10.11) that allows for the formation of the black hole, two basic questions arise next: (i) what *part* of the available centre-of-mass energy E is absorbed inside the black hole and (ii) how *likely* is the creation of a black hole at the first place. In order to answer these questions, we need to study the details of the high-energy particle collision in a strong gravitational background. A theory of quantum gravity could provide the answers; however, such a theory – in a complete, consistent form – is still missing. Over the years the fast-moving, colliding particles have been modelled by gravitational waves, shock waves, and strings in the context of different theories such as general relativity (with or without quantum mechanics) [39–51], string theory [52–57] and topological field theory [58]. The most widely established method is the use of the concept of the Aichelburg–Sexl shock wave [59] that was developed more than 20 years ago in the context of a four-dimensional gravitational theory. An Aichelburg–Sexl shock wave follows from a Schwarzschild line element boosted along the z axis, with a Lorentz factor $\gamma = 1/\sqrt{1-\beta^2}$. In the limit $\gamma \to \infty$, the boosted line element becomes [45–47]

$$ds^2 = -dudv + dx^2 + dy^2 + 4\mu\ln(x^2+y^2)\,\delta(u)\,du^2, \tag{10.12}$$

with $u = t - z$, $v = t + z$ and μ the particle's energy. The above line element is everywhere flat apart from the point $u = 0$ where a discontinuity arises. Therefore, it describes a shock wave located at this point and moving along the $+z$ axis at the speed of light.

We now assume that two Aichelburg–Sexl waves, with their centres at $u = 0$ and $v = 0$, are moving in opposite directions, one along the $+z$ axis and the other along the $-z$. Then, the two shock waves will collide at $u = v = 0$, as we may see in Fig. 10.2. The points at regions I, II and III that lie away from the moving trajectories and the collision point are flat; however, the region IV which forms after the collision is highly non-linear and curved.[3] If, at the union of the two shock waves, a *closed trapped surface* (i.e. a closed two-dimensional spacelike surface on which the outgoing orthogonal null geodesics have positive convergence [60]) or an *apparent horizon* (that is, a closed trapped surface with exactly zero convergence) is formed, then a black hole has been created – since according to the cosmic censorship hypothesis, the apparent horizon either coincides with or lies inside the event horizon [60].

The creation therefore of a black hole is nothing but a boundary value problem. In $D = 4$ dimensions and for a head-on collision ($b = 0$), this problem can be solved analytically. This task was performed by Penrose (1974, unpublished), more than 30 years ago, who found that an apparent horizon is indeed formed with an area equal to $32\pi\mu^2$. This can put a lower bound on the area of the event horizon and thus on the black-hole mass as follows:

$$A_H \equiv 4\pi r_H^2 \geq 32\pi\mu^2 \Rightarrow M_{BH} \equiv \frac{r_H}{2} \geq \frac{1}{\sqrt{2}}(2\mu). \qquad (10.13)$$

From the above one may conclude that at least 71% of the initial energy $E = 2\mu$ of the collision is trapped inside the black hole. Alternatively, one may compute the amount of energy emitted in the form of gravitational waves during the violent collision. This was done in [45–47] where it was found that that amount was of the order of 16%, which raised the percentage of energy absorbed by the black hole to 84% of the initial energy E. In more recent years, numerical analyses, where one [61, 62] or both [63] of the colliding bodies were assumed to be described by a black

Fig. 10.2 Two Aichelburg–Sexl shock waves propagating in opposite directions. The two shock waves collide at $u = v = 0$ and if a closed trapped surface or an apparent horizon is formed, a black hole is created

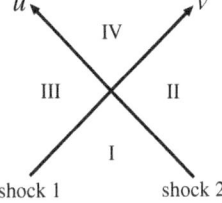

---

[3] The task of finding the exact form of spacetime in region IV involves strong, and thus non-linear, gravitational calculations; until today no answer – analytical or numerical – has been given to this question.

hole, have found that the percentage of energy lost in the form of gravitational waves is in the area of 14%, which is in very good agreement with the results of [45–47].

In the brane-world scenario, the colliding particles need to enter the higher-dimensional regime in order to create a black hole. In that case, every closed trapped surface will be a $(D-2)$ surface instead of a two-dimensional one. Nevertheless, the same procedure for investigating the creation of a black hole can be followed in this case, too. For a head-on collision, the corresponding boundary value problem can be again solved analytically leading to [64]

$$M_{BH} \geq [0.71 \,(\text{for } D = 4) - 0.58 \,(\text{for } D = 11)] \,(2\mu). \tag{10.14}$$

Therefore, as the dimensionality D of spacetime increases, smaller and smaller black holes will be created. On the other hand, if the collision is not head-on, i.e. $b \neq 0$, numerical means have to be used to find the solution of the problem. This was done in [64–66] leading, respectively, to the results

- $D = 4$: $b \leq b_{max} \simeq 0.8 \, r_H$,

- $D = 4 + n$: $b \leq b_{max} \simeq 3 \, 2^{-(n+2)/(n+1)} \, r_H$.

Thus, for a non-head-on collision, a black hole will be created if the impact parameter is smaller than a fraction of the event horizon radius. This fraction is 0.8 in $D = 4$ and increases, reaching asymptotically unity, as D becomes larger.

The impact parameter can offer us a measure of how likely the creation of a black hole is. For a high-energy collision with a non-zero impact parameter b, the production cross-section is found by using the geometric limit:

$$\sigma_{\text{production}} \simeq \pi b^2. \tag{10.15}$$

According to the above, the cross-section for the production of black holes from two fast-moving particles is assumed to be given by the classical formula for the "target" area defined by the impact parameter. One might intuitively think that the formation of a black hole at such high energies would be governed by quantum, rather than classical, effects – in [67, 68], the argument that the production cross-section would be suppressed by an exponential factor involving the Euclidean action of the system was put forward. However, in subsequent studies [69–73] it was shown that the creation of the black hole was a classically allowed process and not a quantum phenomenon; the main contribution to the production cross-section is therefore given by the classical expression (10.15), with the quantum corrections being indeed small as in [67, 68].

Early studies used the approximate expression $\sigma_{\text{production}} \simeq \pi r_H^2$, but today a more precise expression is needed. As we just saw, even in the $D = 4$ case, a black hole will not be created unless the impact parameter is smaller than $0.8 \, r_H$, a result that leads to the more accurate estimate for the production cross-section: $\sigma_{\text{production}} \simeq 0.64 \,(\pi r_H^2)$. Nevertheless, we are not done: novel estimates for the production cross-section have emerged from the study of [74], where the search for the creation of a closed trapped surface was extended in the regime ($u = 0$, $v > 0$) and ($u > 0$, $v = 0$), i.e. in the "future" of the collision point. This extension gave a boost to the

Table 10.3 Black-hole production cross-section as a function of the dimensionality of spacetime [74]

D	4	5	6	7	8	9	10	11
$\sigma_{\text{production}}/(\pi r_H^2)$	0.71	1.54	2.15	2.52	2.77	2.95	3.09	3.20

production cross-section, since in cases where it was previously concluded that no event horizon had been formed at the collision point, now such a surface was found when the extended regime was used instead. Therefore, the state-of-the-art values of the black-hole production cross-section are the ones given in Table 10.3 [74]. For example, the $D = 4$ value of 0.64, in units of πr_H^2, has increased to 0.71, with similar or larger enhancements taking place for the other values of D, too.[4]

Focusing, for a moment, on the geometrical instead of the numerical factor in the expression of the production cross-section, we may write

$$\sigma_{\text{production}} \propto \pi r_H^2 \sim \frac{1}{M_*^2} \left(\frac{E}{M_*} \right)^{2/(n+1)}. \tag{10.16}$$

The above expression gives the dependence of the production cross-section on the centre-of-mass energy of the collision and reveals the enhancement of it with E, a dependence that is not seen in any other SM or beyond the standard model process.[5] The above expression is valid for the production of a black hole out of two elementary, non-composite particles (i.e. partons). The final result for the production cross-section out of two accelerated composite particles, such as protons, follows by properly summing over all pairs of partons that carry enough energy to produce a black hole. This is finally given by [81, 82]

$$\sigma_{\text{production}}^{pp \to BH} = \sum_{ij} \int_{\tau_m}^{1} d\tau \int_{\tau}^{1} \frac{dx}{x} f_i(x) f_j\left(\frac{\tau}{x} \right) \sigma_{\text{production}}^{ij \to BH}, \tag{10.17}$$

where x is the parton-momentum fraction, $\tau = \sqrt{x_i x_j}$, and $f_i(x)$ are the so-called parton distribution functions (PDFs) that determine the fraction of the centre-of-mass energy that is carried by the partons.

---

[4] We should note here that the use of the generalised uncertainty principle has shown to lead to an increase in the minimum amount of energy needed for the creation of a black hole [75]. Similarly, the production cross-section comes out to be suppressed if the charge of the colliding particles exceeds a certain value [76], while the angular momentum of the black hole enhances $\sigma_{\text{production}}$ [77, 78]. Finally, if one assumes the existence of a non-Gaussian point in general relativity and thus a running gravitational coupling, the black-hole production cross-section is greatly suppressed in part of the parameter space [79].

[5] For a black hole produced in a five-dimensional anti-de Sitter spacetime, the above result for the production cross-section holds if we assume that $r_H \ll 1/k$; on the other hand, if $r_H \geq 1/k$, then the corresponding expression for the production cross-section is $\sigma_{\text{production}} \propto \ln^2 E$ [80].

Summing over all possible pairs of partons gives another considerable boost to the production cross-section. One could naively think that by increasing without limit the available centre-of-mass energy E, one could create extremely energetic pairs of partons each one of which would certainly create a black hole. However, the parton distribution functions $f_i(x)$ decrease rapidly with the centre-of-mass energy E, and with them the amount of energy that is passed to (and retained by) the partons. As a result, the production cross-section cannot be indefinitely increased. Numerical calculations that take into account the compositeness of the accelerated particles and the behaviour of PDFs have derived some indicative values for $\sigma_{\text{production}}$ [81, 82]. For example, if we assume that $M_* = 1$ TeV and $D = 10$, then the production cross-section for a black hole with $M_{BH} = 5$ TeV turns out to be $\sigma_{\text{production}} \sim 10^5$ fb, while for a black hole with $M_{BH} = 10$ TeV it is found that $\sigma_{\text{production}} \sim 10$ fb. For beyond the SM processes, the aforementioned values are quite significant – in the first case, the value of $\sigma_{\text{production}}$ amounts to one black hole created per second! Whether LHC will indeed prove to be a black-hole factory remains to be seen.

10.2.3 Black-Hole Properties

We now turn to the properties of the higher-dimensional black holes that may be produced during trans-planckian particle collisions [83, 84]. We will use as a prototype for our discussion the spherically symmetric, neutral black hole described by the Schwarzschild–Tangherlini line element (10.8). Let us start with the value of the horizon radius – how big (or small) are actually these black holes? The value of r_H as a function of the mass of the black hole is given in (10.10). In order to derive some realistic estimates, we assume again $M_* = 1$ TeV and $M_{BH} = 5$ TeV and calculate the value of the horizon as a function of the number of extra dimensions n. These values are presented in Table 10.4. From these we may easily conclude that, in the presence of extra dimensions, in order to create a black hole we only need to access subnuclear distances. To have a measure of comparison, let me note that, in $D = 4$ with $M_* = M_P \simeq 10^{19}$ GeV, the same objective could only be achieved if the two colliding particles came within a distance of 10^{-35} fm!

Drawing from our knowledge of their four-dimensional analogues, we expect that these miniature black holes will go through the following stages during their lifetime [81]. (i) The balding phase: the initially highly asymmetric black hole will

Table 10.4 Horizon radius and temperature of the Schwarzschild–Tangherlini black hole as a function of the number of extra dimensions, for $M_* = 1$ TeV and $M_{BH} = 5$ TeV

n	1	2	3	4	5	6	7
r_H $(10^{-4}$ fm)	4.06	2.63	2.22	2.07	2.00	1.99	1.99
T_H (GeV)	77	179	282	379	470	553	629

shed all quantum numbers and multipole moments apart from its mass M, electromagnetic charge Q and angular momentum J – during this phase, we expect some visible but mainly invisible energy emission. (ii) The spin-down phase: the black hole will start losing its angular momentum via the emission of Hawking radiation through mainly visible channels. (iii) The Schwarzschild phase: after its angular momentum the black hole will now start losing its mass through the emission again of Hawking radiation. (iv) The Planck phase: when the black-hole M_{BH} approaches M_*, it becomes a quantum object whose properties would follow only from a quantum theory of gravity – possible scenarios for this phase are the emission of a few energetic quanta leading to the complete evaporation of the black hole or the formation of a stable "quantum" remnant.

The emission of Hawking radiation [85] is sourced by the non-vanishing temperature of the black hole. This is defined in terms of the black hole's surface gravity k as follows:

$$T_H = \frac{k}{2\pi} = \frac{1}{4\pi} \frac{1}{\sqrt{|g_{tt} g_{rr}|}} \left(\frac{d|g_{tt}|}{dr} \right)_{r=r_H} = \frac{(n+1)}{4\pi r_H} . \tag{10.18}$$

By using again, as an indicative case, the values $M_* = 1$ TeV and $M_{BH} = 5$ TeV, for the fundamental Planck scale and black-hole mass, as well as the values of the horizon radius r_H given in Table 10.4, we may calculate the temperature of the black hole in terms of the number of extra dimensions. These are also given in Table 10.4. We observe that a higher-dimensional black hole, with mass in the range of values that would allow it to be produced at LHC, comes out to have in addition a temperature that would greatly facilitate its detection in present and future experiments – unlike the large astrophysical black holes that are characterised by an extremely low temperature and the majority of primordial black holes that have an extremely high temperature. Finally, let us add that due to the emission of Hawking radiation, the lifetime of a black hole is finite. In the case of a higher-dimensional black hole, this quantity is given by [83]

$$\tau_{(n+4)} \sim \frac{1}{M_*} \left(\frac{M_{BH}}{M_*} \right)^{\frac{(n+3)}{(n+1)}} > \tau_{(4)} . \tag{10.19}$$

For the same values of M_* and M_{BH}, the typical lifetime of the black hole comes out to be $\tau = (1.7 - 0.5) \times 10^{-26}$ s for $n = 1 - 7$. In other words, the produced black hole will evaporate instantly after its creation, and it will do so right in front of our detectors.

That is why we need to study in the greatest possible detail the spectrum of the Hawking radiation emitted by the black hole as this will probably be the main observable effect associated with this gravitational object. Although a purely geometrical property, the temperature of a black hole leads to the emission of thermal radiation similar to that of a black body. The Hawking radiation [85] is therefore a classical phenomenon but with a quantum origin, since classically nothing is allowed to escape from within the black-hole horizon. The emission of radiation from a black hole, four-dimensional and high dimensional alike, can be realised through the creation of a virtual pair of particles just outside the horizon; when the antiparticle happens to fall inside the black hole, the particle can **then** propagate away from the black hole whose mass has decreased due to the negative amount of energy it received. The radiation spectrum is therefore a nearly black-body spectrum with energy emission rate given by an expression of the form [85]

$$\frac{dE(\omega)}{dt} = \frac{|\mathscr{A}(\omega)|^2 \, \omega}{\exp(\omega/T_H) \mp 1} \frac{d\omega}{(2\pi)} \,. \tag{10.20}$$

The quantity $|\mathscr{A}(\omega)|^2$ appearing in the numerator is the absorption probability (or *greybody factor*). Its presence is due to the fact that a particle, propagating in the $(4+n)$-dimensional black-hole background, needs to escape the strong gravitational field, which the black hole creates, to reach the asymptotic observer. In order to see this, we may write the equation of motion of an arbitrary field in the aforementioned background in the form of a Schrödinger-like equation

$$-\frac{d^2\Psi}{dr_*^2} + V(r_*, n, l, \omega, s, \dots)\Psi = \omega^2 \, \Psi \,, \tag{10.21}$$

in terms of the so-called tortoise coordinate $dr_* = \left[1 - \left(\frac{r_H}{r}\right)^{(n+1)}\right]^{-1} dr$. The gravitational barrier $V(r_*, n, l, \omega, s, \dots)$ will reflect some particles back to the black hole while it will allow others to escape to infinity. The rate at which particles and therefore energy are "arriving" at the location of the asymptotic observer will thus be proportional to the transmission (or, absorption, as we will shortly see) probability and thus different from the one for the usual, flat-space blackbody radiation. However, the extra difficulty that the greybody factor introduces in the calculation of the radiation spectrum is compensated by the following fact: the barrier, and consequently the absorption probability, depends on a number of parameters that describe both particle properties (spin s, energy ω, angular momentum numbers $l, m, \dots$) and spacetime properties (number of extra dimensions n, angular momentum of black hole a, cosmological constant Λ, etc.). As a result, the Hawking radiation spectrum, when computed, is bound to be a vital source of information on the emitted particles and gravitational background.

But how does the radiation spectrum follow? For this, we need to do a quantum field theory analysis in curved spacetime. The first step is to define a basis for our fields: in the four-dimensional case, we write [86–88]

$$u_{\omega lm} = \frac{N}{r} e^{-i\omega t} e^{im\varphi} S_{\omega lm}(\theta) R_{\omega lm}(r) \,, \tag{10.22}$$

where N is a normalisation constant, (l, m) the angular momentum numbers with $|m| \leq l$ and $S_{\omega lm}(\theta)$ the spherical harmonics. We also need to define the vacuum state of the theory. The one that describes perfectly the Hawking radiation emission process is the *past Unruh vacuum* $|U^-\rangle$: this state has no incoming radiation from past null infinity $\mathcal{I}^-$ (i.e. far away from the BH at some asymptotic initial time) but modes can "come out" of the black hole.

The gravitational potential V that appears in the equation of motion of the field propagating in the black-hole background has the form of a barrier: it is localised and vanishes at both the horizon and infinity. At these two asymptotic regimes, (10.21) can then be easily solved, and the radial part of the field assumes the forms

$$
R_{\omega l m}^{up}(r) \sim \begin{cases} e^{i\omega r_*} + A^{up} e^{-i\omega r_*}, & r \to r_H \\ B^{up} e^{i\omega r_*}, & r \to \infty \end{cases}. \tag{10.23}
$$

The solution is, as expected, a superposition of free plane waves, where the constants A^{up} and B^{up} can be viewed as the reflection and transmission coefficients.

The fluxes of particles N and energy E emitted by the black hole and measured by an observer at infinity are given by the vacuum expectation values of the radial component of the conserved current J^μ and the (tr)-component of the energy–momentum tensor $T_{\mu\nu}$, respectively, evaluated at infinity [86, 87]:

$$
\frac{d^2}{r^2 \, dt \, d\Omega} \begin{pmatrix} N \\ E \end{pmatrix} = \left\langle U^- \left| \begin{pmatrix} J^r \\ T^{tr} \end{pmatrix} \right| U^- \right\rangle_\infty. \tag{10.24}
$$

Using the asymptotic form (10.23) for the radial part of the field at infinity, and after some algebra, we find

$$
\frac{d^2}{dt \, d\omega} \begin{pmatrix} N \\ E \end{pmatrix} = \frac{1}{2\pi} \sum_l \frac{N_l \, |B^{up}|^2}{\exp(\omega/T_H) \mp 1} \begin{pmatrix} 1 \\ \omega \end{pmatrix}, \tag{10.25}
$$

where $N_l = 2l + 1$ is the multiplicity of states that have the same value of the angular momentum number l, and the ± 1 factor is a statistics factor for fermions and bosons, respectively.

We note that in the numerator of the above expression, it is the transmission probability $|B^{up}|^2$ that appears, as expected. However, one may define an alternative, but equivalent, basis, namely

$$
R_{\omega l m}^{in}(r) \sim \begin{cases} B^{in} e^{-i\omega r_*}, & r \to r_H \\ e^{-i\omega r_*} + A^{in} e^{i\omega r_*}, & r \to \infty \end{cases}. \tag{10.26}
$$

This basis describes modes that originate not from the black hole but from the past null infinity. Now, A^{in} and B^{in} can be viewed as the reflection and absorption coefficients, respectively. As both sets of solutions satisfy the same radial equation, one may easily show that the following relations hold:

$$
1 - |A^{in}|^2 = |B^{in}|^2 \equiv |B^{up}|^2 = 1 - |A^{up}|^2. \tag{10.27}
$$

From the above, we may easily conclude that the transmission probability $|B^{up}|^2$ for the "up" modes originating from inside the black hole is equal to the absorption probability $|B^{in}|^2$ for the "in" modes originating from past null infinity – we denote these two quantities collectively as $|\mathscr{A}(\omega)|^2$ and write

$$\frac{d^2}{dt\,d\omega}\begin{pmatrix} N \\ E \end{pmatrix} = \frac{1}{2\pi}\sum_l \frac{N_l\,|\mathscr{A}(\omega)|^2}{\exp(\omega/T_H)\mp 1}\begin{pmatrix} 1 \\ \omega \end{pmatrix}. \tag{10.28}$$

In the case of a rotating (Kerr) black hole, we may compute three rates: the emission rates of particles N and energy E and the rate of loss of the angular momentum J of the black hole. These are given by the expressions

$$\frac{d^2}{r^2\,dt\,d\Omega}\begin{pmatrix} N \\ E \\ J \end{pmatrix} = \left\langle U^- \left| \begin{pmatrix} J^r \\ T^{tr} \\ T^r_\varphi \end{pmatrix} \right| U^- \right\rangle_\infty. \tag{10.29}$$

The asymptotic solutions for the radial part of the field for either the "up" modes or the "in" modes propagating in a Kerr black-hole background are now given by

$$R^{up}_{\omega lm}(r) \sim \begin{cases} e^{i\tilde\omega r_*} + A^{up}\,e^{-i\tilde\omega r_*}, & r \to r_H \\ B^{up}\,e^{i\omega r_*}, & r \to \infty \end{cases} \tag{10.30}$$

and

$$R^{in}_{\omega lm}(r) \sim \begin{cases} B^{in}\,e^{-i\tilde\omega r_*}, & r \to r_H \\ e^{-i\omega r_*} + A^{in}\,e^{i\omega r_*}, & r \to \infty \end{cases}. \tag{10.31}$$

In the above, the parameter $\tilde\omega$ is defined as

$$\tilde\omega \equiv \omega - m\,\Omega_H = \omega - m\frac{a}{r_H^2 + a^2}, \tag{10.32}$$

where Ω_H is the angular velocity of the rotating black hole and a the angular momentum parameter to be defined later. By using as a basis the "up" modes, that, as we saw, describe more accurately the Hawking radiation emission process, we find the expressions

$$\frac{d^2}{dt\,d\omega}\begin{pmatrix} N \\ E \\ J \end{pmatrix} = \frac{1}{2\pi}\sum_{l,m} \frac{\omega}{\tilde\omega}\frac{|B^{up}|^2}{\exp(\tilde\omega/T_H)\mp 1}\begin{pmatrix} 1 \\ \omega \\ m \end{pmatrix}. \tag{10.33}$$

As in the non-rotating case, we also find that the following relations hold between the coefficients of the asymptotic solutions (10.30) and (10.31) for the two sets of modes

$$\frac{\omega}{\tilde\omega}|B^{up}|^2 = 1 - |A^{up}|^2 \equiv 1 - |A^{in}|^2 = \frac{\tilde\omega}{\omega}|B^{in}|^2, \tag{10.34}$$

leading to the final, simpler formula for the three rates

$$\frac{d^2}{dt\,d\omega}\begin{pmatrix} N \\ E \\ J \end{pmatrix} = \frac{1}{2\pi}\sum_{l,m} \frac{|\mathscr{A}(\omega)|^2}{\exp(\tilde\omega/T_H)\mp 1}\begin{pmatrix} 1 \\ \omega \\ m \end{pmatrix}, \tag{10.35}$$

where now $|\mathscr{A}(\omega)|^2 \equiv = 1 - |A^{up}|^2 \equiv 1 - |A^{in}|^2$. Let us also note that if $\tilde{\omega} = \omega - m\Omega_H < 0$, then from (10.34) the reflection probabilities $|A^{up}|^2$ and $|A^{in}|^2$ can be larger than unity – this happens only for modes with $m > 0$ and signals the effect of *superradiance* [89], where the incident wave "steals" energy from the rotating black hole and escapes with an amplitude larger than the original one.

Let us now introduce a number of additional, spacelike dimensions in our theory. Surprisingly, not much changes in the functional form of the above formulae. The emission rates for a higher-dimensional, rotating black hole will still be given by expressions of the form [77, 78, 90–98]

$$\frac{d^2}{dt\,d\omega}\begin{pmatrix} N \\ E \\ J \end{pmatrix} = \frac{1}{2\pi} \sum_{l,m,j\ldots} \frac{|\mathscr{A}(\omega)|^2}{\exp(\tilde{\omega}/T_H) \mp 1} \begin{pmatrix} 1 \\ \omega \\ m \end{pmatrix}. \qquad (10.36)$$

Where does the difference from the four-dimensional case lie? To start with, the temperature of the black hole will acquire an n-dependence. In addition, the equation of motion of a given field is going to depend on the specific background; therefore, the greybody factor $|\mathscr{A}(\omega)|^2$ that follows by solving the corresponding equation of motion is going to change too. Also, the symmetry and structure of the higher-dimensional spacetime may introduce additional quantum numbers and/or change the multiplicities of states that carry the same sets of quantum numbers.

Another important factor is whether we are considering emission of particles on the brane or in the bulk. Unlike a purely four-dimensional black hole, a higher-dimensional one can emit particles either in the "brane channel" or in the "bulk channel". The species of particles that can be emitted in the bulk are particles that are allowed by the model to propagate in the higher-dimensional spacetime, namely gravitons but also scalar fields that carry no quantum numbers under the SM gauge group. These bulk modes "see" the full $(4 + n)$-dimensional gravitational background and they are invisible to us; therefore, any energy emitted in the bulk will be interpreted as a missing energy signal for a brane observer. On the other hand, the black hole can emit a variety of particles in the "brane channel", namely fermions, gauge bosons and Higgs-like scalars. These brane-localised modes "see" only the projected-on-the-brane four-dimensional gravitational background and they are directly visible to a brane observer; as a result, they are the most interesting emission channel to study from the phenomenological point of view.

10.3 Hawking Radiation Spectra and Observable Signatures

Having discussed the properties of the miniature black holes that may be created during a high-energy particle collision in the context of a low-scale higher-dimensional gravitational theory, we now proceed to discuss in more detail the spectra of the Hawking radiation emitted by these black holes and the information on particle and spacetime properties that we may deduce from them.

10.3.1 The Schwarzschild Phase on the Brane

As we mentioned in the previous section, a black hole emits Hawking radiation during the two intermediate phases of its life, namely the spin-down and the Schwarzschild phases. We will start from the latter one, which although follows the spin-down phase was the first one to be studied due to the simpler form of the line element that describes the gravitational background around it. This is given by the Schwarzschild–Tangherlini solution (10.8) and describes, as we have seen, a spherically symmetric, neutral black hole that has lost all of its angular momentum. For the purpose of studying the emission of Hawking radiation directly on the brane, we will be interested in the brane-localised modes that "see" only the projected-on-the-brane background. In order to derive the latter, we fix the values of all the additional θ_i coordinates, with $i = 2, \ldots, n+1$, introduced to describe the additional spacelike dimensions, to $\frac{\pi}{2}$. Then, the resulting brane background assumes the form

$$ds_4^2 = - \left[1 - \left(\frac{r_H}{r} \right)^{n+1} \right] dt^2 + \left[1 - \left(\frac{r_H}{r} \right)^{n+1} \right]^{-1} dr^2 + r^2 \, d\Omega_2^2 . \qquad (10.37)$$

The above line element describes a four-dimensional black-hole background on the brane which, although resembles a Schwarzschild background, is distinctly different as it carries a non-trivial n-dependence. The horizon radius is still given by (10.10) and its temperature by (10.18) – note that both the horizon radius and the black-hole temperature follow from geometrical arguments involving only the g_{tt} and g_{rr} metric components, and these are not affected by the projection of the $(4+n)$-dimensional line element onto the brane.

 However, the different form of the gravitational background is bound to change the equation of motion of the relevant species of particles, and thus the value of the greybody factor $|\mathscr{A}(\omega)|^2$. In order to study in a combined way the behaviour of fields with spin $s = 0, 1/2$ and 1, a "master" equation of motion with s appearing as a parameter was derived in [84, 90, 91]. For this, we used a factorised ansatz for the wavefunction of the field of the form

$$\Psi_s = e^{-i\omega t} \, e^{im\varphi} \, \Delta^{-s} R_s(r) \, S_{sl}^m(\theta) \qquad (10.38)$$

and employed the Newman–Penrose method [99, 100] that combines multi-component fields with curved gravitational backgrounds. Then, two decoupled equations, one for the radial function $R_s(r)$ and one for the spin-weighted spherical harmonics [101] $S_{sl}^m(\theta)$, were derived having the form

$$\Delta^s \frac{d}{dr} \left(\Delta^{1-s} \frac{dR_s}{dr} \right) + \left[\frac{\omega^2 r^2}{h} + 2i\omega sr - \frac{is\omega r^2 h'}{h} - \lambda_{sl} \right] R_s(r) = 0 \qquad (10.39)$$

and

$$\frac{1}{\sin\theta} \frac{d}{d\theta} \left(\sin\theta \frac{dS_{sl}^m}{d\theta} \right) + \left[-\frac{2ms\cot\theta}{\sin\theta} - \frac{m^2}{\sin^2\theta} + s - s^2 \cot^2\theta + \lambda_{sl} \right] S_{sl}^m(\theta) = 0, \quad (10.40)$$

respectively. In the above, we have defined the function $\Delta \equiv r^2 h \equiv r^2 \left[1 - \left(\frac{r_H}{r} \right)^{n+1} \right]$, while $\lambda_{sl} = l(l+1) - s(s-1)$ is the eigenvalue of the spin-weighted spherical harmonics. The above equations resemble the ones derived by Teukolsky [102, 103] in the background of a purely four-dimensional black-hole background and differ only in the expressions of the functions $h(r)$ and $\Delta(r)$.

The radial equation, from where the value of the greybody factor will follow, may be solved either analytically or numerically. If the analytic approach is chosen [90, 91], an approximation method must be followed according to which (i) we solve the equation of motion in the near-horizon (NR) regime ($r \simeq r_H$) where it takes the form of a hypergeometric equation, (ii) then we solve the equation of motion in the far-field (FF) regime ($r \gg r_H$) where it takes the form of a confluent hypergeometric equation and (iii) finally, we match the two asymptotic solutions in an intermediate zone to guarantee the existence of a smooth solution over the whole radial regime. Once the solution for the radial function $R_s(r)$ is found, we compute the absorption probability (we use the "in" modes as a basis) through the formula

$$|\mathscr{A}(\omega)|^2 \equiv 1 - |\mathscr{R}(\omega)|^2 \equiv \frac{\mathscr{F}_{\text{horizon}}}{\mathscr{F}_{\text{infinity}}}, \tag{10.41}$$

where $\mathscr{R}(\omega)$ is the reflection coefficient and $\mathscr{F}$ the flux of energy towards the black hole.

Whereas the absorption probability is a dimensionless quantity varying between 0 and 1 (in the non-rotating case), a dimensionful quantity may be constructed out of it, namely the absorption cross-section, which is measured in units of the horizon area (πr_H^2) and is defined as [104]

$$\sigma_{\text{abs}}(\omega) = \sum_l \frac{\pi r_H^2}{(\omega r_H)^2} (2l+1) |\mathscr{A}(\omega)|^2. \tag{10.42}$$

By following the approximate method, described above, to solve the radial equation, one may compute the absorption probability and from that the absorption cross-

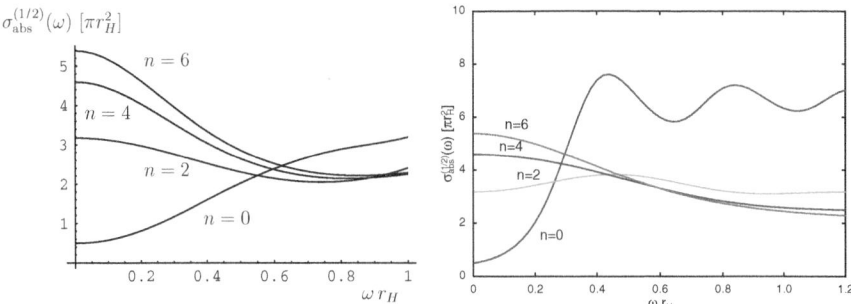

Fig. 10.3 Absorption cross-section for brane-localised fermions evaluated analytically (*left plot*) and numerically (*right plot*)

section. As an indicative case, in Fig. 10.3(left plot) we present the result for $\sigma_{\text{abs}}(\omega)$ for the case of fermions propagating in the projected-on-the-brane black-hole background. As we observe, the horizontal axis does not extend to large values of the energy parameter ωr_H; the reason for this is that during the matching of the two asymptotic solutions, the assumption was made that $\omega r_H \ll 1$, which inevitably restricts the validity of the analytic result to small values of the energy. Therefore, the behaviour of $\sigma_{\text{abs}}(\omega)$, or $\mathscr{A}(\omega)$, for arbitrary values of the energy can only be derived if numerical techniques are employed [105] for the solution of the radial equation. Then, the plot appearing in Fig. 10.3(right plot) can be constructed. The qualitative agreement between the two plots is obvious and one can see that the low-energy behaviour of $\sigma_{\text{abs}}(\omega)$ is accurately reproduced by the analytic result. However, as ωr_H increases, deviations start appearing. To complete the picture, in Figs. 10.4 we present the behaviour of the absorption cross-section for scalars and gauge bosons [105], respectively. What is important in the behaviour of $\sigma_{\text{abs}}(\omega)$ is that (a) it behaves differently for each species of fields and (b) has a rather strong dependence on the number of spacelike dimensions that exist transversely to the brane.

When the (numerically) computed absorption probability and the temperature of the black hole are substituted in the formula for the energy emission rate, we obtain the radiation spectrum [105] that, for the indicative case of fermions, is depicted in Fig. 10.5. The different curves on the plot stand for the differential energy emission rates per unit time and unit frequency for the cases with $n = 0, 1, 2, 4$ and 6 (from bottom to top). We may easily observe that the energy emission rate is greatly enhanced by the number of extra spacelike dimensions, a result that also holds for scalars and gauge bosons. In order to derive the total emissivity, i.e. the energy emitted over the whole frequency regime per unit time, we integrate over ωr_H. The results for all species of brane-localised fields are presented in the first three rows of Table 10.5. From there, we may see that the total emissivity for the SM fields is enhanced up to three or four orders of magnitude with the number of extra dimensions.

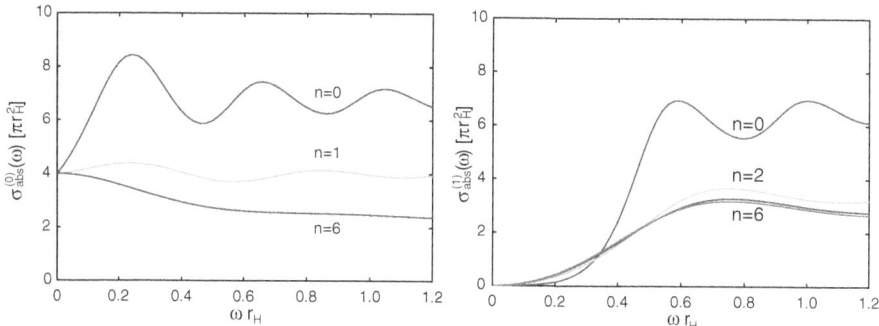

Fig. 10.4 Absorption cross-section for brane-localised scalars (*left plot*) and gauge bosons (*right plot*)

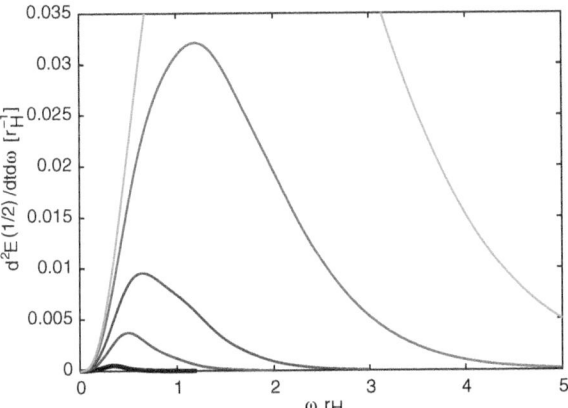

Fig. 10.5 Hawking radiation energy emission rates for brane-localised fermions, for $n = 0, 1, 2, 4$ and 6 (from *bottom* to *top*)

Table 10.5 Total emissivities for brane-localised scalars, fermions and gauge bosons [105] and bulk gravitons [119–121]

n	0	1	2	3	4	5	6	7
Scalars	1.0	8.94	36.0	99.8	222	429	749	1220
Fermions	1.0	14.2	59.5	162	352	664	1140	1830
Gauge bosons	1.0	27.1	144	441	1020	2000	3530	5740
Gravitons	1.0	103	1036	5121	2×10^4	7×10^4	2.5×10^5	8×10^5

We finish this subsection with an interesting observation that applies for the relative emissivities of brane-localised fields. We have already seen that the absorption cross-section, and consequently the absorption probability, has a strong dependence on the spin of the propagating field. One thus expects that different species of particles will have different emission rates. Indeed, this may be seen by putting the emission curves of scalars, fermions and gauge bosons on the same graph. For a purely four-dimensional black hole [106], this is shown in Fig. 10.6(left plot). According

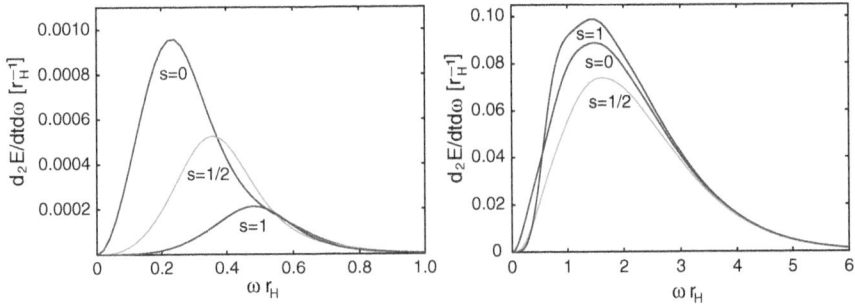

Fig. 10.6 Relative emissivities for brane-localised fields for $n = 0$ (*left plot*) and $n = 6$ (*right plot*)

to this, the dominant type of particles emitted by a black hole in four dimensions is scalars; then come the fermions and finally the gauge bosons. Figure 10.6(right plot) shows the same emission curves but in the case of a ten-dimensional black hole. Here, the gauge bosons are the particles preferably emitted by the black hole, then come the scalars and lastly the fermions. Therefore, the number of extra dimensions determines not only the amount of energy emitted per unit time by the black hole but also the type of the emitted particles.

10.3.2 The Spin-Down Phase on the Brane

We now turn to the phase in the life of the black hole that precedes the Schwarzschild one. This is the spin-down phase during which the black hole has a non-vanishing angular momentum – this is the most generic situation for a black hole created by a non-head-on particle collision. Assuming that the produced black hole has an angular momentum component only along an axis in our three-dimensional space, the line element that describes the gravitational background around such a higher-dimensional black hole is given by the Myers–Perry solution [37]:

$$ds^2 = \left(1 - \frac{\mu}{\Sigma\, r^{n-1}}\right)dt^2 + \frac{2a\mu\sin^2\theta}{\Sigma\, r^{n-1}}\,dt\,d\varphi - \frac{\Sigma}{\Delta}dr^2$$
$$- \Sigma\, d\theta^2 - \left(r^2 + a^2 + \frac{a^2\mu\sin^2\theta}{\Sigma\, r^{n-1}}\right)\sin^2\theta\, d\varphi^2 - r^2\cos^2\theta\, d\Omega_n^2, (10.43)$$

where

$$\Delta = r^2 + a^2 - \frac{\mu}{r^{n-1}}, \qquad \Sigma = r^2 + a^2\cos^2\theta. \tag{10.44}$$

The parameters μ and a that appear in the metric tensor are associated to the black-hole mass and angular momentum, respectively, through the relations

$$M_{BH} = \frac{(n+2)A_{2+n}}{16\pi G}\mu \qquad \text{and} \qquad J = \frac{2}{n+2}aM_{BH}, \tag{10.45}$$

where A_{2+n} is the area of a $(2+n)$-dimensional unit sphere. The horizon radius is found by setting $\Delta(r_H) = 0$ and is found to be $r_H^{n+1} = \mu/(1+a_*^2)$, where we have defined the quantity $a_* \equiv a/r_H$. Finally, the temperature and rotation velocity of this black hole are given by

$$T_H = \frac{(n+1)+(n-1)a_*^2}{4\pi(1+a_*^2)r_H}, \qquad \Omega_H = \frac{a}{(r_H^2+a^2)}. \tag{10.46}$$

Since we are still interested in the emission of brane-localised modes by the black hole, we should first determine the line element on the brane. As in the case of the Schwarzschild phase, this will follow by fixing the values of the "extra" angular coordinates. This results in the disappearance of the $d\Omega_n^2$ part of the metric leaving the remaining unaltered. Then, by employing again the Newman–Penrose method, we compute the two – decoupled again – master equations, one for the radial part of the field and one for the angular part, namely [84, 96]

$$\Delta^{-s} \frac{d}{dr} \left(\Delta^{s+1} \frac{dR_s}{dr} \right) + \left[\frac{K^2 - iKs\Delta'}{\Delta} + 4is\omega r + s\left(\Delta'' - 2\right)\delta_{s,|s|} - \Lambda_{sj}^m \right] R_s = 0$$
(10.47)

and

$$\frac{1}{\sin\theta} \frac{d}{d\theta} \left(\sin\theta \frac{dS_{sj}^m}{d\theta} \right) + \left[-\frac{2ms\cot\theta}{\sin\theta} - \frac{m^2}{\sin^2\theta} + a^2\omega^2\cos^2\theta \right.$$

$$\left. -2a\omega s\cos\theta + s - s^2\cot^2\theta + \lambda_{sj} \right] S_{sj}^m(\theta) = 0.$$
(10.48)

In the above, $S_{sj}^m(\theta)$ are the spin-weighted spheroidal harmonics [107], and we have used the following definitions:

$$K = (r^2 + a^2)\omega - am, \qquad \Lambda_{sj}^m = \lambda_{sj} + a^2\omega^2 - 2am\omega.$$
(10.49)

The angular eigenvalue λ_{sj} does not exist in closed form but it may be computed either analytically, through a power series expansion in terms of the parameter $a\omega$ of the form [108–110]

$$\lambda_{sj} = -s(s+1) + \sum_k f_k^{jms} (a\omega)^k,$$
(10.50)

or numerically [93, 94, 96, 98].

The differential emission rates for the brane-localised modes during the spin-down phase will be given by the four-dimensional formula (10.35) but with the grey-body factor computed from the brane equation of motion (10.47) and the temperature given by (10.46). Despite the complexity of the gravitational background, the absorption probability $|\mathscr{A}(\omega)|^2$ can be again found analytically in the low-energy and low-angular momentum regime. For example, in the case of scalar fields, the dependence of $|\mathscr{A}(\omega)|^2$ on the angular momentum parameter a and number of extra dimensions n is given in Fig. 10.7 [111]. Each curve in the two plots actually consists of two lines: a solid one, representing our analytic result, and a dotted one, representing the numerical result; it is clear that in the low-ω regime, the agreement between the two sets of results is indeed remarkable. A similar agreement is observed for the cases of fermions and gauge bosons [112].

However, for the complete spectrum, we have to retort again to numerical analysis [93–98]. In Fig. 10.8, we present the energy emission rates, for the indicative cases of brane-localised scalars and gauge bosons, again in terms of the angular momentum parameter and number of extra dimensions. It is clear that an increase in any of these two parameters results in the significant enhancement of the energy emission rate. In Table 10.6, we have put together the factors by which the energy emission rates are enhanced, in terms of a and n, for brane-localised scalars [93, 94],

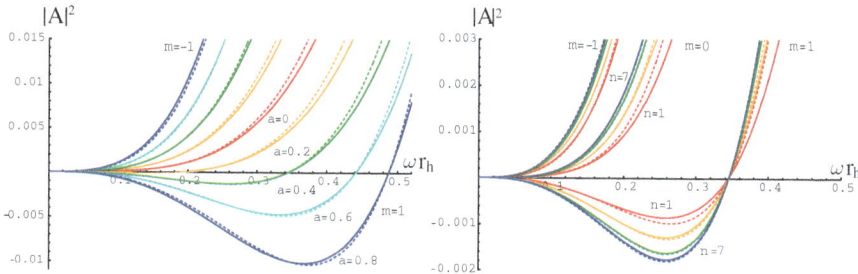

Fig. 10.7 Absorption probabilities for brane-localised scalar fields as a function of the angular momentum parameter a (*left plot*) and number of extra dimensions n (*right plot*)

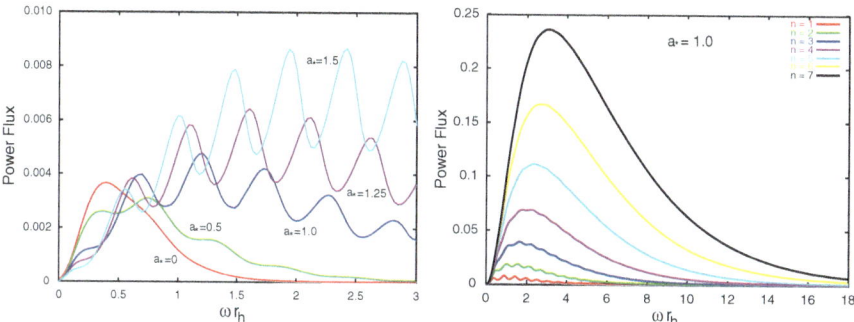

Fig. 10.8 Energy emission rates for brane-localised scalar fields in terms of the angular parameter (*left plot*) and gauge bosons in terms of the number of extra dimensions (*right plot*)

Table 10.6 Enhancement factors for the energy emission rates in terms of the angular momentum parameter and number of extra dimensions

	$(n = 4)$	$a_* = 0$	$a_* = 1.0$	$(a_* = 1)$	$n = 1$	$n = 7$
Scalars		1	≥ 3		1	≥ 100
Fermions		1	6		1	99
Gauge bosons		1	≥ 5		1	≥ 50

gauge bosons [96] and fermions [98]. When the angular momentum parameter increases from 0 to 1, the energy emission rates, for an eight-dimensional black hole, increase by a factor from 3 to 6, whereas, for a black hole with a fixed angular momentum parameter $a_* = 1$, the enhancement factor is of the order of 50–100 when n increases from 1 to 7. If we finally compare the relative emissivities of different species of fields, then, once again, it is the gauge bosons that a higher-dimensional rotating black hole prefers to emit on the brane.

Let us finally comment on a particular feature that the radiation spectra from the spin-down phase in the life of the black hole have. Unlike the line element that

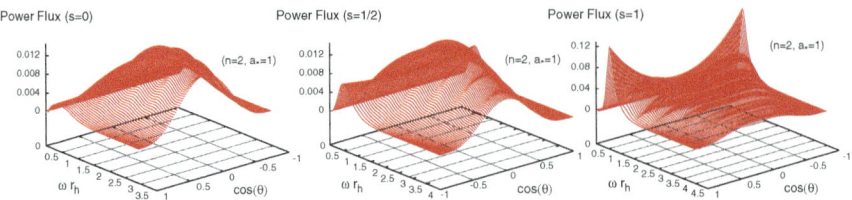

Fig. 10.9 Angular distribution of the energy emission spectra for scalars (*left plot*), fermions (*central plot*) and gauge bosons (*right plot*) for a six-dimensional black hole with $a_* = 1$

describes the background around the black hole during its spherically symmetric Schwarzschild phase, the one for the spin-down phase possesses a preferred axis in space, which is the rotation axis of the black hole. As a result, the radiation spectra of all emitted particles have a non-trivial angular dependence. As an indicative case, in Fig. 10.9, we present the energy emission rates for scalars, fermions and gauge bosons, from a six-dimensional, rotating black hole with $a_* = 1$, as a function of the energy parameter ωr_H and the $\cos(\theta)$ of the angle measured from the rotation axis of the black hole. In all spectra, we observe that most of the energy is emitted along the equatorial plane ($\theta = \pi/2$) as a result of the centrifugal force that is exerted on all species of fields. In the special cases of fermions and gauge bosons, i.e. of particles with non-vanishing spin, there is another effect, that of the spin-rotation coupling, that causes an additional angular dependence in their spectra and aligns the emission along the rotation axis of the black hole – the effect is more dominant for gauge bosons than for fermions, and it dies out as the energy of the emitted particles increases. The angular spectra depicted in Fig. 10.9 follow after solving numerically the angular master equation (10.48) for the value of the spin-weighted spheroidal harmonics $S_{sj}^m(\theta)$ and calculating the differential emission rate

$$\frac{d^3E}{d(\cos\theta)dtd\omega} = \frac{1}{4\pi}\sum_{j=1}^{\infty}\sum_{m=-j}^{j}\frac{\omega\,|\mathcal{A}(\omega)|^2}{\exp(\tilde{\omega}/T_H)-1}\left[\left(S_{|s|j}^m\right)^2 + \left(S_{-|s|j}^m\right)^2\right] \quad (10.51)$$

per unit time, frequency and solid angle for each species of particles [94, 96, 98].

10.3.3 Emission in the Bulk

In the case that higher-dimensional mini black holes can indeed be created during particle collisions, their detection becomes more likely if a significant part of the black-hole energy is channelled, through Hawking radiation emission, into brane fields. Therefore, although the bulk emission will be interpreted as a missing energy signal by a brane observer, we need to know the fraction of the total energy which is lost along this channel. We thus need to study the emission by the black hole of the species of particles that are allowed to propagate in the bulk, that is gravitons and possibly scalar fields. The latter are easier to study as their equation of motion in the higher-dimensional spacetime can be easily found, by generalising its four-dimensional expression, to be

$$\frac{1}{\sqrt{-G}} \partial_M \left[\sqrt{-G}\, G^{MN} \partial_N \Phi \right] = 0, \tag{10.52}$$

where the capital indices take values in the range $(0,1,2,3,\ldots,4+n)$ and G_{MN} is the metric tensor of the higher-dimensional spacetime.

We will study first the Schwarzschild phase, for which more results are available in the literature. In that case, the gravitational background that we need to consider is the higher-dimensional Schwarzschild–Tangherlini one (10.8). By assuming again a factorised ansatz for its wavefunction [105]

$$\Phi(t,r,\theta_i,\varphi) = e^{-i\omega t}\, R_{\omega l}(r)\, \tilde{Y}(\Omega)\,, \tag{10.53}$$

where $\tilde{Y}(\Omega)$ is the higher-dimensional spherical harmonics [113], the equation of motion of the scalar field can reduce to a system of decoupled, radial and angular, equations. From the radial one, we find the absorption probability $|\mathscr{A}(\omega)|^2$ for a bulk scalar field, and finally the radiation spectrum [105]. This is given in Fig. 10.10 in terms of the number of the additional spacelike dimensions n. As in the case of brane emission, the energy emission rate for bulk scalar fields is greatly enhanced as n increases.

Therefore, the question "which scalar channel, bulk or brane, is the most dominant one?" naturally arises. If the black hole has the choice to emit scalar fields both on the brane and in the bulk, which channel is the most effective? In order to answer this question, we need to compute the bulk-to-brane relative emissivity. This follows by integrating the corresponding brane and bulk spectra for scalar fields over the energy parameter ωr_H and computing their ratio. Then, we obtain the values for the bulk-to-brane ratio displayed in Table 10.7 [105]. From these, we see that this ratio becomes smaller than unity as soon as one extra dimension is introduced in the theory, decreases further as n takes intermediate values and increases, while remaining smaller than unity, as n reaches higher (supergravity-inspired) values. Thus, we deduce that, in general, the brane scalar channel is the dominant one; however, for high values of n, the bulk emission becomes indeed significant.

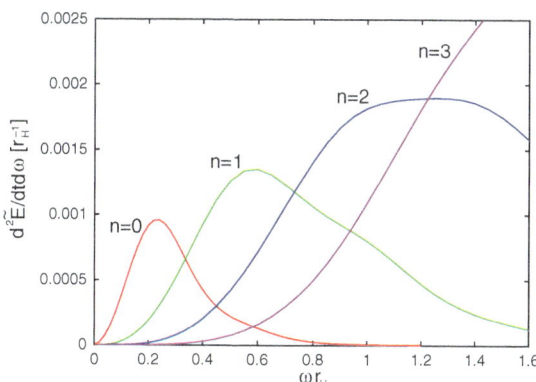

Fig. 10.10 Energy emission rates for bulk scalar fields, as a function of the number of additional spacelike dimensions, for the Schwarzschild phase

Table 10.7 Bulk-to-brane relative emissivity ratio for scalar fields in terms of n

	$n=0$	$n=1$	$n=2$	$n=3$	$n=4$	$n=5$	$n=6$	$n=7$
Bulk/brane	1.0	0.40	0.24	0.22	0.24	0.33	0.52	0.93

The above result gives strong support to the argument presented in [114] where it was argued that most of the energy of a higher-dimensional black hole will be emitted on the brane. The fact that the number of brane-localised degrees of freedom is larger than the bulk ones, combined with the above result that when both channels are available, the black hole still prefers the brane one, solidifies this argument. However, this matter is far from settled since we have not looked yet at one of the most important species of particles that may be emitted by the black hole into the higher-dimensional spacetime, namely the gravitons. If the probability for graviton emission in the bulk comes out to be much higher than the one for lower-spin fields, then the bulk-to-brane balance may be overturned.

The graviton equation of motion in the bulk was derived [115, 116] only a few years ago, in the case of a spherically symmetric, higher-dimensional background. In there, a comprehensive analysis led to Schrödinger-like equations for the three types of gravitational degrees of freedom that one encounters in a higher-dimensional spacetime, namely tensor, vector and scalar ones. In the years that followed, the equations of motion for all three types were studied both analytically [117, 118] and numerically [119–122]. The analytical approaches led to the derivation of the gravitational radiation spectra either in the intermediate [117] or in the low-energy [118] regime. In the latter case, it was shown [118] that, as long as the energy of the emitted particles remain in the lower part of the spectrum, the total bulk graviton emission rate is subdominant to the one for a bulk scalar field, which in turn is subdominant to the one for a brane scalar field. However, a definite answer for the graviton effect on the bulk-to-brane balance can be given only if the complete spectrum for these degrees of freedom is known. This followed from the numerical analysis performed in [119–121]; according to their results, the energy emission rates for gravitons for the Schwarzschild phase of the black hole behave similarly to the ones for the other degrees of freedom, i.e. they are significantly enhanced as the number of additional spacelike dimensions increases. The exact enhancement factors in terms of n appear in the last row of Table 10.5, and a direct comparison is possible: clearly, the bulk graviton emission rate is the one that exhibits the biggest in magnitude enhancement factor.

Similar results follow when the relative emissivities are computed – these are displayed in Table 10.8. Due to the aforementioned enhancement factor, the gravi-

Table 10.8 Relative emissivities for brane-localised standard model fields and bulk gravitons

n	0	1	2	3	4	5	6	7
Scalars	1	1	1	1	1	1	1	1
Fermions	0.55	0.87	0.91	0.89	0.87	0.85	0.84	0.82
Gauge bosons	0.23	0.69	0.91	1.0	1.04	1.06	1.06	1.07
Gravitons	0.053	0.61	1.5	2.7	4.8	8.8	17.7	34.7

tons, from an insignificant part of the total emission in four dimensions, become the dominant type of particles emitted by the black hole as soon as $n \geq 2$. How does this affect the bulk-to-brane energy balance then? Surprisingly, it is not in a position to overturn the dominance of the brane channel. The reason for this is that in the relative emissivities for gravitons displayed in Table 10.8 the total number of gravitational degrees of freedom has already been taken into account. On the other hand, the relative emissivities for the SM fields correspond to individual scalar, fermionic and gauge bosonic degrees of freedoms. When the total number of SM degrees of freedom (not to mention the beyond the SM ones) living on the brane is included in the calculation of the total "brane emissivity", the brane channel turns out to be the dominant one once again.[6]

Have we therefore settled the question of the brane-to-bulk energy balance? Perhaps not. The discussion up to now referred to the Schwarzschild phase in the life of the black hole, and another study needs to be performed for the spin-down phase. The only results available in the literature for the brane-to-bulk ratio in the case of a higher-dimensional, rotating black-hole background are the analytic ones for scalar fields presented in [125]. In Fig. 10.11, we display the ratio of the differential energy emission rate for scalar fields living on the brane over the one for bulk scalar fields from a higher-dimensional black hole with line element given by (10.43). From the left plot of Fig. 10.11, we see that for a black hole with fixed angular momentum ($a_* = 0.5$) the brane-to-bulk ratio remains above unity for all values of n. On the other hand, from the right plot we observe that for a five-dimensional black hole, the same ratio is again larger than unity but it decreases as either the angular momentum of the black hole or the energy of the particle increases. It would be indeed interesting to check whether the brane dominance in the emission of scalar fields persists over the whole frequency regime, especially for large values of a [126].

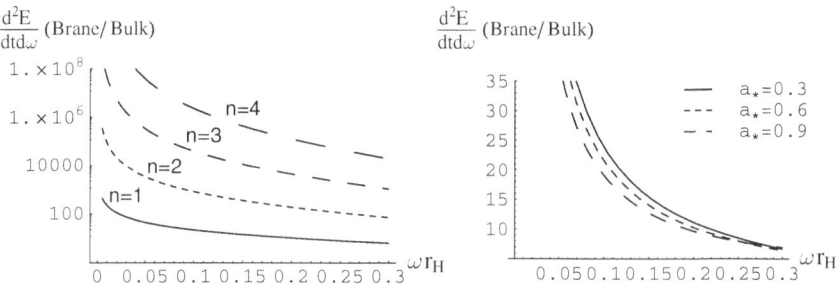

Fig. 10.11 Brane-to-bulk ratio of the differential energy emission rates for scalar fields during the spin-down phase in terms of n (*left plot*) and a (*right plot*)

---

[6] We note that in the presence of higher-derivative curvature terms in the theory, such as the Gauss–Bonnet term, it has been found [123] that the bulk emission might become the dominant one for specific values of the black-hole mass and Gauss–Bonnet coupling constant even for the spherically symmetric Schwarzschild phase. Also, in the case that the model allows for fermions to propagate in the bulk, the bulk-to-brane ratio in the fermionic channel exceeds unity even by an order of magnitude [124].

10.3.4 Deducing Basic Information

Let us now discuss the methodology one should follow, in case we witness the creation of miniature black holes in a collider experiment. In order to deduce any useful information on the fundamental, higher-dimensional theory, we need to compute with the greatest possible accuracy two quantities: the mass of the black hole and its temperature.

During a high-energy collision of composite particles, it is impossible to know which pair of partons led to the creation of the black hole and what its total energy was. Further losses of energy in the form of gravitational or visible radiation during the balding phase complicate things even more. The black-hole mass can therefore be reconstructed only through the measurement of the energy of the particles that appear in the final state after the evaporation of the black hole [82]. Clearly, any missing energy will greatly reduce the efficiency of the method; therefore one needs to focus on events with little or no missing energy. To this end, a cut is imposed on events with missing energy $E > 100$ GeV, so that the black-hole mass resolution is about 4%, i.e. ± 200 GeV if $M_{BH} = 5$ TeV [127].

The temperature of the black hole can be determined by performing a fit on the detected Hawking radiation spectra [82]. Preferably, these spectra should come from events involving only photons and electrons in the final state. The reason is that (a) these events would have a very low background and (b) the energy resolution of these particles is excellent even at high energies.

Once the temperature T_H and mass M_{BH} of the black hole are found with the greatest possible accuracy, one could proceed to determine the dimensionality of spacetime, in other words the value of n. From the temperature–horizon radius relation (10.18), we may write [82]

$$\log(T_H) = -\frac{1}{n+1} \log(M_{BH}) + const. \tag{10.54}$$

Then, the value of n can simply follow by determining the slope of the straight-line fit of the data relating M_{BH} and T_H. The above method is naturally not free of problems – indicatively, we may mention the following:

- The resolution in the measurement of the black-hole mass M_{BH} may not be good.
- The black-hole temperature T_H changes (increases) as a function of time as the evaporation progresses.
- The multiplicity of particles in the final state of the evaporation decreases for high values of n.
- Secondary particles that do not come directly from the evaporating black hole may obscure the spectrum.

We have already discussed the first problem associated with the determination of the black-hole mass. Let us briefly discuss the second one involving the temperature. We consider the special case with fundamental Planck scale given by $M_* = 1$ TeV and number of additional dimensions $n = 2$. We will pretend that we do not know the value of n but rather we are trying to find it through (10.54). We can assume that the temperature of the black hole either remains constant or it increases as the time goes by. Then, the use of (10.54) leads to the two plots, respectively, appearing in Fig. 10.12 [127]. As we see, by fitting the slope of the straight line, we obtain $n = 1.7 \pm 0.3$ in the first case and $n = 3.8 \pm 1.0$ in the second. A realistic model should be in a position to take into account that the temperature of the black hole is indeed increasing as the evaporation progresses but also that the lifetime of the black hole is extremely short. As a result, the real situation should actually be somewhere in between the two cases considered above, and an accurate fitting should be in a position to produce the correct value of n which lies indeed between the two derived values.

The multiplicity of particles emitted by the evaporating black hole depends strongly on the black-hole mass and its temperature – roughly, the first quantity stands for the amount of energy available for emission and the second for the average energy that each emitted particle carries away. More accurately, the number of particles emitted by the black hole is given by the relation [82]

$$\langle N \rangle = \left\langle \frac{M_{BH}}{E} \right\rangle \simeq \frac{M_{BH}}{2T_H} . \tag{10.55}$$

If, given the extremely short lifetime of the black hole, we assume that its mass remains constant and that the black hole evaporates instantly into a number of particles, the multiplicity then depends on the value of T_H. From the entries of Table 10.4, we see that the value of the temperature increases as the number of additional dimensions n increases too. In Fig. 10.13, we display a plot [127] showing the multiplicity of particles emitted from a black hole as a function of M_{BH}, and for various values of n [increasing from 2 (top) to 6 (bottom)] for $M_* = 1$ TeV. From this, it is clear that while for small values of n, a black hole, which might be created at the LHC, can emit up to 25 particles, for large values of n, this number drops at around 10. As a result, the number of data points that we need to construct the $T_H - M_{BH}$ line reduces significantly with n, and with it the accuracy in the determination of its slope. While therefore, by using (10.54), we might be in a position to obtain a rather

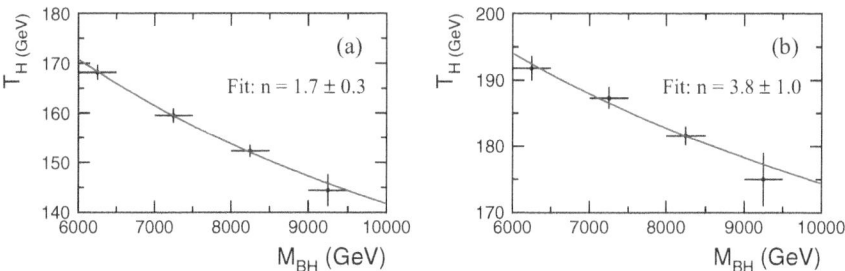

Fig. 10.12 Plots relating the black-hole mass and temperature measurements, and the derived value of n, for constant (*left plot*) and variable (*right plot*) temperatures [127]

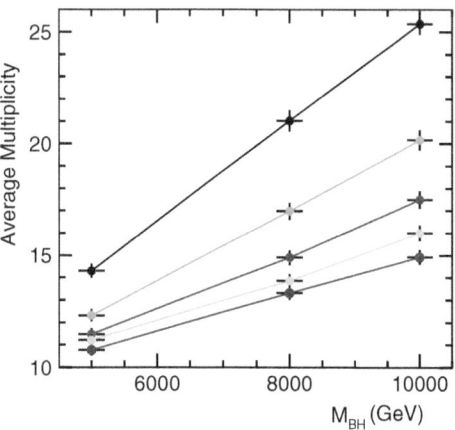

Fig. 10.13 Multiplicity of particles emitted by a black hole as the number of the additional spacelike dimensions n increases from 2 (*top curve*) to 6 (*bottom curve*) [127]

accurate value of n if that lies in the lower part of its range, it might be very difficult to distinguish between the cases with $n = 5$, $n = 6$ or $n = 7$.

Many experiments, looking for beyond the SM physics, have included searches for extra dimensions and miniature black holes in their research programs. At the Large Hadron Collider alone, three collaborations (ALICE, ATLAS and CMS) are planning to do so. But what type of particles and signatures should we expect to see in the detectors? Will we be able to see the Hawking radiation emission spectra that we presented in the previous sections for elementary SM degrees of freedom (the so-called primary particles), or maybe "secondary" composite particles will be detected instead? In order to have a better understanding of the type of particles expected to be seen in the final state, we need a black hole event generator (BHEG) that simulates the black-hole production and decay process given a number of initial conditions. The method followed in a BHEG is roughly the following:

– For a given centre-of-mass energy E of the colliding particles, the black-hole mass M_{BH} is estimated as a fraction of E.
– The theoretically predicted emission rates for the "primary" particles are fed to the BHEG and the "secondary" particle spectra are produced.

At the moment, there are several black hole event generators that have been constructed: CHARYBDIS [128], Catfish [129] and TRUENOIR [130]. For example, the CHARYBDIS generator uses the HERWIG program [131] to handle all the QCD interactions, hadronisation and secondary decays. It also makes specific predictions for the relative emissivities of the different species of SM particles expected to be detected. These are shown in Table 10.9 [127] from where we easily deduce

Table 10.9 Predictions for the relative emissivities of SM fields [127] derived by CHARYBDIS

Type	Quarks	Gluons	Charged leptons	Neutrino	Photons	Z^0	$W^{\pm}$	Higgs
(%)	63.9	11.7	9.4	5.1	1.5	2.6	4.7	1.1

that the dominant type of elementary particles emitted by the black hole should be the quarks. The exact spectrum of emitted particles depends also on what happens during the final phase in the life of the black hole, i.e. whether the black hole evaporates completely by emitting a few energetic particles or a stable remnant is formed [132–137]. For this reason, BHEGs are equipped with an option regarding the nature of the final state of the black hole that can be changed at will leading each time to the corresponding radiation spectra. Finally, any observed deviations from the anticipated behaviour stemming from standard QCD could be considered as additional observable signatures of the black-hole formation. For instance, QCD events with high transverse momentum are expected to become gradually more rare as the energy of the collision increases [33]; on the contrary, black-hole events with high transverse momentum dominate over the QCD events, with this happening at lower energies the smaller the fundamental gravity scale M_* is [138]. In addition, in a standard QCD process, one would expect to see the typical back-to-back di-jet production seen in $p + p$ collisions with a particle distribution peaked at $\Delta\phi = 0$ and π, with $\Delta\phi$ being the difference in the azimuthal coordinate between the two emitted hadrons; as the black hole decays through the emission of individual, sequential "primary" particles that lead to mono-jet events, we expect the back-to-back di-jets to be strongly suppressed in the case of the black-hole formation [138, 139].

If we, therefore, wished to summarise some of the most interesting phenomena associated with the existence of a low-scale gravity and the production of higher-dimensional black holes, we should mention the following (see [140] for a complementary discussion on this):

- Large cross-sections that increase with the centre-of-mass energy of the collision unlike every other SM process – in such a case, the accurate measurement of the cross-section could lead to the value of M_*
- Primary particles emitted by the black hole with a thermal spectrum and a much higher multiplicity than any other SM process
- Energy emission rates and relative emissivities for different species of fields determined by the number of additional spacelike dimensions
- Non-trivial angular distribution in the radiation spectra coming from the spin-down phase
- Comparison of the observed with the predicted spectra could lead to the detection of the final remnant – its presence would increase the multiplicity of particles in the final decay, lower the total transverse momentum by an amount equal to its mass and, if charged, could even be directly detected via an ionising track in the detector
- Events with high transverse momentum, above the expected QCD background
- Strong suppression of back-to-back di-jet events contrary to the expected QCD behaviour
- A significant amount of missing energy – larger than the one for SM or SUSY – due to the emission of weakly interacting particles on the branes and of gravitons or scalars in the bulk.

10.3.5 Schwarzschild–de Sitter Black Holes

We would like to finish the discussion of the properties and fate of higher-dimensional black holes with a brief reference to the class of black holes that are formed in the presence of a positive cosmological constant Λ in the higher-dimensional spacetime. The geometrical background around such a Schwarzschild–de Sitter black hole is given by the line element [36]:

$$ds^2 = -h(r)\,dt^2 + \frac{dr^2}{h(r)} + r^2 d\Omega_{2+n}^2 , \tag{10.56}$$

where

$$h(r) = 1 - \frac{\mu}{r^{n+1}} - \frac{2\kappa_D^2 \Lambda r^2}{(n+3)(n+2)} , \tag{10.57}$$

with $\kappa_D^2 \equiv 8\pi G_D = 8\pi/M_*^{2+n}$ and μ given again by (10.45). The equation $h(r) = 0$ has two real, positive solutions, r_H and r_C standing for the black hole and the cosmological horizon, respectively. The temperature of the black hole is given by the expression [141]

$$T_H = \frac{1}{\sqrt{h(r_0)}} \frac{1}{4\pi r_H} \left[(n+1) - \frac{2\kappa_D^2 \Lambda}{(n+2)} r_H^2 \right], \tag{10.58}$$

where r_0 is the value of the radial coordinate where the metric function $h(r)$ reaches its maximum value – the presence of the factor $1/\sqrt{h(r_0)}$ in the expression for the temperature is necessary for its consistent definition [142]. A similar expression can be written for the temperature T_C corresponding to the cosmological horizon – the fact that $r_C > r_H$ guarantees that $T_C < T_H$; therefore the flow of energy is from the black hole towards the remaining spacetime.

The line element of the gravitational background on the brane follows as before by fixing the values of the additional angular coordinates to $\theta_i = \pi/2$. It is then straightforward to write the equation of motion of a scalar field propagating in the projected-to-the-brane background. By solving the radial part of the equation of motion, we may again determine the absorption probability $|\mathscr{A}(\omega)|^2$, and in turn the energy emission rate. As in the case of a flat spacetime, the energy emission rate is found to be greatly enhanced not only with the number of extra dimensions n but also with the value of the cosmological constant Λ. This enhancement is clearly shown in Fig. 10.14 [141]. What is, however, more important is the fact that, unlike in the case where $\Lambda = 0$, for $\Lambda \neq 0$, the emission curve reaches an asymptotic non-zero value as $\omega \to 0$. This asymptotic value increases with the value of Λ and it might, in principle, be used to "read" the value of the cosmological constant from the observed radiation spectra. This non-zero asymptotic value is due to the fact that, unlike in the case of a flat spacetime, the absorption probability acquires a non-vanishing value when $\omega \to 0$. This is given by the expression [141] (see also [143])

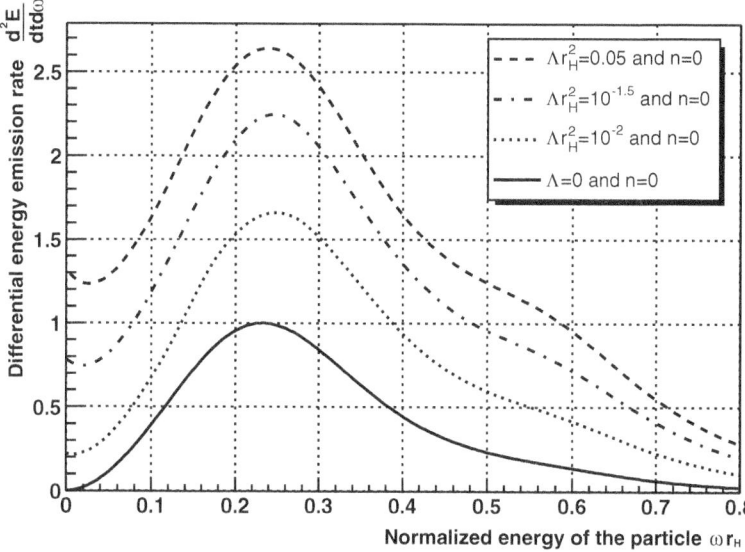

Fig. 10.14 Energy emission rate for a four-dimensional black hole in terms of the cosmological constant Λ

$$|\mathscr{A}(\omega = 0)|^2 = \frac{4r_C^2 r_H^2}{(r_C^2 + r_H^2)^2} \tag{10.59}$$

and is clearly caused by the presence of the cosmological horizon in the theory – in the limit $r_C \to \infty$, the asymptotic value of the absorption probability, and in turn of the energy emission rate, reduces to zero. This effect is independent of the existence of additional spacelike dimensions and should be manifest also in the radiation spectra of four-dimensional primordial black holes.

10.4 Conclusions

In the context of the theories predicting the existence of either Large or Warped Extra Dimensions, a low-scale gravitational theory, characterised by a fundamental Planck scale M_* much smaller than the four-dimensional one M_P, can be realised. This theory becomes accessible as soon as the energy of a given experiment exceeds M_* and manifests itself through a number of strong gravity effects. These effects should be present even at ordinary standard model particle collisions, taking place, for instance, at ground-based colliders. As a matter of fact, it is expected that during collisions with $E > M_*$ we should witness the creation of not point-like particles anymore but of extended heavy objects. One such type of objects is black holes, one of the most fascinating classes of solutions in general relativity.

The Large Hadron Collider at CERN will have a centre-of-mass energy of 14 TeV, i.e. more than an order of magnitude larger than the value of the fundamental Planck scale $M_* = 1$ TeV, suggested by the most optimistic scenarios with extra dimensions. It becomes then a natural place to look for strong gravity effects, and possibly for the creation of black holes. Studies have shown that their production can be realised as long as the energy of the collision exceeds at least the value of 8 TeV. At the same time, the produced black holes are expected to have a mass of at least a few times the value of the fundamental Planck scale if we want the classical theory of general relativity and its predictions to be still applicable. According to the above restrictions, the Large Hadron Collider is found to lie on the edge of both the classical regime and the black-hole creation threshold.

The calculation of the value of the corresponding production cross-section has attracted a great attention over the years. The current results seem to support the claim that this value is significant and that it will lead to the creation of, at least, a few black-hole events per day. In addition, the study of the properties of these higher-dimensional black holes suggests that the presence of extra dimensions greatly facilitates their creation: for instance, the horizon radius of these black holes, although tiny, is orders of magnitude larger compared to the one for a four-dimensional black hole with the same mass.

When it comes to the detection of these events, the terms are also favourable. The most important observable associated with the creation of the black hole will be the emission of Hawking radiation, in the form of elementary particles, as the black hole evaporates. The corresponding radiation spectrum will be centred around the value of the temperature of the black hole, which, for the mass values that would allow their creation at LHC, comes out to be in the range 100–600 GeV. A thorough theoretical study, employing either analytic or numerical techniques, is necessary in order to determine, and thus predict, the exact Hawking radiation spectrum from a decaying black hole. The differential energy emission rates are found to depend on a number of particle and spacetime properties, and thus to encode a valuable amount of information for the gravitational background and for the species of particles emitted. Some of the quantities on which we may deduce information are the number of additional dimensions that exist transversely to the brane, the black-hole angular momentum, the cosmological constant, the spin of the emitted particles and so on.

In order to make a realistic prediction of the radiation spectra and also to model in more detail the dynamical aspects of the production and evaporation process, black hole event generators have been constructed. The exact form of the radiation spectra, together with an additional number of distinct observable signatures, should make the detection of black holes, and thus of the existence itself of the extra dimensions, possible at the Large Hadron Collider. Hopefully, during the coming years, our understanding of particle and gravitational physics, and the fundamental theory that describes them, will be considerably extended beyond the current limits.

Acknowledgments I am grateful to my collaborators (J. March-Russell, C. Harris, A. Barrau, J. Grain, G. Duffy, E. Winstanley, M. Casals, S. Creek, O. Efthimiou, K. Tamvakis and S. Dolan, in chronological order) for their valuable help and inspiration while trying to uncover some of the secrets of the higher-dimensional black holes. I would also like to thank the organisers of the fourth Aegean Summer School on Black Holes for their kind invitation to present these lectures. I finally acknowledge financial support from the UK PPARC Research Grant PPA/A/S/2002/00350 and participation in the RTN Universenet (MRTN-CT-2006035863-1 and MRTN-CT-2004-503369).

References

1. I. Antoniadis, Phys. Lett. B **246**, 377 (1990).
2. P. Horava and E. Witten, Nucl. Phys. B **460**, 506 (1996).
3. P. Horava and E. Witten, Nucl. Phys. B **475**, 94 (1996).
4. J. Lykken, Phys. Rev. D **54**, 3693 (1996).
5. Akama, K.: Lect. Notes Phys. **176**, 267 (1982).
6. V. A. Rubakov and M. E. Shaposhnikov, Phys. Lett. B **125**, 139 (1983).
7. V. A. Rubakov and M. E. Shaposhnikov, Phys. Lett. B **125**, 136 (1983).
8. M. Visser, Phys. Lett. B **159**, 22 (1985).
9. G. W. Gibbons and D. L. Wiltshire, Nucl. Phys. B **287**, 717 (1987).
10. N. Arkani-Hamed, S. Dimopoulos and G. Dvali, Phys. Lett. B **429**, 263 (1998).
11. N. Arkani-Hamed, S. Dimopoulos and G. Dvali, Phys. Rev. D **59**, 086004 (1999).
12. I. Antoniadis, N. Arkani-Hamed, S. Dimopoulos and G. R. Dvali, Phys. Lett. B **4**36, 257 (1998).
13. L. Randall and R. Sundrum, Phys. Rev. Lett. **83**, 3370 (1999).
14. L. Randall and R. Sundrum, Phys. Rev. Lett. **83**, 4690 (1999).
15. P. Abreu et al. [DELPHI Collaboration], Eur. Phys. J. **C17**, 53 (2000).
16. G. Abbiendi et al. [OPAL Collaboration], Eur. Phys. J. **C18**, 253 (2000).
17. D. Acosta et al. [CDF Collaboration], Phys. Rev. Lett. **89**, 281801 (2002).
18. C. D. Hoyle, U. Schmidt, B. R. Heckel, E. G. Adelberger, J. H. Gundlach, D. J. Kapner and H. E. Swanson, Phys. Rev. Lett. **86**, 1418 (2001).
19. D. J. Kapner, T. S. Cook, E. G. Adelberger, J. H. Gundlach, B. R. Heckel, C. D. Hoyle and H. E. Swanson, Phys. Rev. Lett. **98**, 021101 (2007).
20. L. J. Hall and D. R. Smith, Phys. Rev. **D60**, 085008 (1999).
21. S. Cullen and M. Perelstein, Phys. Rev. Lett. **83**, 268 (1999).
22. V. D. Barger, T. Han, C. Kao and R. J. Zhang, Phys. Lett. **B461**, 34 (1999).
23. C. Hanhart, D. Phillips, S. Reddy and M. Savage, Nucl. Phys. **B595**, 335 (2001).
24. C. Hanhart, J. A. Pons, D. R. Phillips and S. Reddy, Phys. Lett. **B509**, 1 (2001).
25. R. Allahverdi, C. Bird, S. Groot Nibbelink and M. Pospelov, Phys. Rev. **D69**, 045004 (2004).
26. S. Hannestad and G. Raffelt, Phys. Rev. Lett. **87**, 051301 (2001).
27. S. Hannestad, Phys. Rev. **D64**, 023515 (2001).
28. S. Hannestad and G. Raffelt, Phys. Rev. Lett. **88**, 071301 (2002).
29. S. Hannestad and G. Raffelt, Phys. Rev. **D67**, 125008 (2003).
30. S. Hannestad and G. Raffelt, Phys. Rev. **D69**, 029901 (2004) (Erratum).
31. M. Gogberashvili, A. S. Sakharov and E. K. G. Sarkisyan, Phys. Lett. **B644**, 179 (2007).
32. L. A. Anchordoqui, J. L. Feng, H. Goldberg and A. D. Shapere, Phys. Rev. **D68**, 104025 (2003).
33. T. Banks and W. Fischler, hep-th/9906038.
34. K. S. Thorne, in *Magic Without Magic*, ed. J. R. Klauder (Freeman, San Francisco, 1972).
35. D. Ida and K. i. Nakao, Phys. Rev. D **66**, 064026 (2002).
36. F. R. Tangherlini, Nuovo Cim. **27**, 636 (1963).
37. R. C. Myers and M. J. Perry, Ann. Phys. **172**, 304 (1986).
38. P. Meade and L. Randall, JHEP **0805**, 003 (2008).
39. K.A. Khan and R. Penrose, Nature **229**, 185 (1971).
40. P. Szekeres, J. Math. Phys. **13**, 286 (1972).
41. T. Dray and G. 't Hooft, Nucl. Phys. **B253**, 173 (1985).

42. T. Dray and G. 't Hooft, Class. Quant. Grav. **3**, 825 (1986).
43. U. Yurtsever, Phys. Rev. **D38**, 1706 (1988).
44. U. Yurtsever, Phys. Rev. **D38**, 1731 (1988).
45. P. D. D'Eath and P. N. Payne, Phys. Rev. **D46**, 658 (1992).
46. P. D. D'Eath and P. N. Payne, Phys. Rev. **D46**, 675 (1992).
47. P. D. D'Eath and P. N. Payne, Phys. Rev. **D46**, 694 (1992).
48. G. 't Hooft, Phys. Lett. **B198**, 61 (1987).
49. G. 't Hooft, Nucl. Phys. **B304**, 867 (1988).
50. G. 't Hooft, Nucl. Phys. **B335**, 138 (1990).
51. S. B. Giddings and M. Srednicki, Phys. Rev. D **77**, 085025 (2008).
52. D. J. Gross and P. F. Mende, Phys. Lett. **B197**, 129 (1987).
53. D. J. Gross and P. F. Mende, Nucl. Phys. **B303**, 407 (1988).
54. D. Amati, M. Ciafaloni and G. Veneziano, Phys. Lett. **B197**, 81 (1987).
55. D. Amati, M. Ciafaloni and G. Veneziano, Int. J. Mod. Phys. **A3**, 1615 (1988).
56. D. Amati, M. Ciafaloni and G. Veneziano, Phys. Lett. **B216**, 41 (1989).
57. S. B. Giddings, D. J. Gross and A. Maharana, Phys. Rev. D **77**, 046001 (2008).
58. H. Verlinde and E. Verlinde, Nucl. Phys. **B371**, 246 (1992).
59. P.C. Aichelburg and R.U. Sexl, Gen. Rel. Grav. **2**, 303 (1971).
60. V. P. Frolov and I. D. Novikov, *Black Hole Physics: Basic Concepts and New Developments* (Kluwer Academic, Dordrecht, Netherlands 1998).
61. V. Cardoso and J. P. S. Lemos, Phys. Lett. **B538**, 1 (2002).
62. V. Cardoso and J. P. S. Lemos, Phys. Rev. **D67**, 084005 (2003).
63. U. Sperhake, V. Cardoso, F. Pretorius, E. Berti and J.A. Gonzalez, arXiv:0806.1738 [gr-qc].
64. D. M. Eardley and S. B. Giddings, Phys. Rev. D **66**, 044011 (2002).
65. H. Yoshino and Y. Nambu, Phys. Rev. D **66**, 065004 (2002).
66. H. Yoshino and Y. Nambu, Phys. Rev. D **67**, 024009 (2003).
67. M. B. Voloshin, Phys. Lett. **B518**, 137 (2001).
68. M. B. Voloshin, Phys. Lett. **B524**, 376 (2002).
69. S. Dimopoulos and R. Emparan, Phys. Lett. **B526**, 393 (2002).
70. E. Kohlprath and G. Veneziano, JHEP **0206**, 057 (2002).
71. S. N. Solodukhin, Phys. Lett. **B533**, 153 (2002).
72. T. G. Rizzo, JHEP **0202**, 011 (2002).
73. S. D. H. Hsu, Phys. Lett. **B555**, 92 (2003).
74. H. Yoshino and V. S. Rychkov, Phys. Rev. **D71**, 104028 (2005).
75. M. Cavaglia, S. Das and R. Maartens, Class. Quant. Grav. **20**, L205 (2003).
76. H. Yoshino and R. B. Mann, Phys. Rev. **D74**, 044003 (2006).
77. D. Ida, K. y. Oda and S. C. Park, Phys. Rev. D **67**, 064025 (2003).
78. D. Ida, K. y. Oda and S. C. Park, Phys. Rev. D **69**, 049901 (2004) (Erratum).
79. B. Koch, Phys. Lett. **B663/4**, 334 (2008).
80. S. B. Giddings, Phys. Rev. D **67**, 126001 (2003).
81. S. B. Giddings and S. Thomas, Phys. Rev. D **65**, 056010 (2002).
82. S. Dimopoulos and G. Landsberg, Phys. Rev. D **87**, 161602 (2001).
83. P. C. Argyres, S. Dimopoulos and J. March-Russell, Phys. Lett. B **441**, 96 (1998).
84. P. Kanti, Int. J. Mod. Phys. A **19**, 4899 (2004).
85. S. W. Hawking, Commun. Math. Phys. **43**, 199 (1975).
86. W. G. Unruh, Phys. Rev. D **10**, 3194 (1974).
87. W. G. Unruh, Phys. Rev. D **14**, 3251 (1976).
88. A. C. Ottewill and E. Winstanley, Phys. Rev. D **62**, 084018 (2000).
89. Y. B. Zel'dovich, JETP Lett. **14**, 180 (1971).
90. P. Kanti and J. March-Russell, Phys. Rev. D **66**, 024023 (2002).
91. P. Kanti and J. March-Russell, Phys. Rev. D **D67**, 104019 (2003).
92. V. Frolov and D. Stojkovic, Phys. Rev. D **67**, 084004 (2003).
93. C. M. Harris and P. Kanti, Phys. Lett. B **633**, 106 (2006).
94. G. Duffy, C. Harris, P. Kanti and E. Winstanley, JHEP **0509**, 049 (2005).
95. D. Ida, K. y. Oda and S. C. Park, Phys. Rev. D **71**, 124039 (2005).

96. M. Casals, P. Kanti and E. Winstanley, JHEP **0602**, 051 (2006).
97. D. Ida, K. y. Oda and S. C. Park, Phys. Rev. D **73**, 124022 (2006).
98. M. Casals, S. R. Dolan, P. Kanti and E. Winstanley, JHEP **0703**, 019 (2007).
99. E. Newman and R. Penrose, J. Math. Phys. **3**, 566 (1962).
100. S. Chandrasekhar, *The Mathematical Theory of Black Holes* (Oxford University Press, New York 1983).
101. J. N. Goldberg, A. J. MacFarlane, E. T. Newman, F. Rohrlich and E. C. Sudarshan, J. Math. Phys. **8**, 2155 (1967).
102. S. A. Teukolsky, Phys. Rev. Lett. **29**, 1114 (1972).
103. S. A. Teukolsky, Astrophys. J. **185**, 635 (1973).
104. S. S. Gubser, I. R. Klebanov and A. A. Tseytlin, Nucl. Phys. **B499**, 217 (1997).
105. C. M. Harris and P. Kanti, JHEP **0310**, 014 (2003).
106. D. N. Page, Phys. Rev. D **13**, 198 (1976).
107. C. Flammer, *Spheroidal Wave Functions* (Stanford University Press, Stanford, USA 1957).
108. A. A. Starobinskii and S. M. Churilov, Sov. Phys.-JETP **38**, 1 (1974).
109. E. D. Fackerell and R. G. Crossman, J. Math. Phys. **18**, 1849 (1977).
110. E. Seidel, Class. Quant. Grav. **6**, 1057 (1989).
111. S. Creek, O. Efthimiou, P. Kanti and K. Tamvakis, Phys. Rev. D **75**, 084043 (2007).
112. S. Creek, O. Efthimiou, P. Kanti and K. Tamvakis, Phys. Rev. D **76**, 104013 (2007).
113. C. Muller, *Lecture Notes in Mathematics: Spherical Harmonics* (Springer-Verlag, Berlin-Heidelberg 1966).
114. R. Emparan, G. T. Horowitz and R. C. Myers, Phys. Rev. Lett. **85**, 499 (2000).
115. H. Kodama and A. Ishibashi, Prog. Theor. Phys. **110**, 701 (2003).
116. H. Kodama and A. Ishibashi, Prog. Theor. Phys. **111**, 29 (2004).
117. A. S. Cornell, W. Naylor and M. Sasaki, JHEP **0602**, 012 (2006).
118. S. Creek, O. Efthimiou, P. Kanti and K. Tamvakis, Phys. Lett. B **635**, 39 (2006).
119. V. Cardoso, M. Cavaglia and L. Gualtieri, Phys. Rev. Lett. **96**, 071301 (2006).
120. V. Cardoso, M. Cavaglia and L. Gualtieri, Phys. Rev. Lett. **96**, 219902 (2006) (Erratum).
121. V. Cardoso, M. Cavaglia and L. Gualtieri, JHEP **0602**, 021 (2006).
122. D. K. Park, Phys. Lett. B **638**, 246 (2006).
123. J. Grain, A. Barrau and P. Kanti, Phys. Rev. D **72**, 104016 (2005).
124. H. T. Cho, A. S. Cornell, J. Doukas and W. Naylor, Phys. Rev. D **77**, 016004 (2008).
125. S. Creek, O. Efthimiou, P. Kanti and K. Tamvakis, Phys. Lett. B **656**, 102 (2007).
126. M. Casals, S. R. Dolan, P. Kanti and E. Winstanley, JHEP **0806**, 071 (2008).
127. C. M. Harris, M. J. Palmer, M. A. Parker, P. Richardson, A. Sabetfakhri and B. R. Webber, JHEP **0505**, 053 (2005).
128. C. M. Harris, P. Richardson and B. R. Webber, JHEP **0308**, 033 (2003).
129. M. Cavaglia, R. Godang, L. Cremaldi and D. Summers, Comput. Phys. Commun. **177**, 506 (2007).
130. G. L. Landsberg, J. Phys. G **32**, R337 (2006).
131. G. Corcella et al., JHEP **0101**, 010 (2001).
132. Y. Aharonov, A. Casher and S. Nussinov, Phys. Lett. B **191**, 51 (1987).
133. T. Banks, A. Dabholkar, M. R. Douglas and M. O'Loughlin, Phys. Rev. D **45**, 3607 (1992).
134. T. Banks and M. O'Loughlin, Phys. Rev. D **47**, 540 (1993).
135. T. Banks, M. O'Loughlin and A. Strominger, Phys. Rev. D **47**, 4476 (1993).
136. S. B. Giddings, Phys. Rev. D **49**, 947 (1994).
137. B. Koch, M. Bleicher and S. Hossenfelder, JHEP **0510**, 053 (2005).
138. T. J. Humanic, B. Koch and H. Stoecker, Int. J. Mod. Phys. E **16**, 841 (2007).
139. B. Koch, M. Bleicher and H. Stoecker, J. Phys. G **34**, S535 (2007).
140. S. B. Giddings, AIP Conf. Proc. **957**, 69 (2007).
141. P. Kanti, J. Grain and A. Barrau, Phys. Rev. D **71**, 104002 (2005).
142. R. Bousso and S. W. Hawking, Phys. Rev. D **54**, 6312 (1996).
143. T. Harmark, J. Natario and R. Schiappa, arXiv:0708.0017 [hep-th].

Part III
Perturbations of Black Holes

Chapter 11
Perturbations and Stability
of Higher-Dimensional Black Holes

H. Kodama

Abstract In this article, I explain the gauge-invariant formulation for perturbations of background spacetimes with untwisted homologous Einstein fibres, which include lots of practically important spacetimes such as static black holes, static black branes and rotating black holes in various dimensions. As applications, we discuss the stability of static black holes in higher dimensions and flat black branes.

11.1 Introduction

Perturbation analysis is a very powerful tool to investigate the dynamical response of a system against small disturbances. In particular, in general relativity whose fundamental equations are quite hard to solve analytically in general due to their non-linearity and strong couplings, perturbation analysis of exact solutions plays crucial roles in physical and astrophysical problems. The most successful example is the perturbative studies of cosmological perturbations, which has in particular provided the foundation for the present structure formation theory and the precise observational cosmology in terms of CMB and gravitational waves.

Another important example is the perturbative studies of black holes. Such an investigation was first systematically done for the Schwarzschild black hole by Regge and Wheeler [1] in 1957. In particular, they succeeded in reducing the Einstein equations for odd-parity perturbations, which is called vector perturbations in the present lecture, to a single master ODE, which is called the Regge–Wheeler equation now. The formulation was extended to even-parity perturbations (scalar perturbations in this lecture) by Zerilli 13 years later [2], and the master equation called the Zerilli equation was derived for such perturbations. Soon later, Teukolsky succeeded in deriving similar master equations for perturbations of the Kerr black hole [3].

The original purpose of these formulations appears to have been to study gravitational emissions from particles plunging into or orbiting around black holes.

H. Kodama (✉)
Cosmophysics Group, IPNS, KEK and the Graduate University of Advanced Studies, 1-1 Oho, Tsukuba 305-0801, Japan
Hideo.Kodama@kek.jp

Kodama, H.: *Perturbations and Stability of Higher-Dimensional Black Holes.* Lect. Notes Phys. **769**, 427–470 (2009)
DOI 10.1007/978-3-540-88460-6_11 ⓒ Springer-Verlag Berlin Heidelberg 2009

However, it was soon recognised [4, 5] that the formulation can be used to study the stability of black holes, which is also a practically important problem in determining the final fate of gravitational collapse. Actually, the asymptotically flat neutral and charged black holes were shown to be stable (for the proof and its historical background, see the excellent book by Chandrasekhar [6]). This together with the uniqueness theorem for black holes in the asymptotically flat electrovac system [7] now provides the basis of the current black hole astrophysics.

These results on four-dimensional black holes are practically sufficient for investigations of low-energy phenomena. However, taking account of the higher dimensionality of the present candidates for the unified theory, it is likely that higher-dimensional black holes are formed in the early universe and in extremely high-energy astrophysical phenomena as well as in particle accelerators. In fact, motivated by this expectation, lots of work has been done on higher-dimensional black holes in various theories, and astonishing discoveries have been obtained. In particular, it is now widely recognised that black hole uniqueness does not hold in higher dimensions, except for static black holes [8, 9]. Furthermore, a full list of regular black holes has not been obtained even in five dimensions [10, 11](cf. [12, 13]).

In this situation, perturbative analysis of exact solutions found so far is expected to be quite useful for the study of stability and uniqueness of higher-dimensional black holes [14]. In particular, it will be a great help if we can reduce the Einstein equations for perturbations of higher-dimensional black holes to decoupled master equations, as in four dimensions. In this lecture, we show that we can really reduce the perturbation equations to decoupled mater equations for some classes of black holes and study the stability with the help of them. We also point out that such a reduction is not always possible.

The remaining part is organized as follows. First, in the next section, we briefly overview the present status of the black-hole stability issue in four and higher dimensions. Then, in Sect. 11.3, we explain the basic aspects of the gauge-invariant formulation for perturbations of a general class of background spacetimes that can be written as a warped product of a lower-dimensional spacetime and an Einstein space. In Sect. 11.4, we apply this formulation to static black holes and discuss their stability. Next, in Sect. 11.5, we study the stability of flat black branes and point out the non-hermitian nature of the perturbation equations of this system. Section 11.6 is devoted to brief summary and discussion.

11.2 Present Status of the Black-Hole Stability Issue

In this section, we briefly overview the present status of the investigations of the stability problem of black holes. It is far from complete.

11.2.1 Four Dimensions

The present status of the stability issue for four-dimensional black holes is summarised as follows:

– Stable

  – Schwarzschild black hole [4, 5, 15, 16]
  – Reissner–Nördstrom black hole [6]
  – AdS/dS neutral/charged black holes [17, 18]
  – Kerr black hole [19]
  – Skyrme black hole (non-unique system) [20–22]

– Unstable

  – YM black hole (non-unique system) [23, 24]
  – Kerr-AdS black hole ($\ell\Omega_h < 1, r_h \ll \ell$) [25].

As is seen from this list, the stability is established for all AF black holes with connected horizon in the Einstein–Maxwell system, except for the charged rotating black hole (Kerr–Newman black hole), for which the perturbation equations have not been reduced to decoupled single master equations for this system.

In the asymptotically adS/dS case, the stability of static black holes have been established with the help of mater equations derived by Cardoso and Lemos [26]. In contrast, in the rotating case, it was conjectured that large Kerr-AdS black holes are stable, while small ones are superradiant unstable [27, 28]. Recently, the conjecture was shown to be true in the limit of slow rotation and small horizon [25].

11.2.2 Higher Dimensions

In contrast to the four-dimensional case, in addition to conventional black holes, there exist different kinds of black objects such as black strings, black branes, Kaluza–Klein black holes, Kaluza–Klein bubbles and black tubes in higher dimensions. The classification of these black objects is far from complete, and rather a little is known about the stability of known solutions. For example, concerning the asymptotically flat/dS/adS black holes and black branes, the present status of the stability issue is summarised as follows:

– Stable

  – AF vacuum static (Schwarzschild–Tangherlini) [17]
  – AF-charged static ($D = 5, 6 - 11$) [18, 29]
  – dS vacuum static ($D = 5, 6, 7 - 11$), dS charged static ($D = 5, 6 - 11$) [17, 18, 29]
  – BPS-charged black branes (in type II SUGRA) [30, 31]

– Unstable

  – AF/adS static black string and AF black branes (non-BPS) [31–39]
  – Rapidly rotating special Kerr-AdS black holes [40]

In this list, the stability of static black holes in higher dimensions ($D > 4$) has been proved analytically only for $D = 5$ in the AF/dS charged case and for $D = 5, 6$

in the dS neutral case, as explained in Sect. 11.4. The stability in other dimensions up to $D = 11$ for these black holes was proved numerically [29]. It is expected that the same result holds for $D > 11$ as well.

In contrast, in the asymptotically adS case, the stability in $D > 4$ is not certain even for neutral static black holes. This is a delicate problem because instability is expected for rotating adS black holes [25, 27, 41–43]. This instability is understood to arise from the combination of superradiance due to a rotating black hole and the time-like nature of the adS infinity. Some people conjectured that this superradiance also invokes instabilities in doubly spinning black rings [44] and Kerr black branes of the form $\text{Kerr}_4 \times \mathbb{R}^p$ [42].

The most impressive result about the stability of black objects in higher dimensions is the discovery of the Gregory–Laflamme instability of black strings/branes [32]. Since then, a large amount of work has been done on the classification of black holes and black strings/branes in the S^1/torus-compactified system and their stability. These researches revealed a rich structure of the phase diagram for such systems as well as new instabilities (for review, see [10, 45]). However, no clear understanding has been obtained about the origin and fate of these instabilities. It is partly because most of the researches were done by numerical methods. In Sect. 11.5, we point out some features that may be obstacles against the analytic approach.

Finally, we have to emphasise that very little is known about the stability of asymptotically flat solutions with rotating black objects. For example, rapidly rotating Myers–Perry solutions were conjectured to suffer from a Gregory–Laflamme type instability [46], but it has not been proved by any exact analysis. Black rings are also expected to be unstable because of their similarity to black string solutions, but no exact proof has been presented.

11.3 Gauge-Invariant Perturbation Theory

In this section, we explain the basic idea and techniques of the gauge-invariant formulation of perturbations [47, 48] for a class of background spacetimes that includes static black hole spacetimes as special case.

11.3.1 Background Solution

We assume that a background spacetime can be locally written as the warped product of a m-dimensional spacetime $\mathcal{N}$ and an n-dimensional Einstein space $\mathcal{K}$ as

$$M^{n+m} \approx \mathcal{N} \times \mathcal{K} \ni (z^M) = (y^a, x^i) \tag{11.1}$$

and has the metric

$$ds^2 = g_{MN}dz^M dz^N = g_{ab}(y)dy^a dy^b + r(y)^2 d\sigma_n^2, \tag{11.2}$$

where $d\sigma_n^2 = \gamma_{ij}dx^i dx^j$ is an n-dimensional Einstein metric on $\mathcal{K}$ satisfying the condition

$$\hat{R}_{ij} = (n-1)K\gamma_{ij}. \tag{11.3}$$

Note that for $n \leq 3$, an Einstein space is automatically a constant curvature space, while for $n > 3$, $\mathcal{K}$ does not have a constant curvature generically.

For this type of spacetimes, we can express the covariant derivative ∇_M, the connection coefficients $\Gamma_{NL}^M(z)$ and the curvature tensor $R_{MNLS}(z)$ in terms of the corresponding quantities for $\mathcal{N}^m$ and $\mathcal{K}^n$. We denote them as $D_a, {}^m\Gamma_{bc}^a(y), {}^m R_{abcd}(y)$ and $\hat{D}_i, \hat{\Gamma}_{jk}^i(x), \hat{R}_{ijkl}(x)$, respectively. For example, the curvature tensor can be expressed as

$$R^a{}_{bcd} = {}^m R^a{}_{bcd}, \quad R^i{}_{ajb} = -\frac{D_a D_b r}{r}\delta_j^i, \quad R^i{}_{jkl} = {}^m R_{abcd} - (Dr)^2(\delta_k^i \gamma_{jl} - \delta_l^i \gamma_{jk}), \tag{11.4}$$

and the non-vanishing components of the Einstein tensor are given by

$$G_{ab} = {}^m G_{ab} - \frac{n}{r}D_a D_b r - \left[\frac{n(n-1)}{2}\frac{K-(Dr)^2}{r^2} - \frac{n}{r}\Box r\right]g_{ab} \tag{11.5a}$$

$$G_j^i = \left[-\frac{1}{2}{}^m R - \frac{(n-1)(n-2)}{2}\frac{K-(Dr)^2}{r^2} + \frac{n-1}{r}\Box r\right]\delta_j^i. \tag{11.5b}$$

From this and the Einstein equations $G_{MN} + \Lambda g_{MN} = \kappa^2 T_{MN}$, it follows that the energy–momentum tensor of the background solution should take the form

$$T_{ab} = T_{ab}(y), \quad T_{ai} = 0, \quad T_j^i = P(y)\delta_j^i. \tag{11.6}$$

11.3.1.1 Examples

This class of background spacetimes includes quite a large variety of important solutions to the Einstein equations in four and higher dimensions.

1. **Robertson–Walker universe**: $m = 1$ and $\mathcal{K}$ is a constant curvature space.

$$ds^2 = -dt^2 + a(t)^2 d\sigma_n^2.$$

The gauge-invariant formulation was first introduced for perturbations of this background by Bardeen [47] and applied to realistic cosmological models by the author [48–50].

2. **Braneworld model**: $m = 2$ (and $\mathcal{K}$ is a constant curvature space). For example, the metric of adS^{n+2} spacetime can be written

$$ds^2 = \frac{dr^2}{1-\lambda r^2} - (1-\lambda r^2)dt^2 + r^2 d\Omega_n^2. \tag{11.7}$$

The gauge-invariant formulation of this background was first discussed by Muko-hyama [51] and then applied to the braneworld model taking account of the junction conditions by the author and collaborators [52].

3. **Higher-dimensional static Einstein black holes**: $m = 2$ and $\mathcal{K}$ is a compact Einstein space. For example, for the Schwarzschild–Tangherlini black hole, $\mathcal{K} = S^n$. In general, the generalised Birkhoff theorem says [18] that the electrovac solutions of the form (11.2) with $m = 2$ to the Einstein equations are exhausted by the Nariai-type solutions such that M is the direct product of a two-dimensional constant curvature spacetime $\mathcal{N}$ and an Einstein space $\mathcal{K}$ with $r = \text{const}$ and the black-hole type solution whose metric is given by

$$ds^2 = \frac{dr^2}{f(r)} - f(r)dt^2 + r^2 d\sigma_n^2; \tag{11.8}$$

$$f(r) = K - \frac{2M}{r^{n-1}} + \frac{Q^2}{r^{2n-2}} - \lambda r^2. \tag{11.9}$$

The gauge-invariant formulation for perturbations was applied to this background to discuss the stability of static black holes by the author and collaborators [17, 18, 53]. This application is explained in the next section.

4. **Black branes**: $m = 2 + k$ and $\mathcal{K} = $ Einstein space. In this case, the spacetime factor $\mathcal{N}$ is the product of a two-dimensional black-hole sector and a k-dimensional brane sector:

$$ds^2 = \frac{dr^2}{f(r)} - f(r)dt^2 + d\mathbf{z} \cdot d\mathbf{z} + r^2 d\sigma_n^2. \tag{11.10}$$

One can also generalise this background to introducing a warp factor in front of the black-hole metric part. The stability of this background for the case in which $\mathcal{K}$ is an Euclidean space is discussed in Sect. 11.5.

5. **Higher-dimensional rotating black hole** (a special Myers–Perry solution): $m = 4$ and $\mathcal{K} = S^n$.

$$ds^2 = g_{rr}dr^2 + g_{\theta\theta}d\theta^2 + g_{tt}dt^2 + 2g_{t\phi}dtd\phi + g_{\phi\phi}d\phi^2 + r^2\cos^2\theta d\Omega_n^2, \tag{11.11}$$

where all the metric coefficients are functions only of r and θ. The stability of this background was studied in [43].

6. **Axisymmetric spacetime**: m is general and $n = 1$.

11.3.2 Perturbations

11.3.2.1 Perturbation Equations

In order to describe the spacetime structure and matter configuration $(\tilde{M}, \tilde{g}, \tilde{\Phi})$ as a perturbation from a fixed background (M, g, Φ), we introduce a mapping F: background $M \rightarrow \tilde{M}$ and define perturbation variables on the fixed background spacetime as follows:

$$h := \delta g = F^* \tilde{g} - g, \quad \phi := \delta \Phi = F^* \tilde{\Phi} - \Phi. \tag{11.12}$$

Then, if these perturbation variables have small amplitudes, the Einstein equations and the other equations for matter can be described by linearised equations well. For example, in terms of the variable $\psi_{MN} = h_{MN} - h g_{MN}/2$, the linearised Einstein equations can be written as

$$\triangle_L \psi_{MN} + \nabla_M \nabla_A \psi_N^A + \nabla_N \nabla_A \psi_M^A - \nabla^A \nabla^B \psi_{AB} g_{MN} + R^{AB} \psi_{AB} g_{MN}$$
$$- R \psi_{MN} = 2 \kappa^2 \delta T_{MN}. \tag{11.13}$$

where $\triangle_L$ is the Lichnerowicz operator defined by

$$\triangle_L \psi_{MN} := -\Box \psi_{MN} + R_{MA} \psi_N^A + R_{NA} \psi_M^A - 2 R_{MANB} \psi^{AB}. \tag{11.14}$$

11.3.2.2 Gauge Problem

For a different mapping F', the perturbation variables defined above change their values, which has no physical meaning and can be regarded as a kind of gauge freedom. Because F and F' are related by a diffeomorphism, the corresponding changes of the variables are identical to the transformation of the variables with respect to the transformation $f = F'^{-1} F$. In the framework of linear perturbation theory, we can restrict considerations to infinitesimal changes of F. Hence, f is expressed in terms of an infinitesimal transformation ξ^M as

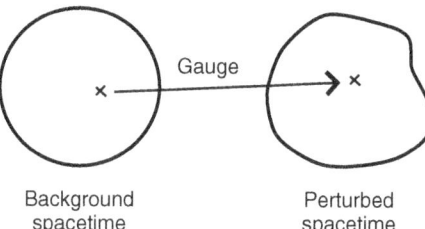

Fig. 11.1 Gauge transformation   Background spacetime   Perturbed spacetime

$$\bar{\delta} x^M = x^M(f(p)) - x^M(p) = \xi^M, \tag{11.15}$$

and the are expressed as

$$\bar{\delta} h_{MN} = -\mathscr{L}_\xi g_{MN} \equiv -\nabla_M \xi_N - \nabla_N \xi_M, \quad \bar{\delta}\phi = -\mathscr{L}_\xi \Phi. \tag{11.16}$$

From its origin, the perturbation equations including the linearised Einstein equations given above are invariant under this gauge transformation.

To be specific, for our background spacetime, the metric perturbation transforms as

$$\bar{\delta}h_{ab} = -D_a\xi_b - D_b\xi_a, \tag{11.17a}$$

$$\bar{\delta}h_{ai} = -r^2 D_a\left(\frac{\xi_i}{r^2}\right) - \hat{D}_i\xi_a, \tag{11.17b}$$

$$\bar{\delta}h_{ij} = -\hat{D}_i\xi_j - \hat{D}_j\xi_i - 2rD^a r\xi_a\gamma_{ij} \tag{11.17c}$$

and the perturbation of the energy–momentum tensor $\tau_{MN} = \delta T_{MN}$ transforms as

$$\bar{\delta}\tau_{ab} = -\xi^c D_c T_{ab} - T_{ac}D_b\xi^c - T_{bc}D_a\xi^c, \tag{11.18a}$$

$$\bar{\delta}\tau_{ai} = -T_{ab}\hat{D}_i\xi^b - r^2 P D_a(r^{-2}\xi_i), \tag{11.18b}$$

$$\bar{\delta}\tau_{ij} = -\xi^a D_a(r^2 P)\gamma_{ij} - P(\hat{D}_i\xi_j + \hat{D}_j\xi_i) \tag{11.18c}$$

In order to remove this gauge freedom, one of the following two approaches is adopted in general:

(i) This method is direct, but it is rather difficult to find relations between perturbation variables in different gauges in general.
(ii) This method describes the theory only in terms of gauge-invariant quantities. Such quantities have non-local expressions in terms of the original perturbation variables in general.

These two approaches are mathematically equivalent, and a gauge-invariant variable can be regarded as some perturbation variable in some special gauge in general. Therefore, the non-locality of the gauge-invariant variables implies that the relation of two different gauges are non-local.

11.3.2.3 Tensorial Decomposition of Perturbations

In this lecture, we focus on the gauge-invariant approach to perturbations and explain that in the class of background spacetimes described above, we can locally construct fundamental gauge-invariant variables with help of harmonic expansions. This construction becomes more transparent if we decompose the perturbation variables into components of specific tensorial types. This decomposition also helps us to divide the coupled set of perturbation equations into decoupled smaller subsets, and in some cases into single master equations.

First of all, note that the basic perturbation variables h_{MN} and τ_{MN} can be classified into the following three algebraic types according to their transformation property as tensors on the n-dimensional space $\mathcal{K}$:

(i) Spatial scalar: h_{ab}, τ_{ab}
(ii) Spatial vector: h_{ai}, τ_i^a
(iii) Spatial tensor: h_{ij}, τ_j^i

Among these, spatial vectors and tensors can be further decomposed into more basic quantities. First, we decompose a vector field v_i on $\mathscr{K}$ into a scalar field $v^{(s)}$ and a transverse vector $v_i^{(t)}$ as

$$v_i = \hat{D}_i v^{(s)} + v_i^{(t)}; \quad \hat{D}_i v^{(t)i} = 0. \tag{11.19}$$

Then, from the relation

$$\hat{\triangle} v^{(s)} = \hat{D}_i v^i, \tag{11.20}$$

the component fields $v^{(s)}$ and $v_i^{(t)}$ can be uniquely determined from v_i up to the ineffective freedom in $v^{(s)}$ to add a constant, provided this Poisson equation has a unique solution on $\mathscr{K}$ up to the same freedom. For example, when $\mathscr{K}$ is compact and closed, this condition is satisfied.

Next, we decompose a symmetric tensor field of rank 2 on $\mathscr{K}$ as

$$t_{ij} = \frac{1}{n} t g_{ij} + \hat{D}_i \hat{D}_j s - \frac{1}{n} \hat{\triangle} s g_{ij} + \hat{D}_i t_j + \hat{D}_j t_i + t_{ij}^{(tt)}; \tag{11.21a}$$

$$\hat{D}_i t^i = 0, \quad t_i^{(tt)i} = 0, \quad \hat{D}_i t_j^{(tt)i} = 0. \tag{11.21b}$$

Here, t is uniquely determined as $t = t_i^i$. Further, from the relations derived from this definition,

$$\hat{\triangle}(\hat{\triangle} + nK)s = \frac{n}{n-1}\left(\hat{D}_i \hat{D}_j t^{ij} - \frac{1}{n}\hat{\triangle} t\right), \tag{11.22a}$$

$$[\hat{\triangle} + (n-1)K]t^i = (\delta_j^i - \hat{D}^i \hat{\triangle}^{-1}\hat{D}_j)(\hat{D}_m t^{jm} - n^{-1}\hat{D}^j t), \tag{11.22b}$$

s and t_i, hence $t_{ij}^{(tt)}$, can be uniquely determined from t_{ij} up to the addition of ineffective zero modes, provided that these Poisson equations have solutions unique up to the same ineffective freedom.

After these decompositions of vectors and tensors to basic components, we can classify these components into the following three types:

(i) Scalar type: $v^i = \hat{D}^i v^{(s)}$, $t_{ij} = \frac{1}{n} t g_{ij} + \hat{D}_i \hat{D}_j s - \frac{1}{n}\hat{\triangle} s g_{ij}$.
(ii) Vector type: $v_i = v_i^{(t)}$, $t_{ij} = \hat{D}_i t_j + \hat{D}_j t_i$.
(iii) Tensor type: $v^i = 0$, $t_{ij} = t_{ij}^{(tt)}$.

We call these types . In the linearised Einstein equations, through the covariant differentiation and tensor-algebraic operations, quantities of different algebraic tensorial types can appear in each equation. However, in the case in which $\mathscr{K}$ is a constant curvature space, perturbation variables belonging to different reduced ten-sorial types do not couple in the linearised Einstein equations if we decompose these perturbation equations into reduced tensorial types as well, because there exists no quantity of the vector or the tensor type in the background except for the metric tensor. The same result holds even in the case in which $\mathcal{K}$ is an Einstein space with non-constant curvature, because the only non-trivial background tensor other than the metric is the Weyl tensor that can only transform a second rank tensor to a second rank tensor.

Here, note that gauge transformations can be also decomposed into reduced ten-sorial types, and the gauge transformation of each type affects only the decomposed perturbation variables of the same reduced tensorial type. Hence, gauge-invariant variables can be constructed in each reduced tensorial types independently.

11.3.3 Tensor Perturbation

Let us start from the tensor-type perturbation, for which the argument is simplest.

11.3.3.1 Tensor Harmonics

We utilise tensor harmonics to expand tensor-type perturbations. They are defined as the basis for second-rank symmetric tensor fields satisfying the following eigenvalue problem:

$$(\hat{\triangle}_L - \lambda_L)\mathbb{T}_{ij} = 0; \quad \mathbb{T}_i^i = 0, \quad \hat{D}_j \mathbb{T}_i^j = 0, \tag{11.23}$$

where $\hat{\triangle}_L$ is the Lichnerowicz operator on $\mathcal{K}$ defined by

$$\hat{\triangle}_L h_{ij} := -\hat{D} \cdot \hat{D} h_{ij} - 2\hat{R}_{ikjl} h^{kl} + 2(n-1)K h_{ij}. \tag{11.24}$$

When $\mathcal{K}$ is a constant curvature space, this operator is related to the Laplace–Beltrami operator by

$$\hat{\triangle}_L = -\hat{\triangle} + 2nK, \tag{11.25}$$

and, $\mathbb{T}_{ij}$ satisfies

$$(\hat{\triangle} + k^2)\mathbb{T}_{ij} = 0; \quad k^2 = \lambda_L - 2nK. \tag{11.26}$$

We use k^2 in the meaning of $\lambda_L - 2nK$ from now on when $\mathcal{K}$ is an Einstein space with non-constant sectional curvature.

The harmonic tensor has the following basic properties:

1. **Identities**: Let T_{ij} be a symmetric tensor of rank 2 satisfying

$$T_i^i = 0, \quad D^j T_{ij} = 0.$$

Then, the following identities hold:

$$2D_{[i}T_{j]k}D^{[i}T^{j]k} = 2D^i(T_{jk}D^{[i}T^{j]k}) + T_{jk}\left[-\triangle T^{jk} + R_l^j T^{lk} + R_i{}^{jk}{}_l T^{il}\right],$$

$$2D_{(i}T_{j)k}D^{(i}T^{j)k} = 2D^i(T_{jk}D^{(i}T^{j)k}) + T_{jk}\left[-\triangle T^{jk} - R_l^j T^{lk} - R_i{}^{jk}{}_l T^{il}\right].$$

On the constant curvature space with sectional curvature K, these identities read

$$2D_{[i}T_{j]k}D^{[i}T^{j]k} = 2D^i(T_{jk}D^{[i}T^{j]k}) + T_{jk}(-\triangle + nK)T^{jk},$$
$$2D_{(i}T_{j)k}D^{(i}T^{j)k} = 2D^i(T_{jk}D^{(i}T^{j)k}) + T_{jk}(-\triangle - nK)T^{jk}.$$

2. **Spectrum**: When $\mathscr{K}$ is a compact and closed space with constant sectional curvature K, these identities lead to the following condition on the spectrum of k^2:

$$k^2 \geq n|K|. \tag{11.27}$$

In contrast, when $\mathscr{K}$ is not a constant curvature space, no general lower bound on the spectrum k^2 is known.
3. When $\mathscr{K}$ is a two-dimensional surface with a constant curvature K, a symmetric second-rank harmonic tensor that is regular everywhere can exist only for $K \leq 0$: for T^2 ($K = 0$), the corresponding harmonic tensor T_{ij} becomes a constant tensor in the coordinate system such that the metric is written $ds^2 = dx^2 + dy^2 (k^2 = 0)$; for $H^2/\Gamma (K = -1)$, a harmonic tensor corresponds to an infinitesimal deformation of the moduli parameters.
4. For $\mathscr{K} = S^n$, the spectrum of k^2 is given by

$$k^2 = l(l+n-1) - 2; \quad l = 2, 3, \cdots, \tag{11.28}$$

11.3.3.2 Perturbation Equations

The metric and energy–momentum perturbations can be expanded in terms of the tensor harmonics as

$$h_{ab} = 0, \quad h_{ai} = 0, \quad h_{ij} = 2r^2 H_T \mathbb{T}_{ij}, \tag{11.29}$$

$$\tau_{ab} = 0, \quad \tau_i^a = 0, \quad \tau_j^i = \tau_T \mathbb{T}_j^i. \tag{11.30}$$

Since the coordinate transformations contain no tensor-type component, H_T and τ_T are gauge invariant by themselves:

$$\xi^M = \bar{\delta}z^M = 0; \quad \bar{\delta}H_T = 0, \ \bar{\delta}\tau_T = 0. \tag{11.31}$$

Only the (i, j)-component of the Einstein equations has the tensor-type component:

$$-\Box H_T - \frac{n}{r} Dr \cdot DH_T + \frac{k^2 + 2K}{r^2} H_T = \kappa^2 \tau_T. \tag{11.32}$$

Here, $\Box = D^a D_a$ is the D'Alembertian in the m-dimensional spacetime $\mathcal{N}$. Thus, the Einstein equations for tensor-type perturbations can be always reduced to the single master equation on our background spacetime.

11.3.4 Vector Perturbation

11.3.4.1 Vector Harmonics

We expand transverse vector fields in terms of the complete set of harmonic vectors defined by the eigenvalue problem

$$(\hat{\triangle} + k^2)\mathbb{V}_i = 0; \quad \hat{D}_i \mathbb{V}^i = 0. \tag{11.33}$$

Tensor fields of the vector-type can be expanded in terms of the harmonic tensors derived from these vector harmonics as

$$\mathbb{V}_{ij} = -\frac{1}{2k}(\hat{D}_i \mathbb{V}_j + \hat{D}_j \mathbb{V}_i). \tag{11.34}$$

They satisfy

$$\left[\hat{\triangle} + k^2 - (n+1)K\right]\mathbb{V}_{ij} = 0, \tag{11.35a}$$

$$\mathbb{V}_i^i = 0, \quad \hat{D}_j \mathbb{V}_i^j = \frac{k^2 - (n-1)K}{2k}\mathbb{V}_i. \tag{11.35b}$$

Here, there is one subtle point; $\mathbb{V}_{ij}$ vanishes when $\mathbb{V}_i$ is a Killing vector. For this mode, from the above relations, we have $k^2 = (n-1)K$. We will see below that the converse holds when $\mathcal{K}$ is compact and closed. We call these modes *exceptional modes*.

Now, we list up some basic properties of the vector harmonics relevant to the subsequent discussions.

1. **Spectrum**: In an n-dimensional Einstein space $\mathcal{K}$ satisfying $R_{ij} = (n-1)Kg_{ij}$, we have

$$2D_{[i}V_{j]}D^{[i}V^{j]} = 2D_i(V_j D^{[i}V^{j]}) + V_j[-\triangle + (n-1)K]V^j \tag{11.36a}$$

$$2D_{(i}V_{j)}D^{(i}V^{j)} = 2D_i(V_j D^{(i}V^{j)}) + V_j[-\triangle - (n-1)K]V^j \tag{11.36b}$$

When $\mathcal{K}$ is compact and closed, from the integration of these over $\mathcal{K}$, we obtain the following general restriction on the spectrum of k^2:

$$k^2 \geq (n-1)|K|. \tag{11.37}$$

Here, when the equality holds, the corresponding harmonic vector becomes a Killing vector for $K \geq 0$ and a harmonic 1-form for $K \leq 0$, respectively.

2. For $\mathscr{K}^n = S^n$, we have

$$k^2 = \ell(\ell + n - 1) - 1, \quad (\ell = 1, 2, \cdots). \tag{11.38}$$

Here, the harmonic vector field $\mathbb{V}_i$ becomes a Killing vector for $l = 1$ and is exceptional.

3. For $K = 0$, the exceptional mode exists only when $\mathscr{K}$ is isometric to $T^p \times \mathscr{C}^{n-p}$, where $\mathscr{C}^{n-p}$ is a Ricci flat space with no Killing vector.

11.3.4.2 Perturbation Equations

Vector perturbations of the metric and the energy–momentum tensor can be expanded in terms of the vector harmonics as

$$h_{ab} = 0, \quad h_{ai} = r f_a \mathbb{V}_i, \quad h_{ij} = 2r^2 H_T \mathbb{V}_{ij}, \tag{11.39a}$$

$$\tau_{ab} = 0, \quad \tau_i^a = r \tau_a \mathbb{V}_i, \quad \tau_j^i = \tau_T \mathbb{V}_j^i. \tag{11.39b}$$

For the vector-type gauge transformation

$$\xi_a = 0, \quad \xi_i = r L \mathbb{V}_i \tag{11.40}$$

the perturbation variables transform as

$$\bar{\delta} f_a = -r D_a \left(\frac{L}{r} \right), \quad \bar{\delta} H_T = \frac{k}{r} L, \quad \bar{\delta} \tau_a = 0, \quad \bar{\delta} \tau_T = 0. \tag{11.41}$$

Hence, we adopt the following combinations as the fundamental gauge-invariant variables for the vector perturbation:

$$\text{generic modes:} \quad \tau_a, \tau_T, F_a = f_a + \frac{r}{k} D_a H_T \tag{11.42}$$

$$\text{exceptional modes:} \quad \tau_a, F_{ab}^{(1)} = r D_a \left(\frac{f_b}{r} \right) - r D_b \left(\frac{f_a}{r} \right) \tag{11.43}$$

. Note that for exceptional modes, $F_a = f_a$ because H_T is not defined.

The reduced vector part of the Einstein equations come from the components corresponding to G_i^a and G_j^i. In terms of the gauge-invariant variables defined above, these equations can be written as follows.

– **Generic modes**:

$$\frac{1}{r^{n+1}} D^b \left(r^{n+1} F_{ab}^{(1)} \right) - \frac{k^2 - (n-1)K}{r^2} F_a = -2\kappa^2 \tau_a,$$

(11.44a)

$$\frac{k}{r^n} D_a (r^{n-1} F^a) = -\kappa^2 \tau_T.$$

(11.44b)

– **Exceptional modes**: $k^2 = (n-1)K > 0$. For these modes, the second of the above equations coming from G^i_j does not exist.

$$\frac{1}{r^{n+1}} D^b \left(r^{n+1} F_{ab}^{(1)} \right) = -2\kappa^2 \tau_a.$$

(11.45)

11.3.5 Scalar Perturbation

11.3.5.1 Scalar Harmonics

Scalar functions on $\mathscr{K}$ can be expanded in terms of the harmonic functions defined by

$$(\hat{\triangle} + k^2)\mathbb{S} = 0.$$

(11.46)

Correspondingly, scalar-type vector and tensor fields can be expanded in terms of harmonic vectors $\mathbb{S}_i$ and harmonic tensors $\mathbb{S}_{ij}$ define by

$$\mathbb{S}_i = -\frac{1}{k} \hat{D}_i \mathbb{S},$$

(11.47a)

$$\mathbb{S}_{ij} = \frac{1}{k^2} \hat{D}_i \hat{D}_j \mathbb{S} + \frac{1}{n} \gamma_{ij} \mathbb{S}.$$

(11.47b)

These harmonic tensors satisfy the following relations:

$$\hat{D}_i \mathbb{S}^i = k \mathbb{S},$$

(11.48a)

$$[\hat{\triangle} + k^2 - (n-1)K]\mathbb{S}_i = 0,$$

(11.48b)

$$\mathbb{S}^i_i = 0, \quad \hat{D}_j \mathbb{S}^j_i = \frac{n-1}{n} \frac{k^2 - nK}{k} \mathbb{S}_i,$$

(11.48c)

$$[\hat{\triangle} + k^2 - 2nK]\mathbb{S}_{ij} = 0.$$

(11.48d)

Note that as in the case of vector harmonics, there are some exceptional modes:

(i) $k = 0$: $\mathbb{S}_i \equiv 0$, $\mathbb{S}_{ij} \equiv 0$.
(ii) $k^2 = nK$ ($K > 0$): $\mathbb{S}_{ij} \equiv 0$.

For scalar harmonics, $k^2 = 0$ is obviously always the allowed lowest eigenvalue. Therefore, the information on the second eigenvalue is important. In general, it is difficult to find such information. However, when $\mathscr{K}^n$ is a compact Einstein space with $K > 0$, we can obtain a useful constraint as follows. Let us define Q_{ij} by

$$Q_{ij} := D_i D_j Y - \frac{1}{n} g_{ij} \triangle Y.$$

Then, we have the identity

$$Q_{ij} Q^{ij} = D^i (D^i Y D_i D_j Y - Y D_i \triangle Y - R_{ij} D^i Y) \tag{11.49}$$
$$+ Y [\triangle (\triangle + (n-1)K)] Y - \frac{1}{n} (\triangle Y)^2.$$

For $Y = \mathbb{S}$, integrating this identity, we obtain the constraint on the second eigenvalue

$$k^2 \geq nK. \tag{11.50}$$

For $\mathscr{K}^n = S^n$, the equality holds because the full spectrum is given by

$$k^2 = \ell(\ell + n - 1), \quad (\ell = 0, 1, 2, \cdots). \tag{11.51}$$

11.3.5.2 Perturbation Equations

The scalar perturbation of the metric and the energy–momentum tensor can be expanded as

$$h_{ab} = f_{ab} \mathbb{S}, \quad h_{ai} = r f_a \mathbb{S}_i, \quad h_{ij} = 2r^2 (H_L \gamma_{ij} \mathbb{S} + H_T \mathbb{S}_{ij}), \tag{11.52a}$$

$$\tau_{ab} = \tau_{ab} \mathbb{S}, \quad \tau_i^a = r \tau_a \mathbb{S}_i, \quad \tau_j^i = \delta P \delta_j^i \mathbb{S} + \tau_T \mathbb{S}_j^i. \tag{11.52b}$$

For the scalar-type gauge transformation

$$\xi_a = T_a \mathbb{S}, \quad \xi_i = r L \mathbb{S}_i, \tag{11.53}$$

these harmonic expansion coefficients for generic modes $k^2(k^2 - nK) > 0$ of a scalar-type perturbation transform as

$$\bar{\delta} f_{ab} = -D_a T_b - D_b T_a, \quad \bar{\delta} f_a = -r D_a \left(\frac{L}{r} \right) + \frac{k}{r} T_a, \tag{11.54a}$$

$$\bar{\delta} H_L = -\frac{k}{nr} L - \frac{D^a r}{r} T_a, \quad \bar{\delta} H_T = \frac{k}{r} L, \tag{11.54b}$$

$$\bar{\delta} \tau_{ab} = -T^c D_c \tau_{ab} - \tau_{ac} D_b T^c - \tau_{bc} D_a T^c, \tag{11.54c}$$

$$\bar{\delta} \tau_a = \frac{k}{r} (\tau_{ab} T^b - P T_a), \quad \bar{\delta}(\delta P) = -T^a D_a P, \quad \bar{\delta} \tau_T = 0. \tag{11.54d}$$

From these we obtain

$$\bar{\delta}X_a = T_a; \quad X_a = \frac{r}{k}\left(f_a + \frac{r}{k}D_aH_T\right). \tag{11.55}$$

Hence, the fundamental gauge invariants can be given by τ_T and the following combinations:

$$F = H_L + \frac{1}{n}H_T + \frac{1}{r}D^a r X_a, \tag{11.56a}$$

$$F_{ab} = f_{ab} + D_a X_b + D_b X_a, \tag{11.56b}$$

$$\Sigma_{ab} = \tau_{ab} + T_b^c D_a X_c + T_a^c D_b X_c + X^c D_c T_{ab}, \tag{11.56c}$$

$$\Sigma_a = \tau_a - \frac{k}{r}(T_a^b X_b - P X_a), \tag{11.56d}$$

$$\Sigma_L = \delta P + X^a D_a P. \tag{11.56e}$$

The scalar part of the Einstein equations comes from G_{ab}, G_{ai} and G_j^i. First, from δG_{ab}, we obtain

$$-\Box F_{ab} + D_a D_c F_b^c + D_b D_c F_a^c + n\frac{D^c r}{r}(-D_c F_{ab} + D_a F_{cb} + D_b F_{ca})$$

$$+ {}^mR_a^c F_{cb} + {}^mR_b^c F_{ca} - 2{}^mR_{acbd}F^{cd} + \left(\frac{k^2}{r^2} - R + 2\Lambda\right)F_{ab} - D_a D_b F_c^c$$

$$-2n\left(D_a D_b F + \frac{1}{r}D_a r D_b F + \frac{1}{r}D_b r D_a F\right)$$

$$-\left[D_c D_a F^{cd} + \frac{2n}{r}D^c r D^d F_{cd} + \left(\frac{2n}{r}D^c D^d r + \frac{n(n-1)}{r^2}D^c r D^d r\right)\right.$$

$$\left. - {}^mR^{cd}\right)F_{cd} - 2n\Box F - \frac{2n(n+1)}{r}Dr \cdot DF + 2(n-1)\frac{k^2 - nK}{r^2}F$$

$$\left. -\Box F_c^c - \frac{n}{r}Dr \cdot DF_c^c + \frac{k^2}{r^2}F_c^c\right]g_{ab} = 2\kappa^2 \Sigma_{ab}. \tag{11.57}$$

Second, from δG_i^a, we obtain

$$\frac{k}{r}\left[-\frac{1}{r^{n-2}}D_b(r^{n-2}F_a^b) + rD_a\left(\frac{1}{r}F_b^b\right) + 2(n-1)D_a F\right] = 2\kappa^2 \Sigma_a. \tag{11.58}$$

Finally, from the trace-free part of δG_j^i, we obtain

$$-\frac{k^2}{2r^2}[2(n-2)F + F_a^a] = \kappa^2 \tau_T, \tag{11.59}$$

and from the trace δG^i_i,

$$-\frac{1}{2}D_aD_bF^{ab} - \frac{n-1}{r}D^arD^bF_{ab}$$

$$+\left[\frac{1}{2}{}^mR^{ab} - \frac{(n-1)(n-2)}{2r^2}D^arD^br - (n-1)\frac{D^aD^br}{r}\right]F_{ab}$$

$$+\frac{1}{2}\Box F^c_c + \frac{n-1}{2r}Dr\cdot DF^c_c - \frac{n-1}{2n}\frac{k^2}{r^2}F^c_c + (n-1)\Box F$$

$$+\frac{n(n-1)}{r}Dr\cdot DF - \frac{(n-1)(n-2)}{n}\frac{k^2-nK}{r^2}F = \kappa^2\Sigma_L. \qquad (11.60)$$

Note that for the exceptional mode with $k^2 = nK > 0$, the third equation does not exist, and for the mode with $k^2 = 0$, the second and the third equations do not exist. For these exceptional modes, the other equations hold without change, but the variables introduced above are not gauge invariant.

Although the energy–momentum conservation equation $\nabla_N T^N_M = 0$ can be derived from the Einstein equations, it is often useful to know its explicit form. For scalar-type perturbations, they are given by the following two sets of equations:

$$\frac{1}{r^{n+1}}D_a(r^{n+1}\Sigma^a) - \frac{k}{r}\Sigma_L + \frac{n-1}{n}\frac{k^2-nK}{kr}\tau_T$$

$$+\frac{k}{2r}(T^{ab}F_{ab} - PF^a_a) = 0, \qquad (11.61a)$$

$$\frac{1}{r^n}D_b\left[r^n(\Sigma^b_a - T^c_aF^b_c)\right] + \frac{k}{r}\Sigma_a - n\frac{D_ar}{r}\Sigma_L$$

$$+n\left(T^b_aD_bF - PD_aF\right) + \frac{1}{2}\left(T^b_aD_bF^c_c - T^{bc}D_aF_{bc}\right) = 0. \qquad (11.61b)$$

11.4 Stability of Static Black Holes

We study the stability of static black holes utilising the gauge-invariant formulation for perturbations explained in the previous section. We consider the static Einstein black hole which corresponds to the case with $m = 2$ of the general background considered in the previous section and has the metric (11.9). The key point is the fact that gauge-invariant perturbation equations can be reduced to decoupled single master equations of the Schrödinger type for any type of perturbations in this background.

11.4.1 Tensor Perturbations

The gauge-invariant equation for tensor perturbations is already given by a single equation for each mode. Assuming that the source term vanishes, it reads

$$-\partial_t H_T^2 + f\partial_r(f\partial_r H_T) - \frac{k^2+2K}{r^2}fH_T = 0. \tag{11.62}$$

Here, note that even if there exist electromagnetic fields, τ_T vanishes because the electromagnetic field is vector like and does not produce a tensor-type quantity in the linear order at least.

With the help of the Fourier transformation with respect to t, i.e., assuming $H_T \propto e^{-i\omega t}$, this equation can be put into the Schrödinger-type eigenvalue problem;

$$\omega^2\Phi = -f\partial_r(f\partial_r\Phi) + V_t\Phi; \quad H_T = r^{-n/2}\Phi(r)e^{-i\omega t}, \tag{11.63}$$

where

$$V_t = \frac{f}{r^2}\left[k^2 + 2K + \frac{nrf'}{2} + \frac{n(n-2)f}{4}\right] \tag{11.64}$$

$$= \frac{f}{r^2}\left[k^2 + \frac{n^2-2n+8}{4}K - \frac{n(n+2)}{4}\lambda r^2 + \frac{n^2M}{2r^{n-1}} - \frac{n(3n-2)Q^2}{4r^{2n-2}}\right].$$

If V_t is non-negative, we can directly conclude the stability. However, it is not so easy to see whether V_t is non-negative or not outside the horizon. This technical difficulty is easily resolved by considering the energy integral

$$E := \int_{r_h}^{r_\infty} dr\left[\frac{1}{f}(\partial_t H_T)^2 + f(\partial_r H_T)^2 + \frac{k^2+2K}{r^2}H_T^2\right]. \tag{11.65}$$

From the equation for H_T, we find that

$$\partial_t E = 2\left[f\partial_t H_T \partial_r H_T\right]_{r_h}^{r_\infty} = 0. \tag{11.66}$$

Hence, in the case $\mathcal{K}$ is a constant curvature space, the condition on the spectrum $k^2 \geq n|K|$ guarantees the positivity of all terms in E, and as a consequence the stability of the system.

11.4.2 Vector Perturbations

11.4.2.1 Master Equation

For vector perturbations, the energy–momentum conservation law is written

$$D_a(r^{n+1}\tau^a) + \frac{m_v}{2k}r^n\tau_T = 0. \tag{11.67}$$

For $m_v \equiv k^2 - (n-1)K \neq 0$, with the help of this equation, the second of the perturbation equations, (11.44b), can be written as

$$D_a(r^{n-1}F^a) = \frac{2\kappa^2}{m_v}D_a(r^{n+1}\tau^a).$$ (11.68)

In the case of $m = 2$, from this it follows that F^a can be written in terms of a variable Ω as

$$r^{n-1}F^a = \epsilon^{ab}D_b\Omega + \frac{2\kappa^2}{m_v}r^{n+1}\tau^a.$$ (11.69)

Further, the first of the perturbation equations, (11.44a), is equivalent to

$$D_a\left(r^{n+1}F^{(1)}\right) - m_v r^{n-1}\epsilon_{ab}F^b = -2\kappa^2 r^{n+1}\epsilon_{ab}\tau^b,$$ (11.70)

where ϵ_{ab} is the two-dimensional Levi–Civita tensor for g_{ab}, and

$$F^{(1)} = \epsilon^{ab}rD_a\left(\frac{F_b}{r}\right) = \epsilon^{ab}rD_a\left(\frac{f_b}{r}\right).$$ (11.71)

Inserting the expression for F_a in terms of Ω into (11.70), we obtain the master equation

$$r^n D_a\left(\frac{1}{r^n}D^a\Omega\right) - \frac{m_v}{r^2}\Omega = -\frac{2\kappa^2}{m_v}r^n\epsilon^{ab}D_a(r\tau_b).$$ (11.72)

Next, for $m_v = 0$, the perturbation variables H_T and τ_T do not exist. The matter variable τ_a is still gauge-invariant, but concerning the metric variables, only the combination $F^{(1)}$ defined in terms of f_a in (11.71) is gauge invariant. In this case, the Einstein equations are reduced to the single equation (11.70), and the energy–momentum conservation law is given by (11.67) without the τ_T term. Hence, τ_a can be expressed in terms of a function $\tau^{(1)}$ as

$$r^{n+1}\tau_a = \epsilon_{ab}D^b\tau^{(1)}.$$ (11.73)

Inserting this expression into (11.70) with $\epsilon^{cd}D_c(F_d/r)$ replaced by $F^{(1)}/r$, we obtain

$$D_a(r^{n+1}F^{(1)}) = -2\kappa^2 D_a\tau^{(1)}.$$ (11.74)

Taking account of the freedom of adding a constant in the definition of $\tau^{(1)}$, the general solution can be written as

$$F^{(1)} = -\frac{2\kappa^2\tau^{(1)}}{r^{n+1}}.$$ (11.75)

Hence, there exists no dynamical freedom in these special modes. In particular, in the source-free case in which $\tau^{(1)}$ is a constant and $K = 1$, this solution corresponds to adding a small rotation to the background static black-hole solution.

11.4.2.2 Neutral Black Holes

For a neutral static Einstein black hole, the master equation for a generic mode can be put into the canonical form as

$$\omega^2 \Phi = -f \partial_r (f \partial_r \Phi) + V_v \Phi; \quad \Omega = r^{n/2} \Phi(r) e^{-i\omega t}, \tag{11.76}$$

where

$$V_v = \frac{f}{r^2} \left[m_v - \frac{nrf'}{2} + \frac{n(n+2)f}{4} \right]$$

$$= \frac{f}{r^2} \left[k^2 + \frac{n(n+2)K}{4} - \frac{n(n-2)}{4} \lambda r^2 - \frac{3n^2 M}{2r^{n-1}} \right]. \tag{11.77}$$

This equation is identical to the Regge–Wheeler equation for $n = 2, K = 1$ and $\lambda = 0$. In this case, we can put V_v into an obviously non-negative form as

$$V_v = \frac{f}{r^2} (m_v + 3f), \tag{11.78}$$

proving the stability of the black hole against vector perturbations (or axial or odd perturbations).

In higher dimensions, the potential V_v is not positive definite anymore and we cannot use this type of argument. However, we can still prove the stability with the help of the conserved energy integral as in the case of tensor perturbations. In the present case, if we define E as

$$E := \int_{r_h}^{r_\infty} \frac{dr}{r^n} \left[\frac{1}{r} (\partial_t \Omega)^2 + f(\partial_r \Omega)^2 + \frac{m_v}{r^2} \Omega^2 \right], \tag{11.79}$$

we have

$$\dot{E} = 2 \left[\frac{f}{r^n} \partial_t \Omega \partial_r \Omega \right]_{r_h}^{r_\infty} = 0. \tag{11.80}$$

Further, all terms of E is non-negative because $m_v \geq 0$. Hence, the stability can be concluded.

11.4.2.3 Charged Black Hole

The formulation for neutral static black holes can be extended to charged static black holes. The final master equations consist of two equations: the extension of the equation for gravitational perturbations with an electromagnetic source and the equation coming from the Maxwell equations [18]:

Fig. 11.2 V_v for $K = 1$, $\lambda = 0$, $l = 2$

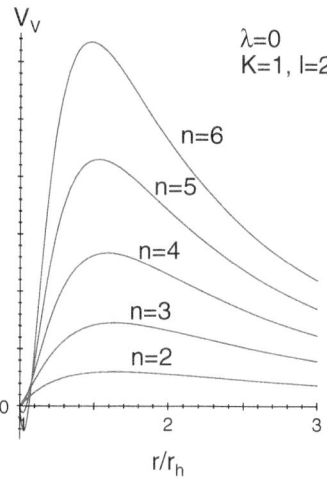

Fig. 11.3 V_v for $K = 1$, $\lambda < 0$, $l = 2$

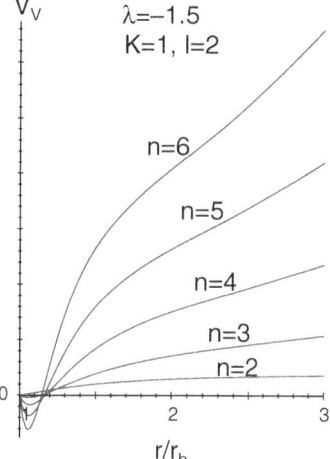

$$r^n D_a \left(\frac{1}{r^n} D^a \Omega \right) \quad -\frac{m_v}{r^2}\Omega = \frac{2\kappa^2 q}{r^2}\mathscr{A}, \tag{11.81a}$$

$$\frac{1}{r^{n-2}} D_a (r^{n-2} D^a \mathscr{A}) \quad -\frac{m_v + 2(n-1)K}{r^2}\mathscr{A} = q\frac{m_v \Omega + 2\kappa^2 q \mathscr{A}}{r^{2n}} \tag{11.81b}$$

where $\mathscr{A}$ is the gauge invariant representing a vector perturbation of the vector potential of the electromagnetic field defined by

$$\delta A_a = 0, \quad \delta A_i = \mathscr{A}\mathbb{V}_i, \tag{11.82}$$

and q is the black-hole charge related to the charge parameter Q in the background metric by

$$Q^2 := \frac{\kappa^2 q^2}{n(n-1)}, \tag{11.83}$$

By taking appropriate combinations, these equations can be transformed to the two decoupled equations

$$-\partial_t^2 \Phi_\pm = (-\partial_{r_*}^2 + V_\pm)\Phi_\pm,$$ (11.84)

where the effective potentials are given by

$$V_\pm = \frac{f}{r^2}\left[m_v + \frac{n(n+2)K}{4} - \frac{n(n-2)}{4}\lambda r^2 + \frac{n(5n-2)Q^2}{4r^{2n-2}} + \frac{\mu_\pm}{r^{n-1}} \right],$$ (11.85)

with

$$\mu_\pm = -\frac{n^2+2}{2}M \pm \Delta; \quad \Delta^2 = (n^2-1)^2 M^2 + 2n(n-1)m_v Q^2.$$ (11.86)

11.4.2.4 S-Deformation

The effective potentials $V_\pm$ are not positive definite as in the neutral case. In the present case, we prove that the system is still stable not by the energy integral method, but rather by a different method, which we call *the S-deformation* [17].

We first explain the basic idea by the eigenvalue equation

$$\omega^2 \Phi = \left(-D^2 + V(r)\right)\Phi,$$ (11.87)

where $D = \partial_{r_*}$. If there exists an unstable mode with $\omega^2 < 0$ and if V is non-negative at horizon and at infinity, we can show Φ falls off sufficiently rapidly at horizon and at infinity. Hence, we obtain the integral identity,

$$\omega^2 \int_{r_h}^{r_\infty} |\Phi|^2 \frac{dr}{f} = \int_{r_h}^{r_\infty} \left[|D\Phi|^2 + V(r)|\Phi|^2 \right] \frac{dr}{f}.$$ (11.88)

If $V(r)$ is non-negative definite, this leads to contradiction and hence proves the stability because the right-hand side is non-negative. In contrast, in the case in which the sign of V is not definite, we cannot say anything about stability from this equation.

In order to treat such a case, let us replace D by $D = \tilde{D} - S(r)$. Then, by partial integrations, we obtain the modified integral identity with D and V replaced by $\tilde{D}$ and $\tilde{V}$ given by

$$\tilde{V} = V + f\frac{dS}{dr} - S^2.$$ (11.89)

Hence, if we can find S such that the modified effective potential $\tilde{V}$ is non-negative, we can establish the stability of the system even when the original potential is not non-negative definite.

For example, by the S-transformation with $S = nf/(2r)$, the effective potentials $V_\pm$ above can be modified into

$$\tilde{V}_{\pm} = V_{\pm} + f\frac{dS}{dr} - S^2 = \frac{f}{r^2}\left[m_v + \frac{1}{r^{n-1}}\left(\frac{3n^2}{2}M + \mu_{\pm}\right)\right]. \tag{11.90}$$

Here, $\tilde{V}_+$ is obviously positive definite. We can also show that $\tilde{V}_-$ is also positive definite. Hence, a charged static Einstein black hole is stable for vector perturbations.

11.4.3 Scalar Perturbations

11.4.3.1 Master Equation

For a static Einstein black-hole background, assuming that $F_{ab}, F \propto e^{-i\omega t}$, we can reduce the whole linearised Einstein equations into a single master equation, as in the case of vector perturbations [53]:

$$\omega^2\Phi = -f\partial_r(f\partial_r\Phi) + V_s\Phi, \tag{11.91}$$

where the master variable Φ is defined as

$$\Phi = \frac{nr^{n/2}}{H}\left(2F + \frac{F_t^r}{i\omega r}\right); \quad H = m + \frac{n(n+1)}{2}x, \tag{11.92}$$

with $m = k^2 - nK$ and $x = 2M/r^{n-1}$, and the effective potential V_s is given by $V_s(r) = \frac{fU(r)}{16r^2H^2}$ with

$$U(r) = -\left[n^3(n+2)(n+1)^2x^2 - 12n^2(n+1)(n-2)mx\right.$$
$$\left. + 4(n-2)(n-4)m^2\right]\lambda r^2 + n^4(n+1)^2x^3$$
$$+ n(n+1)\left[4(2n^2 - 3n + 4)m + n(n-2)(n-4)(n+1)K\right]x^2$$
$$- 12n\left[(n-4)m + n(n+1)(n-2)K\right]mx$$
$$+ 16m^3 + 4Kn(n+2)m^2. \tag{11.93}$$

11.4.3.2 Neutral Black Holes

The above master equation is identical to the Zerilli equation for the four-dimensional Schwarzschild black hole ($n = 2, K = 1$ and $\lambda = 0$). In this case, from

$$V_s = \frac{f}{r^2H^2}\left(m^2(m+2) + \frac{6m^2M}{r} + \frac{36mM^2}{r^2} + \frac{72M^3}{r^3}\right) \geq 0, \tag{11.94}$$

where $m = (l-1)(l+2)(l = 2, 3, \cdots)$, we can easily prove the stability of the black hole. In higher dimensions, however, the effective potential V_s is not positive definite. Hence, an instability may arise.

Fig. 11.4 V_s for $K = 1$, $\lambda = 0$, $l = 2$

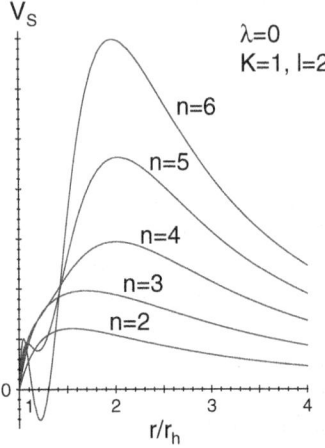

Fig. 11.5 V_s for $K = 1$, $\lambda < 0$, $l = 2$

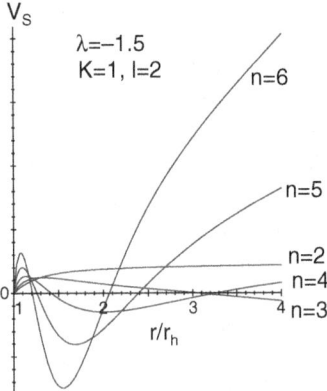

Nevertheless, in the case of $K = 1$ and $\lambda = 0$, i.e., for the Schwarzschild–Tangherlini black hole, we can prove the stability by applying the S-deformation to the energy integral. First, from the above master equation, we obtain

$$E := \int_{r_0}^{r_\infty} \frac{dr}{f} \left[(\partial_t \Phi)^2 + (D\Phi)^2 + V_s \Phi^2 \right], \qquad (11.95)$$

$$\dot{E} = [2f \partial_t \Phi \partial_r \Phi]_{r_h}^{r_\infty} = 0, \qquad (11.96)$$

where $D = f \partial_r$. Next, we replace D to $\tilde{D} = f \partial_r + S$. Then, by partial integration we obtain

$$E = \int_{r_0}^{r_\infty} \frac{dr}{f} \left[(\partial_t \Phi)^2 + (\tilde{D}\Phi)^2 + \tilde{V}_s \Phi^2 \right], \qquad (11.97)$$

where

$$\tilde{V}_s = V_s + f \frac{dS}{dr} - S^2. \qquad (11.98)$$

For example, for

$$S = \frac{f}{h}\frac{dh}{dr}, \quad h \equiv r^{n/2+l-1}\{(l-1)(l+n) + n(n+1)x/2\}. \tag{11.99}$$

we obtain

$$\tilde{V}_s = \frac{f(r)\tilde{Q}(r)}{4r^2\{(l-1)(l+n) + n(n+1)x/2\}}, \tag{11.100}$$

where

$$\tilde{Q}(r) \equiv lx[ln(n+1)x + 2(l-1)\{n^2 + n(3l-2) + (l-1)^2\}]. \tag{11.101}$$

Clearly $\tilde{V}_s > 0$.

11.4.3.3 Charged Black Holes

For charged black holes, we can also reduce the perturbation equations to decoupled single master equations. First, we generalise the master variable Φ for the metric perturbation given in (11.92) by replacing H by

$$H = m + \frac{n(n+1)M}{r^{n-1}} - \frac{n^2Q^2}{r^{2n-2}}. \tag{11.102}$$

Next, we introduce the gauge-invariant variable $\mathscr{A}$ in terms of which the scalar perturbation of the electromagnetic field is expressed as

$$\delta\mathscr{F}_{ab} + D_c(E_0X^c)\epsilon_{ab}\mathbb{S} = \mathscr{E}\epsilon_{ab}\mathbb{S}, \tag{11.103a}$$

$$\delta\mathscr{F}_{ai} - kE_0\epsilon_{ab}X^b\mathbb{S}_i = r\epsilon_{ab}\mathscr{E}^b\mathbb{S}_i, \tag{11.103b}$$

$$\delta\mathscr{F}_{ij} = 0, \tag{11.103c}$$

with

$$\mathscr{E}_a = \frac{k}{r^{n-1}}D_a\mathscr{A}, \quad r^n\mathscr{E} = -k^2\mathscr{A} + \frac{q}{2}(F_c^c - 2nF). \tag{11.104}$$

Then, the Einstein and Maxwell equations for scalar perturbations of a charge Einstein black hole can be reduced to the following two coupled equations [18]:

$$\omega^2\Phi = -\frac{d^2\Phi}{dr_*^2} + V_s\Phi + \frac{\kappa^2qfP_{S1}}{r^{3n/2}H^2}\mathscr{A}, \tag{11.105a}$$

$$\omega^2\mathscr{A} = -r^{n-2}\frac{d}{dr_*}\left(\frac{1}{r^{n-2}}\frac{d\mathscr{A}}{dr_*}\right) + f\left(\frac{k^2}{r^2}\mathscr{A} + \frac{2n^2(n-1)^2Q^2f}{r^{2n}H}\right)\mathscr{A}$$

$$+ f\frac{(n-1)q}{r^{n/2}}\left(\frac{4H^2 - nP_Z}{4nH}\Phi + fr\partial_r\Phi\right). \tag{11.105b}$$

where V_s. P_{S1} and P_Z are the functions of r (see [18] for their exiplict expressions).

As in the case of vector perturbations, we can find linear combinations of $\mathscr{A}$ and Φ, in terms of which these equations are transformed to the decoupled equations

$$\frac{\omega^2}{f}\Phi_\pm = -(f\Phi'_\pm)' + \frac{V_\pm}{f}\Phi_\pm; \quad V_\pm = \frac{fU_\pm}{64r^2H_\pm^2}, \tag{11.106}$$

Here, $H_+ = 1 - n(n+1)\delta x/2, H_- = m + n(n+1)(1+m\delta)x/2$ and δ is a non-negative constant determined from Q by

$$Q^2 = (n+1)^2M^2\delta(1+m\delta). \tag{11.107}$$

The effective potentials $U_\pm$ can be expressed in terms of $x, \lambda r^2, m$ and δ as follows:

$$
\begin{aligned}
U_+ =& \left[-4n^3(n+2)(n+1)^2\delta^2x^2 - 48n^2(n+1)(n-2)\delta x\right.\\
&\left. - 16(n-2)(n-4)\right]\lambda r^2 - \delta^3 n^3(3n-2)(n+1)^4(1+m\delta)x^4\\
&+ 4\delta^2 n^2(n+1)^2\left\{(n+1)(3n-2)m\delta + 4n^2+n-2\right\}x^3\\
&+ 4\delta(n+1)\left\{(n-2)(n-4)(n+1)(m+n^2K)\delta\right.\\
&\left. - 7n^3 + 7n^2 - 14n + 8\right\}x^2\\
&+ \left\{16(n+1)\left(-4m + 3n^2(n-2)K\right)\delta - 16(3n-2)(n-2)\right\}x\\
&+ 64m + 16n(n+2)K,\\
U_- =& \left[-4n^3(n+2)(n+1)^2(1+m\delta)^2x^2 + 48n^2(n+1)(n-2)m(1+m\delta)x\right.\\
&\left. - 16(n-2)(n-4)m^2\right]y - n^3(3n-2)(n+1)^4\delta(1+m\delta)^3x^4\\
&- 4n^2(n+1)^2(1+m\delta)^2\left\{(n+1)(3n-2)m\delta - n^2\right\}x^3\\
&+ 4(n+1)(1+m\delta)\left\{m(n-2)(n-4)(n+1)(m+n^2K)\delta\right.\\
&\left. + 4n(2n^2 - 3n+4)m + n^2(n-2)(n-4)(n+1)K\right\}x^2 \tag{11.108}
\end{aligned}
$$

$$
\begin{aligned}
&- 16m\left\{(n+1)m\left(-4m + 3n^2(n-2)K\right)\delta\right.\\
&\left. + 3n(n-4)m + 3n^2(n+1)(n-2)K\right\}x\\
&+ 64m^3 + 16n(n+2)m^2K. \tag{11.109}
\end{aligned}
$$

By applying the S-deformation to V_+ with

$$S = \frac{f}{h_+}\frac{dh_+}{dr}; \ h_+ = r^{n/2-1}H_+, \tag{11.110}$$

we obtain

$$\tilde{V}_{S+} = \frac{k^2 f}{2r^2 H_+}\left[(n-2)(n+1)\delta x + 2\right]. \tag{11.111}$$

Since this is positive definite, the electromagnetic mode Φ_+ is always stable for any values of K, M, Q and λ, provided that the spacetime contains a regular black hole, although V_+ has a negative region near the horizon when $\lambda < 0$ and Q^2/M^2 is small.

Using a similar transformation, we can also prove the stability of the gravitational mode Φ_- for some special cases. For example, the S-deformation of V_- with

$$S = \frac{f}{h_-}\frac{dh_-}{dr}; \ h_- = r^{n/2-1}H_- \tag{11.112}$$

leads to

$$\tilde{V}_- = \frac{k^2 f}{2r^2 H_-}\left[2m - (n+1)(n-2)(1+m\delta)x\right]. \tag{11.113}$$

For $n = 2$, this is positive definite for $m > 0$. When $K = 1$, $\lambda \geq 0$ and $n = 3$ or when $\lambda \geq 0, Q = 0$ and the horizon is S^4, from $m \geq n+2$ ($l \geq 2$) and the behaviour of the horizon value of x (see [18] for details), we can show that $\tilde{V}_{S-} > 0$. Hence, in these special cases, the black hole is stable with respect to any type of perturbation.

However, for the other cases, $\tilde{V}_{S-}$ is not positive definite for generic values of the parameters. The S-deformation used to prove the stability of neutral black holes is not effective either. Recently, Konoplya and Zhidenko studied the stability of this system for $n > 2$ numerically. They found that if $\lambda \geq 0$, the system is stable for $n \leq 9$, i.e., $D \leq 11$ [29].

11.4.4 Summary of the Stability Analysis

The results of the stability analysis in this section can be summarised in Table 11.1. In this table, D represents the spacetime dimension, $n + 2$. The results for tensor perturbations apply only for maximally symmetric black holes, while those for vector and scalar perturbations are valid for black holes with generic Einstein horizons, except in the case with $K = 1, Q = 0, \lambda > 0$ and $D = 6$.

Note that this is a summary of the analytic study. As we mentioned above, the stability of AF/dS black hole is shown for $D < 12$ numerically.

Table 11.1 Stability of generalised static black holes

		Tensor	Vector	Scalar	
		$\forall Q$	$\forall Q$	$Q=0$	$Q \neq 0$
$K=1$	$\lambda=0$	OK	OK	OK	D=4,5 OK
					D≥6 ?
	$\lambda>0$	OK	OK	D≤6 OK	D=4,5 OK
				D≥7 ?	D≥6 ?
	$\lambda<0$	OK	OK	D=4 OK	D=4 OK
				D≥5 ?	D≥5 ?
$K=0$	$\lambda<0$	OK	OK	D=4 OK	D=4 OK
				D≥5 ?	D≥5 ?
$K=-1$	$\lambda<0$	OK	OK	D=4 OK	D=4 OK
				D≥5 ?	D≥5 ?

11.5 Flat Black Brane

Static flat black brane solutions are perturbatively unstable in contrast to asymptotically simple static black holes discussed in the previous section. This was first shown by Gregory and Laflamme for the s-mode perturbation, i.e., perturbations that is spherically symmetric in the directions perpendicular to the brane [32, 33]. Later on, it was shown that the system has no other unstable modes numerically [37, 38]. These analyses however assumed that the frequency of an unstable mode, if it exists, is pure imaginary. In the static system this assumption may appear to be natural, but it is not the case in reality. In this section, we explain this point explicitly by applying the gauge-invariant formulation in the previous section to this system.

11.5.1 Strategy

Let us rewrite the $(m+n+2)$-dimensional flat black brane solution

$$ds^2 = -f(r)dt^2 + f(r)^{-1}dr^2 + r^2 d\sigma_n^2 + d\boldsymbol{x}^2, \tag{11.114}$$

which is the product of $(n+2)$-dimensional static black-hole solution and the m-dimensional Euclidean space as

$$ds^2 = g_{ab}(y)dy^a dy^b + r^2 d\sigma_n^2, \tag{11.115}$$

with the $(m+2)$-dimensional metric

$$ds_{m+2}^2 = g_{ab}(y)dy^a dy^b = -f(r)dt^2 + f(r)^{-1}dr^2 + d\boldsymbol{x}^2. \tag{11.116}$$

Then, we can classify metric perturbations into tensor, vector and scalar types with respect to the n-dimensional constant curvature space $\mathscr{K}^n$ with the metric $d\sigma_n^2 = \gamma_{ij}(z)dz^i dz^j$ and apply the gauge-invariant formulation developed in the previous chapter to them. Further, since the background spacetime is homogeneous in the brane direction x, for each type of perturbations, we can apply the Fourier transformation with respect to $x = (x^p)$ to the perturbation variable as

$$\delta g_{\mu\nu} = h_{\mu\nu}(t, r, z^i) e^{ik \cdot x}. \tag{11.117}$$

Since the background metric is static, we can further apply the Fourier transformation with respect to t to $h_{\mu\nu}$ if necessary and assume that

$$h_{\mu\nu} \propto e^{-i\omega t}. \tag{11.118}$$

Hence, we can reduce the Einstein equations for perturbations to a set of ODEs with respect to r. In this section, we assume that $\mathcal{K}^n$ is compact.

11.5.2 Tensor Perturbations

The equation for tensor perturbations (11.32) with $\tau_T = 0$ reads for the present system

$$-\partial_t^2 H_T + \frac{f}{r^n} \partial_r(r^n f \partial_r H_T) - f\left(\frac{k_T^2 + 2K}{r^2} + k^2\right) H_T = 0. \tag{11.119}$$

Let us define the energy integral for a tensor perturbation by

$$E := \int_{r_h}^{\infty} dr\, r^n \left[\frac{1}{f}\dot{H}_T^2 + f(H_T')^2 + \left(\frac{k_T^2 + 2K}{r^2} + k^2\right) H_T^2\right]. \tag{11.120}$$

Then, from the perturbation equation, we have $\dot{E} = 2\left[r^n f \dot{H}_T H_T'\right]_{r_h}^{\infty}$. If there exists an unstable solution $H_T \propto e^{-i\omega t}$ with $\text{Im}\,\omega < 0$, it must fall off exponentially at $r \to \infty$ and vanish at the horizon from the above equation, provided that the solution is uniformly bounded. For such a solution, E becomes constant and contradicts the assumed exponential growth because all terms in the energy integral is non-negative definite. Hence, the black brane solution is stable for tensor perturbations.

11.5.3 Vector Perturbations

11.5.3.1 Basic Perturbation Equations

Basic gauge-invariant variables for vector perturbations are given by $F^a(t, r)$ with $a = t, r, p\ (p = 1, \cdots, m)$. Among these components, we decompose the part parallel to the brane, F_p, into the longitudinal component F_k proportional to the wave vector k^p and the transversal components $F_p^{\perp}$ as

$$F_k = ik^p F_p = \partial_p F^p, \quad F_p^\perp = F_p + \frac{ik_p}{k^2} F_k. \tag{11.121}$$

With this decomposition, the perturbation equations (11.44a) and (11.44a) can be written as the four-wave equations

$$\frac{1}{f}\partial_t^2 F_t - \frac{f}{r^n}\partial_r(r^n\partial_r F_t) + \frac{nf+m_v+r^2k^2}{r^2}F_t = \left(f'-\frac{2f}{r}\right)\partial_t F_r, \tag{11.122a}$$

$$\frac{1}{f}\partial_t^2 F^r - \frac{f}{r^{n-2}}\partial_r(r^{n-2}\partial_r F^r) + \frac{2(n-1)f+m_v+k^2r^2}{r^2}F^r = \frac{f'}{f}\partial_t F_t, \tag{11.122b}$$

$$\frac{1}{f}\partial_t^2 F_k - \frac{1}{r^n}\partial_r(r^n f\partial_r F_k) + \frac{rf'+nf+m_v+k^2r^2}{r^2}F_k = \frac{2k^2}{r}F^r, \tag{11.122c}$$

$$\frac{1}{f}\partial_t^2\left(\frac{F^\perp}{r}\right) - \frac{1}{r^{n+2}}\partial_r\left[r^{n+2}f\partial_r\left(\frac{F^\perp}{r}\right)\right] + \left(k^2+\frac{m_v}{r^2}\right)\left(\frac{F^\perp}{r}\right) = 0 \tag{11.122d}$$

and the constraint

$$-\frac{1}{f}\partial_t F_t + \frac{1}{r^{n-1}}\partial_r(r^{n-1}fF_r) + F_k = 0. \tag{11.123}$$

With the help of this constraint, the second of the above can be also written as

$$\frac{1}{f}\partial_t^2 F^r - \frac{1}{r^{n-2}}\partial_r\left(r^{n-2}f\partial_r F^r\right) + \frac{(n-1)(2f-rf')+m_v+k^2r^2}{r^2}F^r = f'F_k. \tag{11.124}$$

Clearly, the transversal part $F_p^\perp$ decouple from the other modes and each component obeys the same single wave equation. Further, each of (F_t, F^r) and (F^r, F_k) obeys a closed set of equations, and the remaining components F_k and F_t, respectively, are directly determined from them with the help of the above constraint equation.

11.5.3.2 Master Equation

Let us take F^r and F_k as fundamental variables and set

$$\Psi := \begin{pmatrix} r^{n/2}F_k \\ (n+1)r^{n/2-1}F^r + r^{n/2}F_k \end{pmatrix}. \tag{11.125}$$

Then, the perturbation equations can be put into the form

$$\omega^2\Psi = \left(-D^2 + V + fA\right)\Psi, \tag{11.126}$$

where V is the scalar potential

$$V = f\left[\frac{m_v}{r^2} + k^2 + \frac{n(n+2)}{4r^2}f\right], \tag{11.127}$$

and A is the matrix potential

$$A = \begin{pmatrix} \dfrac{2k^2}{n+1} + \dfrac{(n+2)f'}{2r} & -\dfrac{2k^2}{n+1} \\ \dfrac{2k^2}{n+1} & -\dfrac{2k^2}{n+1} - \dfrac{n}{2r}f' \end{pmatrix}. \tag{11.128}$$

In order to see whether this set of equations can be reduced into decoupled single equations, we introduce a new vector variable Φ by

$$\Phi = Q\Psi + P\Psi', \tag{11.129}$$

where P and Q are matrix functions of r that are independent of ω. If we require that Φ obeys the equation of the form

$$\Phi'' + (\omega^2 - V - W)\Phi = 0 \tag{11.130}$$

with a diagonal matrix W, we obtain constraints on V and B.

For the exceptional mode with $m_v = 0$, these constraints are satisfied, and we find that for the choice $P = 1$ and

$$Q = \begin{pmatrix} -\dfrac{k^2 r}{n+1} - \dfrac{n+2}{2r}f & \dfrac{k^2 r}{n+1} \\ -\dfrac{k^2 r}{n+1} & \dfrac{k^2 r}{n+1} + \dfrac{n}{2r}f \end{pmatrix}, \tag{11.131}$$

W is given by the diagonal matrix whose entries are

$$W_1 = \frac{n+2}{r^2}f\left(1 - \frac{(n+1)M}{r^{n-1}}\right), \quad W_2 = -\frac{n}{r^2}f\left(1 - \frac{(n+1)M}{r^{n-1}}\right). \tag{11.132}$$

The corresponding equations for Φ decouple to

$$\Phi_i'' + (\omega^2 - V_i)\Phi_i = 0, \tag{11.133}$$

$$V_1 = f\left[k^2 + \frac{n+2}{4r^2}\left(n+4 - \frac{2(3n+2)M}{r^{n-1}}\right)\right], \tag{11.134}$$

$$V_2 = f\left[k^2 + \frac{n}{4r^2}\left(n-2 + \frac{2nM}{r^{n-1}}\right)\right]. \tag{11.135}$$

V_2 is clearly positive and further, in terms of the S-deformation with $S = (n+2)f/(2r)$, V_1 is transformed into $\tilde{V}_1 = k^2 f > 0$. Hence, this system is stable for this exceptional mode.

If we apply the same transformation in the case $m_v \neq 0$, we obtain

$$\left[(f\partial_r)^2 - \frac{2m_v fh}{r(r^2\omega^2 - m_v f)}f\partial_r + \omega^2 - V_0\right]\Phi = \frac{fh}{(n+1)(r^2\omega^2 - m_v f)}B\Phi, \tag{11.136}$$

where $h = 1 - (n+1)M/r^{n-1}$ and

$$V_0 = f\left[\frac{m_v}{r^2} + k^2 + \frac{n^2 + 2n + 4}{4r^2} - \frac{(n^2 + 4n + 2)M}{2r^{n+1}} + \frac{m_v fh}{r^2(r^2\omega^2 - m_v f)}\right], \tag{11.137}$$

$$B = \begin{pmatrix} (n+1)^2\omega^2 + 2m_v k^2 & -2m_v k^2 \\ 2m_v k^2 & -\{(n+1)^2\omega^2 + 2m_v k^2\} \end{pmatrix} \tag{11.138}$$

. Since B is a constant matrix with eigenvalues

$$\lambda = \pm(n+1)\omega\left[(n+1)^2\omega^2 + 4m_v k^2\right]^{1/2}, \tag{11.139}$$

we can reduce the set of equations for Φ to decoupled single second-order ODEs. However, these equations are not useful in the stability analysis because their coefficients depend on ω^2 nonlinearly and have singularities in general.[1]

11.5.3.3 Stability Analysis

Since we cannot find a convenient master equation, let us try to analyse the stability by directly looking into the structure of the set of equations (11.126). The subtle point of this set of equations is that the operator on the right-hand side is not self-adjoint because A is not a hermitian matrix. Therefore, we cannot directly conclude that ω^2 is real.

Allowing for the possible existence of the imaginary part of ω^2, we obtain the following two integral relations from the above equation:

$$\text{Re}\,(\omega^2)(\Psi,\Psi) = \int_{r_h}^{\infty}\frac{dr}{f}\left[(D\Psi_1)^2 + (D\Psi_2)^2 + fU_1|\Psi_1|^2 + fU_2|\Psi_2|^2\right], \tag{11.140a}$$

$$\text{Im}\,(\omega^2)(\Psi,\Psi) = -\frac{4k^2}{n+1}\int_{r_h}^{\infty}dr\text{Im}\,(\bar{\Psi}_1\Psi_2). \tag{11.140b}$$

---

[1] In [38] the author derived a well-behaved single master equation of second-order for the black string background. There, the author took the gauge in which $f_z = 0$ and $H_T = 0$. Such a gauge cannot be realised in general because the gauge transformations of f_z and H_T are given by $\delta f_z = -S\partial_z(L/S)$ and $\delta H_T = k_v L/S$. If we set $f_z = 0$, we cannot change the z-dependence of H_T in general.

Here, $D = fd/dr$ and

$$U_1 = \frac{m_v}{r^2} + \frac{n+3}{n+1}k^2 + \frac{n(n+2)}{4r^2}f + \frac{(n+2)f'}{2r}, \tag{11.141a}$$

$$U_2 = U_1 - \frac{4k^2}{n+1} - \frac{n+1}{r}\frac{f'}{.} \tag{11.141b}$$

By applying the S-deformation with $S = \dfrac{n}{2r}f$ to Ψ_2, the right-hand side of the equation corresponding to $\mathrm{Re}\,(\omega^2)$ is deformed to

$$D\Psi_2 \rightarrow (D+S)\Psi_2, \quad U_2 \rightarrow \frac{m_v}{r^2} + \frac{n-1}{n+1}k^2. \tag{11.142}$$

Therefore, if we assume that ω^2 is real, as is assumed in most work, we can conclude that the system is stable against vector perturbations. However, we cannot exclude the possible existence of an unstable mode with $\mathrm{Im}\,(\omega^2) \neq 0$.

11.5.4 Scalar Perturbations

11.5.4.1 Perturbation Variables

The gauge-invariant variable set F_{ab} in the general formulation can be decomposed into the scalar, vector and tensor parts by their transformation behaviour with respect to the brane coordinates as

Scalar part: $F_{tt}, F_{tr}, F_{rr}, F_{kt}, F_{kr}, F_{kk}, F_\perp$.
Vector part: $F_{\perp pt}, F_{\perp pr}, F_{\perp pk}$.
Tensor part: $F_{\perp p \perp q}$.

Here,

$$F_{ka} = \partial^p F_{pa}/(ik) = (k^p/k)F_{pa}, \tag{11.143a}$$

$$F_{\perp pa} = F_{pa} - (k_p/k^2)k^q F_{qa} = F_{pa} - (k_p/k)F_{ka}, \tag{11.143b}$$

$$F_{kk} = (k^p k^q/k^2)F_{pq}, \quad F_\perp = F_p^p - F_{kk}, \tag{11.143c}$$

$$F_{\perp p \perp q} = F_{\perp pq} - (k_q/k)F_{\perp pk} - \frac{1}{d-1}F_\perp(\delta_{pq} - k_p k_q/k^2). \tag{11.143d}$$

The remaining gauge-invariant variable F in the general formulation also belongs to the scalar part. Note that the vector and tensor parts do not exist for the black string background.

11.5.4.2 S-Mode

First, we consider the exceptional mode with $k_s = 0$, which is often called the S-mode. For this exceptional mode, the general gauge-invariant variables reduce to $F_{ab} = f_{ab}$ and $F = H_L$ due to the non-existence of corresponding harmonic vectors and tensors. These variables are not gauge invariant and subject to the gauge transformation law

$$\delta H_L = -\frac{f}{r} T_r, \tag{11.144a}$$

$$\bar{\delta} f_{tt} = 2i\omega T_t + f f' T_r, \quad \bar{\delta} f_{tr} = i\omega T_r - f(T_t/f)',$$

$$\bar{\delta} f_{rr} = -2T_r' - (f'/f) T_r, \tag{11.144b}$$

$$\bar{\delta} f_{tk} = i\omega T_k - ik T_t, \quad \bar{\delta} f_{rk} = -T_k' - ik T_r, \quad \bar{\delta} f_{kk} = -2ik T_k \tag{11.144c}$$

$$\bar{\delta} f_{\perp pt} = i\omega T_{\perp p}, \quad \bar{\delta} f_{\perp pr} = -T_{\perp p}', \quad \bar{\delta} f_{\perp pk} = -ik T_{\perp p}, \tag{11.144d}$$

$$\bar{\delta} f_\perp = 0, \quad \bar{\delta} f_{\perp p \perp q} = 0. \tag{11.144e}$$

In particular, we have

$$\bar{\delta} \left(f_{tk} + \frac{\omega}{2k} f_{kk} \right) = -ik T_t, \quad \bar{\delta} \left(f_{rk} + \frac{i}{2k} f_{kk}' \right) = -ik T_r. \tag{11.145}$$

From these, we can construct the following five gauge invariants for the scalar part:

$$r^{2-n} X = -F_\perp, \tag{11.146a}$$

$$r^{2-n} (X - Y) = F_r' - 2r F' - \left(\frac{rf'}{f} - 2 \right) F, \tag{11.146b}$$

$$r^{2-n} Z = F_t^r + \frac{if^2}{2\omega} (F_t^t)' + i\omega r F - \frac{if^2}{2\omega} \left(\frac{rf'}{f} F \right)', \tag{11.146c}$$

$$r^{2-n} V^t = F_k^t + \frac{k}{2\omega} \left(F_t^t - \frac{rf'}{f} F \right) - \frac{\omega}{2kf} F_{kk}, \tag{11.146d}$$

$$r^{2-n} V^r = F_k^r - ik r F + \frac{if}{2k} F_{kk}' \tag{11.146e}$$

For the vector part, we adopt the following two gauge invariants

$$r^{2-n} W_p^t = F_{\perp p}^t - \frac{\omega}{kf} F_{\perp pk}, \quad r^{2-n} W_p^r = F_{\perp p}^r + \frac{if}{k} F_{\perp pk}'. \tag{11.147}$$

Tensor Part

First, we study the stability in the tensor part. The perturbation variable of this part, $F_{\perp p \perp q}$, follows the closed equation

$$-f(r^n f F'_{\perp p \perp q})' + (k^2 f - \omega^2)r^n F_{\perp p \perp q} = 0. \tag{11.148}$$

From this we obtain the integral relation

$$\omega^2 \int_{r_h}^{\infty} dr_* r^n |F_{\perp p \perp q}|^2 = \int_{r_h}^{\infty} dr r^n \left[f |F'_{\perp p \perp q}|^2 + k^2 |F_{\perp p \perp q}|^2 \right]$$

$$- \left[r^n f \bar{F}^{\perp p \perp q} F'_{\perp p \perp q} \right]_{r_h}^{\infty}. \tag{11.149}$$

If there exists an unstable mode with $\omega = \omega_1 + i\omega_2$ ($\omega_2 > 0$), a solution that is bounded at the horizon behaves as $F_{\perp p \perp q} \sim e^{-i\omega r_*}$ near the horizon. Next, at infinity, the solution behaves

$$F_{\perp p \perp q} \sim \frac{1}{r^{(n-1)/2}} Z_\nu(\sqrt{\omega^2 - k^2}\,r) \sim r^{-n/2} \exp(\pm i\sqrt{\omega^2 - k^2}\,r). \tag{11.150}$$

Therefore, for an unstable mode that is uniformly bounded, the boundary term in the above integral relation vanishes and the integral at the left-hand side converges. This implies that $\omega^2 > 0$ and leads to contradiction.

Vector Part

Next, for the vector part, we obtain the following two equations for the gauge-invariant variables W_p^t and W_p^r:

$$-i\omega \left[W'_{\perp p}{}' - \left(\frac{n-2}{r} + \frac{f'}{f} \right) W^t_{\perp p} \right] + \frac{k^2 f - \omega^2}{f^2} W^r_{\perp p} = 0, \tag{11.151a}$$

$$-i\omega W^t_{\perp p} + W^r_{\perp p}{}' + \frac{2}{r} W^r_{\perp p} = 0. \tag{11.151b}$$

Therefore, we can set

$$W^r_{\perp p} = r^{-2}\Phi, \quad i\omega W^t_{\perp p} = r^{-2}\Phi', \tag{11.152}$$

and the perturbation equations can be reduced to the following single master equation for Φ;

$$-f(r^{-n}f\Phi')' + (k^2 f - \omega^2)r^{-n}\Phi = 0. \tag{11.153}$$

By the same argument for the tensor part, we can show that this equation does not have a uniformly bounded solution with $\text{Im}(\omega) > 0$.

Scalar Part

Finally for the scalar part, the perturbation equations gives the closed first-order set
of equations for X, Y, Z, V^t,

$$X' = \frac{1}{k^2 r H f^2}\left[r^2\omega^4 - \omega^2\left\{ k^2 r^2 f + n - n(n+1)x + \frac{3n^2 + 2n - 1}{4}x^2 \right\} \right.$$

$$\left. - \left(2 + \frac{n-5}{2}x \right)k^2 H f \right]X + \frac{1}{k^2 r f H}\left\{ n\omega^2\left(1 - \frac{n+1}{2}x \right) + k^2 H^2 \right\}Y$$

$$+ \frac{2i\omega}{k^2 f^2 H}(n\omega^2 - k^2 H)Z + \frac{\omega}{krfH}\left\{ 2\omega^2 r^2 + (n-1)xH \right\}V^t, \qquad (11.154\text{a})$$

$$Y' = \frac{1}{k^2 r f^2 H}\left[r^2\omega^4 - \omega^2\left\{ 2k^2 r^2 f + n - n(n+1)x + \frac{3n^2 + 2n - 1}{4}x^2 \right\} \right.$$

$$\left. + r^2 k^4 f^2 - \left\{ n - (n^2 + 1)x + \frac{(n+1)^2}{4}x^2 \right\}k^2 f \right]X$$

$$+ \frac{1}{k^2 r f H}\left[n\omega^2\left(1 - \frac{n+1}{2}x \right) + (n-1)k^2\left\{ n - \frac{5n}{2}x + \frac{3(n+1)}{4}x^2 \right\} \right]Y$$

$$+ \frac{2in\omega}{k^2 f^2 H}(\omega^2 - k^2 f)Z + \frac{\omega}{krfH}\left\{ 2r^2\omega^2 - 2k^2 r^2 f + (n-1)xH \right\}V^t,$$

$$(11.154\text{b})$$

$$Z' = - i\frac{(n-1)^2 x}{2\omega r^2}X + \frac{i}{2r^2\omega}\left\{ r^2\omega^2 + (n-1)^2 x \right\}Y$$

$$- \frac{2}{rf}\left(1 - \frac{n+1}{2}x \right)Z + ikfV^t, \qquad (11.154\text{c})$$

$$(V^t)' = \frac{\omega}{2k^3 f^3 H}\left[-r^2\omega^4 + \omega^2\left\{ 2k^2 r^2 f + n - n(n+1)x + \frac{3n^2 + 2n - 1}{4}x^2 \right\} \right.$$

$$\left. - r^2 k^4 f^2 - k^2 f\left\{ n - 2n^2 x + \frac{5n^2 + 2n - 3}{4}x^2 \right\} \right]X$$

$$+ \frac{\omega}{2rk^3 f^2 H}\left[-n\omega^2\left(1 - \frac{n+1}{2}x \right) \right.$$

$$\left. + k^2\left\{ n - n(n+1)x + \frac{3n^2 + 2n - 1}{4}x^2 \right\} \right]Y$$

$$+ \frac{i}{k^3 f^3 H}\left\{ -n\omega^4 + k^2\omega^2\left(2n - \frac{3n+1}{2}x \right) - k^4 f H \right\}Z$$

$$+ \frac{1}{rk^2 f^2 H}\left\{ -r^2\omega^4 + \omega^2\left(k^2 r^2 f - \frac{n-1}{2}xH \right) \right.$$

$$+ (n-2)k^2 f^2 H \right\}V^t. \qquad (11.154\text{d})$$

and the expression for V^r in terms of these quantities,

$$V^r = \frac{i(\omega^2 + k^2 f)}{2nk^3 r f^2} \left[((\omega^2 + k^2 f)r^2 + nf^2)X + nf^2Y \right.$$

$$\left. + \frac{2in\omega^3 r}{\omega^2 + k^2 f}Z + 2\omega k r^2 f V^t \right]. \tag{11.155}$$

Here, $x = 2M/r^{n-1}$.

From these equations, we find that X obeys the closed second-order ODE

$$-f(fX')' + (n-4)\frac{f^2}{r}X'$$

$$+ \left[-\omega^2 + f\left(k^2 + \frac{n-2}{r^2}\{1 + (n-2)x\} \right) \right] X = 0, \tag{11.156}$$

which can be put into the canonical form in terms of Φ defined by

$$X = r^{n/2-2}\Phi \tag{11.157}$$

as

$$-f(f\Phi')' + \left[-\omega^2 + f\left\{ k^2 + \frac{n}{4r^2}(n-2+nx) \right\} \right] \Phi = 0. \tag{11.158}$$

It is clear that this equation does not have an unstable mode.

Next, let us define the new variable Ω by

$$\Omega := PX + nf\left(1 - \frac{n+1}{2}x \right)Y + 2in\omega rZ + 2k\omega r^2 f V^t; \tag{11.159}$$

$$P := \left[\frac{n+1}{2n}x - \frac{(n-1)x}{2k^2 r^2}\left(n - \frac{n+1}{2}x \right) \right] \omega^2 r^2 + \frac{n-1}{2n}xk^2 r^2$$

$$-n + n(n+1)x - (3n^2 + 2n - 1)x^2, \tag{11.160}$$

Then, we find that Ω satisfies a closed second-order ODE mod $X = 0$:

$$-f(f\Omega')' + Af\Omega' + (-\omega^2 + V_\Omega)\Omega = BX, \tag{11.161}$$

where

$$A = \frac{f}{4r^3 gH} \left[\{4n^2 + 2(n+1)(n-2)x\}k^2 r^2 \right.$$

$$\left. + n^2(n-1)x\{3(n+1)x - 2(n+2)\} \right], \tag{11.162a}$$

$$V_\Omega = \frac{f}{8r^4 g^2 H} \left[2\{2n - (n+1)x\}^2 k^4 r^4 \right.$$

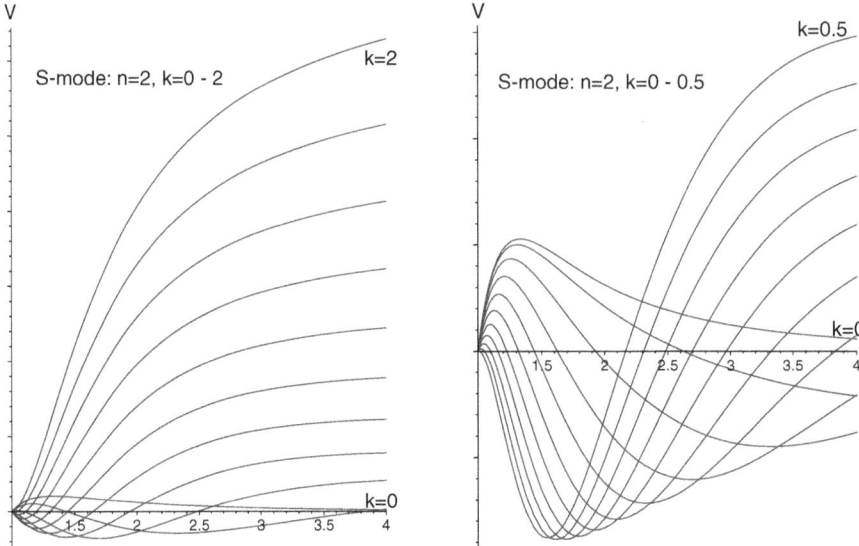

Fig. 11.6 The effective potential for S-modes

$$+ \left\{ 8n^2(n+2) - 4n(n+2)(3n^2+n+2)x \right.$$

$$+ 2n(n+1)(8n^2+5n+5)x^2 - (n+2)(3n-1)(n+1)^2x^3 \right\} k^2 r^2$$

$$+ n^2(n-1)x \left\{ n(n+1)^2x^3 - 3(3n-1)(n+1)x^2 \right.$$

$$\left. + 4(2n^2+2n-1)x - 4n^2 \right\} \right], \tag{11.162b}$$

$$B = \frac{fg}{nr^2H} \left\{ (n+1)\omega^2 - k^2 \right\} \left\{ -2k^2r^2(1-nx) \right.$$

$$\left. + n(n-1)x(n-x) \right\}, \tag{11.162c}$$

$$H := k^2 + \frac{n(n-1)}{2r^2}x, \quad g := n - \frac{n+1}{2}x. \tag{11.162d}$$

By the transformation

$$\Omega = r^{n/2}gH\Psi, \tag{11.163}$$

we can put this equation into the canonical form

$$-f(f\Psi')' + (-\omega^2 + V)\Psi = r^{n/2}gHBX, \tag{11.164}$$

where $V = fU/H^2$ with

$$U = k^6 + \frac{(n+4)k^4}{4r^2}(n+2-3nx) - \frac{n(n-1)k^2}{4r^4}\Big\{3n(n+2)$$

$$-(2n^2+3n+4)x\Big\}x + \frac{n^3(n-1)^2}{16r^6}x^2(n-2+nx). \qquad (11.165)$$

This potential has a deep negative region for $0 < k < k_n$ with some constant k_n dependent on n. It has been shown by numerical calculations [37, 38] that the eigenvalue ω^2 becomes negative for some range $0 < k < k_c$, as first pointed out by Gregory and Laflamme using a different reduction [32].

11.5.4.3 Generic Scalar Perturbation

Tensor Part

The tensor part of generic scalar-type perturbations obeys the decoupled second-order ODE

$$-f(r^n f F'_{\perp p \perp q})' + \Big(-\omega^2 + k^2 f + \frac{n+m}{r^2}f\Big)r^n F_{\perp p \perp q} = 0. \qquad (11.166)$$

It is obvious that this equation has no unstable mode.

Vector Part

In terms of the gauge-invariant fundamental variables

$$V_t = r^{n-2}F_{t \perp p}, \quad V_r = r^{n-2}F_{r \perp p}, \quad V_k = r^{n-2}F_{k \perp p}, \qquad (11.167)$$

the perturbation equations for the vector part are expressed as

$$-(fV_r)' - i\omega f^{-1}V_t - ikV_k = 0, \qquad (11.168a)$$

$$i\frac{\omega}{f}\Big(V_t' - \frac{n-2}{r}fV_t\Big) + \Big(-\frac{\omega^2}{f} + k^2 + \frac{n+m}{r^2}\Big)V_r$$

$$+ik\Big(V_k' - \frac{n-2}{r}V_k\Big) = 0, \qquad (11.168b)$$

$$-r^{n-4}(r^{4-n}fV_k')' + \Big(-\frac{\omega^2}{f} + \frac{n+m}{r^2} + \frac{n-2}{r}f' + \frac{n-2}{r^2}f\Big)V_k$$

$$-\frac{k\omega}{f}V_t + ik\Big((fV_r)' + \frac{2f}{r}V_r\Big) = 0, \qquad (11.168c)$$

$$-r^{n-4}f(r^{4-n}V_t')' + \left(\frac{m+n}{r^2} + \frac{n-2}{r^2}\right)V_t - i\omega f\left(V_r' + \frac{2}{r}V_r\right)$$
$$+k\omega V_k = 0. \tag{11.168d}$$

By eliminating V_t and introducing the new variables Y and Z by

$$\Phi = \begin{pmatrix} Z \\ Y \end{pmatrix}; \quad V_k = -ir^{n/2-2}Z, \quad V_r = f^{-1}r^{n/2-1}Y, \tag{11.169}$$

this set of equations are reduced to a set of two ODEs,

$$D^2\Phi - \left\{-\omega^2 + k^2 f + \frac{n+m}{r^2}f + \frac{n(n-2)}{4}f^2\right\}\Phi = A\Phi; \tag{11.170}$$

$$A = \begin{pmatrix} \dfrac{nf'}{2r}f & -2kf \\ -\dfrac{kf'}{r}f & -\dfrac{(n-2)f'}{2r}f \end{pmatrix}. \tag{11.171}$$

This set of equations has the same structure as that for vector perturbations and can be shown to have no unstable mode if ω^2 is real.

Scalar Part

Finally, we discuss the scalar part of the generic scalar-type perturbation. Utilising one of the Einstein equations

$$E_T \equiv 2(n-2)F + F_a^a = 0, \tag{11.172}$$

the basic perturbation variables can be expressed in terms of X, Y, Z, V^t, V^r, S and Ψ as

$$\tilde{F}_t^t = X + 2\tilde{F} - k^2 fV^t, \quad \tilde{F}_r^r = Y + 2\tilde{F}, \quad \tilde{F}_t^r = i\omega Z, \tag{11.173a}$$
$$\tilde{F}_k^r = ikV^r, \quad \tilde{F}_k^t = \omega kV^t, \quad \tilde{F}_{kk} = S + \omega^2 V^t + 2\tilde{F}, \tag{11.173b}$$
$$2(n+1)\tilde{F} = -\Psi - X - Y - S - (\omega^2 - k^2 f)V^t, \quad \tilde{F}_\perp = \Psi. \tag{11.173c}$$

Here, $\tilde{Q} = r^{n-2}Q$ in general.

In terms of these variables, the Einstein equations can be reduced to the decoupled single equation for Ψ,

$$-r^{-n}f(r^n f\Psi')' + \left[-\omega^2 + \left(k^2 + \frac{n+m}{r^2}\right)f\right]\Psi = 0. \tag{11.174}$$

and the regular first-order set of ODEs for X, Y, Z, V^t, V^r and S,

$$Z' = X, \tag{11.175a}$$

$$X' = \frac{n-2}{r}X + \left(\frac{f'}{f} - \frac{2}{r}\right)Y + \frac{1}{f}\left(-\frac{\omega^2}{f} + k^2 + \frac{m+n}{r^2}\right)Z$$

$$+k^2 f'V', \tag{11.175b}$$

$$Y' = \frac{f'}{2f}(X - Y) + \frac{\omega^2}{f^2}Z + \frac{k^2}{f}\left(V^r - \frac{ff'}{2}V'\right), \tag{11.175c}$$

$$(V^r)' = -S, \tag{11.175d}$$

$$S' = \frac{n-2}{r}S - \frac{2}{r}Y + \omega^2\frac{f'}{f}V' + \frac{1}{f}\left(\frac{\omega^2}{f} - k^2 - \frac{n+m}{r^2}\right)V^r \tag{11.175e}$$

$$k^2 r^2 f' f^2 (V')' = \left[2\omega^2 r^2 + (n-1)x\left(n - \frac{n+1}{2}x\right)\right]X$$

$$+ \left[2\omega^2 r^2 - 2(k^2 r^2 + n + m)f + 2n - 4nx + \frac{(n+1)^2}{2}x^2\right]Y$$

$$+ \frac{1}{r}\left[-2n\omega^2 r^2 + (n-1)x(k^2 r^2 + n + m)\right]Z$$

$$- (n-1)k^2 x\left(2 + \frac{n-5}{2}x\right)fV' - 2k^2 r\left(n - \frac{n+1}{2}x\right)V^r$$

$$- 2k^2 r^2 fS, \tag{11.175f}$$

where $x = 2M/r^{n-1}$.

If we define X_1, X_2, X_3 by

$$X_1 = Z, \quad X_2 = r(X + Y) - nZ + k^2 rfV',$$

$$X_3 = -\left(1 + \frac{rf'}{2nf}\right)[r(X + Y) - nZ] - \frac{k^2 r^2}{2n}f'V', \tag{11.176}$$

and introduce Φ by

$$\Phi := \begin{pmatrix} r^{-n/2}X_1 \\ r^{-n/2+1}f^{-1}X_2 \\ r^{-n/2}X_3 \end{pmatrix}, \tag{11.177}$$

we can reduce the above set of first-order ODEs to the set of second-order ODEs of the normal eigenvalue type as

$$\omega^2 \Phi = (-D^2 + V_0 + W)\Phi. \tag{11.178}$$

Here, V_0 is the scalar potential

$$V_0 = \frac{f}{4r^2}\left[4(m + k^2 r^2) + n^2 - 2n + n(n+4)x\right], \tag{11.179}$$

and W is the following matrix of rank 3:

$$W_{11} = 0, \quad W_{12} = \frac{(n^2 - 1)xf^2}{nr^3}, \quad W_{13} = \frac{2f^2}{r^2}, \tag{11.180a}$$

$$W_{21} = \frac{\{4 - 2(n+1)x\} k^2 r^2 - 2(n-1)mx - n(n^2 - 1)x}{rf}, \tag{11.180b}$$

$$W_{22} = \frac{nf^2}{r^2}, \quad W_{23} = 0, \tag{11.180c}$$

$$W_{31} = \frac{1}{2nr^2 f} \left[2(n-1)x(n-2+x)r^2 k^2 + \{4n + 2n(n-5)x + 2(n+1)x^2\} m \right.$$
$$\left. + n(n+1)x \{2n^2 - (2n^2 + 3n - 1)x + n(n+1)x^2\} \right], \tag{11.180d}$$

$$W_{32} = \frac{(n^2 - 1)x \{2 - (n+1)x\} f}{2nr^3}, \quad W_{33} = \frac{(n+1) \{2 - (n+1)x\} f}{r^2}. \tag{11.180e}$$

Unfortunately, it is not possible to analyse the stability of this second-order system by an analytic method, partly because it is not a self-adjoint system. However, all numerical calculations done by various authors have found no evidence of instability for this system [37, 38].

11.6 Summary and Discussion

In this lecture, we have explained the gauge-invariant formulation for perturbations of a class of background solutions to the Einstein equations that include various practically important spacetimes as special cases. Then, we have illustrated its power by applying it to the stability problem of static black holes in higher dimensions and flat black branes.

These two systems have one important common feature in addition to staticity that the background spacetime is of the cohomogeneity one. That is, the isotropy group of the spacetime has orbits with codimension one, and roughly speaking, the spacetime is inhomogeneous only in one direction, say r. In this case, the perturbation equations for the system can be automatically reduced to a set of ODEs for functions of r with the help of the harmonic expansion. This applies to a rotating black hole case as well [54].

There exists however one crucial difference between the two systems. In the static black-hole case, the perturbation equations can be reduced to decoupled single second-order ODE, and the stability problem is formulated as an eigenvalue problem for the corresponding self-adjoint operator. In contrast, in the black brane case, it appears to be impossible to reduce all the perturbation equations to decoupled single master equations. Further, the eigenvalue problem for the stability issue cannot be put in the self-adjoint form even if we allow for a multi-component expression, except for some special modes. Nevertheless, numerical calculations indicate that there exists no eigen-mode with an imaginary frequency [37, 38]. There must be a profound reason behind this result.

The gauge-invariant formulation developed in Sect. 11.3 can be also applied to spacetimes whose cohomogeneity dimension is greater than one. Although this implies that the formulation can be applied to perturbations of rotating black holes in higher dimensions such as the Myers–Perry solution [55] and its generalisation to non-vanishing cosmological constant [56], it may not be practically useful in most case, because we obtain a couple set of partial differential equations in a reduced spacetime with dimensions smaller than the original one. However, in some special cases, we obtain a single PDE that is separable to ODEs. For example, for a Kerr(-AdS) black hole that rotates in a two-dimensional plane, we can classify perturbations into tensor, vector and scalar types as in the static case, and among these, the perturbation equation for the tensor-type perturbation is separable [43]. There may exist other cases in which similar phenomena happen.

Acknowledgments   The author thanks the organisers of the Fourth Aegean Summer School, especially Prof. Elefteris Papantonopoulos for giving him the chance to deliver a lecture at the beautiful island Lesvos. This work is partly supported by Grants-in-Aid for Scientific Research from JSPS (No. 18540265).

References

1. T. Regge and J. Wheeler, Phys. Rev. **108**, 1063–1069 (1957).
2. F. Zerilli, Phys. Rev. Lett. **24**, 737–738 (1970).
3. S. Teukolsky, Phys. Rev. Lett. **29**, 1114–1118 (1972).
4. C. Vishveshwara, Phys. Rev. D **1**, 2870–2879 (1970).
5. R. Price, Phys. Rev. D **5**, 2419–2438 (1972).
6. S. Chandrasekhar, *The Mathematical Theory of Black Holes* (Clarendon Press, New York 1983).
7. M. Heusler, *Black Hole Uniqueness Theorems* (Cambridge University Press, Cambridge 1996).
8. R. Emparan, JHEP **03**, 064 (2004).
9. H. Kodama, J. Korean Phys. Soc. **45**, S68–76 (2004).
10. B. Kol, Phys. Rep. C **422**, 119–165 (2006).
11. H. Elvang, R. Emparan and P. Figueras, JHEP **0705**, 056 (2007).
12. S. Hollands and S. Yazadjiev, arXiv:0711.1722 [gr-qc] (2007).
13. Y. Morisawa, S. Tomizawa and Y. Yasui, arXiv:0710.4600 [hep-th] (2007).
14. H. Kodama, Prog. Theor. Phys. **112**, 249–274 (2004).
15. R. M. Wald, J. Math. Phys. **20**, 1056–1058 (1979).
16. R. M. Wald, J. Math. Phys. **21**, 218 (1980).
17. A. Ishibashi and H. Kodama, Prog. Theor. Phys. **110**, 901–919 (2003).
18. H. Kodama and A. Ishibashi, Prog. Theor. Phys. **111**, 29–73 (2004).
19. B. Whiting, J. Math. Phys. **30**, 1301–1305 (1989).
20. M. Heusler, S. Droz and N. Straumann, Phys. Lett. B **271**, 61–67 (1991).
21. M. Heusler, S. Droz and N. Straumann, Phys. Lett. B **285**, 21–26 (1992).
22. M. Heusler, N. Straumann and Z. -H. Zhou, Helv. Phys. Acta **66**, 614–632 (1993).
23. N. Straumann and Z. -H. Zhou, Phys. Lett. B **237**, 353–356 (1990).
24. Zhou, Z. -H. and N. Straumann, Nucl. Phys. B **360**, 180–96 (1991).
25. V. Cardoso, O. Dias and S. Yoshida, Phys. Rev. D **74**, 044008 (2006).
26. V. Cardoso and J. P. S. Lemos, Phys. Rev. D **64**, 084017 (2001).
27. S. Hawking and H. Reall, Phys. Rev. D **61**, 024014 (1999).
28. V. Cardoso and O. Dias, Phys. Rev. D **70**, 084011 (2004).
29. R. Konoplya and A. Zhidenko, Nucl. Phys. B **777**, 182–202 (2007).

30. R. Gregory and R. Laflamme, Nucl. Phys. B **428**, 399–434 (1994).
31. T. Hirayama, G. Kang and Y. Lee, Phys. Rev. D **67**, 024007 (2003).
32. R. Gregory and R. Laflamme, Phys. Rev. Lett. **70**, 2837–2840 (1993).
33. R. Gregory and R. Laflamme, Phys. Rev. D **51**, 305–309 (1995).
34. R. Gregory, Class. Quantum Grav. **17**, L125–32 (2000).
35. T. Hirayama and G. Kang, Phys. Rev. D **64**, 064010 (2001).
36. G. Kang, J. Korean Phys. Soc. **45**, S86–S89 (2004).
37. S. Seahra, C. Clarkson and R. Maartens, Phys. Rev. Lett. **94**, 121302 (2005).
38. H. Kudoh, Phys. Rev. D **73**, 104034 (2006).
39. Y. Brihaye, T. Delsate and E. Radu, arXiv: 0710.4034[hep-th] (2007).
40. H. Kunduri, J. Lucietti and H. Reall, Phys. Rev. D **74**, 084021 (2006).
41. V. Cardoso, O. Dias, J. Lemos and S. Yoshida, Phys. Rev. D **70**, 044039 (2004).
42. V. Cardoso and S. Yoshida, JHEP **0507**, 009 (2005).
43. H. Kodama, arXiv:0711.4184 [hep-th] (2007).
44. O. Dias, Phys. Rev. D **73**, 124035 (2006).
45. T. Harmark, V. Niarchos and N. Obers, hep-th/0701022 (2007).
46. R. Emparan and R. C. Myers, JHEP **0309**, 025 (2003).
47. J. Bardeen, Phys. Rev. D **22**, 1882–905 (1980).
48. H. Kodama and M. Sasaki, Prog. Theor. Phys. Suppl. **78**, 1–166 (1984).
49. H. Kodama, Prog. Theor. Phys. **71**, 946–959 (1984).
50. H. Kodama, Prog. Theor. Phys. **73**, 674–682 (1985).
51. S. Mukohyama, Phys. Lett. B **473**, 241 (2000).
52. H. Kodama, A. Ishibashi and O. Seto, Phys. Rev. D **62**, 064022 (2000).
53. H. Kodama and A. Ishibashi, Prog. Theor. Phys. **110**, 701–722 (2003).
54. K. Murata and J. Soda, arXiv:0710.0221[hep-th] (2007).
55. R. Myers and M. Perry, Ann. Phys. **172**, 304–347 (1986).
56. G. W. Gibbons, H. Lü, D. N. Page and C. N. Pope, J. Geom. Phys. **53**, 49–73 (2005).

Chapter 12
Analytic Calculation of Quasi-Normal Modes

G. Siopsis

Abstract We discuss the analytic calculation of quasi-normal modes of various types of perturbations of black holes both in asymptotically flat and in anti-de Sitter spaces. We obtain asymptotic expressions and also show how corrections can be calculated perturbatively. We pay special attention to low-frequency modes in anti-de Sitter space because they govern the hydrodynamic properties of a gauge theory fluid according to the AdS/CFT correspondence. The latter may have experimental consequences for the quark-gluon plasma formed in heavy ion collisions.

12.1 Introduction

To many practitioners of quantum gravity the black hole plays the role of a soliton, a non-perturbative field configuration that is added to the spectrum of particle-like objects only after the basic equations of their theory have been put down, much like what is done in gauge theories of elementary particles, where Yang–Mills equations with small coupling constants determine the small-distance structure, and solitons and instantons govern the large-distance behavior.

Such an attitude however is probably not correct in quantum gravity. The coupling constant increases with decreasing distance scale which implies that the smaller the distance scale, the stronger the influences of "solitons". At the Planck scale it may well be impossible to disentangle black holes from elementary particles.

– G. 't Hooft

Quasi-normal modes (QNMs) describe small perturbations of a black hole which is a thermodynamical system whose (Hawking) temperature and entropy are given in terms of its global characteristics (total mass, charge and angular momentum). They are obtained by solving a wave equation for small fluctuations subject to the conditions that the flux be ingoing at the horizon and outgoing at asymptotic infinity. These boundary conditions in general lead to a discrete spectrum of complex frequencies whose imaginary part determines the decay time of the small fluctuations

G. Siopsis (✉)
Department of Physics and Astronomy, The University of Tennessee, Knoxville,
TN 37996-1200, USA
siopsis@tennessee.edu

Siopsis G.,: *Analytic Calculation of Quasi-Normal Modes.* Lect. Notes Phys. **769**, 471–508 (2009)
DOI 10.1007/978-3-540-88460-6_12   © Springer-Verlag Berlin Heidelberg 2009

$$\text{Im}\,\omega = \frac{1}{\tau}.$$ (12.1)

There is a vast literature on quasi-normal modes and we make no attempt to review it. Instead, we concentrate on obtaining analytic expressions for quasi-normal modes of various black-hole perturbations of interest. One can rarely obtain analytic expressions in closed form. Instead, we discuss techniques which allow one to calculate the spectrum perturbatively starting with an asymptotic regime (e.g., high or low overtones). In asymptotically flat space, we discuss the cases of four-dimensional Schwarzschild and Kerr black holes. Generalization to higher-dimensional spacetimes does not present substantially new calculational challenges. However, we should point out that the case of a rotating black hole is considerably harder than the Schwarzschild case.

We also discuss asymptotically AdS spaces and obtain the spectrum as a perturbative expansion around high overtones. At leading order the frequencies are proportional to the radius of the horizon. When expanding around low overtones, one in general obtains an additional frequency which is inversely proportional to the horizon radius. Thus for large black holes there is a gap between the lowest frequency and the rest of the spectrum of quasi-normal modes. We pay special attention to the lowest frequencies because they govern the behavior of the gauge theory fluid on the boundary per the AdS/CFT correspondence. The latter may have experimental consequences pertaining to the formation of the quark-gluon plasma in heavy ion collisions.

12.2 Flat Spacetime

We start with a study of QNMs in asymptotically flat spacetimes. We discuss scalar perturbations of Schwarzschild and Kerr black holes in four dimensions.

12.2.1 Schwarzschild Black Holes

The metric of a Schwarzschild black hole in four dimensions is

$$ds^2 = -f(r)\,dt^2 + \frac{dr^2}{f(r)} + r^2 d\Omega^2\,,\quad f(r) = 1 - \frac{2GM}{r}\,.$$ (12.2)

The Hawking temperature is

$$T_H = \frac{1}{8\pi GM} = \frac{1}{4\pi r_0}\,,$$ (12.3)

where $r_0 = 2GM$ is the radius of the horizon.

A spin-j perturbation of frequency ω is governed by the radial equation

$$-f(r)\frac{d}{dr}\left(f(r)\frac{d\Psi}{dr}\right)+V(r)\Psi = \omega^2\Psi\,, \tag{12.4}$$

where $V(r)$ is the "Regge–Wheeler" potential

$$V(r) = f(r)\left(\frac{\ell(\ell+1)}{r^2}+\frac{(1-j^2)r_0}{r^3}\right)\,. \tag{12.5}$$

The spin is $j = 0, 1, 2$ for scalar, electromagnetic and gravitational perturbations, respectively. It is advantageous to avoid integer values of j throughout the discussion and only take the limit $j \to$ integer at the end of the calculation.

By defining the "tortoise coordinate"

$$r_* = \int \frac{dr}{f(r)} = r + r_0 \ln\left(\frac{r}{r_0}-1\right)\,, \tag{12.6}$$

the wave equation may be brought into a Schrödinger-like form,

$$-\frac{d^2\Psi}{dr_*^2} + V(r(r_*))\Psi = \omega^2\Psi \tag{12.7}$$

to be solved along the entire real r_*-axis. At both ends the potential vanishes ($V \to 0$ as $r_* \to \pm\infty$) therefore the solutions behave as $\Psi \sim e^{\pm i\omega r_*}$. For QNMs, we demand

$$\Psi \sim e^{\mp i\omega r_*}\,,\quad r_* \to \pm\infty\,. \tag{12.8}$$

assuming $\mathrm{Re}\,\omega > 0$.

12.2.1.1 Limit $\ell \to \infty$

In this case it suffices to consider the potential near its maximum. Expanding around the maximum of the potential ($V_0'(r_{max}) = 0$) [1],

$$r_{max} = \frac{3}{2}r_0 + \mathcal{O}(1/\ell)\,, \tag{12.9}$$

we obtain

$$V_0[r(r_*)] \approx \alpha^2 - \beta^2(r_* - r_*(r_{max}))^2\,, \tag{12.10}$$

where

$$\alpha^2 = \frac{4}{27}\left(\ell+\frac{1}{2}\right)r_0^2 + \mathcal{O}(1/\ell)\,,\quad \beta^2 = \frac{16}{729}\left(\ell+\frac{1}{2}\right) + \mathcal{O}(1/\ell)\,. \tag{12.11}$$

The solutions to the wave equation are

$$\Psi_n = H_n(\sqrt{i\beta}x)e^{i\beta x^2/2} \ , \quad n = 0,1,2,\dots \tag{12.12}$$

where H_n are Hermite polynomials. The corresponding eigenvalues are

$$\omega_n = \frac{2}{3\sqrt{3}\,r_0}\left\{\ell + \frac{1}{2} + i\left(n + \frac{1}{2}\right)\right\} + \mathcal{O}(1/\ell) \ . \tag{12.13}$$

This result is in agreement with the standard WKB approach [2].

12.2.1.2 Limit $n \to \infty$

The asymptotic form of QNMs for large n is

$$\frac{\omega_n}{T_H} = (2n+1)\pi i + \ln 3 \tag{12.14}$$

independent of the angular momentum quantum number ℓ. This form was first derived numerically [3–7] and subsequently confirmed analytically [8]. The large imaginary part of the frequency (Im ω_n) makes the numerical analysis cumbersome but is easy to understand because the spacing of frequencies is $2\pi i T_H$ which is the same as the spacing of poles of a thermal Green function on the Schwarzschild black-hole background. On the other hand the real part (Re ω_n) is small. Its analytical value was first proposed by Hod [9].

The analytical derivation of the asymptotic form (12.14) of QNMs by Motl and Neitzke [8] offered a new surprise because it heavily relied on the black-hole singularity. It is intriguing that the unobservable region beyond the horizon influences the behavior of physical quantities.

We shall calculate the asymptotic formula for QNMs including first-order corrections [10] by solving the wave equation perturbatively for arbitrary spin of the wave. We shall obtain agreement with results from numerical analysis for gravitational and scalar waves [5, 11] and WKB analysis for gravitational waves [12].

Let

$$\Psi = e^{-i\omega r_*}f(r_*) \ . \tag{12.15}$$

We have $f(r_*) \sim 1$ as $r_* \to +\infty$ and near the horizon, $f(r_*) \sim e^{2i\omega r_*}$ (as $r_* \to -\infty$). Let us continue r analytically into the complex plane and define the boundary condition at the horizon in terms of the monodromy of $f(r_*(r))$ around the singular point $r = r_0$,

$$\mathcal{M}(r_0) = e^{-4\pi\omega r_0} \tag{12.16}$$

along a contour running counterclockwise. We may deform the contour in the complex r-plane so that it either lies beyond the horizon (Re$r < r_0$) or at infinity ($r \to \infty$). The monodromy only gets a contribution from the segment lying beyond the horizon.

It is convenient to change variables to

$$z = \omega(r_* - i\pi r_0) = \omega(r + r_0 \ln(1 - r/r_0)) \,, \tag{12.17}$$

(where we chose a branch such that $z \to 0$ as $r \to 0$). The potential can be written as a series in $\sqrt{z}$,

$$V(z) = -\frac{\omega^2}{4z^2}\left(1 - j^2 + \frac{3\ell(\ell+1)+1-j^2}{3}\sqrt{-\frac{2z}{\omega r_0}} + \cdots\right) \tag{12.18}$$

which is a formal expansion in $1/\sqrt{\omega}$.

Now deform the contour defining the monodromy so that it gets mapped onto the real axis in the z-plane. Near the singularity $z = 0$,

$$z \approx -\frac{\omega}{2r_0}r^2 \,. \tag{12.19}$$

Choose a contour in the r-plane so that near $r = 0$, the positive and negative real axes in the z-plane are mapped onto

$$\arg r = \pi - \frac{\arg\omega}{2} \,, \quad \arg r = \frac{3\pi}{2} - \frac{\arg\omega}{2} \,, \tag{12.20}$$

in the r-plane, respectively. These segments form a $\pi/2$ angle (independent of $\arg\omega$).

To avoid the $r = 0$ singularity, go around an arc of angle $3\pi/2$ which corresponds to an angle of 3π around $z = 0$ in the z-plane.

Considering the black-hole singularity ($r = 0$), we note that there are two solutions,

$$f_\pm(r) = r^{1\pm j}Z_\pm(r) \,, \tag{12.21}$$

where $Z_\pm$ are analytic functions of r. Going around an arc of angle of $3\pi/2$, we obtain

$$f_\pm(e^{3\pi i/2}r) = e^{3\pi(1\pm j)i/2} f_\pm(r), \tag{12.22}$$

which is an *exact* result.

To proceed further, we need to relate the behavior of the wavefunction near the black-hole singularity to its behavior at large r in the complex r-plane. To this end, we shall solve the wave equation perturbatively, thus writing the wavefunction as a perturbation series in $1/\sqrt{\omega}$.

At zeroth order, the wave equation reads

$$\frac{d^2\Psi^{(0)}}{dz^2} + \left(\frac{1-j^2}{4z^2} + 1\right)\Psi^{(0)} = 0 \,. \tag{12.23}$$

Two linearly independent solutions are

$$f_\pm^{(0)}(z) = e^{iz}\Psi_\pm^{(0)} = e^{iz}\sqrt{\frac{\pi z}{2}}J_{\pm j/2}(z) \tag{12.24}$$

in terms of Bessel functions. We deduce the behavior at infinity ($z \to \infty$)

$$f_{\pm}^{(0)}(z) \sim e^{iz}\cos(z - \pi(1 \pm j)/4) \ . \tag{12.25}$$

The boundary conditions imply $f(z) \sim$ const. as $z \to \infty$ along the positive real axis in the z-plane. Therefore, we ought to adopt the linear combination

$$f^{(0)} = f_{+}^{(0)} - e^{-\pi ji/2} f_{-}^{(0)} \sim e^{iz}\sqrt{z}H_{j/2}^{(1)}(z) \tag{12.26}$$

(in terms of a Hankel function). As $z \to \infty$, we obtain

$$f^{(0)}(z) \sim -e^{-\pi(1+j)i/4}\sin(\pi j/2) \tag{12.27}$$

a constant, as desired.

Going along the 3π arc around $z = 0$ in the z-plane, we have

$$f^{(0)}(e^{3\pi i}z) = e^{3\pi(1+j)i/2}\left(f_{+}^{(0)}(z) - e^{-7\pi ji/2} f_{-}^{(0)}(z)\right) \ . \tag{12.28}$$

As $z \to \infty$,

$$f^{(0)}(z) \sim e^{-\pi(1+j)i/4}\sin(3\pi j/2) + e^{\pi(1-j)i/4}\sin(2\pi j)e^{2iz} \ . \tag{12.29}$$

The monodromy to zeroth order is

$$\mathscr{M}(r_0) = -\frac{\sin(3\pi j/2)}{\sin(\pi j/2)} = -(1 + 2\cos(\pi j)) \ , \tag{12.30}$$

leading to a discrete set of complex frequencies (QNMs) [8]

$$\frac{\omega_n}{T_H} = (2n+1)\pi i + \ln(1 + 2\cos(\pi j)) + \mathscr{O}(1/\sqrt{n}) \ . \tag{12.31}$$

Next, we calculate the first-order correction to the above expression [10]. Expanding the wavefunction in $1/\sqrt{\omega}$,

$$\Psi = \Psi^{(0)} + \frac{1}{\sqrt{-\omega r_0}}\Psi^{(1)} + \mathscr{O}(1/\omega) \tag{12.32}$$

the first-order correction obeys

$$\frac{d^2\Psi^{(1)}}{dz^2} + \left(\frac{1-j^2}{4z^2} + 1\right)\Psi^{(1)} = \sqrt{-\omega r_0}\,\delta V \Psi^{(0)} \ , \tag{12.33}$$

where

$$\delta V(z) = \frac{1-j^2}{4z^2} + \frac{1}{\omega^2}V[r(z)] \ . \tag{12.34}$$

Two linearly independent solutions are

$$\Psi_{\pm}^{(1)}(z) = \mathscr{C}\Psi_{+}^{(0)}(z)\int_0^z \Psi_{-}^{(0)}\delta V \Psi_{\pm}^{(0)} - \mathscr{C}\Psi_{-}^{(0)}(z)\int_0^z \Psi_{+}^{(0)}\delta V \Psi_{\pm}^{(0)} \ , \tag{12.35}$$

where $\mathscr{C} = \dfrac{\sqrt{-\omega r_0}}{\sin(\pi j/2)}$ and the integral is along the positive real axis on the z-plane ($z > 0$). We obtain the large-z behavior

$$\Psi_{\pm}^{(1)}(z) \sim c_{-\pm}\cos(z - \pi(1+j)/4) - c_{+\pm}\cos(z - \pi(1-j)/4) , \tag{12.36}$$

where

$$c_{\pm\pm} = \mathscr{C}\int_0^{\infty}\Psi_{\pm}^{(0)}\delta V\Psi_{\pm}^{(0)} . \tag{12.37}$$

To obtain the small-z behavior, expand

$$\delta V(z) = -\frac{3\ell(\ell+1)+1-j^2}{6\sqrt{-2\omega r_0}}z^{-3/2} + \mathscr{O}(1/\omega) . \tag{12.38}$$

It follows that

$$\Psi_{\pm}^{(1)} = z^{1\pm j/2}G_{\pm}(z) + \mathscr{O}(1/\omega) , \tag{12.39}$$

where $G_{\pm}$ are even analytic functions of z.

For the desired behavior as $z \to \infty$, define

$$\Psi = \Psi_{+}^{(0)} + \frac{1}{\sqrt{-\omega r_0}}\left\{\Psi_{+}^{(1)} - e^{-\pi ji/2}\Psi_{-}^{(1)} + e^{-\pi ji/2}\xi\Psi_{-}^{(0)}\right\} + \cdots , \tag{12.40}$$

where $\xi \sim \mathscr{O}(1)$ and dots represent terms of order higher than $\mathscr{O}(1/\sqrt{\omega})$. By demanding $\Psi \sim e^{-iz}$ as $z \to +\infty$, we fix

$$\xi = \xi_{+} + \xi_{-} , \quad \xi_{+} = c_{++}e^{\pi ji/2} - c_{+-} , \quad \xi_{-} = c_{--}e^{-\pi ji/2} - c_{+-} . \tag{12.41}$$

Then the requirement $f(z) = e^{iz}\Psi(z) \sim \text{const.}$ as $z \to \infty$ yields

$$f(z) \sim -e^{-\pi(1+j)i/4}\sin(\pi j/2)\left\{1 - \frac{\xi_{-}}{\sqrt{-\omega r_0}}\right\} . \tag{12.42}$$

In the neighborhood of the black-hole singularity (around $z = 0$), going around a 3π arc, we obtain

$$\Psi_{\pm}^{(1)}(e^{3\pi i}z) = e^{3\pi(2\pm j)i/2}\Psi_{\pm}^{(1)}(z) , \tag{12.43}$$

therefore

$$\Psi(e^{3\pi i}z) = \Psi^{(0)}(e^{3\pi i}z)$$
$$- e^{3\pi ji/2}\frac{1}{\sqrt{-\omega r_0}}\left\{\Psi_{+}^{(1)}(z) - e^{-7\pi ji/2}(\Psi_{-}^{(1)}(z) - i\xi\Psi_{-}^{(0)}(z))\right\} . \tag{12.44}$$

As $z \to \infty$ along the real axis,

$$f(z) \sim e^{-\pi(1+j)i/4}\sin(3\pi j/2)\left\{1 - \frac{1}{\sqrt{-\omega r_0}}A\right\}$$
$$+ e^{\pi(1-j)i/4}\sin(2\pi j)\left\{1 - \frac{1}{\sqrt{-\omega r_0}}B\right\}e^{2iz} ,$$

where

$$A = \frac{i-1}{2} e^{\pi j i/2} \left(\xi_+ + i\xi_- - \xi \cot(3\pi j/2) \right) \tag{12.45}$$

and B is not needed for our purposes. The monodromy to this order reads

$$\mathcal{M}(r_0) = -\frac{\sin(3\pi j/2)}{\sin(\pi j/2)} \left\{ 1 + \frac{i-1}{2\sqrt{-\omega r_0}} e^{\pi j i/2} \left(\xi_- - \xi_+ + \xi \cot(3\pi j/2) \right) \right\}, \tag{12.46}$$

leading to the QNM frequencies [10]

$$\frac{\omega_n}{T_H} = (2n+1)\pi i + \ln(1 + 2\cos(\pi j)) + \frac{e^{\pi j i/2}}{\sqrt{n+1/2}} \left(\xi_- - \xi_+ + \xi \cot(3\pi j/2) \right)$$
$$+ \mathcal{O}(1/n), \tag{12.47}$$

which includes the $\mathcal{O}(1/\sqrt{n})$ correction to the $\mathcal{O}(1)$ asymptotic expression (12.31).

For an explicit expression, use

$$\mathcal{J}(\nu,\mu) \equiv \int_0^\infty dz z^{-1/2} J_\nu(z) J_\mu(z) = \frac{\sqrt{\pi/2}\,\Gamma(\frac{\nu+\mu+1/2}{2})}{\Gamma(\frac{-\nu+\mu+3/2}{2})\Gamma(\frac{\nu+\mu+3/2}{2})\Gamma(\frac{\nu-\mu+3/2}{2})}. \tag{12.48}$$

We obtain

$$c_{\pm\pm} = \pi \frac{3\ell(\ell+1) + 1 - j^2}{12\sqrt{2}\,\sin(\pi j/2)} \mathcal{J}(\pm j/2, \pm j/2) \tag{12.49}$$

therefore

$$\xi_- - \xi_+ + \xi \cot(3\pi j/2) = (1-i) \frac{3\ell(\ell+1) + 1 - j^2}{24\sqrt{2}\pi^{3/2}} \frac{\sin(2\pi j)}{\sin(3\pi j/2)}$$
$$\times \Gamma^2(1/4)\,\Gamma(1/4 + j/2)\,\Gamma(1/4 - j/2), \tag{12.50}$$

where we also used the identity $\Gamma(y)\Gamma(1-y) = \frac{\pi}{\sin(\pi y)}$. This expression has a well-defined finite limit as $j \to$ integer.

For scalar waves, let $j \to 0^+$. We obtain

$$\frac{\omega_n}{T_H} = (2n+1)\pi i + \ln 3 + \frac{1-i}{\sqrt{n+1/2}} \frac{\ell(\ell+1) + 1/3}{6\sqrt{2}\pi^{3/2}} \Gamma^4(1/4) + \mathcal{O}(1/n), \tag{12.51}$$

which is in agreement with numerical results [11].

For gravitational waves, we let $j \to 2$ and obtain

$$\frac{\omega_n}{T_H} = (2n+1)\pi i + \ln 3 + \frac{1-i}{\sqrt{n+1/2}} \frac{\ell(\ell+1) - 1}{18\sqrt{2}\pi^{3/2}} \Gamma^4(1/4) + \mathcal{O}(1/n), \tag{12.52}$$

which is in agreement with the results from a WKB analysis [12] as well as numerical analysis [5].

12.2.2 Kerr Black Holes

Extending the above discussion to rotating (Kerr) black holes is not straightforward. Bohr's correspondence principle

$$\delta M = \hbar \mathrm{Re}\,\omega \,, \tag{12.53}$$

and the first law of black-hole mechanics

$$\delta M = T_H \delta S_{BH} + \Omega \delta J \,, \tag{12.54}$$

imply the asymptotic expression [9]

$$\mathrm{Re}\,\omega = T_H \ln 3 + m\Omega \,, \tag{12.55}$$

where m is the azimuthal eigenvalue of the wave, and Ω is the angular velocity of horizon. In deriving the above, we identified $\delta S_{BH} \equiv \ln 3$ [13]. Even though the above result has the correct limit as $\Omega \to 0$ (in agreement with the Schwarzschild expression (12.14)), it is in conflict with numerical results [14] indicating $\mathrm{Re}\,\omega \approx m\Omega$.

To resolve the above contradiction, we shall obtain an analytic solution to the wave (Teukolsky [15]) equation which will be valid for asymptotic modes bounded from above by $1/a$, where

$$a = \frac{J}{M} \,, \tag{12.56}$$

with J being the angular momentum and M the mass of the Kerr black hole. The calculation will be valid for $a \ll 1$ which includes the Schwarzschild case ($a = 0$) [16]. Our results will confirm Hod's expression (12.55) and not necessarily contradict numerical results (the latter may still be valid in the asymptotic regime $1/a \lesssim \omega$). In the Schwarzschild limit ($a \to 0$) the range of frequencies extends to infinity and our expression reduces to the expected form (12.14).

The metric of a Kerr black hole is

$$ds^2 = - \left(1 - \frac{2Mr}{\Sigma}\right) dt^2 + \frac{4Mar\sin^2\theta}{\Sigma}\, dtd\phi + \frac{\Sigma}{\Delta}\, dr^2$$
$$+ \Sigma d\theta^2 + \sin^2\theta \left(r^2 + a^2 + \frac{2Ma^2 r\sin^2\theta}{\Sigma}\right) d\phi^2 \,, \tag{12.57}$$

where $\Sigma = r^2 + a^2\cos^2\theta$, $\Delta = r^2 - 2Mr + a^2 = (r - r_-)(r - r_+)$ and we have set Newton's constant $G = 1$. The angular velocity of the horizon and Hawking temperature, respectively, are

$$\Omega = \frac{a}{2Mr_+} \,, \quad T_H = \frac{1 - r_-/r_+}{8\pi M} \,. \tag{12.58}$$

12.2.2.1 Massless Perturbations

Massless perturbations are governed by the Teukolsky wave equation [15]

$$
\left(\frac{(r^2+a^2)^2}{\Delta} - a^2\sin^2\theta\right)\frac{\partial^2\Psi}{\partial t^2} + \frac{4Mar}{\Delta}\frac{\partial^2\Psi}{\partial t\partial\phi} + \left(\frac{a^2}{\Delta} - \frac{1}{\sin^2\theta}\right)\frac{\partial^2\Psi}{\partial\phi^2}
$$

$$
- \frac{1}{\Delta^s}\frac{\partial}{\partial r}\left(\Delta^{s+1}\frac{\partial\Psi}{\partial r}\right) - 2s\left(\frac{M(r^2-a^2)}{\Delta} - r - ia\cos\theta\right)\frac{\partial\Psi}{\partial t}
$$

$$
- \frac{1}{\sin\theta}\frac{\partial}{\partial\theta}\left(\sin\theta\frac{\partial\Psi}{\partial\theta}\right) - 2s\left(\frac{a(r-M)}{\Delta} + \frac{i\cos\theta}{\sin^2\theta}\right)\frac{\partial\Psi}{\partial\phi} + (s^2\cot^2\theta - s)\Psi = 0 \,,
$$

$$\tag{12.59}$$

where $s = 0, -1, -2$ for scalar, electromagnetic and gravitational perturbations, respectively. Writing the wavefunction in the form

$$
\Psi = e^{-i\omega t}e^{im\phi}S(\theta)f(r) \tag{12.60}
$$

we obtain the angular equation

$$
\frac{1}{\sin\theta}(\sin\theta\, S')' + \left(a^2\omega^2\cos^2\theta - \frac{m^2}{\sin^2\theta} - 2a\omega s\cos\theta - \frac{2ms\cos\theta}{\sin^2\theta} - s^2\cot^2\theta\right)S
$$

$$
= -(A+s)S \,, \tag{12.61}
$$

where A is the separation constant (eigenvalue) and the radial equation

$$
\frac{1}{\Delta^s}(\Delta^{s+1}f')' + V(r)f = (A+a^2\omega^2)f \,, \tag{12.62}
$$

where the potential is given by

$$
V(r) = \frac{(r^2+a^2)^2\omega^2 - 4aMr\omega m + a^2m^2 + 2ia(r-M)ms - 2iM(r^2-a^2)\omega s}{\Delta}
$$

$$
+ 2ir\omega s \,. \tag{12.63}
$$

Let us simplify the notation by placing the horizon at $r = 1$, i.e., by setting

$$
2M = 1+a^2 \,, \quad r_- = a^2 \,, \quad r_+ = 1 \tag{12.64}
$$

and solve the two wave equations by expanding in a. We shall keep terms up to $\mathcal{O}(a)$ assuming ω is large but bounded from above by $1/a$, $(1 \lesssim \omega \lesssim 1/a)$. Thus ω is in an intermediate range which becomes asymptotic in the Schwarzschild limit $a \to 0$.

The solutions to the angular equation to lowest order are spin-weighted spherical harmonics with eigenvalue

$$
A = \ell(\ell+1) - s(s+1) + \mathcal{O}(a\omega) \,. \tag{12.65}
$$

Near the horizon ($r \rightarrow 1$),

$$f(r) \sim (r-1)^{\lambda} \ , \quad \lambda = i(\omega - am) + \mathcal{O}(1/\omega) \ . \tag{12.66}$$

At infinity ($r \rightarrow \infty$), $f(r) \sim e^{i\omega r}$. Introducing the "tortoise coordinate"

$$z = \omega r + (\omega - am)\ln(r-1) \ , \tag{12.67}$$

the boundary conditions read

$$f(z) \sim e^{\pm iz} \ , \quad z \rightarrow \pm\infty \ . \tag{12.68}$$

From the boundary condition at the horizon we deduce the monodromy for the function $\mathscr{F}(z) \equiv e^{iz} f(z)$ (notice that $\mathscr{F} \sim$ const. as $z \rightarrow +\infty$) around the singular point $r = 1$,

$$\mathscr{M}(1) = e^{4\pi(\omega - am)} + \mathcal{O}(a^2) \ . \tag{12.69}$$

To express the radial equation in terms of the tortoise coordinate, define

$$f(r) = \Delta_0^{-s/2} \frac{R(r)}{\sqrt{r(\omega r - am)}} \ , \tag{12.70}$$

$\Delta_0 = r(r-1)$ (note $\Delta = \Delta_0 + \mathcal{O}(a^2)$). Inverting $z = z(r)$,

$$r = \sqrt{-\frac{2z}{\omega}} + \mathcal{O}(1/\omega) \ , \tag{12.71}$$

the radial equation to lowest order in $1/\sqrt{\omega}$ in terms of R reads

$$\frac{d^2R}{dz^2} + \left\{ 1 + \frac{3is}{2z} + \frac{4-s^2-4iams}{16z^2} \right\} R = 0 \tag{12.72}$$

to be solved along the entire real axis. This is Whittaker's equation. The solutions may be written as

$$M_{\kappa,\pm\mu}(x) = e^{-x/2} x^{\pm\mu+1/2} M(\frac{1}{2} \pm \mu - \kappa, 1 \pm 2\mu, x) \ , \tag{12.73}$$

where $\kappa = \dfrac{3s}{4}$, $\mu^2 = \dfrac{s(s+4iam)}{16}$, $M_{\kappa,\pm\mu}$ is Kummer's function (also called Φ) and we set $x = 2iz$. We need to introduce Whittaker's function

$$W_{\kappa,\mu}(x) = \frac{\Gamma(-2\mu)}{\Gamma(\frac{1}{2} - \mu - \kappa)} M_{\kappa,\mu}(x) + \frac{\Gamma(2\mu)}{\Gamma(\frac{1}{2} + \mu - \kappa)} M_{\kappa,-\mu}(x) \ , \tag{12.74}$$

due to its clean asymptotic behavior,

$$W_{\kappa,\mu}(x) \sim e^{-x/2} x^{\kappa} (1 + \mathcal{O}(1/x)) \ , \quad |x| \rightarrow \infty \ . \tag{12.75}$$

We may compute the monodromy by deforming the contour as before. Going around an arc of angle 3π, we have

$$M_{\kappa,\pm\mu}(e^{3\pi i}x) = -ie^{\pm 3\pi i\mu}M_{-\kappa,\pm\mu}(x) , \qquad (12.76)$$

where we used $M(a,b,-x) = e^{-x}M(b-a,b,x)$, therefore

$$W_{\kappa,\mu}(e^{3\pi i}x) = -ie^{3\pi i\mu}\frac{\Gamma(-2\mu)}{\Gamma(\frac{1}{2}-\mu-\kappa)}M_{-\kappa,\mu}(x) - ie^{-3\pi i\mu}\frac{\Gamma(2\mu)}{\Gamma(\frac{1}{2}+\mu-\kappa)}M_{-\kappa,-\mu}(x) . \qquad (12.77)$$

To find the asymptotic behavior, we need

$$M_{-\kappa,\mu}(x) = \frac{\Gamma(1+2\mu)}{\Gamma(\frac{1}{2}+\mu+\kappa)}e^{-i\pi\kappa}W_{\kappa,\mu}(e^{i\pi}x) + \frac{\Gamma(1+2\mu)}{\Gamma(\frac{1}{2}+\mu-\kappa)}e^{-i\pi(\frac{1}{2}+\mu+\kappa)}W_{-\kappa,\mu}(x) . \qquad (12.78)$$

As $|x| \to \infty$, we obtain

$$W_{\kappa,\mu}(e^{3\pi i}x) \sim Ae^{x/2}x^{\kappa} + Be^{-x/2}x^{-\kappa} , \qquad (12.79)$$

where

$$A = -ie^{3\pi i\mu}\frac{\Gamma(-2\mu)}{\Gamma(\frac{1}{2}-\mu-\kappa)}\frac{\Gamma(1+2\mu)}{\Gamma(\frac{1}{2}+\mu+\kappa)}e^{-\pi i\kappa} + (\mu \to -\mu) , \qquad (12.80)$$

and B is not needed for our purposes. After some algebra, we deduce

$$A = -(1+2\cos\pi s) + \mathcal{O}(a^2) , \qquad (12.81)$$

where we used the identities $\Gamma(1-x)\Gamma(x) = \frac{\pi}{\sin\pi x}$, $\Gamma(\frac{1}{2}+x)\Gamma(\frac{1}{2}-x) = \frac{\pi}{\cos\pi x}$. The monodromy around $r = 1$ is

$$\mathcal{M}(1) = e^{4\pi(\omega-ma)} = A , \qquad (12.82)$$

therefore [16]

$$\mathrm{Re}\,\omega = \frac{1}{4\pi}\ln(1+2\cos\pi s) + ma + \mathcal{O}(a^2) , \qquad (12.83)$$

in agreement with Hod's formula for gravitational waves ($s = -2$) in the small-a limit (in which $\Omega \approx a$, $T_H \approx \frac{1}{4\pi}$). However, it should be emphasized that these are not asymptotic values of QNMs but bounded from above by $1/a$.

12.2.2.2 Massive Perturbations

The case of massive perturbations is interesting because it reveals instabilities. As is well-known, the Schwarzschild spacetime is stable against all kinds of perturbations, massive or massless which makes the Schwarzschild geometry appropriate to study astrophysical objects. On the other hand, Kerr spacetime is stable against massless perturbations but not against massive bosonic fields [17]. The instability timescale is much larger than the age of the Universe so the problem is not expected to have observable consequences. Nevertheless, the study of instabilities is an important subject and QNMs provide an indispensable tool.

For a massive scalar of mass μ, the radial wave equation reads

$$\frac{d}{dr}\left(\Delta\frac{dR}{dr}\right) + \left\{\frac{\omega^2(r^2+a^2)^2 - 4aMm\omega r + m^2a^2}{\Delta} - \mu^2 r^2 - a^2\omega^2 - \ell(\ell+1)\right\}R = 0, \tag{12.84}$$

We are interested in solving this equation for a small mass and low frequencies $(\mu, \omega \ll 1/M)$ [17].

Away from the horizon $(r \gg M)$, we may approximate by

$$\frac{d^2}{dr^2}(rR) + \left[-k^2 + \frac{2M\mu^2}{r} - \frac{\ell(\ell+1)}{r^2}\right]rR = 0 \ , \quad k^2 = \mu^2 - \omega^2 \ . \tag{12.85}$$

The solution to this equation is given in terms of a confluent hypergeometric function,

$$R(r) = (2kr)^\ell e^{-kr} U(\ell+1 - M\mu^2/k, 2(\ell+1), 2kr) \ . \tag{12.86}$$

Near the horizon $(r \ll \ell/|k|)$, we may approximate by

$$z(z+1)\frac{d}{dz}\left[z(z+1)\frac{dR}{dz}\right] + \left[P^2 - \ell(\ell+1)z(z+1)\right]R = 0 \ , \tag{12.87}$$

where $P = \frac{am - 2Mr_+\omega}{r_+ - r_-}$, $z = \frac{r - r_+}{r_+ - r_-}$. The solution to this equation is given in terms of a hypergeometric function,

$$R(z) = \left(\frac{z}{z+1}\right)^{iP} F(-\ell, \ell+1; 1-2iP; z+1) \ . \tag{12.88}$$

Matching the two expressions in the overlap region $(M \ll r \ll \ell/|k|)$, we obtain the frequencies

$$\omega_n \approx \mu + i\gamma_n \ , \quad n \in \mathbb{N} \ , \tag{12.89}$$

where

$$\gamma_n = \mathscr{C}_{\ell n}\mu(\mu M)^{4(\ell+1)}\frac{am}{M - 2\mu r_+}\prod_{j=1}^{\ell}\left[j^2\left(1 - \frac{a^2}{M^2}\right) + \left(\frac{am}{M} - 2\mu r_+\right)^2\right] \ , \tag{12.90}$$

and $\mathscr{C}_{\ell n} = \dfrac{2^{2(2\ell+1)}(2\ell+1+n)!(\ell!)^2}{(\ell+1+n)^{2(\ell+2)}(2\ell+1)^2 n!((2\ell)!)^4}$.

For $m > 0$, we have $\gamma_n > 0$ yielding an instability. For the fastest growing mode (with $\ell = 1, m = 1, n = 2$ (2p state)) we have

$$\tau = \frac{1}{\gamma} = \frac{24}{a\mu^2(\mu M)^7} \qquad (12.91)$$

which is generally large.

Notice that there is no instability in the Schwarzschild limit ($a \to 0$) and for massless perturbations ($\mu \to 0$); in both cases, $\gamma \to 0$ and therefore the lifetime $\tau \to \infty$.

12.2.3 Half-Integer Spin

In the case of a perturbation of half-integer spin we need to solve the Teukolsky equation [18] with potential

$$V(r) = f(r)\left(\frac{\ell(\ell+1)}{r^2} + \frac{1}{r^3}\right) + \frac{2i\omega j}{r} - \frac{3i\omega j}{r^2} + \frac{j^2}{4r^4} , \qquad (12.92)$$

where j is the spin of the perturbing field (e.g., $j = 1/2$ for Dirac fermion). We shall set $r_0 = 1$, for simplicity, so $f(r) = 1 - \frac{1}{r}$.

Expanding around the black-hole singularity $z = \omega r_* = 0$,

$$\frac{1}{\omega^2}V(z) = \frac{3ij}{2z} - \frac{4-j^2}{16z^2} + \frac{\mathscr{A}}{\omega^{1/2}z^{3/2}} + \mathscr{O}(1/\omega) , \quad \mathscr{A} = \frac{\ell(\ell+1) + \frac{1-j^2}{3}}{2\sqrt{2}} , \qquad (12.93)$$

we obtain the zeroth-order wave equation

$$\frac{d^2\Psi}{dz^2} + \left[1 - \frac{3ij}{2z} - \frac{4-j^2}{16z^2}\right]\Psi = 0 , \qquad (12.94)$$

whose solutions are the Whittaker functions

$$\Psi_{\pm}^{(0)}(z) = M_{\lambda,\pm\mu}(-2iz) , \quad \lambda = \frac{3j}{4} , \quad \mu = \frac{j}{4} . \qquad (12.95)$$

The calculation of the monodromy as before leads to the modes [18]

$$\frac{\omega_n}{T_H} = -(2n+1)\pi i + \ln(1 + 2\cos\pi j) + \mathscr{O}(1/\sqrt{n}) \qquad (12.96)$$

in agreement with the result for integer spin (which came from the Regge–Wheeler equation). For a Dirac fermion, $j = 1/2$, so asymptotically, the real part vanishes.

The first-order correction may also be calculated as before [19]. The result is

$$\frac{\omega_n}{T_H} = -(2n+1)\pi i + \ln(1 + 2\cos\pi j)$$

$$- \frac{2i}{\sqrt{-in/2}}\sin 4\pi\mu \frac{\bar{b}_+A_-B_- + \bar{b}_-A_+B_+}{e^{-4\pi i\mu}A_+B_- - e^{4\pi i\mu}A_-B_+} + \mathscr{O}(1/n) , \qquad (12.97)$$

where

$$\bar{b}_{\pm} = \frac{\mathscr{A}}{4\mu} \int_0^\infty \frac{dz}{z^{3/2}} M_{\lambda,\pm\mu}(-2iz) M_{\lambda,\pm\mu}(-2iz) \tag{12.98}$$

and $A_{\pm} = \frac{\Gamma(1\pm2\mu)}{\Gamma(\frac{1}{2}\pm\mu+\lambda)} e^{i\pi(\frac{1}{2}\pm\mu-\lambda)}$, $B_{\pm} = \frac{\Gamma(1\pm2\mu)}{\Gamma(\frac{1}{2}\pm\mu-\lambda)} e^{-i\pi\lambda}$. This result appears to be a complicated function of j, so let us look at specific cases.

For $j = 1/2$ (Dirac fermions), we obtain

$$\frac{\omega_n}{T_H} = -(2n+1)\pi i + \frac{1+i}{2\sqrt{n}} \left(\ell + \frac{1}{2}\right)^2 \Gamma^2\left(\frac{1}{4}\right) + \mathscr{O}(1/n). \tag{12.99}$$

which is in good agreement with numerical data [19].

For $j = 3/2$, we find

$$\frac{\omega_n}{T_H} = -(2n+1)\pi i + \mathscr{O}(1/n), \tag{12.100}$$

so there are no first-order corrections to the spectrum.

For $j = 5/2$, we have

$$\frac{\omega_n}{T_H} = -(2n+1)\pi i + \frac{1+i}{\sqrt{2n}} \mathscr{A} \Gamma^2\left(\frac{1}{4}\right) + \mathscr{O}(1/n). \tag{12.101}$$

etc.

All of the above spectra agree with the general expression we obtained for integer spin using the Regge–Wheeler equation. The relation of the latter to the Teukolsky equation is worth exploring further.

12.3 Anti-de Sitter Spacetime

According to the AdS/CFT correspondence, QNMs of AdS black holes are expected to correspond to perturbations of the dual Conformal Field Theory (CFT) on the boundary. The establishment of such a correspondence is hindered by difficulties in solving the wave equation governing the various types of perturbation. In three dimensions one obtains a hypergeometric equation which leads to explicit analytic expressions for the QNMs [20, 21]. In five dimensions one obtains a Heun equation and a derivation of analytic expressions for QNMs is no longer possible. On the other hand, numerical results exist in four, five and seven dimensions [22–24].

12.3.1 Scalar Perturbations

To find the asymptotic form of QNMs, we need to find an approximation to the wave equation valid in the high-frequency regime. In three dimensions the resulting wave equation will be an exact equation (hypergeometric equation). In five dimensions, we shall turn the Heun equation into a hypergeometric equation which will lead to an analytic expression for the asymptotic form of QNM frequencies in agreement with numerical results.

12.3.1.1 AdS$_3$

In three dimensions the wave equation for a massless scalar field is

$$\frac{1}{R^2 r} \partial_r \left(r^3 \left(1 - \frac{r_0^2}{r^2} \right) \partial_r \Phi \right) - \frac{R^2}{r^2 - r_0^2} \partial_t^2 \Phi + \frac{1}{r^2} \partial_x^2 \Phi = 0 . \qquad (12.102)$$

Writing the wavefunction in the form

$$\Phi = e^{i(\omega t - px)} \Psi(y), \quad y = \frac{r_0^2}{r^2} , \qquad (12.103)$$

the wave function becomes

$$y^2(y-1)\left((y-1)\Psi'\right)' + \hat{\omega}^2 y\Psi + \hat{p}^2 y(y-1)\Psi = 0 \qquad (12.104)$$

to be solved in the interval $0 < y < 1$, where

$$\hat{\omega} = \frac{\omega R^2}{2r_0} = \frac{\omega}{4\pi T_H}, \quad \hat{p} = \frac{pR}{2r_0} = \frac{p}{4\pi R T_H} . \qquad (12.105)$$

For QNMs, we are interested in the solution

$$\Psi(y) = y(1-y)^{i\hat{\omega}} {}_2F_1(1+i(\hat{\omega}+\hat{p}), 1+i(\hat{\omega}-\hat{p}); 2; y) , \qquad (12.106)$$

which vanishes at the boundary ($y \to 0$). Near the horizon ($y \to 1$), we obtain a mixture of ingoing and outgoing waves,

$$\Psi \sim A_+(1-y)^{-i\hat{\omega}} + A_-(1-y)^{+i\hat{\omega}} , \quad A_\pm = \frac{\Gamma(\pm 2i\hat{\omega})}{\Gamma(1 \pm i(\hat{\omega}+\hat{p}))\Gamma(1 \pm i(\hat{\omega}-\hat{p}))} .$$

Setting $A_- = 0$, we deduce the quasi-normal frequencies

$$\hat{\omega} = \pm \hat{p} - in , \quad n = 1, 2, \dots \qquad (12.107)$$

which form a discrete spectrum of complex frequencies with $\text{Im}\,\hat{\omega} < 0$.

12.3.1.2 AdS$_5$

Restricting attention to the case of a large black hole, the massless scalar wave equation reads

$$\frac{1}{r^3}\partial_r(r^5 f(r)\,\partial_r\Phi) - \frac{R^4}{r^2 f(r)}\partial_t^2\Phi - \frac{R^2}{r^2}\nabla^2\Phi = 0 \ , \quad f(r) = 1 - \frac{r_0^4}{r^4} \ . \quad (12.108)$$

Writing the solution in the form

$$\Phi = e^{i(\omega t - \mathbf{p}\cdot\mathbf{x})}\Psi(y) \ , \quad y = \frac{r^2}{r_0^2} \ , \tag{12.109}$$

the radial wave equation becomes

$$(y^2 - 1)\left(y(y^2 - 1)\Psi'\right)' + \left(\frac{\hat{\omega}^2}{4}y^2 - \frac{\hat{p}^2}{4}(y^2 - 1)\right)\Psi = 0 \ . \tag{12.110}$$

For QNMs, we are interested in the analytic solution which vanishes at the boundary and behaves as an ingoing wave at the horizon. The wave equation contains an additional (unphysical) singularity at $y = -1$, at which the wavefunction behaves as $\Psi \sim (y+1)^{\pm\hat{\omega}/4}$. Isolating the behavior of the wavefunction near the singularities $y = \pm 1$,

$$\Psi(y) = (y-1)^{-i\hat{\omega}/4}(y+1)^{\pm\hat{\omega}/4}F_\pm(y) \ , \tag{12.111}$$

we shall obtain two sets of modes with the same $\mathrm{Im}\,\hat{\omega}$, but opposite $\mathrm{Re}\,\hat{\omega}$.

$F_\pm(y)$ satisfies the Heun equation

$$y(y^2 - 1)F\pm'' + \left\{\left(3 - \frac{i\pm1}{2}\hat{\omega}\right)y^2 - \frac{i\pm1}{2}\hat{\omega}y - 1\right\}F_\pm'$$

$$+ \left\{\frac{\hat{\omega}}{2}\left(\pm\frac{i\hat{\omega}}{4}\mp 1 - i\right)y - (i\mp 1)\frac{\hat{\omega}}{4} - \frac{\hat{p}^2}{4}\right\}F_\pm = 0 \quad (12.112)$$

to be solved in a region in the complex y-plane containing $|y| \geq 1$ which includes the physical regime $r > r_h$.

For large $\hat{\omega}$, the constant terms in the polynomial coefficients of F' and F are small compared with the other terms, therefore they may be dropped. The wave equation may then be approximated by a hypergeometric equation

$$(y^2 - 1)F_\pm'' + \left\{\left(3 - \frac{i\pm1}{2}\hat{\omega}\right)y - \frac{i\pm1}{2}\hat{\omega}\right\}F_\pm' + \frac{\hat{\omega}}{2}\left(\pm\frac{i\hat{\omega}}{4}\mp 1 - i\right)F_\pm = 0 \ , \tag{12.113}$$

in the asymptotic limit of large frequencies $\hat{\omega}$. The acceptable solution is

$$F_0(x) = {}_2F_1(a_+, a_-; c; (y+1)/2) \ , \quad a_\pm = 1 - \frac{i\pm1}{4}\hat{\omega}\pm 1 \ , \quad c = \frac{3}{2}\pm\frac{1}{2}\hat{\omega} \ . \tag{12.114}$$

For proper behavior at the boundary ($y \to \infty$), we demand that F be a *polynomial*, which leads to the condition

$$a_+ = -n \ , \ n = 1, 2, \ldots \tag{12.115}$$

Indeed, it implies that F is a polynomial of order n, so as $y \to \infty$, $F \sim y^n \sim y^{-a_+}$ and $\Psi \sim y^{-i\hat{\omega}/4} y^{\pm\hat{\omega}/4} y^{-a_+} \sim y^{-2}$, as expected.

We deduce the quasi-normal frequencies [25]

$$\hat{\omega} = \frac{\omega}{4\pi T_H} = 2n(\pm 1 - i) \ , \tag{12.116}$$

in agreement with numerical results.

It is perhaps worth mentioning that these frequencies may also be deduced by a simple monodromy argument [25]. Considering the monodromies around the singularities, if the wavefunction has no singularities other than $y = \pm 1$, the contour around $y = +1$ may be unobstructedly deformed into the contour around $y = -1$, which yields

$$\mathcal{M}(1)\mathcal{M}(-1) = 1 \ . \tag{12.117}$$

Since the respective monodromies are $\mathcal{M}(1) = e^{\pi\hat{\omega}/2}$ and $\mathcal{M}(-1) = e^{\mp i\pi\hat{\omega}/2}$, using Im $\hat{\omega} < 0$, we deduce $\hat{\omega} = 2n(\pm 1 - i)$, in agreement with our resultl above.

12.3.2 Gravitational Perturbations

Next we consider gravitational perturbations of AdS Schwarzschild black holes of arbitrary size in d dimensions. We shall derive analytic expressions for the asymptotic spectrum [26] including first-order corrections [27]. Our results will be in good agreement with numerical results.

The metric is

$$ds^2 = -f(r)dt^2 + \frac{dr^2}{f(r)} + r^2 d\Omega_{d-2}^2 \ , \quad f(r) = \frac{r^2}{R^2} + 1 - \frac{2\mu}{r^{d-3}} \ . \tag{12.118}$$

The radial wave equation can be cast into a Schrödinger-like form,

$$-\frac{d^2\Psi}{dr_*^2} + V[r(r_*)]\Psi = \omega^2\Psi \ , \tag{12.119}$$

in terms of the tortoise coordinate defined by

$$\frac{dr_*}{dr} = \frac{1}{f(r)} \ . \tag{12.120}$$

The potential V for the various types of perturbation has been found by Ishibashi and Kodama [28]. For scalar, vector and tensor perturbations, we obtain, respectively,

$$V_S(r) = \frac{f(r)}{4r^2} \left[\ell(\ell+d-3) - (d-2) + \frac{(d-1)(d-2)\mu}{r^{d-3}} \right]^{-2}$$

$$\times \left\{ \frac{d(d-1)^2(d-2)^3\mu^2}{R^2 r^{2d-8}} - \frac{6(d-1)(d-2)^2(d-4)[\ell(\ell+d-3)-(d-2)]\mu}{R^2 r^{d-5}} \right.$$

$$+ \frac{(d-4)(d-6)[\ell(\ell+d-3)-(d-2)]^2 r^2}{R^2} + \frac{2(d-1)^2(d-2)^4\mu^3}{r^{3d-9}}$$

$$+ \frac{4(d-1)(d-2)(2d^2-11d+18)[\ell(\ell+d-3)-(d-2)]\mu^2}{r^{2d-6}}$$

$$+ \frac{(d-1)^2(d-2)^2(d-4)(d-6)\mu^2}{r^{2d-6}}$$

$$- \frac{6(d-2)(d-6)[\ell(\ell+d-3)-(d-2)]^2\mu}{r^{d-3}}$$

$$- \frac{6(d-1)(d-2)^2(d-4)[\ell(\ell+d-3)-(d-2)]\mu}{r^{d-3}}$$

$$\left. + 4[\ell(\ell+d-3)-(d-2)]^3 + d(d-2)[\ell(\ell+d-3)-(d-2)]^2 \right\}, \quad (12.121)$$

$$V_V(r) = f(r) \left\{ \frac{\ell(\ell+d-3)}{r^2} + \frac{(d-2)(d-4)f(r)}{4r^2} - \frac{rf'''(r)}{2(d-3)} \right\}, \quad (12.122)$$

$$V_T(r) = f(r) \left\{ \frac{\ell(\ell+d-3)}{r^2} + \frac{(d-2)(d-4)f(r)}{4r^2} + \frac{(d-2)f'(r)}{2r} \right\}. \quad (12.123)$$

Near the black-hole singularity ($r \sim 0$),

$$V_T = -\frac{1}{4r_*^2} + \frac{\mathscr{A}_T}{[-2(d-2)\mu]^{\frac{1}{d-2}}} r_*^{-\frac{d-1}{d-2}} + \dots, \quad \mathscr{A}_T = \frac{(d-3)^2}{2(2d-5)} + \frac{\ell(\ell+d-3)}{d-2},$$

$$(12.124)$$

$$V_V = \frac{3}{4r_*^2} + \frac{\mathscr{A}_V}{[-2(d-2)\mu]^{\frac{1}{d-2}}} r_*^{-\frac{d-1}{d-2}} + \dots, \quad \mathscr{A}_V = \frac{d^2-8d+13}{2(2d-15)} + \frac{\ell(\ell+d-3)}{d-2}$$

$$(12.125)$$

and

$$V_S = -\frac{1}{4r_*^2} + \frac{\mathscr{A}_S}{[-2(d-2)\mu]^{\frac{1}{d-2}}} r_*^{-\frac{d-1}{d-2}} + \cdots, \quad (12.126)$$

where

$$\mathscr{A}_S = \frac{(2d^3 - 24d^2 + 94d - 116)}{4(2d-5)(d-2)} + \frac{(d^2-7d+14)[\ell(\ell+d-3)-(d-2)]}{(d-1)(d-2)^2}.$$

$$(12.127)$$

We have included only the terms which contribute to the order we are interested in.
We may summarize the behavior of the potential near the origin by

$$V = \frac{j^2 - 1}{4r_*^2} + \mathscr{A} \, r_*^{-\frac{d-1}{d-2}} + \cdots \tag{12.128}$$

where $j = 0$ (2) for scalar and tensor (vector) perturbations.

On the other hand, near the boundary (large r),

$$V = \frac{j_\infty^2 - 1}{4(r_* - \bar{r}_*)^2} + \cdots \ , \quad \bar{r}_* = \int_0^\infty \frac{dr}{f(r)} \ , \tag{12.129}$$

where $j_\infty = d - 1$, $d - 3$ and $d - 5$ for tensor, vector and scalar perturbations, respectively.

After rescaling the tortoise coordinate ($z = \omega r_*$), the wave equation to first order becomes

$$\left(\mathscr{H}_0 + \omega^{-\frac{d-3}{d-2}} \mathscr{H}_1 \right) \Psi = 0 \ , \tag{12.130}$$

where

$$\mathscr{H}_0 = \frac{d^2}{dz^2} - \left[\frac{j^2 - 1}{4z^2} - 1 \right] \ , \quad \mathscr{H}_1 = -\mathscr{A} \, z^{-\frac{d-1}{d-2}} \ . \tag{12.131}$$

By treating $\mathscr{H}_1$ as a perturbation, we may expand the wave function

$$\Psi(z) = \Psi_0(z) + \omega^{-\frac{d-3}{d-2}} \Psi_1(z) + \cdots \tag{12.132}$$

and solve the wave equation perturbatively.

The zeroth-order wave equation,

$$\mathscr{H}_0 \Psi_0(z) = 0 \ , \tag{12.133}$$

may be solved in terms of Bessel functions,

$$\Psi_0(z) = A_1 \sqrt{z} J_{\frac{j}{2}}(z) + A_2 \sqrt{z} N_{\frac{j}{2}}(z) \ . \tag{12.134}$$

For large z, it behaves as

$$\Psi_0(z) \sim \sqrt{\frac{2}{\pi}} \left[A_1 \cos(z - \alpha_+) + A_2 \sin(z - \alpha_+) \right]$$

$$= \frac{1}{\sqrt{2\pi}} (A_1 - iA_2) e^{-i\alpha_+} e^{iz} + \frac{1}{\sqrt{2\pi}} (A_1 + iA_2) e^{+i\alpha_+} e^{-iz} \ ,$$

where $\alpha_\pm = \frac{\pi}{4}(1 \pm j)$.

At the boundary ($r \to \infty$), the wavefunction ought to vanish, therefore the acceptable solution is

$$\Psi_0(r_*) = B \sqrt{\omega(r_* - \bar{r}_*)} \, J_{\frac{j_\infty}{2}}(\omega(r_* - \bar{r}_*)) \ . \tag{12.135}$$

Indeed, $\Psi \to 0$ as $r_* \to \bar{r}_*$, as desired.

Asymptotically (large z), it behaves as

$$\Psi(r_*) \sim \sqrt{\frac{2}{\pi}} B \cos\left[\omega(r_* - \bar{r}_*) + \beta\right] , \quad \beta = \frac{\pi}{4}(1 + j_\infty) . \tag{12.136}$$

We ought to match this to the asymptotic form of the wavefunction in the vicinity of the black-hole singularity along the Stokes line $\mathrm{Im}\, z = \mathrm{Im}\,(\omega r_*) = 0$. This leads to a constraint on the coefficients A_1 and A_2,

$$A_1 \tan(\omega\bar{r}_* - \beta - \alpha_+) - A_2 = 0 . \tag{12.137}$$

By imposing the boundary condition at the horizon

$$\Psi(z) \sim e^{iz} , \quad z \to -\infty , \tag{12.138}$$

we obtain a second constraint. To find it, we need to analytically continue the wavefunction near the black-hole singularity ($z = 0$) to negative values of z. A rotation of z by $-\pi$ corresponds to a rotation by $-\pi/d - 2$ near the origin in the complex r-plane. Using the known behavior of Bessel functions

$$J_\nu(e^{-i\pi}z) = e^{-i\pi\nu}J_\nu(z) , \quad N_\nu(e^{-i\pi}z) = e^{i\pi\nu}N_\nu(z) - 2i\cos\pi\nu J_\nu(z) , \tag{12.139}$$

for $z < 0$ the wavefunction changes to

$$\Psi_0(z) = e^{-i\pi(j+1)/2}\sqrt{-z}\left\{\left[A_1 - i(1 + e^{i\pi j})A_2\right] J_{\frac{j}{2}}(-z) + A_2 e^{i\pi j} N_{\frac{j}{2}}(-z)\right\} , \tag{12.140}$$

whose asymptotic behavior is given by

$$\Psi \sim \frac{e^{-i\pi(j+1)/2}}{\sqrt{2\pi}}\left[A_1 - i(1 + 2e^{j\pi i})A_2\right] e^{-iz} + \frac{e^{-i\pi(j+1)/2}}{\sqrt{2\pi}}\left[A_1 - iA_2\right] e^{iz} . \tag{12.141}$$

Therefore we obtain a second constraint

$$A_1 - i(1 + 2e^{j\pi i})A_2 = 0 . \tag{12.142}$$

The two constraints are compatible provided

$$\begin{vmatrix} 1 & -i(1 + 2e^{j\pi i}) \\ \tan(\omega\bar{r}_* - \beta - \alpha_+) & -1 \end{vmatrix} = 0 , \tag{12.143}$$

which yields the quasi-normal frequencies [26]

$$\omega\bar{r}_* = \frac{\pi}{4}(2 + j + j_\infty) - \tan^{-1}\frac{i}{1 + 2e^{j\pi i}} + n\pi . \tag{12.144}$$

The first-order correction to the above asymptotic expression may be found by standard perturbation theory [27]. To first order, the wave equation becomes

$$\mathcal{H}_0 \Psi_1 + \mathcal{H}_1 \Psi_0 = 0 . \tag{12.145}$$

The solution is

$$\Psi_1(z) = \sqrt{z} N_{\frac{1}{2}}(z) \int_0^z dz' \frac{\sqrt{z'} J_{\frac{1}{2}}(z') \mathcal{H}_1 \Psi_0(z')}{\mathcal{W}}$$
$$- \sqrt{z} J_{\frac{1}{2}}(z) \int_0^z dz' \frac{\sqrt{z'} N_{\frac{1}{2}}(z') \mathcal{H}_1 \Psi_0(z')}{\mathcal{W}} , \tag{12.146}$$

where $\mathcal{W} = 2/\pi$ is the Wronskian.

The wavefunction to first order reads

$$\Psi(z) = \{A_1[1 - b(z)] - A_2 a_2(z)\} \sqrt{z} J_{\frac{1}{2}}(z) + \{A_2[1 + b(z)] + A_1 a_1(z)\} \sqrt{z} N_{\frac{1}{2}}(z) , \tag{12.147}$$

where

$$a_1(z) = \frac{\pi \mathcal{A}}{2} \omega^{-\frac{d-3}{d-2}} \int_0^z dz' \, z'^{-\frac{1}{d-2}} J_{\frac{1}{2}}(z') J_{\frac{1}{2}}(z')$$
$$a_2(z) = \frac{\pi \mathcal{A}}{2} \omega^{-\frac{d-3}{d-2}} \int_0^z dz' \, z'^{-\frac{1}{d-2}} N_{\frac{1}{2}}(z') N_{\frac{1}{2}}(z')$$
$$b(z) = \frac{\pi \mathcal{A}}{2} \omega^{-\frac{d-3}{d-2}} \int_0^z dz' \, z'^{-\frac{1}{d-2}} J_{\frac{1}{2}}(z') N_{\frac{1}{2}}(z') .$$

and $\mathcal{A}$ depends on the type of perturbation.

Asymptotically, it behaves as

$$\Psi(z) \sim \sqrt{\frac{2}{\pi}} [A_1' \cos(z - \alpha_+) + A_2' \sin(z - \alpha_+)] , \tag{12.148}$$

where
$$A_1' = [1 - \bar{b}] A_1 - \bar{a}_2 A_2 , \quad A_2' = [1 + \bar{b}] A_2 + \bar{a}_1 A_1 \tag{12.149}$$

and we introduced the notation

$$\bar{a}_1 = a_1(\infty) , \quad \bar{a}_2 = a_2(\infty) , \quad \bar{b} = b(\infty) . \tag{12.150}$$

The first constraint is modified to

$$A_1' \tan(\omega \bar{r}_* - \beta - \alpha_+) - A_2' = 0 . \tag{12.151}$$

Explicitly,

$$[(1 - \bar{b}) \tan(\omega \bar{r}_* - \beta - \alpha_+) - \bar{a}_1] A_1 - [1 + \bar{b} + \bar{a}_2 \tan(\omega \bar{r}_* - \beta - \alpha_+)] A_2 = 0 . \tag{12.152}$$

To find the second constraint to first order, we need to approach the horizon. This entails a rotation by $-\pi$ in the z-plane. Using

$$a_1(e^{-i\pi}z) = e^{-i\pi\frac{d-3}{d-2}}e^{-i\pi j}a_1(z) \,,$$

$$a_2(e^{-i\pi}z) = e^{-i\pi\frac{d-3}{d-2}}\left[e^{i\pi j}a_2(z) - 4\cos^2\frac{\pi j}{2}a_1(z) - 2i(1+e^{i\pi j})b(z)\right] \,,$$

$$b(e^{-i\pi}z) = e^{-i\pi\frac{d-3}{d-2}}\left[b(z) - i(1+e^{-i\pi j})a_1(z)\right] \,.$$

in the limit $z \to -\infty$ we obtain

$$\Psi(z) \sim -ie^{-ij\pi/2}B_1\cos(-z-\alpha_+) - ie^{ij\pi/2}B_2\sin(-z-\alpha_+) \,, \qquad (12.153)$$

where

$$\begin{aligned}
B_1 &= A_1 - A_1 e^{-i\pi\frac{d-3}{d-2}}[\bar{b} - i(1+e^{-i\pi j})\bar{a}_1] \,, \\
&\quad - A_2 e^{-i\pi\frac{d-3}{d-2}}\left[e^{+i\pi j}\bar{a}_2 - 4\cos^2\frac{\pi j}{2}\bar{a}_1 - 2i(1+e^{+i\pi j})\bar{b}\right] \\
&\quad - i(1+e^{i\pi j})\left[A_2 + A_2 e^{-i\pi\frac{d-3}{d-2}}[\bar{b} - i(1+e^{-i\pi j})\bar{a}_1] + A_1 e^{-i\pi\frac{d-3}{d-2}}e^{-i\pi j}\bar{a}_1\right] \\
B_2 &= A_2 + A_2 e^{-i\pi\frac{d-3}{d-2}}[\bar{b} - i(1+e^{-i\pi j})\bar{a}_1] + A_1 e^{-i\pi\frac{d-3}{d-2}}e^{-i\pi j}\bar{a}_1 \,.
\end{aligned}$$

Therefore the second constraint to first order reads

$$[1 - e^{-i\pi\frac{d-3}{d-2}}(i\bar{a}_1 + \bar{b})]A_1 - [i(1+2e^{i\pi j}) + e^{-i\pi\frac{d-3}{d-2}}((1+e^{i\pi j})\bar{a}_1 + e^{i\pi j}\bar{a}_2 - i\bar{b})]A_2 = 0 \,. \qquad (12.154)$$

Compatibility of the two first-order constraints yields

$$\begin{vmatrix} 1 + \bar{b} + \bar{a}_2\tan(\omega\bar{r}_* - \beta - \alpha_+) & i(1+2e^{i\pi j}) + e^{-i\pi\frac{d-3}{d-2}}((1+e^{i\pi j})\bar{a}_1 + e^{i\pi j}\bar{a}_2 - i\bar{b}) \\ (1-\bar{b})\tan(\omega\bar{r}_* - \beta - \alpha_+) - \bar{a}_1 & 1 - e^{-i\pi\frac{d-3}{d-2}}(i\bar{a}_1 + \bar{b}) \end{vmatrix}$$
$$= 0 \,. \qquad (12.155)$$

leading to the first-order expression for quasi-normal frequencies,

$$\begin{aligned}
\omega\bar{r}_* &= \frac{\pi}{4}(2 + j + j_\infty) + \frac{1}{2i}\ln 2 + n\pi \\
&\quad - \frac{1}{8}\left\{6i\bar{b} - 2ie^{-i\pi\frac{d-3}{d-2}}\bar{b} - 9\bar{a}_1 + e^{-i\pi\frac{d-3}{d-2}}\bar{a}_1 + \bar{a}_2 - e^{-i\pi\frac{d-3}{d-2}}\bar{a}_2\right\} \,,
\end{aligned}$$

where

$$\bar{a}_1 = \frac{\pi\mathscr{A}}{4}\left(\frac{n\pi}{2\bar{r}_*}\right)^{-\frac{d-3}{d-2}}\frac{\Gamma(\frac{1}{d-2})\Gamma(\frac{j}{2} + \frac{d-3}{2(d-2)})}{\Gamma^2(\frac{d-1}{2(d-2)})\Gamma(\frac{j}{2} + \frac{d-1}{2(d-2)})}$$

$$\bar{a}_2 = \left[1 + 2\cot\frac{\pi(d-3)}{2(d-2)}\cot\frac{\pi}{2}\left(-j + \frac{d-3}{d-2}\right)\right]\bar{a}_1$$

$$\bar{b} = -\cot\frac{\pi(d-3)}{2(d-2)}\,\bar{a}_1 \,.$$

Thus the first-order correction is $\sim \mathscr{O}(n^{-\frac{d-3}{d-2}})$.

The above analytic results are in good agreement with numerical results [29] (see [27] for a detailed comparison).

12.3.3 Electromagnetic Perturbations

The electromagnetic potential in four dimensions is

$$V_{\mathsf{EM}} = \frac{\ell(\ell+1)}{r^2} f(r) \,. \tag{12.156}$$

Near the origin,

$$V_{\mathsf{EM}} = \frac{j^2 - 1}{4r_*^2} + \frac{\ell(\ell+1)r_*^{-3/2}}{2\sqrt{-4\mu}} + \cdots \,, \tag{12.157}$$

where $j = 1$. Therefore we have a vanishing potential to zeroth order. To calculate the QNM spectrum we need to include first-order corrections from the outset. Working as with gravitational perturbations, we obtain the QNMs

$$\omega \bar{r}_* = n\pi - \frac{i}{4}\ln n + \frac{1}{2i}\ln\left(2(1+i)\mathscr{A}\sqrt{\bar{r}_*}\right) \,, \quad \mathscr{A} = \frac{\ell(\ell+1)}{2\sqrt{-4\mu}} \,. \tag{12.158}$$

Notice that the first-order correction behaves as $\ln n$, a fact which may be associated with gauge invariance.

As with gravitational perturbations, the above analytic results are in good agreement with numerical results [29] (see [27] for a detailed comparison).

12.4 AdS/CFT Correspondence and Hydrodynamics

A second unexpected connection comes from studies carried out using the Relativistic Heavy Ion Collider, a particle accelerator at Brookhaven National Laboratory. This machine smashes together nuclei at high energy to produce a hot, strongly interacting plasma. Physicists have found that some of the properties of this plasma are better modelled (via duality) as a tiny black hole in a space with extra dimensions than as the expected clump of elementary particles in the usual four dimensions of spacetime. The prediction here is again not a sharp one, as the string model works much better than expected. String-theory skeptics could take the point of view that it is just a mathematical spinoff. However, one of the repeated lessons of physics is unity - nature uses a small number of principles in diverse ways. And so the quantum gravity that is manifesting itself in dual form at Brookhaven is likely to be the same one that operates everywhere else in the universe.

– Joe Polchinski

There is a correspondence between $\mathscr{N} = 4$ Super Yang–Mills (SYM) theory in the large N limit and type-IIB string theory in AdS$_5 \times$ S^5 (AdS/CFT correspondence).

In the low-energy limit, string theory is reduced to classical supergravity and the AdS/CFT correspondence allows one to calculate all gauge field-theory correlation functions in the strong coupling limit leading to non-trivial predictions on the behavior of gauge theory fluids. For example, the entropy of $\mathcal{N} = 4$ SYM theory in the limit of large 't Hooft coupling is precisely 3/4 its value in the zero coupling limit.

The long-distance (low-frequency) behavior of any interacting theory at finite temperature must be described by fluid mechanics (hydrodynamics). This leads to a universality in physical properties because hydrodynamics implies very precise constraints on correlation functions of conserved currents and the stress–energy tensor. Their correlators are fixed once a few transport coefficients are known.

12.4.1 Hydrodynamics

To study hydrodynamics of the gauge theory fluid, suppose it possesses a conserved current j^μ. For simplicity, let us set the chemical potential $\mu = 0$, so that in thermal equilibrium the charge density $\langle j^0 \rangle = 0$. The retarded thermal Green function is given by

$$G^R_{\mu\nu}(\omega, q) = -i \int d^4 x\, e^{-iq \cdot x}\, \theta(t) \langle [j_\mu(x), j_\nu(0)] \rangle, \qquad (12.159)$$

where $q = (\omega, \boldsymbol{q})$, $x = (t, \boldsymbol{x})$. It determines the response of the system to a small external source coupled to the current. For small ω and $|\boldsymbol{q}|$, the external perturbation varies slowly in space and time. Then a macroscopic hydrodynamic description for its evolution is possible [30].

For a charged density obeying the diffusion equation

$$\partial_0 j^0 = D \nabla^2 j^0, \qquad (12.160)$$

where D is the diffusion constant with dimension of length, we obtain an overdamped mode with dispersion relation

$$\omega = -iDq^2, \qquad (12.161)$$

The corresponding retarded Green function has a pole at $\omega = -iDq^2$ in the complex ω-plane.

Another important conserved current is the stress–energy tensor $T^{\mu\nu}$. Its conservation law may be written as

$$\begin{aligned} \partial_0 \tilde{T}^{00} + \partial_i T^{0i} &= 0, \\ \partial_0 T^{0i} + \partial_j \tilde{T}^{ij} &= 0, \end{aligned} \qquad (12.162)$$

where

$$\tilde{T}^{00} = T^{00} - \rho, \qquad \rho = \langle T^{00} \rangle,$$

$$\tilde{T}^{ij} = T^{ij} - p\delta^{ij} = -\frac{1}{\rho + p}\left[\eta\left(\partial_i T^{0j} + \partial_j T^{0i} - \frac{2}{3}\delta^{ij}\partial_k T^{0k}\right) + \zeta\delta^{ij}\partial_k T^{0k}\right],$$

$$(12.163)$$

and ρ (p) is the energy density (pressure) of the fluid, η (ζ) is its shear (bulk) viscosity.

One obtains two types of eigenmodes, the shear modes which consist of transverse fluctuations of the momentum density T^{0i}, with a purely imaginary eigenvalue

$$\omega = -iDq^2 , \qquad D = \frac{\eta}{\rho + p}, \tag{12.164}$$

and a sound wave due to simultaneous fluctuations of the energy density T^{00} and the longitudinal component of momentum density T^{0i}, with dispersion relation

$$\omega = u_s q - \frac{i}{2}\frac{1}{\rho + p}\left(\zeta + \frac{4}{3}\eta\right)q^2 , \qquad u_s^2 = \frac{\partial p}{\partial \rho}. \tag{12.165}$$

In a conformal field theory, the stress-energy tensor is traceless, so

$$\rho = 3p , \qquad \zeta = 0 , \qquad u_s = \frac{1}{\sqrt{3}} . \tag{12.166}$$

12.4.2 Branes

To understand the gravitational side of the AdS/CFT correspondence, consider a non-extremal 3-brane which is a solution of type-IIB low-energy equations of motion. In the near-horizon limit $r \ll R$ where R is the AdS radius, the metric becomes

$$ds_{10}^2 = \frac{(\pi TR)^2}{u}\left(-f(u)dt^2 + dx^2 + dy^2 + dz^2\right) + \frac{R^2}{4u^2 f(u)}du^2 + R^2 d\Omega_5^2 , \tag{12.167}$$

where $T = \frac{r_0}{\pi R^2}$ is the Hawking temperature, and we have defined $u = \frac{r_0^2}{r^2}$, $f(u) = 1 - u^2$. The horizon corresponds to $u = 1$ whereas spatial infinity is at $u = 0$.

According to the gauge theory/gravity correspondence, the above background metric with non-extremality parameter r_0 is dual to $\mathcal{N} = 4$ $SU(N)$ SYM at finite temperature T in the limit of $N \to \infty$, $g_{YM}^2 N \to \infty$. For the retarded Green function

$$G_{\mu\nu,\lambda\rho}(\omega,\boldsymbol{q}) = -i\int d^4x\, e^{-iq\cdot x}\,\theta(t)\langle[T_{\mu\nu}(x), T_{\lambda\rho}(0)]\rangle. \tag{12.168}$$

we deduce by considering an appropriate perturbation of the background metric [30],

$$G_{xy,xy}(\omega,\boldsymbol{q}) = -\frac{N^2 T^2}{16}\left(i2\pi T\omega + q^2\right). \tag{12.169}$$

leading to the shear viscosity of strongly coupled $\mathscr{N} = 4$ SYM plasma (Kubo formula)

$$\eta = \lim_{\omega \to 0} \frac{1}{2\omega} \int dt \, d\mathbf{x} \, e^{i\omega t} \, \langle [T_{xy}(x), T_{xy}(0)] \rangle = \frac{\pi}{8} N^2 T^3 \,. \tag{12.170}$$

Other correlators may also be found by different perturbations of the metric. One obtains

$$G_{tx,tx}(\omega, \mathbf{q}) = \frac{N^2 \pi T^3 \mathbf{q}^2}{8(i\omega - \mathscr{D}\mathbf{q}^2)} + \cdots \,,$$

$$G_{tx,xz}(\omega, \mathbf{q}) = -\frac{N^2 \pi T^3 \omega |\mathbf{q}|}{8(i\omega - \mathscr{D}\mathbf{q}^2)} + \cdots \,,$$

$$G_{xz,xz}(\omega, \mathbf{q}) = \frac{N^2 \pi T^3 \omega^2}{8(i\omega - \mathscr{D}\mathbf{q}^2)} + \cdots \,, \tag{12.171a}$$

where $\mathscr{D} = \frac{1}{4\pi T}$.

From the above results, one may deduce the viscosity η. Indeed, recall from hydrodynamics $\mathscr{D} = \frac{\eta}{\rho + p}$. Using the entropy

$$s = \frac{3}{4} s_0 = \frac{\pi^2}{2} N^2 T^3 \,, \tag{12.172}$$

where s_0 is the entropy at zero coupling, and the thermodynamic equations $s = \frac{\partial P}{\partial T}$, $\rho = 3p$, we deduce $\rho + p = \frac{\pi^2}{2} N^2 T^4$, therefore

$$\eta = \frac{\pi}{8} N^2 T^3 \,, \quad \frac{\eta}{s} = \frac{1}{4\pi} \,, \tag{12.173}$$

which agrees with the Kubo formula. It should be pointed out that there is no agreement unless $s = \frac{3}{4} s_0$, a fact which is still poorly understood.

The above result for the viscosity is based on the gravity dual of the gauge theory fluid and should correspond to its strong coupling regime. At weak coupling, one obtains by a direct calculation

$$\frac{\eta}{s} \gg \frac{1}{4\pi} \,. \tag{12.174}$$

Thus the viscocity coefficient η varies as a function of the 't Hooft coupling,

$$\eta = f_\eta(g_{\mathrm{YM}}^2 N) N^2 T^3 \,, \tag{12.175}$$

where $f_\eta(x) \sim \dfrac{1}{-x^2 \ln x}$ for $x \ll 1$ and $f_\eta(x) = \dfrac{\pi}{8}$ for $x \gg 1$.

12.4.3 Schwarzschild Black Holes

In the metric considered above, the horizon was flat. This corresponds to the limit of a large black hole. For a black hole of finite size, the horizon is generally a sphere. Then the boundary of spacetime is $S^3 \times \mathbb{R}$. This may be conformally mapped onto a flat Minkowski space. Then by holographic renormalization, the AdS$_5$-Schwarzschild black hole is dual to a spherical shell of plasma on the four-dimensional Minkowski space which first contracts and then expands (conformal soliton flow) [31].

Quasi-normal modes govern the properties of this plasma with long-lived modes (i.e., of small Im ω) having the most influence. For example, one obtains the ratio

$$\frac{v_2}{\delta} = \frac{1}{6\pi} \mathrm{Re} \frac{\omega^4 - 40\omega^2 + 72}{\omega^3 - 4\omega} \sin \frac{\pi\omega}{2} , \qquad (12.176)$$

where $v_2 = \langle \cos 2\phi \rangle$ evaluated at $\theta = \frac{\pi}{2}$ (mid-rapidity) and averaged with respect to the energy density at late times; $\delta = \frac{\langle y^2 - x^2 \rangle}{\langle y^2 + x^2 \rangle}$ is the eccentricity at time $t = 0$. Numerically, $\frac{v_2}{\delta} = 0.37$, which compares well with the result from RHIC data, $\frac{v_2}{\delta} \approx 0.323$ [32].

Another observable is the thermalization time which is found to be

$$\tau = \frac{1}{2|\mathrm{Im}\,\omega|} \approx \frac{1}{8.6 T_{\mathrm{peak}}} \approx 0.08 \text{ fm/c} , \quad T_{\mathrm{peak}} = 300 \text{ MeV} , \qquad (12.177)$$

not in agreement with the RHIC result $\tau \sim 0.6$ fm/c [33], but still encouragingly small. For comparison, the corresponding result from perturbative QCD is $\tau \gtrsim 2.5$ fm/c [34, 35].

The above results motivate the calculation of low-lying QNMs. Earlier, we calculated analytically the asymptotic form of QNMs for large black holes. We obtained frequencies which were proportional to the horizon radius r_0. We found an infinite spectrum, however we missed the lowest frequencies which are inversely proportional to r_0. The latter are important in the understanding of the hydrodynamic behavior of the gauge theory fluid via the AdS/CFT correspondence.

12.4.3.1 Vector Perturbations

We start with vector perturbations and work in a d-dimensional Schwarzschild background. It is convenient to introduce the coordinate [36]

$$u = \left(\frac{r_0}{r} \right)^{d-3} . \qquad (12.178)$$

The wave equation becomes

$$-(d-3)^2 u^{\frac{d-4}{d-3}} \hat{f}(u) \left(u^{\frac{d-4}{d-3}} \hat{f}(u) \Psi' \right)' + \hat{V}_V(u)\Psi = \hat{\omega}^2 \Psi \;, \quad \hat{\omega} = \frac{\omega}{r_0} \;, \quad (12.179)$$

where prime denotes differentiation with respect to u and we have defined

$$\hat{f}(u) \equiv \frac{f(r)}{r^2} = 1 - u^{\frac{2}{d-3}} \left(u - \frac{1-u}{r_0^2} \right) \;, \qquad (12.180)$$

$$\hat{V}_V(u) \equiv \frac{V_V}{r_0^2} = \hat{f}(u) \left\{ \hat{L}^2 + \frac{(d-2)(d-4)}{4} u^{-\frac{2}{d-3}} \hat{f}(u) - \frac{(d-1)(d-2)\left(1+\frac{1}{r_0^2}\right)}{2} u \right\}$$

$$(12.181)$$

where $\hat{L}^2 = \frac{\ell(\ell+d-3)}{r_0^2}$.

First let us consider the large black-hole limit $r_0 \to \infty$ keeping $\hat{\omega}$ and $\hat{L}$ fixed (small). Factoring out the behavior at the horizon ($u = 1$)

$$\Psi = (1-u)^{-i\frac{\hat{\omega}}{d-1}} F(u) \;, \qquad (12.182)$$

the wave equation simplifies to

$$\mathscr{A} F'' + \mathscr{B}_{\hat{\omega}} F' + \mathscr{C}_{\hat{\omega},\hat{L}} F = 0 \;, \qquad (12.183)$$

where

$$\mathscr{A} = -(d-3)^2 u^{\frac{2d-8}{d-3}} \left(1 - u^{\frac{d-1}{d-3}} \right)$$

$$\mathscr{B}_{\hat{\omega}} = -(d-3)[d-4-(2d-5)u^{\frac{d-1}{d-3}}]u^{\frac{d-5}{d-3}} - 2(d-3)^2 \frac{i\hat{\omega}}{d-1} \frac{u^{\frac{2d-8}{d-3}}\left(1-u^{\frac{d-1}{d-3}}\right)}{1-u}$$

$$\mathscr{C}_{\hat{\omega},\hat{L}} = \hat{L}^2 + \frac{(d-2)[d-4-3(d-2)u^{\frac{d-1}{d-3}}]}{4} u^{-\frac{2}{d-3}}$$

$$- \frac{\hat{\omega}^2}{1-u^{\frac{d-1}{d-3}}} + (d-3)^2 \frac{\hat{\omega}^2}{(d-1)^2} \frac{u^{\frac{2d-8}{d-3}}\left(1-u^{\frac{d-1}{d-3}}\right)}{(1-u)^2}$$

$$-(d-3)\frac{i\hat{\omega}}{d-1} \frac{[d-4-(2d-5)u^{\frac{d-1}{d-3}}]u^{\frac{d-5}{d-3}}}{1-u} - (d-3)^2 \frac{i\hat{\omega}}{d-1} \frac{u^{\frac{2d-8}{d-3}}\left(1-u^{\frac{d-1}{d-3}}\right)}{(1-u)^2} \;.$$

We may solve this equation perturbatively by separating

$$(\mathscr{H}_0 + \mathscr{H}_1)F = 0 \qquad (12.184)$$

where

$$\mathscr{H}_0 F \equiv \mathscr{A} F'' + \mathscr{B}_0 F' + \mathscr{C}_{0,0} F \;,$$
$$\mathscr{H}_1 F \equiv (\mathscr{B}_{\hat{\omega}} - \mathscr{B}_0)F' + (\mathscr{C}_{\hat{\omega},\hat{L}} - \mathscr{C}_{0,0})F \;.$$

Expanding the wavefunction perturbatively,

$$F = F_0 + F_1 + \cdots \tag{12.185}$$

at zeroth order the wave equation reads

$$\mathcal{H}_0 F_0 = 0 , \tag{12.186}$$

whose acceptable solution is

$$F_0 = u^{\frac{d-2}{2(d-3)}} , \tag{12.187}$$

being regular at both the horizon ($u = 1$) and the boundary ($u = 0$, or $\Psi \sim r^{-\frac{d-2}{2}} \to 0$ as $r \to \infty$). The Wronskian is

$$\mathcal{W} = \frac{1}{u^{\frac{d-4}{d-3}}\left(1 - u^{\frac{d-1}{d-3}}\right)} \tag{12.188}$$

and another linearly independent solution is

$$\check{F}_0 = F_0 \int \frac{\mathcal{W}}{F_0^2} , \tag{12.189}$$

which is unacceptable because it diverges at both the horizon ($\check{F}_0 \sim \ln(1 - u)$ for $u \approx 1$) and the boundary ($\check{F}_0 \sim u^{-\frac{d-4}{2(d-3)}}$ for $u \approx 0$, or $\Psi \sim r^{\frac{d-4}{2}} \to \infty$ as $r \to \infty$).

At first order the wave equation reads

$$\mathcal{H}_0 F_1 = -\mathcal{H}_1 F_0 , \tag{12.190}$$

whose solution may be written as

$$F_1 = F_0 \int \frac{\mathcal{W}}{F_0^2} \int \frac{F_0 \mathcal{H}_1 F_0}{\mathcal{A}\mathcal{W}} . \tag{12.191}$$

The limits of the inner integral may be adjusted at will because this amounts to adding an arbitrary amount of the unacceptable solution. To ensure regularity at the horizon, choose one of the limits of integration at $u = 1$ rendering the integrand regular at the horizon. Then at the boundary ($u = 0$),

$$F_1 = \check{F}_0 \int_0^1 \frac{F_0 \mathcal{H}_1 F_0}{\mathcal{A}\mathcal{W}} + \text{regular terms} . \tag{12.192}$$

The coefficient of the singularity ought to vanish,

$$\int_0^1 \frac{F_0 \mathcal{H}_1 F_0}{\mathcal{A}\mathcal{W}} = 0 , \tag{12.193}$$

which yields a constraint on the parameters (dispersion relation)

$$\mathbf{a}_0 \hat{L}^2 - i\mathbf{a}_1 \hat{\omega} - \mathbf{a}_2 \hat{\omega}^2 = 0 . \tag{12.194}$$

After some algebra, we arrive at

$$\mathbf{a}_0 = \frac{d-3}{d-1} \quad , \quad \mathbf{a}_1 = d-3 .$$

(12.195)

The coefficient $\mathbf{a}_2$ may also be found explicitly for each dimension d, but it cannot be written as a function of d in closed form. It does not contribute to the dispersion relation at lowest order. E.g., for $d = 4,5$, we obtain, respectively

$$\mathbf{a}_2 = \frac{65}{108} - \frac{1}{3}\ln 3 \quad , \quad \frac{5}{6} - \frac{1}{2}\ln 2 .$$

(12.196)

Equation (12.194) is quadratic in $\hat{\omega}$ and has two solutions,

$$\hat{\omega}_0 \approx -i\frac{\hat{L}^2}{d-1} \quad , \quad \hat{\omega}_1 \approx -i\frac{d-3}{\mathbf{a}_2} + i\frac{\hat{L}^2}{d-1} .$$

(12.197)

In terms of the frequency ω and the quantum number ℓ,

$$\omega_0 \approx -i\frac{\ell(\ell+d-3)}{(d-1)r_0} \quad , \quad \frac{\omega_1}{r_0} \approx -i\frac{d-3}{\mathbf{a}_2} + i\frac{\ell(\ell+d-3)}{(d-1)r_0^2} .$$

(12.198)

The smaller of the two, ω_0, is inversely proportional to the radius of the horizon and is not included in the asymptotic spectrum. The other solution, ω_1, is a crude estimate of the first overtone in the asymptotic spectrum, nevertheless it shares two important features with the asymptotic spectrum: it is proportional to r_0 and its dependence on ℓ is $\mathcal{O}(1/r_0^2)$. The approximation may be improved by including higher-order terms. This increases the degree of the polynomial in the dispersion relation (12.194) whose roots then yield approximate values of more QNMs. This method reproduces the asymptotic spectrum derived earlier albeit not in an efficient way.

To include finite size effects, we shall use perturbation theory (assuming $1/r_0$ is small) and replace $\mathcal{H}_1$ by

$$\mathcal{H}_1' = \mathcal{H}_1 + \frac{1}{r_0^2}\mathcal{H}_H ,$$

(12.199)

where

$$\mathcal{H}_H F \equiv \mathcal{A}_H F'' + \mathcal{B}_H F' + \mathcal{C}_H F .$$

(12.200)

The coefficients may be easily deduced by collecting $\mathcal{O}(1/r_0^2)$ terms in the exact wave equation. We obtain

$$\mathcal{A}_H = -2(d-3)^2 u^2 (1-u)$$

$$\mathcal{B}_H = -(d-3)u\left[(d-3)(2-3u) - (d-1)\frac{1-u}{1-u^{\frac{d-1}{d-3}}}u^{\frac{d-1}{d-3}}\right]$$

$$\mathcal{C}_H = \frac{d-2}{2}\left[d-4 - (2d-5)u - (d-1)\frac{1-u}{1-u^{\frac{d-1}{d-3}}}u^{\frac{d-1}{d-3}}\right] .$$

Interestingly, the zeroth order wavefunction F_0 is an eigenfunction of $\mathscr{H}_H$,

$$\mathscr{H}_H F_0 = -(d-2)F_0 , \tag{12.201}$$

therefore the first-order finite-size effect is a simple shift of the angular momentum operator

$$\hat{L}^2 \rightarrow \hat{L}^2 - \frac{d-2}{r_0^2} . \tag{12.202}$$

The QNMs of lowest frequency are modified to

$$\omega_0 = -i\frac{\ell(\ell+d-3)-(d-2)}{(d-1)r_0} + \mathcal{O}(1/r_0^2) . \tag{12.203}$$

For $d=4,5$, we have respectively,

$$\omega_0 = -i\frac{(\ell-1)(\ell+2)}{3r_0} \quad , \quad -i\frac{(\ell+1)^2-4}{4r_0} \quad , \tag{12.204}$$

in agreement with numerical results [29, 31].

According to the AdS/CFT correspondence, dual to the AdS Schwarzschild black hole is a gauge theory fluid on the boundary of AdS ($S^{d-2} \times \mathbb{R}$). Consider the fluid dynamics ansatz

$$u_i = \mathscr{K} e^{-i\Omega\tau} \mathbb{V}_i , \tag{12.205}$$

where u_i is the (small) velocity of a point in the fluid, and $\mathbb{V}_i$ a vector harmonic on S^{d-2}. Demanding that this ansatz satisfy the standard equations of linearized hydrodynamics, one arrives at a constraint on the frequency of the perturbation Ω which yields [37]

$$\Omega = -i\frac{\ell(\ell+d-3)-(d-2)}{(d-1)r_0} + \mathcal{O}(1/r_0^2) , \tag{12.206}$$

in perfect agreement with its dual counterpart.

12.4.3.2 Scalar Perturbations

Next we consider scalar perturbations which are calculationally more involved but phenomenologically more important because their spectrum contains the lowest frequencies. For a scalar perturbation we ought to replace the potential $\hat{V}_V$ by

$$\hat{V}_S(u) = \frac{\hat{f}(u)}{4}\left[\hat{m} + \left(1+\frac{1}{r_0^2}\right)u\right]^{-2}$$

$$\times \left\{ d(d-2)\left(1+\frac{1}{r_0^2}\right)^2 u^{\frac{2d-8}{d-3}} - 6(d-2)(d-4)\hat{m}\left(1+\frac{1}{r_0^2}\right)u^{\frac{d-5}{d-3}} \right.$$

$$+ (d-4)(d-6)\hat{m}^2 u^{-\frac{2}{d-3}} + (d-2)^2\left(1+\frac{1}{r_0^2}\right)^3 u^3$$

$$+ 2(2d^2 - 11d + 18)\hat{m}\left(1 + \frac{1}{r_0^2}\right)^2 u^2$$

$$+ \frac{(d-4)(d-6)\left(1 + \frac{1}{r_0^2}\right)^2}{r_0^2} u^2 - 3(d-2)(d-6)\hat{m}^2\left(1 + \frac{1}{r_0^2}\right)u$$

$$- \frac{6(d-2)(d-4)\hat{m}\left(1 + \frac{1}{r_0^2}\right)}{r_0^2} u + 2(d-1)(d-2)\hat{m}^3 + d(d-2)\frac{\hat{m}^2}{r_0^2}\Bigg\},$$

$$(12.207)$$

where $\hat{m} = 2\frac{\ell(\ell+d-3)-(d-2)}{(d-1)(d-2)r_0^2} = \frac{2(\ell+d-2)(\ell-1)}{(d-1)(d-2)r_0^2}$.

In the large black-hole limit $r_0 \to \infty$ with $\hat{m}$ fixed (small), the potential simplifies to

$$\hat{V}_S^{(0)}(u) = \frac{1 - u^{\frac{d-1}{d-3}}}{4(\hat{m}+u)^2}\left\{ d(d-2)u^{\frac{2d-8}{d-3}} - 6(d-2)(d-4)\hat{m}u^{\frac{d-5}{d-3}} \right.$$

$$+ (d-4)(d-6)\hat{m}^2 u^{-\frac{2}{d-3}} + (d-2)^2 u^3$$

$$\left. + 2(2d^2 - 11d + 18)\hat{m}u^2 - 3(d-2)(d-6)\hat{m}^2 u + 2(d-1)(d-2)\hat{m}^3 \right\}.$$

$$(12.208)$$

The wave equation has an additional singularity due to the double pole of the scalar potential at $u = -\hat{m}$. It is desirable to factor out the behavior not only at the horizon but also at the boundary and the pole of the scalar potential,

$$\Psi = (1 - u)^{-i\frac{\hat{\omega}}{d-1}} \frac{u^{\frac{d-4}{2(d-3)}}}{\hat{m}+u} F(u).$$

$$(12.209)$$

Then the wave equation reads

$$\mathscr{A} F'' + \mathscr{B}_{\hat{\omega}} F' + \mathscr{C}_{\hat{\omega}} F = 0.$$

$$(12.210)$$

where

$$\mathscr{A} = -(d-3)^2 u^{\frac{2d-8}{d-3}}\left(1 - u^{\frac{d-1}{d-3}}\right)$$

$$\mathscr{B}_{\hat{\omega}} = -(d-3)u^{\frac{2d-8}{d-3}}\left(1 - u^{\frac{d-1}{d-3}}\right)\left[\frac{d-4}{u} - \frac{2(d-3)}{\hat{m}+u}\right]$$

$$-(d-3)\left[d - 4 - (2d-5)u^{\frac{d-1}{d-3}}\right]u^{\frac{d-5}{d-3}} - 2(d-3)^2 \frac{i\hat{\omega}}{d-1}\frac{u^{\frac{2d-8}{d-3}}\left(1 - u^{\frac{d-1}{d-3}}\right)}{1-u}$$

$$\mathscr{C}_{\hat{\omega}} = -u^{\frac{2d-8}{d-3}}\left(1-u^{\frac{d-1}{d-3}}\right)\left[-\frac{(d-2)(d-4)}{4u^2} - \frac{(d-3)(d-4)}{u(\hat{m}+u)} + \frac{2(d-3)^2}{(\hat{m}+u)^2}\right]$$

$$-\left[\left\{d-4-(2d-5)u^{\frac{d-1}{d-3}}\right\}u^{\frac{d-5}{d-3}} + 2(d-3)\frac{i\hat{\omega}}{d-1}\frac{u^{\frac{2d-8}{d-3}}\left(1-u^{\frac{d-1}{d-3}}\right)}{1-u}\right]$$

$$\times\left[\frac{d-4}{2u} - \frac{d-3}{\hat{m}+u}\right]$$

$$-(d-3)\frac{i\hat{\omega}}{d-1}\frac{\left[d-4-(2d-5)u^{\frac{d-1}{d-3}}\right]u^{\frac{d-5}{d-3}}}{1-u} - (d-3)^2\frac{i\hat{\omega}}{d-1}\frac{u^{\frac{2d-8}{d-3}}\left(1-u^{\frac{d-1}{d-3}}\right)}{(1-u)^2}$$

$$+\frac{\hat{V}_S^{(0)}(u)-\hat{\omega}^2}{1-u^{\frac{d-1}{d-3}}} + (d-3)^2\frac{\hat{\omega}^2}{(d-1)^2}\frac{u^{\frac{2d-8}{d-3}}\left(1-u^{\frac{d-1}{d-3}}\right)}{(1-u)^2}.$$

We shall define zeroth-order wave equation as $\mathscr{H}_0 F_0 = 0$, where

$$\mathscr{H}_0 F \equiv \mathscr{A} F'' + \mathscr{B}_0 F'. \qquad (12.211)$$

The acceptable zeroth-order solution is

$$F_0(u) = 1 \qquad (12.212)$$

which is plainly regular at all singular points ($u = 0, 1, -\hat{m}$). It corresponds to a wavefunction vanishing at the boundary ($\Psi \sim r^{-\frac{d-4}{2}}$ as $r \to \infty$).

The Wronskian is

$$\mathscr{W} = \frac{(\hat{m}+u)^2}{u^{\frac{2d-8}{d-3}}\left(1-u^{\frac{d-1}{d-3}}\right)}, \qquad (12.213)$$

and an unacceptable solution is $\check{F}_0 = \int \mathscr{W}$. It can be written in terms of hypergeometric functions. For $d \geq 6$, it has a singularity at the boundary, $\check{F}_0 \sim u^{-\frac{d-5}{d-3}}$ for $u \approx 0$, or $\Psi \sim r^{\frac{d-6}{2}} \to \infty$ as $r \to \infty$. For $d = 5$, the acceptable wavefunction behaves as $r^{-1/2}$ whereas the unacceptable one behaves as $r^{-1/2}\ln r$. For $d = 4$, the roles of F_0 and $\check{F}_0$ are reversed, however the results still valid because the correct boundary condition at the boundary is a Robin boundary condition [36, 37]. Finally, we note that $\check{F}_0$ is also singular (logarithmically) at the horizon ($u = 1$).

Working as in the case of vector modes, we arrive at the first-order constraint

$$\int_0^1 \frac{\mathscr{C}_{\hat{\omega}}}{\mathscr{A}\mathscr{W}} = 0, \qquad (12.214)$$

because $\mathscr{H}_1 F_0 \equiv (\mathscr{B}_{\hat{\omega}} - \mathscr{B}_0)F_0' + \mathscr{C}_{\hat{\omega}}F_0 = \mathscr{C}_{\hat{\omega}}$. This leads to the dispersion relation

$$\mathbf{a}_0 - \mathbf{a}_1 i\hat{\omega} - \mathbf{a}_2 \hat{\omega}^2 = 0. \qquad (12.215)$$

After some algebra, we obtain

$$\mathbf{a}_0 = \frac{d-1}{2} \frac{1+(d-2)\hat{m}}{(1+\hat{m})^2} \quad , \quad \mathbf{a}_1 = \frac{d-3}{(1+\hat{m})^2} \quad , \quad \mathbf{a}_2 = \frac{1}{\hat{m}} \{1 + O(\hat{m})\} . \quad (12.216)$$

For small $\hat{m}$, the quadratic equation has solutions

$$\hat{\omega}_0^\pm \approx -i\frac{d-3}{2} \hat{m} \pm \sqrt{\frac{d-1}{2}} \hat{m} , \quad (12.217)$$

related to each other by $\hat{\omega}_0^+ = -\hat{\omega}_0^{-*}$, which is a general symmetry of the spectrum.

Finite size effects at first order amount to a shift of the coefficient $\mathbf{a}_0$ in the dispersion relation

$$\mathbf{a}_0 \rightarrow \mathbf{a}_0 + \frac{1}{r_0^2}\mathbf{a}_H . \quad (12.218)$$

After some tedious but straightforward algebra, we obtain

$$\mathbf{a}_H = \frac{1}{\hat{m}} \{1 + O(\hat{m})\} . \quad (12.219)$$

The modified dispersion relation yields the modes

$$\hat{\omega}_0^\pm \approx -i\frac{d-3}{2} \hat{m} \pm \sqrt{\frac{d-1}{2}} \hat{m} + 1 . \quad (12.220)$$

In terms of the quantum number ℓ,

$$\omega_0^\pm \approx -i(d-3) \frac{\ell(\ell+d-3)-(d-2)}{(d-1)(d-2)r_0} \pm \sqrt{\frac{\ell(\ell+d-3)}{d-2}} , \quad (12.221)$$

in agreement with numerical results [31].

Notice that the imaginary part is inversely proportional to r_0, as in vector case. In the scalar case, we also obtained a finite real part independent of r_0. It yields the speed of sound $v_s = \frac{1}{\sqrt{d-2}}$ which is the correct value in the presence of conformal invariance.

Turning to the implications of the above results for the AdS/CFT correspondence, we may perturb the gauge theory fluid on the boundary of AdS ($S^{d-2} \times \mathbb{R}$) using the ansatz

$$u_i = \mathcal{K}e^{-i\Omega\tau}\nabla_i\mathbb{S} , \quad \delta p = \mathcal{K}'e^{-i\Omega\tau}\mathbb{S} , \quad (12.222)$$

where u_i is the (small) velocity of a point in the fluid and δp is a pressure perturbation. They are both given in terms of $\mathbb{S}$, a scalar harmonic on S^{d-2}. Demanding that this ansatz satisfy the equations of linearized hydrodynamics, one obtains a frequency of perturbation Ω in perfect agreement with our analytic result [36, 37].

12.4.3.3 Tensor Perturbations

Finally, for completeness we discuss the case of tensor perturbations. Unlike the other two cases, the asymptotic spectrum of tensor perturbations is the entire spectrum. To see this, note that in the large black-hole limit, the wave equation reads

$$-(d-3)^2(u^{\frac{2d-8}{d-3}} - u^3)\Psi'' - (d-3)[(d-4)u^{\frac{d-5}{d-3}} - (2d-5)u^2]\Psi'$$
$$+ \left\{ \hat{L}^2 + \frac{d(d-2)}{4}u^{-\frac{2}{d-3}} + \frac{(d-2)^2}{4}u - \frac{\hat{\omega}^2}{1 - u^{\frac{d-1}{d-3}}} \right\}\Psi = 0 .$$

For the zeroth-order equation, we may set $\hat{L} = 0 = \hat{\omega}$. The resulting equation may be solved exactly. Two linearly independent solutions are ($\Psi = F_0$ at zeroth order)

$$F_0(u) = u^{\frac{d-2}{2(d-3)}} , \quad \check{F}_0(u) = u^{-\frac{d-2}{2(d-3)}} \ln\left(1 - u^{\frac{d-1}{d-3}}\right) . \tag{12.223}$$

Neither behaves nicely at both ends ($u = 0, 1$). Therefore both are unacceptable which makes it impossible to build a perturbation theory to calculate small frequencies which are inversely proportional to r_0. This negative result is in agreement with numerical results [29, 31] and in accordance with the AdS/CFT correspondence. Indeed, there is no ansatz that can be built from tensor spherical harmonics $\mathbb{T}_{ij}$ satisfying the linearized hydrodynamic equations, because of the conservation and tracelessness properties of $\mathbb{T}_{ij}$.

12.5 Conclusion

We discussed the calculation of analytic asymptotic expressions for quasi-normal modes of various perturbations of black holes in asymptotically flat as well as anti-de Sitter spaces. We also showed how perturbative corrections to the asymptotic expressions can be systematically calculated.

In view of the AdS/CFT correspondence, in AdS spaces we concentrated on low-frequency modes because they govern the hydrodynamic behavior of the gauge theory fluid which is dual to the black hole. Thus, these modes provide a powerful tool in understanding the hydrodynamics of a gauge theory at strong coupling. They may lead to experimental consequences pertaining to the quark-gluon plasma produced in heavy ion collisions at RHIC and the LHC.

Acknowledgments This research was supported in part by the US Department of Energy under grant DE-FG05-91ER40627.

References

1. V. Ferrari and B. Mashhoon, Oscillations of a black hole. Phys. Rev. Lett. **52**, 1361 (1984).
2. R. A. Konoplya, Quasinormal behavior of the D-dimensional Schwarzshild black hole and higher order WKB approach. Phys. Rev. D **68**, 024018 (2003).

3. S. Chandrasekhar and S. Detweiler, Proc. R. Soc. London, Ser. A **344**, 441 (1975).
4. E. W. Leaver, Proc. R. Soc. London, Ser. A **402**, 285 (1985).
5. H. -P. Nollert, Quasinormal modes of Schwarzschild black holes: The determination of quasi-normal frequencies with very large imaginary parts. Phys. Rev. D **47**, 5253 (1993).
6. N. Andersson, On the asymptotic distribution of quasinormal-mode frequencies for Schwarzschild black holes. Class. Quant. Grav. **10**, L61 (1993).
7. A. Bachelot and A. Motet-Bachelot, Annales Poincarè Phys. Theor. **59**, 3 (1993).
8. L. Motl and A. Neitzke, Asymptotic black hole quasinormal frequencies. Adv. Theor. Math. Phys. **7**, 2 (2003).
9. S. Hod, Bohr's correspondence principle and the area spectrum of quantum black holes. Phys. Rev. Lett. **81**, 4293 (1998).
10. S. Musiri and G. Siopsis, Perturbative calculation of quasi-normal modes of Schwarzschild black holes. Class. Quant. Grav. **20**, L285–L291 (2003).
11. E. Berti and K. D. Kokkotas, Asymptotic quasinormal modes of Reissner-Nordström and Kerr black holes. Phys. Rev. D **68**, 044027 (2003).
12. A. Maasen van den Brink, WKB analysis of the Regge-Wheeler equation down in the frequency plane. J. Math. Phys. **45**. 327 (2004)
13. J. D. Bekenstein and V. F. Mukhanov, Spectroscopy of the quantum black hole. Phys. Lett. B **360**, 7 (1995).
14. E. Berti, V. Cardoso, K. D. Kokkotas and H. Onozawa, Highly damped quasinormal modes of Kerr black holes. Phys. Rev. D **68**, 124018 (2003).
15. S. A. Teukolsky, Rotating black holes: Separable wave equations for gravitational and electromagnetic perturbations. Phys. Rev. Lett. **29**, 1114 (1972).
16. S. Musiri and G. Siopsis, On quasi-normal modes of Kerr black holes. Phys. Lett. B **579**, 25–30 (2004).
17. S. Detweiler, Klein-Gordon equation and rotating black holes. Phys. Rev. D **22**, 2323 (1980).
18. I. B. Khriplovich and G. Yu. Ruban, Quasinormal modes for arbitrary spins in the Schwarzschild background. Int. J. Mod. Phys. D **15**, 879–894 (2006).
19. S. Musiri and G. Siopsis, Perturbative calculation of quasi-normal modes of arbitrary spin in Schwarzschild spacetime. Phys. Lett. B **650**, 279 (2007).
20. V. Cardoso and J. P. S. Lemos, Scalar, electromagnetic and Weyl perturbations of BTZ black holes: Quasi normal modes. Phys. Rev. D **63**, 124015 (2001).
21. D. Birmingham, I. Sachs and S. N. Solodukhin, Conformal field theory interpretation of black hole quasi-normal modes. Phys. Rev. Lett. **88**, 151301 (2002).
22. G. T. Horowitz and V. E. Hubeny, Quasinormal modes of AdS black holes and the approach to thermal equilibrium. Phys. Rev. D **62**, 024027 (2000).
23. A. O. Starinets, Quasinormal modes of near extremal black branes. Phys. Rev. D **66**, 124013 (2002).
24. R. A. Konoplya, On quasinormal modes of small Schwarzschild-Anti-de-Sitter black hole. Phys. Rev. D **66**, 044009 (2002).
25. S. Musiri and G. Siopsis, Asymptotic form of quasi-normal modes of large AdS black holes. Phys. Lett. B **576**, 309–313 (2003).
26. J. Natário and R. Schiappa, On the classification of asymptotic quasi-normal frequencies for d-dimensional black holes and quantum gravity, hep-th/0411267.
27. S. Musiri, S. Ness and G. Siopsis, Perturbative calculation of quasi-normal modes of AdS Schwarzschild black holes. Phys. Rev. D **73**, 064001 (2006).
28. A. Ishibashi and H. Kodama, A master equation for gravitational perturbations of maximally symmetric black holes in higher dimensions. Prog. Theor. Phys. **110** 701 (2003).
29. V. Cardoso, R. A. Konoplya and J. P. S. Lemos, Quasi-normal frequencies of Schwarzschild black holes in AdS space-times: A complete study on the asymptotic behavior. Phys. Rev. D **68**, 044024 (2003).
30. G. Policastro, D. T. Son and A. O. Starinets, From AdS/CFT correspondence to hydrodynamics. JHEP **0209**, 043 (2002).
31. J. J. Friess, S. S. Gubser, G. Michalogiorgakis and S. S. Pufu, Expanding plasmas and quasi-normal modes of anti-de sitter black holes. arXiv:hep-th/0611005.

32. PHENIX Collaboration and A. Adare et al., Scaling properties of azimuthal anisotropy in Au+Au and Cu+Cu collisions at $\sqrt{s_{NN}} = 200$ GeV. arXiv:nucl-ex/0608033.

33. P. Arnold, J. Lenaghan, G. D. Moore and L. G. Yaffe, Apparent thermalization due to plasma instabilities in quark-gluon plasma. Phys. Rev. Lett. **94** 072302 (2005).

34. R. Baier, A. H. Mueller, D. Schiff and D. T. Son, "Bottom-up" thermalization in heavy ion collisions. Phys. Lett. B. **502**, 51–58 (2001)

35. D. Molnar and M. Gyulassy, Saturation of elliptic flow at RHIC: Results from the covariant elastic parton cascade model MPC. Nucl. Phys. A **697**, 495–520 (2002).

36. G. Siopsis, Low frequency quasi-normal modes of AdS black holes. JHEP **0705**, 042 (2007).

37. G. Michalogiorgakis and S. S. Pufu, Low-lying gravitational modes in the scalar sector of the global AdS4 black hole. JHEP **0702**, 023 (2007).

Index